I draw from

MW00607858

physical, mental and spiritual strength. This book is about as mental as one can get... Enjoy!!

Jill S. Heaton

The Mojave Desert

The Mojave Desert

Ecosystem Processes and Sustainability

EDITED BY

ROBERT H. WEBB, LYNN F. FENSTERMAKER, JILL S. HEATON,
DEBRA L. HUGHSON, ERIC V. MCDONALD, DAVID M. MILLER

FOREWORD BY

CHARLES WILKINSON

UNIVERSITY OF NEVADA PRESS  RENO & LAS VEGAS

The publication of this book was made possible in part by the generous support of the California Desert Managers Group.

University of Nevada Press, Reno, Nevada 89557 USA

Manufactured in the United States of America

Library of Congress Cataloging-in-Publication Data [to come]
The Mojave Desert : ecosystem processes and sustainability / edited by Robert H. Webb . . . [et al.] ; foreword by Charles Wilkinson.
p. cm.
Includes bibliographical references and index.
ISBN 978-0-87417-776-3 (hardcover : alk. paper)
1. Desert ecology—California—Mojave Desert. 2. Endangered ecosystems—California—Mojave Desert. I. Webb, Robert H.
QH105.C2M65 2009
577.5409794'95—dc22 2008043584

The paper used in this book is a recycled stock made from 30 percent post-consumer waste materials, certified by FSC, and meets the requirements of American National Standard for Information Sciences—Permanence of Paper for Printed Library Materials, ANSI/NISO Z39.48-1992 (R2002). Binding materials were selected for strength and durability.

First Printing
18 17 16 15 14 13 12 11 10 09
5 4 3 2 1

CONTENTS

ILLUSTRATIONS

FIGURES

Figures with asterisks (*) also appear as color plates following page 162

TABLES

APPENDICES

FOREWORD

CHARLES WILKINSON

In the gray of a crisp desert dawn five years ago, I drove east out of Furnace Creek, having completed my work in Death Valley National Park. It had been an intense two days. I was mediating negotiations between the National Park Service and the Timbisha Shoshone Tribe. The Shoshone settlement in Furnace Creek, old adobes and battered trailers, signaled them as the squatters that they were under the law. These residents, by 12,000 years the oldest society in the valley, wanted land from a government that never gave away parkland.

The tensions lifted during the two days of talks. John Reynolds, the Park Service regional director, was a good man, open-minded. Pauline Estevez, the tribal chair in her seventies, tough and traditional, held a well-warranted suspicion but she listened and grew guardedly optimistic. Maybe these federal people would be different. There was no resolution—that would take many more meetings—but it was a good start.

I had seen Death Valley during this trip, but only at a distance for both negotiating sessions had gone late, and it was then dinner and off to bed. Over the years, I had gotten out on the land in the west and south Mojave, but this was my first trip to this part of the desert. My flight out of Las Vegas was hours away, so I had the chance to get up early and see some country.

During the talks, Pauline had made many statements about the land, about how when she was a girl the people wintered on the valley floor and moved into the higher country during the summers. She was precise, the way Indian people are, about places, animals, and vegetation. One place she mentioned was Ash Meadows. Her family loved it there. The water, the rabbits, the mesquite beans.

So I decided to stop off at Ash Meadows. I knew of it through a Supreme Court case that curtailed groundwater pumping that affected Devil's Hole and its resident population of pupfish. That interested me because I've come to think of lawsuits over public lands as much in terms of place as law. Even more basically, though, Pauline had moved me and I wanted to know more about her life.

I turned left onto the Ash Meadows dirt road, then took a right toward Jackrabbit Spring. I had no idea what to expect. Probably a burbling up of water and a slim outlet trickle. I parked, walked over, and—like many before me, all the way back to Pauline's oldest ancestors—stood stunned, without words. The pool, ten to twenty feet across, six to eight feet deep, pure, clear beyond the saying, the low mesquite, ash, and grass on the banks, the pupfish and dace finning down below. No words.

Of course, Jackrabbit Spring is just one of many miracles of the Mojave. Other springs. Other tribes. The blue and gold carpet on Death Valley's floor in some springs. The Panamints. The desert tortoises, so ancient. The fallen-down cabins that someone somehow survived in. The Joshua trees. The rises and sets of the sun. The space. The hard work it takes to know the desert. The crisp, tart taste of wildness. The time the Mojave gives you to think.

The desert has always been that way. It has always been that way for the Shoshone. But it has not always been that way for us.

For us, "desert" was long a pejorative term. In *The Grapes of Wrath,* John Steinbeck, conjuring images of the Valley of the Shadow of Death, wrote of how Tom Joad had exulted when they finally crossed the terrifying Mojave: "Jesus, the desert's past! Pa, Al, for Christ sakes! The desert's past!" But it turned out that Granma had died and Ma knew it, though she kept it to herself. "I tol' Granma we couldn't help her. The fambly had to get acrost. I tol' her, tol' her when she was a-dyin.' We couldn't stop in the desert." It was a wasteland, a hell. Frank Norris ended his novel, *McTeague,* to evoke ultimate despair, in Death Valley with McTeague handcuffed to a dead man. Without a key. In the summer.

Non-Indians first came in numbers to the Mojave in the 1850s, spillovers from the great rush in the Sierra, in search of mineral wealth. They soon had official carte blanche from the Hardrock Mining Law of 1872: prospecting and extraction required backbreaking labor but the minerals and the land over them were free for the taking. The miners picked over an amazingly large part of the Mojave's vast and difficult terrain, taking out everything from gold to borax to gravel. But they were there for business, not living, and most of them moved on to more appealing locales when the deposits played out.

The same dynamic was at work with the Small Tract Act of 1938, which doled out five-acre lots in the southern and western Mojave for a few dollars each to promote new settlements. It didn't work. The land was too stark, too hostile. Most of the homesteaders fled, leaving their husks of dwellings behind with no takers.

World War II and the Big Buildup that followed changed everything, just as it did all across the Southwest, which has boomed more than four times over, from 8 million people to 34 million, since the war. Level, wide-open expanses of land for military installations was one of the big draws. This was perfect terrain for aerial and ground maneuvers, ideal for tanks, antiaircraft guns, and bombers. Edwards Air Force Base was the first, beginning as a bombing range in the 1930s and made a permanent base in 1942. Then came Fort Irwin, China Lake, Barstow, the Marine Corps Combat Center, Nellis Air Force Base, and the Nevada Test Site. The largest of all was the Desert Training Center (DTC), west of Needles, under the command of General George Patton, which offered intensive training in an area of 45300 km^2 and held a population of nearly 200,000. Unlike the others, the DTC was decommissioned.

Today the military installations encompass more than 2.59 million hectares. It remains the dominant intensive land use in the Mojave Desert. The Bureau of Land Management administers 2.57 million hectares and the National Park Service 2.91 million hectares.

The military reasons—including size and remoteness—for locating in the desert were valid. But mark it down that our societal disdain for deserts was critical to our decision to locate so much military might there. The Mojave was a useless place, a lifeless place, a wasteland, the most logical ground to bomb, grind down with heavy vehicles, and dump. That's what it was, a dump.

Then, after the war, the Big Buildup brought large-scale permanent population to the Mojave. In the western Mojave, Edwards Air Force Base in Antelope Valley attracted, in addition to soldiers and their families, all manner of defense-related contractors and manufacturers. Antelope Valley then sprouted subdivisions to accommodate the new arrivals. The Cold War jumpstarted other Mojave communities and, it needs to be mentioned, cities not far beyond: Phoenix, Salt Lake City, and the Southern California coastal urban areas. At the far eastern edge of the Mojave, an urban behemoth was born. Because the census set a cutoff of 2,500 as the minimum population to qualify as a city, Las Vegas did not even appear in the census until 1930. By the end of the war, though it might technically be defined as a city, it was only a town, with just 45,000 souls. Of course, since then Las Vegas has received that many new residents in a month.

Although this has changed somewhat, we need to appreciate how this postwar surge of residential population held a perception of the desert little different than the early miners, Small Tract homesteaders, and military personnel. The citizens of Antelope Valley, Las Vegas, and the other expanding population centers liked the desert just fine—so long as it came fully equipped with air conditioning, Kentucky bluegrass lawns, golf courses, public water fountains, and opportunities for swimming—for which Jackrabbit Spring over on Ash Meadows did not qualify.

We have taken a great toll on this arid landscape's aquifers. They are in decline in Las Vegas, Antelope, Apple, Cadiz, and Victor valleys as well as elsewhere. We have seen some subsidence of land surfaces. So far, the main response has been long-range transport of water supplies rather than conservation or moratoria on subdivision construction. So our disdain for the desert has been accompanied by a related attitude, our refusal to acknowledge what a desert *is*—and first on the list is that a desert is a place without much water.

To be sure, there have always been people, in addition to the Indians, who understood what the Mojave is and loved it with all their hearts and souls. Mary Austin wrote *The Land of Little Rain,* her book-song to Owens Valley. Joseph Wood Krutch was another. Wallace Stegner helped broaden us: "You have to get over the color green; you have to quit associating beauty with gardens and lawns; you have to get used to an inhuman scale. . . ." And, at least as fundamentally, thousands of impassioned rock hounds, botanists, artists, hikers, and plain citizens knew and respected this dry, white-hot, rocky, scratchy ground.

But it may be that a person irreverent, iconoclastic, and bombastic in the extreme—wild eyed and fire breathing doesn't begin to describe him—was the one who most caused us as a society to take a better and longer look at the deserts, to try to open our minds and understand. And that would be, of course, one Edward Paul Abbey. He dealt mostly with the Colorado Plateau desert but he also knew and wrote about the Sonoran and Mojave deserts—and composed the final chapters of his masterpiece, *Desert Solitaire,* in Ash Meadows.

At its core, *Desert Solitaire* is about teaching. Abbey instructs us about what the desert is and the profound impacts it can have on us. But much of the desert is counterintuitive. You have to spend time at it, work at it, get down on your hands and knees. He gave us *particularity,* vivid descriptions of desert plants, animals, minerals, air, and geologic formations. Consider this, one of his many descriptions of rocks: "The various forms of chalcedony, for example, are strewn liberally over the dismal clay hills along Salt Creek. Here you will find tiny crystals of garnet embedded in a matrix of mica schist—almandite or 'common garnet.' Fragments of quartzite are everywhere, some containing pure quartz crystals. You might find a geode: a lump of sandstone the size and shape of an ostrich egg, or sometimes much larger; slice it through with a diamond wheel and you may find inside a glittering treasure trove of crystals. A treasure not in money but in beauty." And this about midday, after a long passage where Abbey explains that everything—lizards, spiders, red ants, birds, flowers, coyotes, *everything*—shuts down in the heat of noon. "Noontime here is like a drug. The light is psychedelic, that dry electric air narcotic. To me the desert is stimulating, exciting, exacting; I feel no temptation to sleep or to relax into occult dreams but rather, an opposite effect which sharpens and heightens vision, touch, hearing, taste and smell. Each stone, each plant, each grain of sand exists in and for itself with a clarity that is undimmed by any suggestion of a different realm. *Claritas, integritas, veritas.* Only the sunlight holds things together. Noon is the crucial hour: the desert reveals itself nakedly and cruelly, with no meaning but its own existence."

By the 1980s, Abbey, and the deserts that so inflamed his passions, had literally millions of readers. But whatever the specific influence Abbey or anyone else may have had, it is certain that by the 1980s the public's affection for the natural qualities of the desert had risen noticeably and it had grown intense.

Another desert resident had a sad and poignant story to tell in the 1980s and it too gripped the public. An enormously successful animal, in substantially its present form since the late Paleozoic, the desert tortoise found its stability, its staying power—the work of 200 million years—thrown into crisis in the click of a moment. Research scientists explained the reasons. Out of affection, we had collected too many tortoises. Much worse, we had debilitated their habitat: We crushed—with ORV wheels, tank treads, and cattle hooves—desert wildflowers, pads of the prickly pear, cactus flowers, and grasses that these grazing animals need. Our garbage attracted ravens that in turn fed on tortoise eggs and juveniles. Our civilization inadvertently introduced upper respiratory infections that ravaged the population.

This crisis—the tortoise is now listed as a threatened species under the Endangered Species Act—had a powerful impact on the public. These little animals, stolid and silent as the desert itself, gave us a vivid and profound reason to protect the Mojave. The country was finally ready to take action on behalf of the desert.

The origins of the 1994 California Desert Protection Act went back a good thirty years. Research done by BLM and university scientists showed the toll our desert had taken on soils and vegetation, and the animals that depend on them. Congress took a beginning step in the 1976 Federal Land Policy and Management Act (FLPMA) by requiring a special plan for what it called the California Desert Conservation Area. FLPMA prompted still more research and gave the BLM more enforcement authority. But it was the 1994 act that marked the nation's first comprehensive national statement on the worth of deserts and the necessity of preserving the naturalness, remoteness, and wildness of these once-scorned lands.

Congress created the Mojave National Preserve and upgraded the status of Joshua Tree and Death Valley from national monuments to national parks. It set aside more than seven million acres as wilderness, most of it in the Mojave Desert. It is by any standard one of our grandest national statements in favor of the land. We had finally heeded Stegner's call to get over the color green.

The power of the desert landscape, the integrity of good scientific research, and the passion of a people, even when lodged in truly landmark legislation, have only gone so far. We have achieved sustainability over parts of the Mojave, but the future of the larger part of this desert landscape is still very much in question. And now some thoughts about the future of sustainability, science, and the Mojave.

In recent times, natural resource management has emphasized planning and collaboration. Both are important elements of modern natural resources policy, but both are processes—and policy must always have a substantive goal. Otherwise, we are rudderless.

In Aldo Leopold's terms, a land ethic requires the preservation of "land health." As he wrote in *A Sand County Almanac,* "Health is the capacity of the land for self-renewal. Conservation is our effort to understand and preserve this capacity." And "the most important characteristic of an organism [including the land organism] is that capacity for internal self-renewal known as health."

Today we use the term sustainability to articulate our commitment to long-term land and water health. Broadly put, it calls upon us to meet the needs of our generation without compromising the ability of future generations to meet their own needs. At its core, sustainability is a moral idea, premised on intergenerational equity, the ethical obligation to those still to come, a blood commitment that they will have the same or greater bounty with which we have been blessed. We should strive to live within our ecological means because failing to do so means passing on a burden of sacrifice and loss to other people, people who actually will inhabit our places. Though we do not know these people, passing the burden to them would show no respect, love, or morality, precisely the qualities our species values so.

These are high-sounding words, the kind of talk that causes some people to say that sustainability is vague, all puff and no substance. But it is worthwhile to look closely at sustainability and at some of the many different venues where people are hard at work putting sustainability into practice, on the ground, as a working policy.

I think of sustainability as operating on two levels. First, sustainability has great appeal as a broad societal objective—as a symbol, as a statement of some of the fundamental values we hold as a people. Sustainability calls to us because it combines the philosophical and moral force of fairness to future generations with the practical edge of being necessary to our economic and social well-being. In this broad, symbolic sense, sustainability embodies a shared national goal in much the same way that freedom and equality do. Such broad formulations—idealistic and never fully attainable, yet undeniable in their essential truth—are critical for setting an agreed-upon context for making public choices on difficult and contentious issues.

I referred to people who say sustainability is vague, and in many ways it is. It is a young concept. Freedom and equality were young once, too, but both, while maintaining their symbolic force, have also matured into specific programs, including voting rights acts, the right to peacefully protest, fair housing laws, and scores of others.

Sustainability, like freedom and equality, also operates on a more down-to-earth level and has been gaining specificity quite quickly. We recognize three aspects of sustainability—ecological, economic, and cultural. We increasingly focus on a place, usually defined by natural, rather than political, boundaries. The objective of sustainability—what it is we will strive to sustain—is expansive. We mean to sustain the traditional commodity-oriented outputs—megawatts, board feet, acre-feet, animal unit months, visitor days, and so forth—but we refuse to define our future in those terms alone. Modern sustainability goes further and also seeks to sustain such things as wolves and tortoises; lesser-known beings such as voles and Mojave tarplant; good rafting water, long vistas, and archaeological sites; tribal, ranch, and farm communities; and solitude, beauty, and wonder.

I had the occasion to see the definition of sustainability sharpened even further.

In 1997, Agriculture Secretary Glickman appointed me to the Committee of Scientists, a panel authorized by the National Forest Management Act and charged to make recommendations for the Forest Service planning regulations. There were some semantic shenanigans here. Secretary Glickman included political science as a science—a stretch, but so far so good—and law as a branch of political science. Uh, oh: hence my appointment. The real scientists on the panel, ten of them including Norm Johnson, the chair, the hydrologist Bob Bechta, and the ecologist Barry Noon, made it blue ribbon.

Our fundamental recommendations involved the mission of the Forest Service, which we believed should be sustainability, with ecological, economic, and cultural components. The real sharpening, though, came from our view that the first step in federal land management must be to satisfy the requirement of ecological sustainability before turning to the economic and social components. The reason is that without sustaining the ecology—providing clean air and water, stable soils, and wildlife habitat—the economic returns (for example, grazing and the economic aspects of recreation) and cultural benefits (such as community stability, recreation, and beauty) cannot be sustained.

The Committee of Scientists' recommendations were accepted by the Forest Service and built into the 2000 regulations. This was apparently the first time that the primacy of ecological sustainability, which I believe represents the future of land and water management, has been written into law. The new administration revised the regulations, first making the analysis in the 2000 regulations optional then repealing it altogether in 2005. But, as I say, give the idea of the primacy of ecological sustainability time. It will come back into use in the future.

Yet the truest way to understand what sustainability means comes not from these kinds of descriptions, though they help, but from seeing how sustainability has actually been implemented in real places. It is through real-world efforts that you best understand and define sustainability and how it differs from traditional approaches, such as multiple use.

Take the Northwest Forest Plan, President Clinton's 1993 initiative in the federal, tribal, and private forests west of the crest of the Cascade Range. One aim is to sustain ancient forests, salmon, and other species, including the spotted owl. The plan also strives to sustain scaled-back but stable timber and commercial fishing industries. This comprehensive plan may be the most extensive effort in sustainability ever undertaken and the people of the Northwest are well familiar with it. Some oppose it. Some like it. But you don't hear many say that the Northwest Forest Plan is vague.

Two other examples—in addition to the several efforts I have already alluded to—are useful to show that we comprehend the specifics of sustainability, not so much in broad classroom-type formulations, as in real efforts in real places.

In the Greater Yellowstone Ecosystem, we have employed the device of ecosystem management, one of the methods we can use to further the goal of sustainability. You can see sustainability there as a concrete, working policy. At Yellowstone, the things we have decided to sustain are evidenced by such programs as grizzly bear recovery; bison, moose, cutthroat trout, and elk management; fire policy; coordination of information on campsite vacancies throughout the ecosystem; restrictions on road construction for logging and mining; protection of the geothermal resources; land exchanges; and wolf reintroduction. Some of these policies overlap, and all are being changed and modified to reflect new data, but you can quite clearly define what we are trying to sustain, and the ways we are going about it, in Yellowstone.

Compare this to Grand Canyon National Park, where sustainability must be fundamentally different than at Yellowstone or in the Northwest because the *place* is different. Yellowstone's lodgepoles are the Canyon's piñon-juniper; Yellowstone's geysers, the Canyon's exposed geology; Yellowstone's grizzlies, the Plateau's humpback chub; Yellowstone's fires, the Canyon's flash floods.

The broad principles of modern sustainability have been applied in many discrete ways in the Grand Canyon. All of these can and will be improved upon, but look what people at the Grand Canyon have accomplished in recent years. The attempt to save endemic fish species through the Endangered Species Act. The effort to sustain the fish (both the endemic species and introduced trout), the riparian areas, and the quality of recreational raft trips through the Grand Canyon Protection Act. The program to restore air clarity through cleanups of coal-fired power plants and through the work of the Grand Canyon Visibility Transport Commission. The determination to achieve economic sustainability by continuing to manage the Colorado River watershed to provide significant amounts of municipal and irrigation water and electrical energy. The beginning efforts, which will stiffen, to preserve archaeological sites. The regime to limit airplane over-flights in order to protect the solitude. The new Grand Canyon General Management Plan, which deals with overcrowding in the national park by sharply limiting automobile traffic. The implicit, but still firm, decision to sustain and preserve the living river and deep canyon walls and all their 1.7 billion years of world history by never, despite all the many proposals over the years, plugging the Grand Canyon with any dam.

And so in the Mojave, we can see the specific distinctive qualities—mostly different than in the places just mentioned—that need to be sustained. They include the tortoise, many other animals and many plants, the soils, and, in this land of little rain, the aquifers under assault and the associated springs, limestone formations, and saline seeps that support many endemic and rare plant species. The role of science is foundational. Scientists have developed ways, and are refining them, to measure land health—ecological sustainability—and to alert us to the threats and the trends. Scientists did not write the California Desert Protection Act, but their work proved its necessity and gave the people's movement its spine of integrity.

The Mojave, assuming that land health will be maintained, also has economic activities to be sustained. They include ranching, mining operations that protect the land, and recreation—including ORV travel where it is well planned. As for cultural sustainability, Pauline Estevez and the Timbisha Shoshone did finally succeed in negotiating with the Park Service and obtaining federal legislation granting a 121-hectare reservation at Furnace Creek and more than 2430 hectares of BLM land in traditional areas on the perimeter of Death Valley National Park.

No one can say whether we will be using the term sustainability generations from now. My guess, though, is that they will and that it will be the overarching framework for our relationship with the natural world for our careers and beyond. That is not because sustainability holds out some automatic solution, but because it offers a sensible way, at once idealistic and practical, for public bodies and plain citizens to conceive of, and build toward, a fair and promising future, toward sustaining places in a full and vibrant sense.

There are many people determined to put the West up for sale and development, but during the post-War era the West also has gained a sense of itself. Now we understand what we have here in this sacred place. And what we have to lose. For the first time ever, westerners all across the region are complaining, and loudly, about growth and how it is tearing away at everything the West is. For the first time, westerners understand the desert.

This is not an easy time to be a scientist assessing land health—ecological sustainability—yet I will remind you of some history. Beginning in the 1970s, we embarked on a national environmental program far beyond anything ever conceived before. It led to flagship statutes—the National Environmental Policy Act, the Clean Air Act, the Clean Water Act, the Endangered Species Act—and the creation of more than 40 million hectares of wilderness, led by the Alaska and California Desert acts. The centrality of science cannot be overstated: the work of scientists has been historic.

Has it been harder in recent years? Yes. Will it be hard in the next few years? Yes. But please we should do two things, and have faith in them. First, in the short term, we should reaffirm the intrinsic value of science, the gathering and analyses of data, and aim for some progress, even if it be limited. Second, and different, we should reflect on the indisputable trends over the past four decades and believe, for it is true, that we will return to the progressive advances of those decades. For in time the policies will again reflect the passion and particularity of Edward Abbey, the wisdom of Aldo Leopold, the courage of Rachel Carson, and the long-haul understanding of the American West of Wallace Stegner. We revere the desert now, the rare gift of the springs; the tenacity of the Indian people who have made it here for so long; the jerky gait toward the cactus bud that the stony tortoise has used for hundreds of millennia; the scraggly, sun-bleached creosote bush giving notice that it shall hang on through yet another drought; and the elixir of sun and shadows and silence and solitude that we can find nowhere else.

So, in a time when budgets are slim and the pressures keep building, we should all appreciate that the skills and values of scientists are respected and needed across the region and that what scientists research and say and do will help shape how westerners treat the land and wild places and wild things that, in the last analysis, constitute the soul of the American West.

ACKNOWLEDGMENTS

The 2004 gathering of the Third Mojave Desert Science Symposium, focused on the endlessly fascinating Mojave, has been much on my mind ever since Bob Webb extended the honor of giving the keynote address in 2003. I want to thank my research assistant, Christy McCann, for working with me on this foreword.

PREFACE

The first Mojave Desert Science Symposium was convened November 7–8, 1992, in Riverside, California, with a goal of fostering communication between scientists and land managers, and particularly focused on the area of the Mojave National Preserve and its surroundings of the Eastern Mojave Desert of California. The second Mojave Desert Science Symposium, held February 25–27, 1999, at the University of Nevada at Las Vegas, had a much broader agenda in terms of both space and time, encompassing the entire Mojave Desert three-state region. As a result of the broad recognition of the importance of these symposia, the ad hoc steering committee, comprised primarily of scientists representative of federal and state governments, determined that the Mojave Desert Science Symposium should be a regular meeting held at 4- to 5-year intervals. The precedent set by the second symposium carried over to the third one, which is the subject of this book.

One of the goals of the Mojave Desert Science Symposium series is to broadly disseminate current scientific knowledge of Mojave Desert ecological processes and resources. This book addresses that goal and is a specific product of the third symposium, "Continuing to Explore the Interface Between Science and Resource Management," which was held at the University of Redlands in Redlands, California, on November 16–18, 2004. Over 200 scientists and land managers attended this interdisciplinary meeting, which continued a tradition of discussing the Mojave Desert's ecosystem processes and the implications of new and emerging science for resource management. The meeting and its large attendance also reinforced the ongoing nature of these symposia and the importance of scientific investigations to the management of this desert region.

The symposium began with three keynote addresses, the first presented by a legal and historical scholar of the American West, the second by a research scientist, and the third by a group of land managers. These presentations described major scientific, management, and political issues that affect the Mojave Desert, and set the tone for the remainder of the meeting. The symposium was organized around four themes, each with an invited keynote speaker followed by several invited presentations, and with an accompanying poster session—about 70 posters were presented. The four themes were: Threats to the Mojave Desert, Ecosystem Monitoring, Natural Recovery or Active Restoration, and Scales and Sustainability. We rearranged these four themes into the four sections of this book: Regional Threats to the Mojave Desert; Road Effects on the Mojave Desert; The Role of Soils for Plant Communities; and Recovery, Restoration, and Ecosystem Monitoring.

We have incorporated presentations from the symposium into this book, in some cases combining presentations to obtain a broader perspective, and inviting new contributions to fill gaps in the information presented during the symposium. We created a different organization from the symposium, choosing to group similar-themed chapters together to form some state-of-the-art collections on a subject of interest to scientists and land managers. The chapters were written to be accessible to these two groups who are concerned specifically with the Mojave Desert or with ecological processes in desert regions worldwide. A total of thirty-six scientists and land managers contributed to the twenty-six chapters and bridging material included here.

In keeping with the symposium tradition of state-of-the-art science and its interface with resource management, each chapter underwent a rigorous scientific peer review. The editors required a minimum of two peer reviews for each chapter—with one reviewer from within the pool of book authors and one from without—as well as an editorial review that dealt with technical content and presentation. Where necessary, chapters received additional peer review. In addition, the entire book was reviewed by two anonymous reviewers. Both review processes were very positive experiences that have strengthened this book.

We acknowledge a number of symposium and book sponsors, including the U.S. Geological Survey, the Sweeny Granite Mountains Desert Research Center of the University of California, the Desert Research Institute, the National Park Service, the U.S. Fish and Wildlife Service, the Bureau of Land Management, Edwards Air Force Base, and the Redlands Institute of the University of Redlands. We would like to specifically acknowledge the cooperation and assistance of the Western National Parks Association and the California Desert Managers Group. The Third Mojave Desert Science Symposium would not have been possible without the organizational and logistical support of Sue Husch from QBC Events. We thank the more than 40 reviewers who contributed their time to making this a better, more scientifically accurate, book. Tina Kister of the U.S. Geological Survey performed a magnificent job of copyediting and proofreading this book, and we owe a special thanks to Jayne Belnap for donating Tina's time for this effort. Sophie Baker of the Desert Research Institute assisted with the chapter editing. Ben Yetman helped us design the logo for the symposium. Erin Aldrich of the University of Nevada, Reno, did an excellent job of reformatting and standardizing the maps presented in this book. Peter Griffiths and Diane Boyer helped prepare the many illustrations and photographs. Finally, we want to express our sincere gratitude to the chapter authors for sharing their knowledge of the Mojave Desert.

The Mojave Desert

Introduction

ROBERT H. WEBB, JILL S. HEATON, MATTHEW L. BROOKS,
AND DAVID M. MILLER

In this book, we discuss a number of threats and ecosystem processes that affect large parts of or the entire Mojave Desert, culminating in a discussion of the potential for recovery from severe disturbances. Previous information syntheses have concentrated on results from relatively small areas of this desert subject to intensive and sustained research (e.g., Rundel and Gibson 1996), or were focused narrowly on specific management issues (e.g., off-highway vehicles) (Webb and Wilshire 1983). Here, we consider some very broad processes and threats that affect the whole Mojave Desert or differentially affect its subregions. We begin with a delineation of those subregions, updating previous work using recent geographical information system (GIS) analyses to create distinctive maps of this desert.

MOJAVE DESERT SUBREGIONS

Several subdivisions of the Mojave Desert have been proposed, including the generally accepted map presented in Rowlands et al. (1982). Rowlands et al. (1982) proposed very broad, indistinct boundaries, some of which may be appropriate given the gradational nature of transitions among subregions and between the Mojave Desert and its surrounding areas. Most of the proposed changes to that map deal with specific details of where the boundaries should be, and because of gradational boundaries, these changes are mostly cosmetic.

Subregions of the Mojave Desert typically are defined by topography and seasonality of precipitation (Rowlands 1995; Hereford et al. 2004, 2006; Redmond *this volume*). Using these characteristics, the subdivisions in the northern and western Mojave Desert are easily defined, but the southeastern edges grade into the Sonoran Desert. Here, we adapt the subdivisions originally proposed by Bailey (1995, 2005) and modified by The Nature Conservancy (Grossman et al. 1998) as a part of the Ecoregional Assessments of North America (fig. 0.1; see also color plate). These are the six specific subregions that we refer to in this book, and these differ from other subregions, such as the species-specific recovery units used in the Desert Tortoise Recovery Plan (Berry 1997).

Northern Mojave Desert

The Northern Mojave includes much of the broad ecotone that transitions into the Great Basin, but it also includes the most inhospitable terrain associated with Death Valley. This subregion is constrained on the west by the Sierra Nevada and on the north by several mountain ranges and passes. A northward increase in elevation in the Owens Valley (fig. 0.1; see also color plate) provides a finger in the northwest. This subregion includes the mountain ranges and valleys surrounding Death Valley, most notably the Panamint, Grapevine, and Funeral Mountains, and Eureka and Saline Valleys, which are wholly or partially included within Death Valley National Park. These geographic features typify the Mojave Desert in regards to topography, vegetation, climate, and geology. The southeastern boundary of this subregion is poorly defined, primarily by the Spring Mountains northwest of Las Vegas and a series of mountain ranges south of Death Valley.

Growing conditions such as temperature extremes, rainfall totals, and potential evapotranspiration rates vary more dramatically in this than in any other Mojave Desert subregion. This variation results in the highest plant species richness of the five regions (1,025 species) (Rowlands et al. 1982). Most of the rainfall occurs during winter months, which may also limit the establishment of warm-season, nonnative grasses that require summer rainfall. The northern limit is based on the distribution of iconic Mojave Desert species, particularly *Larrea tridentata-Ambrosia dumosa* assemblages (creosote bush-white bursage) and *Yucca brevifolia* (Joshua tree). As a result, the Nevada Test Site (Smith et al. *this volume;* Webb et al., Long-term data, *this volume*) northwest of Las Vegas, is considered to be part of this subregion.

Eastern Mojave Desert

The Eastern Mojave interfaces with the Great Basin to the north, the Colorado Plateau to the east, and the Sonoran Desert to the south, and therefore is defined by topography and the northern limits of iconic species. The subregion, along with the Northern Mojave, contains most of the high-elevation topography in the Mojave Desert. Although local elevations include widely variable gradients, growing conditions do not vary considerably across this subregion. One of the major distinguishing characteristics of the Eastern Mojave is its relatively high amount of summer rainfall.

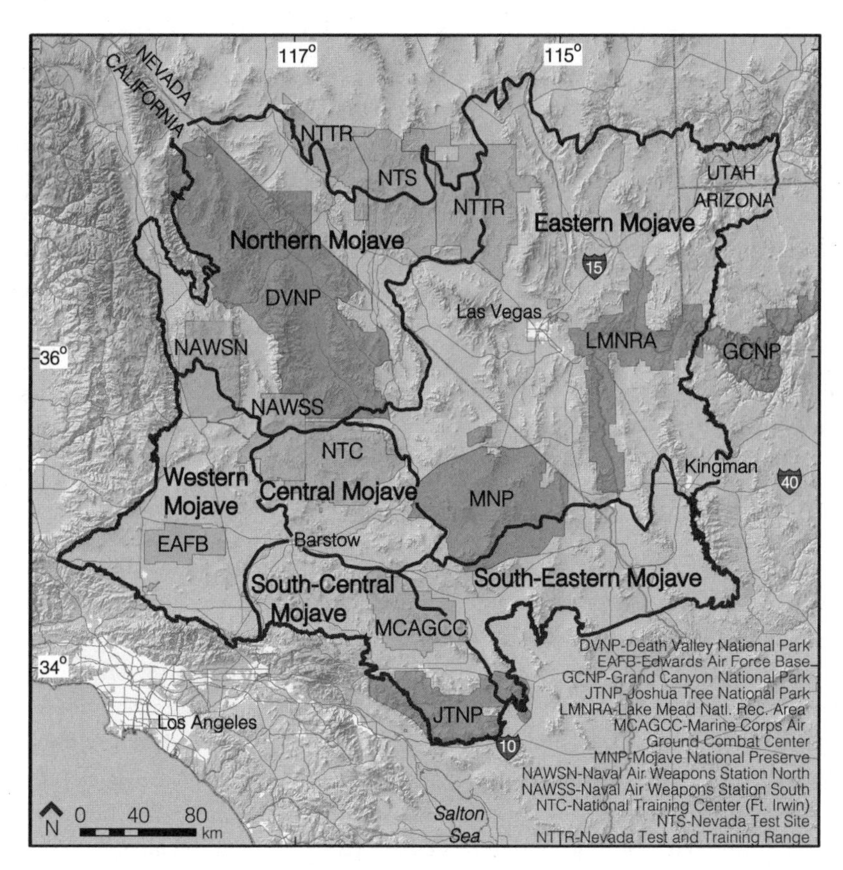

Fig. 0.1. Map of the subregion boundaries within the Mojave Desert ecoregion, with national park and military reserve boundaries. The ecoregion and subregions are based on Bailey (1995) and Environment Canada (Wiken 1986) and were provided by The Nature Conservancy (http://conserveonline.org/workspaces/rmcr.gis/index/ecoregion/md). (See color plate following page 162.)

This subregion encompasses most of the Nevada and Utah parts of the Mojave Desert (fig. 0.1; see also color plate). To the northeast, this subregion includes the *Larrea* assemblages in the vicinity of St. George, Utah, as well as the entire Gold Butte-Pakoon Basin east of the Grand Canyon and the ridgeline of the Virgin Mountains southwest of St. George in Utah and northwestern Arizona. To the east, the Virgin Mountains and the Grand Wash Cliffs make a natural topographic boundary. To the south and southwest, the boundary is more vaguely defined and driven primarily by vegetation differences that may be due to the influence of the summer monsoon. The Mojave National Preserve, which is discussed extensively in this book (e.g., Vogel and Hughson *this volume;* Miller et al. *this volume;* Bedford et al. *this volume*) is situated near the southwestern edge of the Eastern Mojave Desert subregion.

South-Eastern Mojave Desert

Of the six subregions, the South-Eastern Mojave Desert is defined least well by topography. To the east, the Hualapai Mountains of northwestern Arizona, south of Kingman, form a natural topographic boundary, but to the south, topographic boundaries are vague at best. Here, floral elements of the Sonoran and Mojave deserts intermix to form a diffuse boundary, and this may be one of the reasons why Rowlands et al. (1982) did not recognize this subregion and instead included it within the Sonoran Desert. However, the occurrence of *Y. brevifolia* on the eastern corner of this subregion provides some justification for its inclusion in the Mojave Desert. In the southeastern corner of this subregion, *Y. brevifolia* mixes with *Carnegia gigantea* (saguaro), creating a strange floristic brew of iconic species of both the Mojave and Sonoran deserts.

South-Central Mojave Desert

The South-Central Mojave Desert encompasses the southern tip of the Mojave Desert where it interfaces with the San Bernardino Mountains to the southwest and the Sonoran Desert to the southeast and east. The seasonal distribution of rainfall shifts steadily from predominantly winter rainfall (94% winter) near Victorville on its western edge to a more equitable winter-summer rainfall ratio (64% winter) near Twentynine Palms on its eastern edge (Rowlands 1980). The South-Central and Western Mojave subregions are close to the greater Los Angeles metropolitan area, and this proximity makes them most susceptible to atmospheric fallout from pollutants transported from that urban area (Allen et al. *this volume*).

The southern boundary of this subregion reflects the influence of the Colorado Desert subdivision of the Sonoran Desert, as exemplified in the southern part of Joshua Tree National Park. To the south, the heavily populated Coachella Valley with its extensive subdivisions and roads has significant issues with habitat fragmentation (Barrows and Allen *this volume*), which are also issues in this subregion. Several floristic differences justify this separation from the Sonoran Desert; to the north, *Y. brevifolia* is common, and to the south *Olneya tesota* (ironwood), *Psorothamnus spinosa* (smoketree), and *Cercidium microphyllum* (littleleaf palo verde) are common. On the northwest, this subregion has a boundary roughly defined by the course of the Mojave River; on the southwest, by the Transverse Ranges; on the north and northeast, by a series of prominent low valleys stretching from Barstow. To the southeast and east, this region interfaces imperceptibly into the Sonoran Desert.

Central Mojave Desert

The Central Mojave is the only subregion that does not directly interface with adjacent bioregions outside of the Mojave Desert. This subregion encompasses the lower reaches of the Mojave River drainage, including a series of relatively low-elevation playas that are the terminus to this major drainage system, and a series of broad valleys northward that terminate at high elevations. Its separation from major regional ecotones, along with its relatively moderate topography, results in the lowest ecological variability of the six subregions of the Mojave Desert. Another indication of its low environmental variability is the fact that it is estimated to contain the fewest vascular plant species of all the regions (458 species), with the Western Mojave the next lowest (663 species) (Rowlands et al. 1982).

Western Mojave Desert

The Western Mojave is bounded by the San Gabriel Mountains to the south, the Tehachapi and Sierra Nevada Mountains to the west, and the Great Basin and Owens Valley to the north. This subregion is perhaps the best defined of the six subregions, primarily because the Transverse Ranges and Sierra Nevada determine the boundaries on the south and northwest. Rainfall mostly occurs during winter (97%, Rowlands 1980; 82%, Hereford et al. 2004). The vegetation in these regions grades from *Y. brevifolia* into pinyon-juniper woodland. To the east and northeast, prominent mountain ranges also form natural boundaries. This part of the Mojave Desert has low relief and the highest percentage of winter rainfall (Hereford et al. 2004), and the uniseasonal precipitation regime in this subregion was once mistakenly used to characterize the entire Mojave Desert.

OVERVIEW OF THIS BOOK

In this book, we address a number of specific threats to the Mojave Desert, as well as the potential for natural recovery and active restoration, with an emphasis on well-recognized management issues with local effects (e.g., roads) that can scale up to regional threats, combined with recently emphasized regional threats (climate variability and change). The ongoing threat of introduction and expansion of nonnative species is treated with separate chapters on more xerophytic species in uplands and species that are invasive in riparian areas. We report on some new work on the influence of geomorphic processes and substrate on ecosystem function and services, particularly with regard to common perennial plant species and their root systems. We address issues of recovery and restoration, separating natural processes of healing from severe disturbances with active methods of reestablishing plant cover. Finally, we discuss the importance of long-term data to an understanding of change in the Mojave Desert.

ACKNOWLEDGMENTS

The summary of the various ecoregions of the Mojave Desert was inspired by the work of Peter Rowlands and others (Rowlands et al. 1982, Rowlands 1995). We obtained additional geospatial data from The Nature Conservancy. We thank Jeffrey Lovich for reviewing this introductory material.

REFERENCES

Bailey, R. G. 1995. Description of the ecoregions of the United States. Miscellaneous Publication No. 1391, with separate map, scale 1:7,500,000. Second Edition. U.S. Department of Agriculture, Forest Service, Washington, D.C.

Bailey, R. G. 2005. Identifying ecoregion boundaries. *Environmental Management* **34**: S14–S26.

Berry, K. H. 1997. The Desert Tortoise Recovery Plan: an ambitious effort to conserve biodiversity in the Mojave and Colorado deserts of the United States. Pages 430–440 *in* J. V. Abema, editor. *Proceedings of the 1993 International Conference on Conservation, Restoration, and Management of Tortoises and Turtles, July 11–16, 1993, Purchase, New York.* New York Turtle and Tortoise Society, New York, New York.

Grossman, D. H., D. Faber-Langendoen, A. S. Weakley, M. Anderson, P. Bourgeron, R. Crawford, K. Goodin, S. Landaal, K. Metzler, K. Patterson, M. Pyne, M. Reid, and L. Sneddon. 1998. International classification of ecological communities: terrestrial vegetation of the United States. Volume one. The national vegetation classification system: development, status, and applications. The Nature Conservancy, Arlington, Virginia.

Hereford, R., R. H. Webb, and C. I. Longpré. 2004. Precipitation history of the Mojave Desert region, 1893–2001. U.S. Geological Survey Fact Sheet No. 117–03. Reston, Virginia.

Hereford, R., R. H. Webb, and C. Longpré. 2006. Precipitation history and ecosystem response to multidecadal precipitation variability in the Mojave Desert and vicinity, 1893–2001. *Journal of Arid Environments* **67**:13–34.

Rowlands, P. G. 1980. Effects of disturbance on desert soils, vegetation, and community processes with emphasis on off-road vehicles: a critical review. U.S. Department of the Interior, Bureau of Land Management, Riverside, California.

Rowlands, P. G. 1995. Regional bioclimatology of the California Desert. Pages 95–134 *in* J. Latting and P. G. Rowlands, editors. *The California Desert: an introduction to natural resources and man's impact.* Volume one. University of California Riverside Press, Riverside, California.

Rowlands, P. G., H. Johnson, E. Ritter, and A. Endo. 1982. The Mojave Desert. Pages 103–162 *in* G. L. Bender, editor. *Reference handbook on the deserts of North America.* Greenwood Press, Westport, Connecticut.

Rundel, P. W., and A. C. Gibson. 1996. *Ecological communities and processes in a Mojave Desert ecosystem: Rock Valley, Nevada.* Cambridge University Press, New York, New York.

Webb, R. H., and H. G. Wilshire, editors. 1983. *Environmental effects of off-road vehicles.* Springer-Verlag Publishers, New York, New York.

Wiken, E. B. 1986. *Terrestrial ecozones of Canada.* Ecological Land Classification Series 19. Environmental Canada, Hull, Quebec Canada.

PART I

Regional Threats to the Mojave Desert

LYNN F. FENSTERMAKER, ERIC V. MCDONALD,
AND ROBERT H. WEBB

Management of desert ecosystems demands a broad understanding of desert organisms, the environment that supports them, and the complex processes that occur over time and space. History has shown that improper management of arid lands often results in radical changes in ecosystem structure and function, both within the region and neighboring ecosystems (e.g., dust transport, changes in plant community composition, and changes in hydrology). These environmental changes may induce social and economic instability over time. In a brief history of the Middle East, Hillel (2006) showed how early settlement and degradation of arid lands lead to tribal battles, which created the basis for some of the current political unrest in that region.

Threats can be caused by a wide variety of human-induced actions and interactions of these anthropogenic forces with climate variations. Threats are cumulative, interactive, and synergistic; the causative agents, and changes brought about by these agents, interact in complex, often unpredictable ways that may either reduce or heighten the impacts, but almost always make their resolutions and detection very difficult. Science can best contribute to management of threats through: (1) basic research on ecosystem function, (2) evaluation of ecosystem disturbance, and (3) design of holistic inventory and monitoring programs.

Threats to the Mojave Desert ecosystem span many different spatial and temporal scales. The spatial scales range from local (e.g., land use practices), to regional (e.g., atmospheric deposition from pollution sources to the southwest), to global (e.g., large-scale climatic processes and fluctuations). Similarly, the temporal scale ranges from minutes to eons. To further confound the issue, our perception of what constitutes a threat changes over time and is based upon our cultural or social values. For example, a hard-rock miner in the late 1800s would not have the same perception of threats as we do today. Similarly, threats may emerge in the future that are not currently recognized.

One of the difficulties facing resource managers today is the need to understand potential changes in ecosystem function that may occur in response to regional-scale changes, particularly climate fluctuations or change. Managers typically lack information on specific ecosystem attributes and how they are related to ecosystem processes and function. This section provides some examples of how the potential for regional change is on the interface between science and management. Basic research may identify problems that otherwise might go unnoticed, particularly ones that may loom large in the future (e.g., the impact of climate change on invasive species).

While basic research is generally not instigated by management directives, in each of the studies presented in this chapter, the results became very important to management practices. As discussed by Redmond (*this volume*) and Smith et al. (*this volume*), climatic fluctuations, particularly drought, and trace-gas composition of the atmosphere have pervasive ecosystem effects. Drought frequency may well prove to be the dominant force altering floristic systems in this desert, influencing both undisturbed ecosystems (Webb et al., Long-Term Data, *this volume*), as well as the course of natural recovery from severe disturbance.

As discussed in Hughson (*this volume*), human population growth, fueled by nonsustainable imports of energy and food, threatens both terrestrial ecosystems and water supplies, and raises complex questions concerning the potential for regional sustainability. Land uses external to the ecosystem influence nonnative plant invasions (Brooks *this volume*) and lead to nutrient influxes transported in the atmosphere. Atmospheric deposition of nutrients can affect competition between native and nonnative annual species (Allen et al. *this volume*). One lesson learned from these studies is that discussions of threats to the Mojave Desert ecosystem should first start at the largest scale, taking into account both natural climatic variability (Hereford et al. 2004, 2006; Redmond *this volume*) and potential anthropogenically driven climatic changes (Smith et al. *this volume*).

One effect of human population increases in the Mojave Desert is increased rates of dispersal by nonnative species. In upland areas, nonnative plants can compete with native species, and often comprise significant fractions of total plant productivity (Brooks *this volume*). Where their productivity is particularly high, nonnative annual grasses can alter fuel loadings, fire behavior, and fire regimes, creating an invasive plant/fire regime cycle (Brooks et al. 2004; Brooks and Minnich 2006). Their most significant effects on fire regimes have occurred in the Eastern subregion of the Mojave Desert, specifically at the upper ecotones of creosote bush scrub and throughout the blackbrush zone (Brooks and Matchett 2006). Nonnative species in riparian areas, one of the most scarce and valued habitats within the Mojave Desert ecoregion, is a significant threat of high interest to land managers, both because of direct competition with natives and the potential for nonnatives to alter fire regimes (Dudley *this volume*). Increases in nonnative species are both a symptom and a cause of ecological degradation. Nonnative plants have direct effects by competing with natives for resources and indirect effects by altering habitat structure. To many concerned with the interface of science and management, the increases in nonnative vegetation are the largest threats to this ecoregion because nonnative species can fundamentally change ecosystem processes.

Finally, basic research on regional threats serves education. Our public lands, especially in a desert environment that is home to—and in close proximity to—millions of human beings, represent the perfect classrooms and laboratories for educating future generations. Basic research is fundamental in raising public interest and awareness by unveiling the ecological stories, large and small, that help raise awareness of the need for land management practices designed to sustain our resources on public lands. Sharing the wealth of information gained from science is the foundation of the Mojave Desert Science Symposium, and it forms a baseline of knowledge that helps to direct applied research and the quality of stewardship of the Mojave Desert.

REFERENCES

Brooks, M. L., and J. R. Matchett. 2006. Spatial and temporal patterns of wildfires in the Mojave Desert, 1980–2004. *Journal of Arid Environments* **67**:148–164.

Brooks, M. L., and R. A. Minnich. 2006. Southeastern deserts bioregion. Pages 391–414 *in* N.G. Sugihara, J. W. van Wagtendonk, K. E. Shaffer, J. Fites-Kaufman, and A. E. Thode, editors. *Fire in California's Ecosystems.* University of California Press, Berkeley, California.

Brooks, M. L., C. M. D'Antonio, D. M. Richardson, J. Grace, J. J. Keeley, J. DiTomaso, R. Hobbs, M. Pellant, and D. Pyke. 2004. Effects of invasive alien plants on fire regimes. *BioScience* 54:677–688.

Hereford, R., R. H. Webb, and C. I. Longpré. 2004. Precipitation history of the Mojave Desert region, 1893–2001. U.S. Geological Survey Fact Sheet No. 117–03. Reston, Virginia.

Hereford, R., R. H. Webb, and C. Longpré. 2006. Precipitation history and ecosystem response to multidecadal precipitation variability in the Mojave Desert region, 1893–2001. *Journal of Arid Environments* **67**:13–34.

Hillel, D. 2006. Uses and abuses of soil and water resources: an historical review. Presentation at the Eighteenth World Congress of Soil Science, July 9–15, 2006, Philadelphia, Pennsylvania.

ONE

Historic Climate Variability in the Mojave Desert

KELLY T. REDMOND

The Mojave Desert extends eastward to approximately the northwestern boundary of Arizona with Nevada; northward to about the latitude of Beatty, Nevada; westward to the eastern front of the Sierra Nevada; and southward to about the southern side of Joshua Tree National Park and Interstate 10. This 152,000 km^2 area is approximately centered on Pahrump, Nevada, or Shoshone, California (see fig. 0.1, Webb et al., Introduction, *this volume*).

The Mojave is considered a "cold" or winter desert, because about 50%–70% of its annual precipitation typically falls during the cool season from November through March, with additional precipitation in October or April in some locations (Hereford et al. 2004, 2006). As noted by Hereford et al. (2004), the winter contribution to annual precipitation is generally greater in the western portions of the Mojave than in the eastern portions. Most of the remainder falls during the Southwest monsoon period from July through September. Annual precipitation ranges from 50–130 mm in valleys and from 250 to as much as 750 mm in the numerous isolated mountain ranges (fig. 1.1; see also color plate). Higher elevations receive more precipitation because their presence can initiate convection and cloudiness, and enhance existing precipitation systems, and because less evaporation occurs in the very low humidity air during the shorter fall to the ground.

With its valleys lying generally between 600 and 1500 m in elevation, this desert does not, in general, reach the withering temperatures seen in the lower elevation Sonoran Desert, which adjoins to the south. A few lower elevation locations are hot by any standard, notably Death Valley, which is at and slightly below sea level; this is the hottest location in North America and, by just a bare margin,

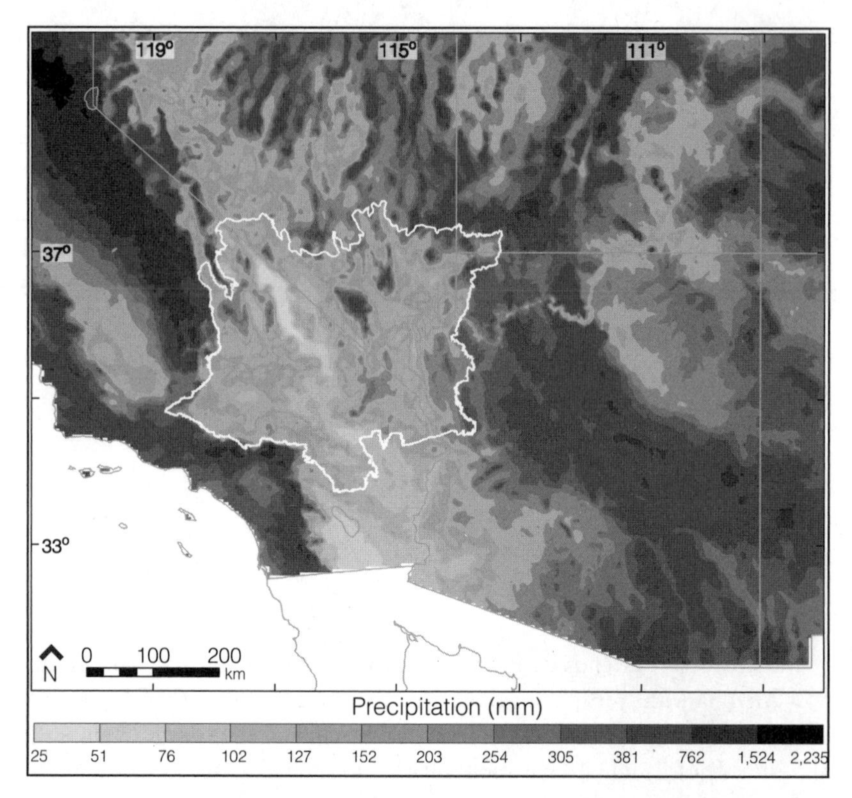

Fig. 1.1. Map depicting mean annual precipitation in the southwestern United States (1961–1990), with warmer colors corresponding to drier locations. The units are millimeters and were converted from inches, and the displayed precipitation range is approximately 50 mm at Death Valley to a maximum of 2235 mm in the central Sierra Nevada (from the Western Regional Climate Center and the PRISM group of the Oregon Climate Service, http://www.wrcc.dri.edu/precip.html). (See color plate following page 162.)

the world. Figure 1.2 (see also color plate) shows the 1971–2000 average monthly maximum temperature in the warmest month (July).

SPECIAL PECULIARITIES OF DESERT CLIMATES

In a typical year, measurable precipitation (at least 0.25 mm, the resolution of a National Weather Service standard rain gauge) occurs 20–25 days a year in most parts of the Mojave Desert, with 26–30 days in a few locations. The wettest of these days brings about a quarter of the annual precipitation at most sites. Statistics from a representative site are quite illustrative. Based on measurements from 1951–2000, the Las Vegas Airport averages 91 hours per year with measurable rain (1.0% of the hours in a year), and 30 of these hours bring only the minimum measurable amount of 0.25 mm. June averages just 1.4 hours with measurable rain,

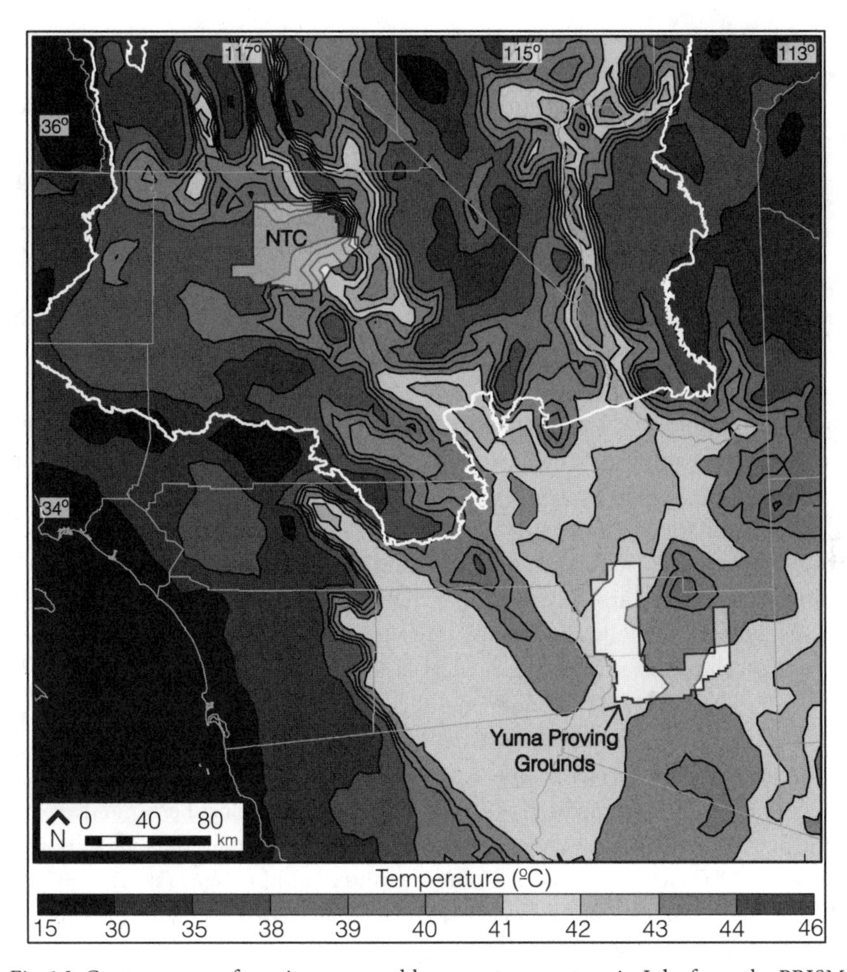

Fig. 1.2. Contour map of maximum monthly mean temperature in July, from the PRISM group. The brightest colors represent the warmest locations. The locations of Fort Irwin National Training Center (NTC) and Yuma Proving Grounds are included for reference. (See color plate following page 162.)

January averages 14.2 hours, and November through March account for a total of 58.4 hours. During the monsoon period, July experiences 5.5 total hours of precipitation, with 6.0 hours in August and 5.4 hours in September. The mean precipitation intensity when rain is falling (snow is very rare in valley locations) ranges from a winter minimum of 0.81 mm hr^{-1} in December to summer maximums of 2.01 and 1.98 mm hr^{-1} in July and August, respectively.

During this 50-year period, the wettest hour recorded in December brought 4.8 mm, the wettest hour in January brought 5.1 mm, and the wettest hour in February brought 7.6 mm. These values clearly show the steadier stratiform precipitation that occurs during the cool season. The wettest hour during this half of the twentieth century occurred in July, with 31.2 mm. On average, the wettest hour of a typical year brings 7.4 mm at the Las Vegas airport, or about 7% of the annual precipitation. This contrasts with Yuma, Arizona, in the Sonoran Desert, which averages 51 hours of precipitation annually, with the wettest hour bringing about 12% of the annual total.

Storms in the Mojave Desert are very spotty, especially during the convective season. Obviously, gages cannot be located everywhere, and storm dimensions are often small, with typical cell diameters of 2–5 km. Ground-based sampling will thus miss many events except where gauge density is high. Over the period from 2000–2005, several National Weather Service spotters in the well-sampled metropolitan Las Vegas area measured daily amounts of 63.5–88.9 mm during the monsoon season—the latter is more than three-quarters of the annual mean precipitation at the Las Vegas airport.

These representative statistics demonstrate that most life in the desert spends much of its time in "waiting" mode, enduring the 99% of hours with no moisture falling from the sky and anticipating, genetically or consciously, the remaining 1% that brings the vital rain. During those few hours when precipitation at last occurs, much happens, to both biological systems and to landscapes. In their own style, Mary Austin (1903) and Edward Abbey (1968) painted vivid and memorable descriptions of these salient characteristics of desert climate.

Because of high surface temperatures in summer, much of the moisture that falls from the sky is immediately recycled back into the atmosphere and does not contribute to groundwater recharge. At the Climate Reference Network station at Stovepipe Wells in Death Valley, California, the ground surface (skin) temperature maximum in July 2005 averaged 14.7°C warmer than the standard eye-height (1.5–2.0 m) maximum air temperature of 47.2°C. Such high temperatures readily evaporate light rains; summer precipitation must be heavy and sustained to cool the surface enough so that some water can make its way into the subsoil. However, such heavy precipitation is also likely to run off in significant quantities. In contrast, temperatures are much cooler during winter precipitation events, which are also of longer duration, so the available moisture is much more able to recharge the soil column. In this sense, not all precipitation is equal in terms of its effects on biological and hydrological systems. Hogan et al. (2004) address this subject more thoroughly.

Diurnal temperature fluctuations—warmer in the daytime and cooler at night—are much larger near the surface. The monthly average of the diurnal range of hourly temperatures in the above example at Death Valley was 12.4°C at eye height and 28.9°C at the surface. This can be understood by considering the flows of energy to and from the surface. During daylight, energy transfer to the atmosphere at the air-soil interface is through molecular diffusion, and the surface skin temperature must be considerably warmer than that of the atmosphere to provide a sufficiently high energy transfer rate. Under clear skies, water vapor is the most important source of infrared radiation emitted downward by the atmosphere. In desert locations this energy flow is much reduced and cannot balance the upward emission from the warm surface. As soon as the daytime solar input ceases, the surface rapidly loses energy through radiative processes. This greatly favors the rapid formation of surface inversions (warmer air temperatures at higher locations in the atmosphere) soon after nightfall.

Deserts are noted for their low humidity, but in summer the Mojave is affected by the intrusion of moist air with the monsoon, and the moisture content of the air rises sharply. During the summer the Great Basin Desert to the north typically has lower absolute and relative humidity than does the Mojave. During this time the air in the Sonoran Desert to the south is moister than in the Mojave, but both deserts can become very muggy, with dew-point temperatures near 21°C and higher. The arrival of the monsoon in the Mojave Desert is typically rather reliable, and occurs within a week of its usual July 3–7 onset in Tucson, Arizona, and seldom is more than two weeks early or late. The operational definition for monsoon arrival in Tucson is the first day of the first three-day period in summer when the dew point is 12.2°C or higher. Monsoon onset statistics can be found at the National Weather Service—Tucson web site. General overviews of the North American Monsoon System are provided by Adams and Comrie (1997) and Higgins et al. (1997).

The monsoon often arrives in a very definite manner, but its departure is halting and episodic and can often only be determined in retrospect. As ocean temperatures continue to warm, the eastern Pacific Ocean hurricane season becomes active. The eastern Pacific, which had 1971–2005 averages of 15.3 tropical storms (sustained winds of at least 56.3 km hr^{-1}), 8.8 hurricanes (winds at least 121 km hr^{-1}), and 4.2 major hurricanes (winds at least 177 km hr^{-1}, Climate Prediction Center Internet Team 2006a), is more active than the Atlantic, which had 1950–2005 averages of 11.0 tropical storms, 6.2 hurricanes, and 2.7 major hurricanes (Climate Prediction Center Internet Team 2006b). Though these storms seldom survive to reach the southwestern United States land mass, their ocean dissipation phase often produces moisture plumes that stream into the Southwest, carrying sufficient water vapor to initiate significant widespread rainfall and flooding in all but the largest stream basins. Tropical storm statistics in this region have been summarized comprehensively by Smith (1986). The causes of year-to-year variations in the timing and strength of the monsoon are not well understood, and were the subject of the North American Monsoon Experiment (Higgins et al. 1997; North American Monsoon Experiment Science Team 2001).

Desert winds have particularly strong diurnal cycles in speed and sometimes direction, especially in the warm season. In concert with low humidity and high

Fig. 1.3. Photograph of dead and dormant plants in the Anza-Borrego Desert Park, California, August 2003, illustrating drought pruning during a severe dry spell. *Larrea tridentata* (creosote bush) is in the foreground, *Ambrosia dumosa* (white bursage) is in the midground (hemispheric shrub), and vegetation probably includes *Opuntia acanthocarpa* (staghorn cholla) as well.

Fig. 1.4. Photograph of *Yucca brevifolia* (Joshua tree) depicting the effects of severe drought in Joshua Tree National Park, California, during August 2003.

temperatures, windy conditions promote rapid loss of exposed liquid water. The strong vertical air mixing associated with high afternoon temperatures cause winds to increase in the afternoon. For example, the 1987–2005 average July wind speed at a height of 6.1 m ranged from 6.0 km hr^{-1} at 6 a.m. to 23.0 km hr^{-1} at 3 p.m. (Local Standard Time) at the somewhat elevated (1400 m) Bureau of Land Management Remote Automated Weather Station at Kane Springs in southern Nevada. For these years, average gust speeds ranged from a mean of 13.8 km hr^{-1} at 6 a.m. to 45 km hr^{-1} at 2 and 3 p.m. Winds were near calm (< 4.8 km hr^{-1}) 25–33% of the hours just before sunrise, and about 1%–2% of the hours in early-mid afternoon. Valley stations, such as the Community Environmental Monitoring Site in Beatty, Nevada (979 m elevation), often show somewhat quieter conditions. Wind speeds there (3.0 m height, 2000–2005) ranged from 5.0 km hr^{-1} during the early morning to about 11.4 km hr^{-1} in mid-afternoon in July, and gusts ranged from 10.0–23.2 km hr^{-1}. Wind speeds were less than 4.8 km hr^{-1} about 2% of the time in early afternoon, and about half the hours overnight. These sites reflect typical summer desert diurnal wind behavior. Winds tend to maintain higher speeds at night on ridgetops than on nearby valley floors. Because wind greatly facilitates mixing and vapor transfer, in order to survive desert vegetation must adapt to the drying influences implied by these statistics.

DROUGHT AND FLOOD IN DRY CLIMATES

Deserts are dry by definition, as demonstrated by the predominant colors, tan or beige, and greens that are usually muted. Long intervals without significant precipitation are not uncommon. Death Valley has gone 385 consecutive days, encompassing a full calendar year, without measurable rain (December 29, 1988 to January 18, 1990) and has had six intervals without measurable precipitation lasting more than half a year since the monitoring station was relocated in 1961. Las Vegas has gone 150 days without measurable precipitation, and a 63-day dry stretch occurs about once a year. Thus, drought can be difficult to recognize. Plants are hardy and adapted to these conditions, but eventually even they will succumb to persistent precipitation shortages (figs. 1.3, 1.4, and 1.5). With two main precipitation seasons, the loss of one season in the Mojave Desert might not seem overly serious if precipitation in the other season is adequate. However, the winter season appears to be more important to groundwater recharge, which generally is much more efficient during the cool portion of the year. Trees especially respond more to winter than to summer precipitation.

Tree rings and other paleoclimate indicators show that droughts that last several years are not uncommon in this area, that severe multi-year droughts occur at least once or twice a century, and that the region is capable of experiencing even longer droughts, lasting up to a decade or two (e.g., Grissino-Mayer 1996). It is important to distinguish between summer droughts and winter droughts, because

Fig. 1.5. Photograph of *Opuntia bigelovii* (Teddy bear cholla), which are adapted to extreme drought, during a severely dry episode, Anza-Borrego Desert State Park, California, August 2003.

their causative mechanisms may be quite different. Ironically, it now appears that single-year and multi-year droughts in this driest portion of North America are controlled, in large part, by conditions in the wettest part of the planet, the Pacific and Indian oceans (Hoerling and Kumar 2003; Schubert et al. 2004a, 2004b).

Because dryness is the defining characteristic of deserts, it seems almost paradoxical that water is the major agent of landscape change in the Mojave Desert. The rare heavy precipitation events described earlier can lead to spectacular floods on three different scales (see Webb and Betancourt 1992). At one spatial scale, heavy summer monsoon thunderstorms can cover several square kilometers and produce rapid runoff on steep topographic gradients. This runoff can pile up and arrive as what is often described as a "wall of water" in normally dry washes, flowing downstream until the dry and sandy streambed absorbs most of the flood. At the next larger scale, copious moisture from tropical storms near Baja California can produce soaking rains over hundreds of square kilometers in late summer/early autumn, typically once or twice a decade somewhere in the Southwest. The resulting runoff can produce significant floods in medium-sized drainages. At the largest scale, the biggest rivers flood in winter, when the jet stream dips to low latitudes to bring abundant subtropical moisture for periods of 1–3 days or more. It is virtually impossible for the physical conditions capable of generating such floods to occur in summer.

On all temporal and spatial scales, runoff and flooding does not occur until sandy soils and streambeds become saturated. Heavy precipitation must persist long enough for this saturation of the soil column to occur and initiate runoff. Precipitation is rare enough—and persistent precipitation is even less common—that the associated floods, particularly on large scales, are uncommon as well. On a smaller scale, in a typical monsoon season, dozens to hundreds of very small but locally disruptive floods likely occur each summer and simply go unnoted. The ubiquitous presence of very wide streambeds, which seem incongruous in their desert surroundings, provides the most salient testimony to the highly skewed statistics of desert hydrology.

Temporal Variability of Climate

The short instrumental record of a little more than a century shows the high relative precipitation variability (i.e., as percentage of annual mean) expected in arid regions. Hereford et al. (2004) showed precipitation variation in the Mojave Desert region for the past century for annual warm and cool seasons. Areal average precipitation for the Southern Nevada Climate Division, located within this region, through water year 2006 (October–September) is shown annually (fig. 1.6), for cool seasons (fig. 1.7), and warm seasons (fig. 1.8), in order to show the very severe drought that became established at the beginning of the twenty-first century. The winter of 2004–2005 was the wettest (316 mm), and provided 56% more rain than the second wettest winter, 1940–1941, a strong El Niño year. The wet 2004–2005 winter followed almost immediately after the driest water year, 2001–2002, when a mere 27 mm of precipitation fell in 12 months. Exceptionally dry conditions then returned in the winter of 2005–2006. Such sequences of extreme variation are more common in desert climates than in any other, and indeed amount to a defining characteristic and are sure to have ecological consequences for many years.

Annual mean calendar year temperature is shown in figure 1.9. This shows a warming in the region starting approximately in the late 1970s. A prior warm period occurred in the first third of the twentieth century. High temperatures in the winter of 1933–34 and calendar year 1934 have long stood out in the West's climate records, and only recently have these individual years been matched. Multiyear average temperatures (fig. 1.9) have climbed well beyond prior values. One constituent station, at the Las Vegas airport, shows urban warming, but the divisional numbers are from 18–20 climate stations, most of them rural. From much other evidence the recent warming is almost certainly real rather than artifact, and is discussed further in Smith et al. (*this volume*).

Precipitation shows strong interannual variability. Winter precipitation in the

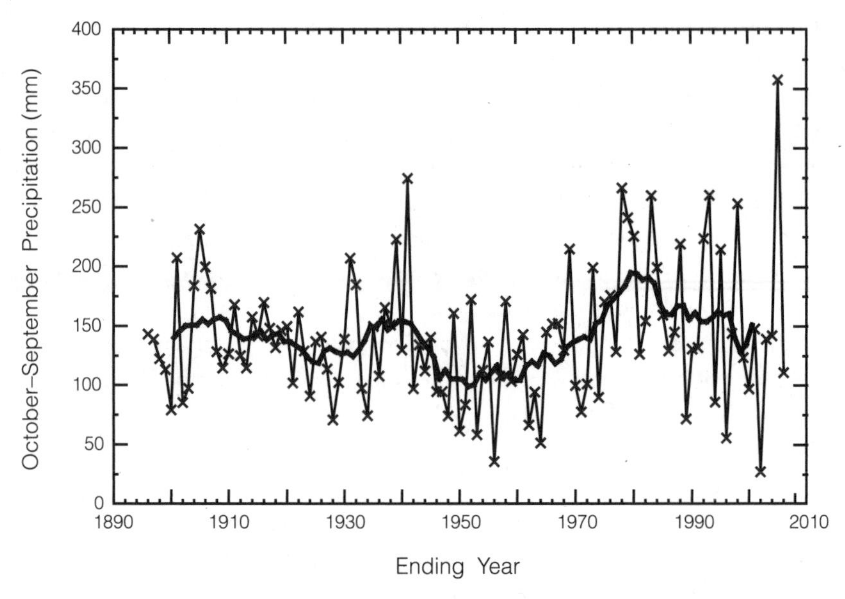

Fig. 1.6. Graph showing water year (October–September) precipitation for the Southern Nevada Climate Division from 1895–1896 through 2004–2005, with a bold line representing the ten-year running mean.

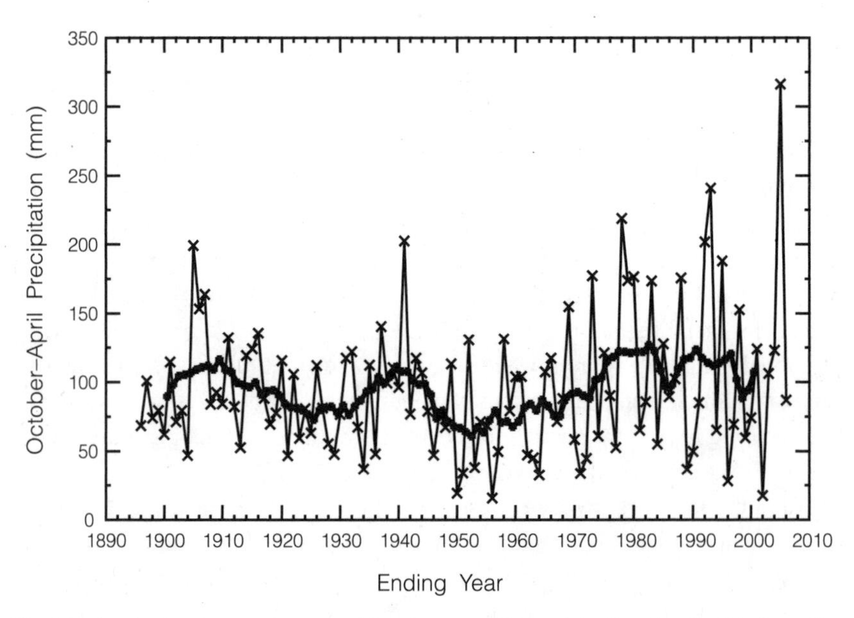

Fig. 1.7. Graph showing cool-season (October–April) precipitation for the Southern Nevada Climate Division from 1895–1896 through 2005–2006, with a bold line representing the ten-year running mean.

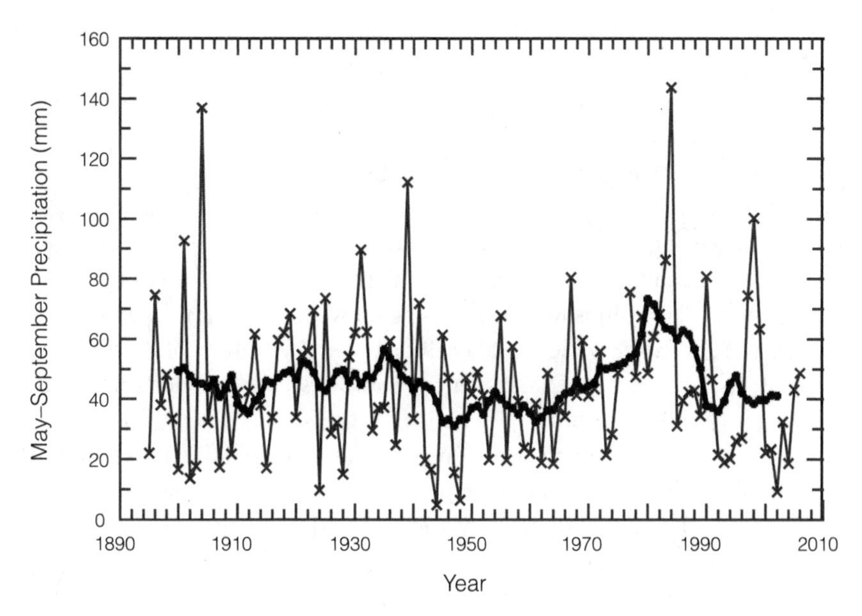

Fig. 1.8. Graph showing warm-season (May–September) precipitation for the Southern Nevada Climate Division from 1895 through 2005, with a bold line representing the ten-year running mean (from Western Regional Climate Center).

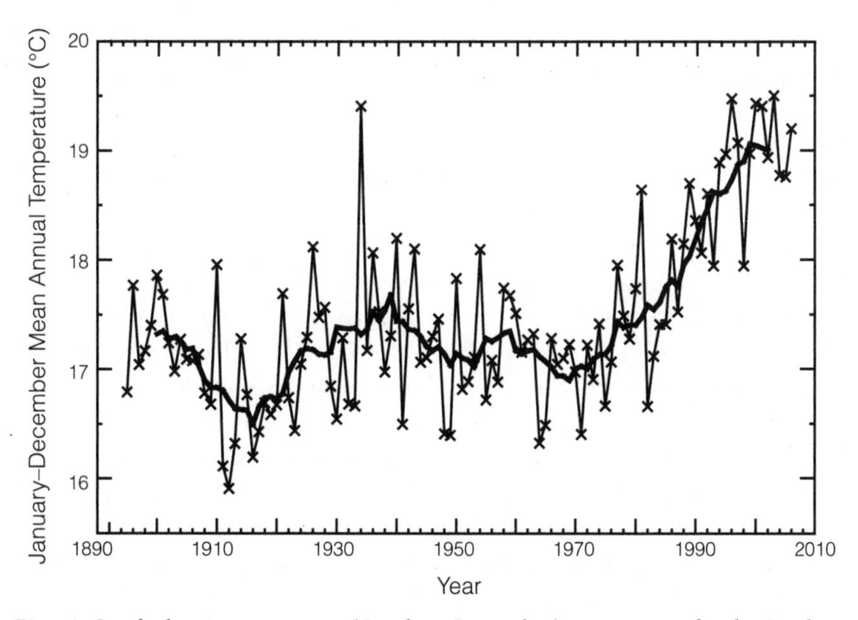

Fig. 1.9. Graph showing water year (October–September) temperature for the Southern Nevada Climate Division from 1895–1896 through 2004–2005, with a bold line representing the ten-year running mean (from Western Regional Climate Center).

Southwest has been shown to relate significantly to the ENSO phase (El Niño/ Southern Oscillation). When El Niño is present (negative Southern Oscillation Index), the likelihood of a wet winter is increased in the Mojave Desert. When La Niña is present (positive Southern Oscillation Index) there is very low likelihood of a wetter-than-average winter (e.g., Redmond and Koch 1991). El Niño and La Niña also influence the likelihood of rainy days in winter, and the day-to-day persistence of rainy patterns (Woolhiser et al. 1993). These effects are greatly magnified in streamflow records (Cayan et al. 1999). The initial statistical ENSO-precipitation relationships found in this region were later reproduced with numerical models, showing the causal chain between the equatorial Pacific Ocean temperature and the storm track over the Southwest, and provided the crucial linkages to streamflow records.

To a considerable extent, the winter precipitation time series reflects long-term fluctuations in the frequency of El Niño and La Niña. A major shift was noted in the Pacific climate in 1976 (Ebbesmeyer et al. 1991), after which El Niño became much more common, and La Niña was seldom seen for 20 years. The 1976 shift coincided with a change in the Southwest from a regime characterized by drier winters to one characterized by wetter winters. During the twentieth century, the Mojave Desert experienced several such shifts in winter precipitation regimes, with a typical duration of 1–3 decades. Similar variations are seen in North Pacific Ocean sea surface temperature patterns, a behavior known as the Pacific Decadal Oscillation (PDO) (Mantua et al. 1997; Mantua and Hare 2002). The physical basis for this "oscillation" has not been firmly established, and it is now frequently referred to as Pacific Decadal Variability (PDV). Likewise, the physical basis for how these ocean variations south of Alaska affect winter precipitation in the southwestern deserts has yet to be firmly established. However, the empirical connection remains quite intriguing.

Recent studies (McCabe et al. 2004; Gray et al. 2004) have suggested a connection between precipitation patterns in the Mojave Desert and in parts of the Colorado River Basin and patterns of variability in the Atlantic Ocean that Enfield et al. (2001) referred to as the Atlantic Multidecadal Oscillation (AMO). Correlations with AMO are highest in the northeastern portion of the Mojave Desert, and are on a time scale of 1–3 decades. The association is of interest, but the physical mechanisms that might produce such a relationship have yet to be described.

As noted above, multiyear droughts occur at least once or twice a century. These must originate with some part of the climate system that retains long-term memory, and the oceans are the logical candidate. Hoerling and Kumar (2003) show that the extended drought early in the twenty-first century in the American Southwest and in a latitude belt that also covers the Near East (Iraq to Afghanistan) can be traced to ocean conditions in the western Pacific and the Indian oceans, as well as the simultaneous occurrence of La Niña in the central and eastern equatorial Pacific. Schubert et al. (2004a, 2004b) likewise show that the Dust Bowl drought (1930s) can be largely explained by ocean conditions in the tropical Pacific and Indian oceans. Seager et al. (2005) show how ocean conditions can lead to prolonged wet and dry episodes in interior North America. This same ocean region has likely contributed to lengthy Southwest droughts.

PALEOCLIMATE

The lack of moisture in deserts greatly slows the decay process and facilitates preservation of evidence of past climate events, in the form of tree rings (living and dead), playa layers, sediments in river beds, packrat middens, and others. Studies have shown that the Mojave Desert and other parts of the Southwest experience significant multiyear droughts approximately 10–20 times per millennium (e.g., Grissino-Mayer 1996; Ni et al. 2002). These are reflected in streamflow reconstructions of the Colorado River (Woodhouse et al. 2006; Woodhouse and Lukas 2006). Some of the water that ends up in playas originates near or beyond the outskirts of the desert itself. For example, the Mojave River begins in mountains close to the ocean just north of Los Angeles but ends in the Mojave Desert near Baker, California, and exhibits the extreme hydrologic variability characteristic of deserts and of southern coastal California (Cayan et al. 2003). From 1931–2004, measured Mojave River peak annual flow at Barstow, California, was zero in 29 of 73 years, but reached 1821 $m^3 s^{-1}$ in March 1938. Enzel et al. (1989, 1992), Ely et al. (1993), and Enzel (1990, 1992) have documented numerous high stands (greatest depth within a long interval of time) of terminal lakes resulting from wet episodes in the Mojave River Basin, with the latest major wet interval about 400 years ago. Reconstructed flow on the Verde River in Arizona shows that the statistics of extreme flows derived from twentieth century records are not representative of all hundred-year episodes of the past 1,400 years (Redmond et al. 2002), information of value for engineering applications as well as ecological understanding.

CLIMATE CHANGE

A slow warming has been noted in the region commencing in the late 1970s to early 1980s (fig. 1.9). Dettinger (2005) has summarized the results of a half dozen climate models using several greenhouse gas emission scenarios from 1900 to 2100, for a location just north of Reno, Nevada; results are also representative of southern Nevada. These multiple model runs collectively show a departure from a baseline case of constant greenhouse gas, which begins to appear about 1980, with the envelope of projections clearing the climate baseline about 2005–2010. This corresponds well with the observed trends shown in figure 1.9. This is part of a general trend toward warming noted throughout the West of about 1°C since 1975–1980 (Redmond, 2007). Confidence in these temperature records is increased by the consistency between them and independent evidence from changes in the dates of spring "pulses" in streamflow and in the dates of phenological stages of lilacs and honeysuckles (Cayan et al. 2001), and the implied temperature changes. Weiss and Overpeck (2005) have noted similar changes in the nearby Sonoran Desert. Warming has been somewhat more pronounced at night than during daylight hours, and in the period from 2000 to 2006, a number of cities in the southern half of the West have experienced all-time record-high individual minimum temperatures. The mean annual temperature over an extended period from 2000–2006 is well above average for the western United States. Expressed in terms of the inherent observed variability, the Southwest has shown the largest anomalies in the continental United States during these years (National Research Council 2007).

The same set of models and scenarios show no projected systematic change in precipitation (Dettinger 2005) in the midlatitudes of the western states. In the Mojave Desert, the annual precipitation projections tend more toward drier than wetter, but by no more than 5%–10% different from present. Observed precipitation does appear to exhibit more interannual variability in the last 30 years of the record (fig. 1.6), particularly in winter (fig. 1.7). Thus, these results do not appear to be inconsistent with the expectations arising from the ensemble of models and greenhouse gas emission scenarios. Though there is significant effort under way to better understand the summer monsoon (North American Monsoon Experiment Science Team 2001), the controlling mechanisms for interannual and long-term variability are not known very well at present, and there seems little basis for assuming any particular climate change projection for summer precipitation. Warming with no change in precipitation is generally considered equivalent to an effective reduction in precipitation, since moisture-loss mechanisms increase as temperature increases. With respect to ecosystem responses to precipitation changes, attention should not be restricted simply to annual totals, but should also take into account the overall character of precipitation, i.e., the frequency of occurrence, average and extreme intensity during events, timing and duration of precipitation episodes, season of occurrence, and the overall frequency distribution of amount and intensity. Desert biology is finely tuned to the entire spectrum of precipitation characteristics, rare though precipitation might be.

IMPLICATIONS FOR SUSTAINABILITY

Many aspects of desert systems have been altered directly or indirectly by human behavior over the decades, and activities are under way to restore these systems toward their prior condition. The changes in climate observed in the last 30 years, if continued as projected, have implications for restoration efforts and for approaches to sustainability. If the climatic background changes sufficiently, the systems of interest may be close enough to transition points in phase space (the collection of indicators that characterize the state of a biological community and its environment) such that prior "equilibrium" conditions are no longer as relevant as before, or are perhaps unreachable. An important question is whether the goal is restoration of *process* or of *outcome*. Presumably, if the original processes can operate within the approximate historic range of physical environmental conditions, systems of interest that have experienced disturbance can eventually evolve toward the characteristics they exhibited in the past (in the absence of hysteresis effects). However, if background environmental conditions become sufficiently different from that history such that these processes operate in a different manner, or if new processes begin to operate, then those prior outcomes might only be attained with significant continuing expenditure of effort and resources, or, in more extreme cases, might not be attainable at all. That is, past statistics may be less applicable to future applications. We do not understand these systems well enough, at present, to make reliable or accurate quantitative deterministic predictions, but we may be in a position to evaluate the probabilities of different outcomes given different strategies.

Because there are so many interacting elements in these systems—many more than in atmospheric systems—there is strong rationale for following an approach that has, in recent years, become standard in atmospheric prediction. The ensemble approach, which entails running a model repeatedly with slightly different starting values that reflect the observational uncertainty of initial conditions, was originally suggested by Epstein (1969) and followed up by Leith (1974). This technique has been widely adopted in the past decade for weather and climate predictions and is now standard practice (Tracton and Kalnay 1993, many others since). Extensions of this method could perturb the representation of internal dynamics to account for imperfections in how well particular processes or constraints are known or are treated. The outcome of this process is a trajectory in state space with an associated probability envelope. There are numerous issues relating to various trade-offs in how this is done, beyond the scope of this discussion. However, such an approach helps avoid locking onto a single particular solution and seems a more suitable match to the fuzziness and uncertainty that are inherent in most decision-making processes.

NEEDS FOR CLIMATE MONITORING

Atmospheric processes in desert regions span a wide range of space and time scales and have special and often difficult statistical properties, many of which are discussed by Warner (2004), among others. Of particular note is the spatial spottiness of summer convective precipitation, with frequently huge downpours in close proximity to areas that receive nothing. It may require many years for these differences to "even out." To capture these effects requires dense networks of gauges, many of which might go long periods recording little or no precipitation. But when precipitation does occur, high time resolution is of great value. Since it is often difficult to justify such in situ networks for long periods because of the expenditure, we must resort to taking advantage of clusters of stations, such as those at the Nevada Test Site or in urban flood control districts in desert cities (such as Las Vegas or Phoenix/Maricopa County), or those used for research purposes (such as the Owens Valley Terrain-induced Rotor Experiment, Grubisic and Kuettner 2004). Because precipitation is rare, these must operate for many years to obtain adequate data for statistical analysis.

Similar statements can be made for stream gauges in streambeds that are usually dry, but that, on occasion, run wildly. In general, however, long-term consistent records (50 years or more) are not common. Hereford et al. (2004) identified 25 stations in the area they considered as the Mojave Desert—about one station per 6000 km^2, or an average spacing of about 80 km. At a given elevation, where precipitation climatologies and statistical properties should be similar, one can to a certain extent "swap space for time," making the assumption that multiple points in space are sufficiently separated and quasi-independent that their individual time series segments could be concatenated to provide an equivalently long single time series.

Mountains in general and desert mountains in particular are not well sampled by current networks (Consortium for Integrated Climate Research in Western Mountains 2006). In a given area, precipitation amount increases with elevation, but there may additionally be changes in other precipitation characteristics, such as seasonality, intensity, duration of events, time between events, and spatial extent. Because precipitation in elevated areas disproportionately contributes to subsurface recharge, more dense observation networks are needed in such settings.

For studies attempting to link biological behavior to climate (e.g., Smith et al. *this volume;* Webb et al., Long-term Data, *this volume*), climate time series must be spatially and temporally representative, and this usually implies close physical proximity of monitoring stations. The times between significant atmospheric events can be long, and much can happen in a short time. This means that monitoring equipment needs to be in a constant state of readiness, and able to work at high sampling rates.

For precipitation, remote sensing (such as the National Weather Service Doppler radar) can provide spatial and temporal coverage, but in dry desert air intense thundershowers can leave little surface precipitation if cloud bases are high, and for each event the reflectivity-intensity relationship often needs to be calibrated separately with the aid of a live surface network. When air has abundant moisture (during monsoon surges, for example), such relationships are more apt to resemble those used in other parts of the country. Mountain blockage hinders coverage, and false echoes from military radar evasion chaff (tiny strips of aluminum-coated glass fibers) can lead to misinterpretation.

A strategy that seems to make sense is one that entails a large-scale network with uniform station spacing at tens of kilometers, supplemented with a few dense networks of multiple physical (meteorological and hydrological) and ecological measurements. The placement and use of these stations should be tied in to programmatic goals of natural resource management agencies, research studies, and when possible, operational agencies with hazard-management responsibilities. Such practical considerations are invaluable for ensuring sustained attention to maintenance that is required if we are to develop the long-term records essential for understanding the behavior of physical and biological systems.

ACKNOWLEDGMENTS

I would like to thank Robert Webb, Eric McDonald, Lynn Fenstermaker, Richard Hereford, Kyle House, Yehouda Enzel, Jon Skindlov, and anonymous reviewers for their suggestions and assistance, and for numerous conversations on various aspects of desert climate. Portions of this work were supported by the Western Regional Climate Center and related National Oceanic and Atmospheric Administration projects.

REFERENCES

Abbey, E. 1968. *Desert solitaire.* Ballantine Books, New York, New York.

Adams, D. K., and A. C. Comrie. 1997. The North American monsoon. *Bulletin of the American Meteorological Society* **78**:2197–2213.

Austin, M. 1903. *The land of little rain.* 1997 reprint. Penguin Books, New York, New York.

Cayan, D. R., K. T. Redmond, and L. G. Riddle. 1999. ENSO and hydrologic extremes in the western United States. *Journal of Climate* **12**:2881–2893.

Cayan, D. R., S. A. Kammerdiener, M. D. Dettinger, J. M. Caprio, and D. H. Peterson. 2001. Changes in the onset of spring in the western United States. *Bulletin of the American Meteorological Society* **82**:399–415.

Cayan, D. R., M. D. Dettinger, K. T. Redmond, G. J. McCabe, N. Knowles, and D. H. Peterson. 2003. The transboundary setting of California's water and hydropower systems—Linkages between the Sierra Nevada, Columbia River, and Colorado River hydroclimates. Pages 237–262 *in* H. F. Diaz and B. J. Morehouse, editors. *Climate and Water: Transboundary challenges in the Americas.* Kluwer Academic Publishers, Dordrecht, the Netherlands.

Climate Prediction Center Internet Team. 2006a. Background information: East Pacific hurricane season. Online. National Weather Service Climate Prediction Center, Camp Springs, Maryland. http://www.cpc.ncep.noaa.gov/products/Epac_hurr/background_information.html.

Climate Prediction Center Internet Team. 2006b. Background information: the North Atlantic hurricane season. Online. National Weather Service Climate Prediction Center, Camp Springs, Maryland. http://www.cpc.noaa.gov/products/outlooks/background_information.shtml.

Consortium for Integrated Climate Research in Western Mountains (CIRMOUNT). 2006. Mapping new terrain: climate change and America's West. Miscellaneous Publication No. PSW-MISC-77. Online. U.S. Department of Agriculture, Forest Service, Pacific Southwest Research Station, Albany, California. http://www.fs.fed.us/psw/cirmount/publications/pdf/new_terrain.pdf.

Dettinger, M. D. 2005. From climate-change spaghetti to climate-change distributions for 21st Century California. Online. San Francisco Estuary and Watershed Science **3**: Article 4. http://repositories.cdlib.org/jmie/sfews/vol3/iss1/art4.

Ebbesmeyer, C. C., D. R. Cayan, D. R. McClain, F. H. Nichols, D. H. Peterson, and K. T. Redmond. 1991. 1976 step in Pacific climate: forty environmental changes between 1968–1975 and 1977–1984. Pages 115–126 *in* J. L. Betancourt and V. L. Tharp, editors. *Proceedings of the 7th Annual Pacific Climate (PACLIM) Workshop, April 10–13 1990, Asilomar, California.* Interagency Ecological Studies Program Technical Report No. 26. California Department of Water Resources, Sacramento, California.

Ely, L. L., Y. Enzel, V. R. Baker, and D. R. Cayan. 1993. A 5000-year record of extreme floods and climate change in the southwestern United States. *Science* **262**:410–412.

Enfield, D. B., A. M. Mestas-Nuñez, and P. J. Trimble. 2001. The Atlantic multidecadal oscillation and its relation to rainfall and river flows in the continental U.S. *Geophysical Research Letters* **28**:2077–2080.

Enzel, Y. 1990. Hydrology of a large closed arid watershed as a basis for paleohydrological and paleoclimatological studies in the Mojave River drainage system, southern California. Ph.D. dissertation. University of New Mexico, Albuquerque, New Mexico.

Enzel, Y. 1992. Flood frequency of the Mojave River and the formation of late Holocene playa lakes, southern California. *The Holocene* **2**:11–18.

Enzel, Y., W. J. Brown, R. Y. Anderson, L. D. McFadden, and S. G. Wells. 1992. Short-duration Holocene lakes in the Mojave River drainage basin, southern California. *Quaternary Research* **38**:60–73.

Enzel, Y., D. R. Cayan, R. Y. Anderson, and S. G. Wells. 1989. Atmospheric circulation during Holocene lake stands in the Mojave Desert: evidence of regional climate change. *Nature* **341**:44–47.

Epstein, E.S. 1969. Stochastic dynamic predictions. *Tellus* **21**:739–759.

Gray, S. T., J. L. Graumlich, J. L. Betancourt, and G. T. Pederson. 2004. A tree-ring based reconstruction of the Atlantic Multidecadal Oscillation since 1567 A.D. *Geophysical Research Letters* **31**:L12205.

Grissino-Mayer, H. D. 1996. A 2129-year reconstruction of precipitation for northwestern New Mexico, USA. Pages 191–204 *in* J. S. Dean, D. M. Meko, and T. W. Swetnam, editors. *Tree rings, environment, and humanity: proceedings of the international conference, May 17–21, 1994, Tucson, Arizona.* Radiocarbon, Tucson, Arizona.

Grubisic, V., and J. P. Kuettner. 2004. Sierra Rotors and Terrain-induced Rotors Experiment (T-REX). Online. Eleventh Conference on Mountain Meteorology and the Annual Mesoscale Alpine Program (MAP), June 20–25, 2004, Bartlett, New Hampshire. American Meteorological Society, Boston, Massachusetts. http://ams.confex.com/ams/11Mountain/techprogram/paper_77384.htm.

Hereford, R., R. H. Webb, and C. I. Longpré. 2004. Precipitation history of the Mojave Desert region, 1893–2001. U.S. Geological Survey Fact Sheet No. 117–03. Reston, Virginia.

Hereford, R., R. H. Webb, and C. Longpré. 2006. Precipitation history and ecosystem response to multidecadal precipitation variability in the Mojave Desert and vicinity, 1893–2001. *Journal of Arid Environments* **67**:13–34.

Higgins, R. W., Y. Yao, and X. L. Wang. 1997. Influence of the North American monsoon system on the U.S. summer precipitation regime. *Journal of Climate* **10**:2600–2622.

Hoerling, M., and A. Kumar. 2003. The perfect ocean for drought. *Science* **299**:691–694.

Hogan, J. F., F. M. Phillips, and B. R. Scanlon, editors. 2004. Groundwater recharge in a desert environment: the southwestern United States. American Geophysical Union, Water Science and Applications Series. Volume nine. American Geophysical Union, Washington, D.C.

Leith, C. E. 1974. Theoretical skill of Monte Carlo forecasts. *Monthly Weather Review* **102**:409–418.

Mantua, N. J., and S. R. Hare. 2002. The Pacific Decadal Oscillation. *Journal of Oceanography* **58**:35–44.

Mantua, N. J., S. R. Hare, Y. Zhang, J. M. Wallace, and R. C. Francis. 1997. A Pacific interdecadal climate oscillation with impacts on salmon production. *Bulletin of the American Meteorological Society* **78**:1069–1079.

McCabe, G. J., M. A. Palecki, and J. L. Betancourt. 2004. Pacific and Atlantic Ocean influences on multidecadal drought frequency in the United States. *Proceedings of the National Academy of Sciences* **101**:4136–4141.

Ni, F., T. Cavazos, M. K. Hughes, A. C. Comrie, and G. Funkhouser. 2002. Cool-season precipitation in the southwestern USA since AD 1000: comparison of linear and nonlinear techniques for reconstruction. *International Journal of Climatology* **22**:1645–1662.

National Research Council. 2007. *Colorado River Basin water management: evaluating and adjusting to hydroclimatic variability.* National Academy Press, Washington, D.C.

North American Monsoon Experiment Science Team. 2001. North American Monsoon Experiment (NAME): science and implementation plan. Online. National Oceanic and Atmospheric Administration, Washington, D.C. http://www.clivar.org/organization/vamos/Publications/name.pdf. Accessed October 19, 2005.

Redmond, K. T. 2007. Climate variability and change as a backdrop for western resource management. *In* L. Joyce, R. Haynes, R. White, R. J. Barbour, editors. *Bringing climate change into natural resource management.* Gen. Tech. Rep. PNW-GTR-706. U.S. Department of Agriculture Forest Service, Pacific Northwest Research Station, Portland, Oregon.

Redmond, K. T., Y. Enzel, P. K. House, and F. Biondi. 2002. Climate variability and flood frequency at decadal to millennial time scales. Pages 21–45 *in* P. K. House, R. H. Webb, and V. R. Baker, editors. *Ancient floods, modern hazards: principles and applications of paleoflood hydrology.* Water Science and Application Monograph 5. American Geophysical Union, Washington, D.C.

Redmond, K. T., and R. W. Koch. 1991. Surface climate and streamflow variability in the western United States and their relationship to large-scale circulation indexes. *Water Resources Research* **27**:2381–2399.

Schubert, S. D., M. J. Suarez, P. J. Pegion, R. D. Koster, and J. T. Bacmeister. 2004a. On the cause of the 1930s Dust Bowl. *Science* **303**:1855–1859.

Schubert, S. D., M. J. Suarez, P. J. Pegion, R. D. Koster, and J. T. Bacmeister. 2004b. Causes of long-term drought in the U.S. Great Plains. *Journal of Climate* **17**:485–503.

Seager, R., Y. Kushnir, C. Herweijer, N. Naik, and J. Velez. 2005. Modeling of tropical forcing of persistent droughts and pluvials over western North America: 1856–2000. *Journal of Climate* **18**:4068–4091.

Smith, W. 1986. The effects of eastern North Pacific tropical cyclones on the southwestern United States. NOAA Technical Memorandum No. NWS WR-197. National Oceanic and Atmospheric Administration, National Weather Service, Salt Lake City, Utah.

Tracton, M. S., and E. Kalnay. 1993. Operational ensemble prediction at the National Meteorological Center: practical aspects. *Weather Forecasting* **8**:378–398.

Warner, T. T. 2004. *Desert meteorology.* Cambridge University Press, New York, New York.

Webb, R. H., and J. L. Betancourt. 1992. Climatic variability and flood frequency of the Santa Cruz River, Pima County, Arizona. U.S. Geological Survey Water Supply Paper No. 2379. U.S. Government Printing Office, Washington, D.C.

Weiss, J. L., and J. T. Overpeck. 2005. Is the Sonoran Desert losing its cool? *Global Change Biology* **11**:2065–2077.

Woodhouse, C. A., S. T. Gray, and D. M. Meko. 2006. Updated streamflow reconstructions for the Upper Colorado River basin. *Water Resources Research* **42**:W05415. doi:10.1029/2005WR004455.

Woodhouse, C. A., and J. J. Lukas. 2006. Multi-century tree-ring reconstructions of Colorado streamflow for water resource planning. *Climatic Change* **78**:298–315. doi: 10.1007/s10584-006-9055-0.

Woolhiser, D. A., T. O. Keefer, and K. T. Redmond. 1993. Southern oscillation effects on daily precipitation in the southwestern United States. *Water Resources Research* **29**:1287–1295.

TWO

Effects of Global Change on Mojave Desert Ecosystems

STANLEY D. SMITH, THERESE N. CHARLET,
LYNN F. FENSTERMAKER, AND BETH A. NEWINGHAM

The human species is embarking on a global experiment in which we will, over the next century, observe how our planet responds to dramatic human-induced changes in climate. Through the burning of fossil fuels, intensive agriculture, and widespread fire and deforestation, we are injecting large quantities of greenhouse gases and nitrogenous compounds into the atmosphere, thereby changing global climate and ecosystem function. Through widespread alterations in land use, we are also influencing important feedbacks between the atmosphere and biosphere that have unpredictable consequences for global climate. Furthermore, we are moving species to new locations on the globe at an unprecedented rate, which is consistently proving to have profoundly negative consequences on native ecosystems. Together, these factors—elevated carbon dioxide (CO_2) concentrations, climate change, increased nitrogen (N) deposition, land use change, and the introduction of invasive species—constitute the phenomenon termed "global change."

There are several reasons to examine the effects of global change in the Mojave Desert. First, global change is a process that will affect all regions of the world, even remote deserts. Second, the southwestern United States is experiencing unprecedented levels of human population growth (Hughson *this volume*). Land use decisions related to that growth must be calibrated in the context of global change, taking into account, for example, the potential impacts on water resources in the region. Third, and probably most important, global change is so pervasive and has such great potential as a driver of biodiversity and ecosystem change that literally all future biological and ecological research will have to be considered in a global-change context if our goal is to provide sound advice for users of scientific information. Studies in biogeography, conservation biology, ecosystem restoration, and other scientific disciplines must take into account the potential for global change to alter the predictions and/or results of those efforts.

In this chapter, we will discuss various types of global change, with specific reference to anticipated changes in arid regions such as the Mojave Desert. We will then present the results of a major global change experiment in the Mojave Desert of southern Nevada and discuss how global change may alter the Mojave Desert in both ecological and land management contexts.

GLOBAL CHANGE AND DESERTS

The Mojave Desert Ecosystem: Water and Nitrogen Feedbacks

Desert ecosystems typically exhibit low annual net primary production (ANPP—the annual net growth of vegetation) and standing biomass (Noy-Meir 1973). The low ANPP of desert ecosystems has been attributed to low annual rainfall, low soil nutrient content, and the inability of perennial vegetation to maintain plant cover and leaf area due to recurring drought conditions (Smith et al. 1997). As a result, climate and land use changes that affect water and nitrogen inputs are anticipated to have important consequences for the structure and function of desert ecosystems. Over the past century, unsustainable land use practices have resulted in a shift from semiarid grassland to desert scrub vegetation in many areas. This process, called desertification, results in even lower ANPP and greater spatial variation in soil resources (Schlesinger et al. 1990).

Worldwide, desertification has been increasing at an alarming pace (Dregne 1991), so that arid and semiarid areas now occupy about 47% of the earth's land surface (Lal 2004). In contrast to land use, other global change factors, such as elevated CO_2, increased rainfall, and higher rates of N deposition, should each increase ANPP, and thus could potentially counteract the desertification process. Therefore, anthropogenic global change can have both positive and negative influences on deserts and on the desertification process. How increases in CO_2, changes in climate, and the disturbance of soil surfaces through accelerated land use will combine to alter the structure and function of desert ecosystems is an open and extremely relevant question. It is one that we must investigate if we wish to predict how desert and semi-desert regions will potentially be transformed in a rapidly changing global environment.

The Mojave Desert is of considerable interest with respect to global change. It contains the driest habitats in North America, and thus most closely approximates the world's great arid deserts (Smith et al. 1997). The Mojave is a warm, arid desert with most rainfall occurring in the winter, and it is often exposed to hard winter frosts. Long summer dry seasons and moderately cold winters result in an open, low-stature vegetation physiognomy dominated by evergreen and drought-deciduous shrubs, with fewer grasses than in the summer-rainfall eastern deserts and fewer trees than in the more subtropical southern deserts. The Mojave Desert is apparently the oldest of the four deserts located within the United States; the Great Basin, Chihuahuan, and northern Sonoran Desert are of more recent origin (Smith et al. 1997). Therefore, changes detected in the Mojave Desert may be a realistic harbinger of long-term changes that may occur in the world's great deserts as the global environment changes in response to human activities.

Elevated Atmospheric Carbon Dioxide

Over the past 250 years, atmospheric CO_2 has risen in concentration from approximately 280 ppm (μmol CO_2 mol^{-1} air) to 380 ppm (Keeling and Whorf 2002). It is predicted that levels will continue to increase for at least the rest of this century, reaching 550 ppm by 2050 and 700 ppm (an approximate doubling of today's concentration) by 2100 (Houghton et al. 2001). In addition to its effects on global climate (discussed later), elevated CO_2 affects two key plant physiological processes: (1) the rate of photosynthetic CO_2 assimilation, which especially increases in C_3 plants; and (2) stomatal conductance and plant water loss (transpiration) per unit leaf area, which decreases (Ellsworth et al. 2004, Nowak et al. 2004a). In combination, the resulting rise in photosynthesis and drop in transpiration substantially increases plant water-use efficiency (carbon gain per unit water loss). Because water-use efficiency should increase to a larger extent than does photosynthesis alone, water-limited ecosystems have been predicted to exhibit the most positive productivity response to elevated CO_2 (Strain and Bazzaz 1983). Indeed, in a survey of the world's primary biome types, Melillo et al. (1993) predicted that desert ecosystems would show the largest increase in ANPP (approximately 50%) of any biome in response to a doubling of CO_2 concentration. However, since elevated CO_2 will, in turn, increase transpirational surface area (due to higher growth rates), extrapolation of this leaf-level physiological response to landscape-scale changes has been more problematic. Even so, deserts are widely thought to be highly responsive to changes in atmospheric CO_2.

Climate Change

Anthropogenic increases in CO_2 have altered, and will continue to alter, the global climate (Schneider 1989; Kacholia and Reck 1997). Temperature increases in the 1990s were unprecedented in at least the past two millennia in the Northern Hemisphere (Mann and Jones 2003), and there is strong consensus that by the middle of this century increased levels of CO_2 and other greenhouse gases will result in a 1.5–4.5°C increase in global air temperatures compared with preindustrial times (Watson et al. 1996; Crowley 2000). Based on analyses of temperature trends over the past century, which indicate nighttime minima are rising more than daytime maxima (Karl et al. 1995; Easterling et al. 1997), projections are that the Mojave Desert will experience significant reductions in freeze events but only moderate increases in heat stress. Freezing stress establishes the northern distributional limit of many species of subtropical origin at the Sonoran-Mojave ecotone (Turnage and Hinckley 1938), and many warm-desert species such as *Larrea tridentata* at the Mojave-Great Basin ecotone (Beatley 1974; Pockman and Sperry 1997). Nocturnal warming could, therefore, relax these biogeographical constraints and result in the northern migration of many species, as has been observed in numerous regions of the globe (Walther 2003).

Global warming is also likely to alter synoptic weather patterns and change precipitation regimes, which, in turn, could have dramatic effects on ecosystems (Peters and Lovejoy 1992; Clark and Fastie 1998). Several global climate models (GCMs) predict that a doubling of atmospheric CO_2 concentration and its accompanying warming effect will significantly increase both winter and summer rainfall, or possibly just summer rainfall, in the southwestern desert region (Weltzin et al. 2003). It is also predicted that the frequency of El Niño events will increase (Tsonis et al. 2005), as will short-duration extreme precipitation events (Karl et al. 1995; Kunkel et al. 2003; Kim 2005). However, there is great variation among GCM simulations, with some predicting potentially drier conditions in the western U.S. (Smith et al. 2005), so at the moment we are not able to confidently predict specific precipitation regimes in the future. Collectively, shifts in precipitation (amount and seasonality) could have profound impacts on the structure and function of Mojave Desert ecosystems. Paleobotanical evidence indicates that the Mojave Desert region changed from a more mesic desert grassland to a more arid desert scrub approximately 5,000 years ago (Spaulding and Graumlich 1986). This transition coincided with the southern displacement of the "monsoon front" (the northward extension of a predictable monsoon in North America), which significantly decreased summer rainfall in the Mojave region. Since desert perennial life forms show differential ability to utilize summer rains (Ehleringer et al. 1991), significant shifts in both species and life-form composition would be expected to occur if summer rainfall were to increase in the region. Similarly, increases in winter rainfall, which tend to more effectively recharge deep soil layers than do summer rains, would act to significantly increase plant growth, cover, and ANPP.

Since the pioneering work of Noy-Meir (1973), desert ecologists have understood that the structure and function of arid systems are largely driven by large precipitation or pulse events, and that the adaptive traits and life histories of desert organisms are most closely approximated by a "pulse-reserve" paradigm (Reynolds et al. 2004). This paradigm predicts that desert life forms will respond in a burst of activity after heavy rain events, rather than the more typical incremental response seen in moister regions, and then quickly return to a quiescent status (seed in annuals, dormancy in perennials). This is because the most critical soil resources for plants—water and nitrogen—tend to be delivered to plants via short, high-intensity events rather than as a continuous resource in the growing season (Loik et al. 2004). Although rainfall in deserts can be delivered as pulses in all seasons, it is particularly the case in the summer (Redmond *this volume*). Within the warm deserts (Mojave, Sonoran, and Chihuahuan), the west-to-east gradient of increasing summer rainfall is accompanied by an increase in the frequency of large (> 5 mm) rainfall events. These large rainfall events activate many perennial plants by breaking drought-induced dormancy in the summer months (Huxman et al. 2004). The Mojave Desert has much lower mean summer rainfall but almost the same frequency of small (< 5 mm) rainfall events as the Chihuahuan Desert (Huxman et al. 2004), which is classified as a summer-rainfall desert (Smith et al. 1997). In contrast to large pulse events, small pulse events result in shallow soil wetting fronts, which tend to activate only near-surface microbes. This results in short bursts of carbon (C) loss from the system (Huxman et al. 2004) and stress to biological soil crusts due to rapid wetting and drying cycles (Belnap et al. 2004; Barker et al. 2005). Therefore, as climate changes, it will be of critical importance in the Mojave Desert ecosystem whether potential increases in summer rainfall occur primarily via an increase in large or small rainfall events. An increase in large pulse events should dramatically increase plant cover and productivity, while an increase primarily in small pulse events could have a wide range of deleterious consequences for the ecosystem, such as the demise of many biological crust organisms (Barker et al. 2005). Of perhaps equal importance is interannual variation in total rainfall. As with pulse events, the Mojave Desert shows greater interannual variation in total rainfall than the more eastern warm deserts (Davidowitz 2002), and so the magnitude to which climate change alters interannual variation in rainfall will also have important implications for ecosystem function and the distribution of species.

Nitrogen Deposition and Land Use Change

Deserts are well known to be nutrient-limited ecosystems. The continuing rapid growth and urbanization of metropolitan areas in the Southwest, as exemplified by Los Angeles and, more recently, Las Vegas, will inevitably lead to increasing N deposition in the Mojave Desert. Industrialized regions of the eastern U.S. have already experienced increased N inputs, primarily due to wet deposition (Aber et al. 1998). However, wet deposition is low in the western U.S., and N primarily enters desert ecosystems via dry deposition (dust), a process which historically has been very difficult to quantify. Even so, we now know that levels of N deposition in the vicinity of major metropolitan areas in the West can be quite high, in some cases similar to levels in heavily industrialized regions of the U.S. (approaching 40 kg ha^{-1} y^{-1}) (Fenn et al. 2003; Galloway et al. 2004). Because plant production in deserts is limited by N (Smith et al. 1997), increases in N deposition will likely result in higher plant production, particularly if accompanied by increases in precipitation and CO_2 concentration.

In addition to rapid urbanization, the deserts of the Southwest have experienced intense land use pressures such as overgrazing by domestic livestock and feral animals, such as horses, and damage due to off-road vehicle use (Webb 1982). These land use practices have resulted in the destruction of biological soil crusts, which historically have lived on the surface of desert soil over large areas (Belnap et al. 2003). These crusts undergo N fixation and can therefore act as important conduits for N input into desert ecosystems (Evans and Johansen 1999). Consequently, crust disturbance can have profound effects on desert fertility (Evans and Ehleringer 1993; Belnap 2003) and they can take hundreds of years to recover (Belnap and Eldridge 2003). Although there are strong indications that increases in summer rainfall, particularly in small events, have a deleterious effect on biological soil crusts (Belnap et al. 2004; Barker et al. 2005), it is not clear how changing precipitation patterns and increased anthropogenic N deposition will affect intact crusts, or interact with degraded desert surfaces to alter the functioning of desert ecosystems. This is a relatively unknown aspect of global change and could have important implications for the Mojave Desert region.

Invasive Species

An important change taking place in North American deserts is the introduction and spread of invasive species. These nonindigenous species have had deleterious consequences on the structure and function of these ecosystems. Two primary examples are *Tamarix ramosissima* (saltcedar), which has expanded to dominate most warm desert floodplain environments (Busch and Smith 1995), and *Bromus* spp., annual grasses that have invaded much of the Intermountain Region (Mack 1981). These annual grasses, particularly *Bromus tectorum* (cheatgrass) of the Great Basin and *Bromus madritensis* ssp. *rubens* (red brome) of the Mojave Desert, are now ubiquitous components of regional vegetation (Hunter 1991; Brooks 1999a). Years of above-average precipitation have caused a buildup of *Bromus* litter and standing dead plants. This has resulted in an increase in summer wildfires, which largely eliminate the above-ground biomass of the ecosystem and induce widespread shrub mortality (Brooks 1999b). Thus, *Bromus* adds a novel disturbance regime to these ecosystems by increasing episodic wildfires (Brooks et al. 2004). Where frequent fires have occurred, *Bromus* has replaced the longer-lived perennials of the desert community, thereby intensifying the fire cycle. Overgrazing, drought, and the destruction of surface crusts have all accelerated the invasion of *Bromus* into the Intermountain Region, and global change threatens to further aggravate this invasion (Sage 1996). Such a transition, from desert scrub, with moderate diversity and intricate feedbacks between its flora and fauna, to an annual grassland dominated by exotic species, would have major ecosystem-level implications for the region. The listing of the Southwestern Willow Flycatcher (*Empidomax traillii* ssp. *extimus*) as an endangered species in

Tamarix-dominated areas in 1995, and the potential listing of the Sage Grouse (*Centrocercus urophasianus*) in *Bromus*-dominated areas, are examples of the negative impact of invasive species on wildlife biodiversity.

One interesting aspect of plant invasions that has recently come to light is that they have been occurring throughout the past century, and not in a static world but in a world of increasing atmospheric CO_2 concentration and climate change. Indeed, Weltzin et al. (2003) convincingly argued that changes in atmospheric CO_2 and subsequent climate change have actually facilitated biological invasions in the recent past. In support of this argument, a study of multiple species of invasive annuals from the midwestern U.S. (Ziska 2003) found that historical increases in atmospheric CO_2 (from 280 to 370 ppm) had a stronger stimulatory effect on plant growth and seed production than simulated future increases in CO_2 (from 370 to 700 ppm). Therefore, we cannot just discuss and study global change that may occur in the future; we must also consider that we are in a climate that is changing at the present time, and that the communities and ecosystems we are now studying may have already begun to show strong, potentially nonlinear, shifts in response to global change.

Why Study Deserts?

Global change research over the past two decades has tended to concentrate on understanding how other biomes, most notably forests and tundra, will respond to global change. This is probably because forests are the major terrestrial reservoirs for biodiversity, and also because forests and arctic tundra are widely assumed to constitute the primary terrestrial "carbon sinks," areas with more C flowing in than out. However, the global community is starting to realize that deserts also harbor important levels of biodiversity, and may represent a larger C sink than was previously thought (Pacala et al. 2001; Lal 2004). Furthermore, it is important, from a global perspective, to understand how global change will influence the desertification process—will desertification speed up or slow down as CO_2 increases and climate changes? Another important reason to study the response of deserts to global change is that deserts represent stressful environments where the effects of global change may be magnified. From a strictly scientific point of view, this is reason enough to try and elucidate how these unique systems may respond to the myriad of global changes anticipated in the future.

CASE STUDY: THE NEVADA DESERT RESEARCH CENTER

Site Descriptions

The Nevada Desert Research Center (NDRC) is a network of two global change study sites located within the boundaries of the Nevada Test Site (NTS) in southern Nevada (145 km north of Las Vegas). The NDRC sites are used to experimentally examine how global change may potentially affect Mojave Desert ecosystems.

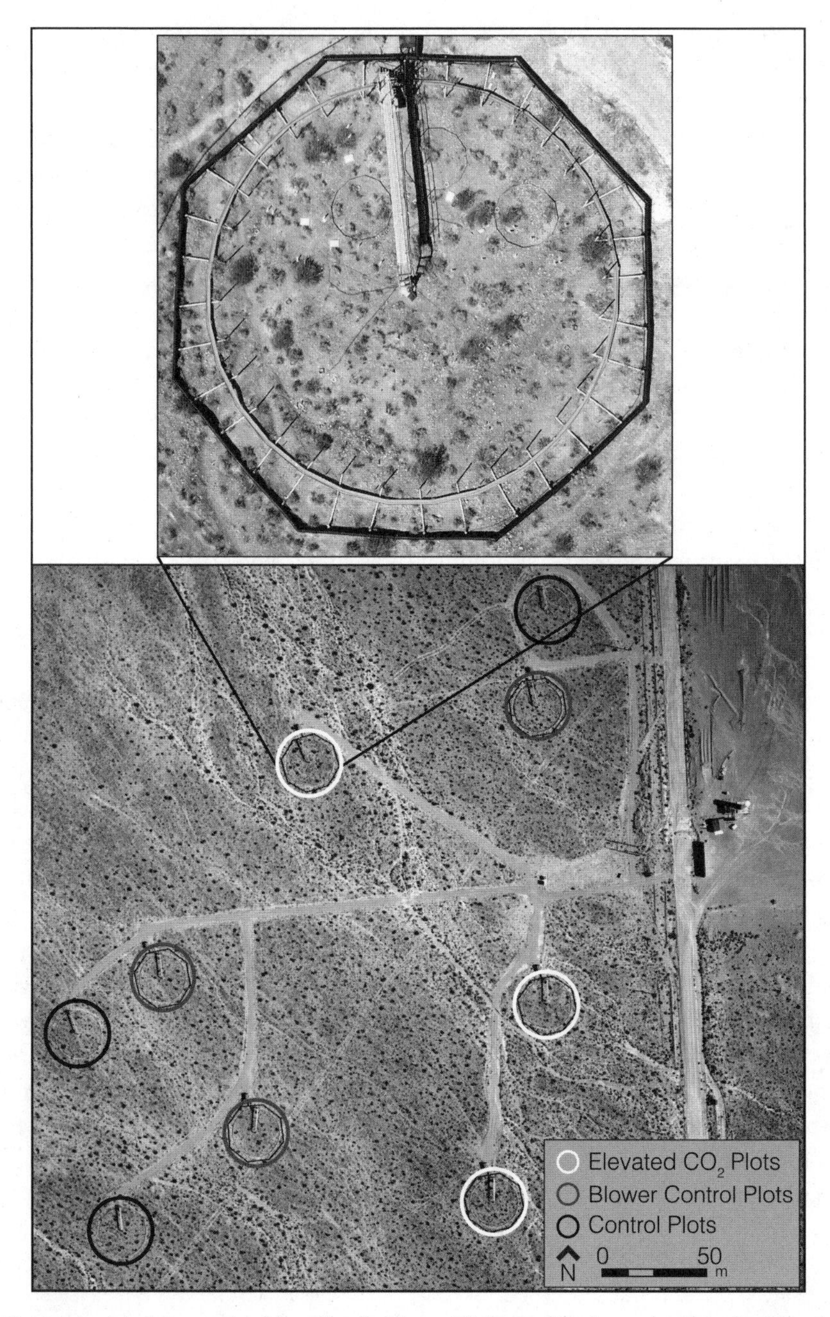

Fig. 2.1. Aerial photograph of the Nevada Desert FACE Facility, located within the Nevada Test Site, depicting the distribution of experimental plots and their respective treatments. A close-up aerial view (*top*) displays the infrastructure and platform used for nondestructive measurements within one of the elevated CO_2 plots. (See color plate following page 162.)

The NDRC is made up of the Nevada Desert Free-Air CO_2 Enrichment (FACE) Facility (36° 49' N, 115° 55' W, 965 m elevation) and the Mojave Global Change Facility (36° 46' N, 115° 59' W, 970 m elevation). The two study sites are located on Frenchman Flat, a broad alluvial fan (bajada) where vegetation is dominated by the evergreen shrub *Larrea tridentata* (creosote bush), the drought-deciduous shrubs *Ambrosia dumosa* (white bursage) and *Lycium* spp. (wolfberry), and *Pleuraphis rigida* (big galleta), a bunchgrass. Depending on rainfall, up to 75 species of winter annuals may also occur, including the exotic annual grass *Bromus madritensis* ssp. *rubens,* which is rapidly invading the NTS (Hunter 1991). Soils at the sites are gravelly sands that have well-developed biological soil crusts made up of bacteria, algae, mosses, and lichens. By the process of fixation, these organisms make atmospheric N available to other organisms (Evans and Ehleringer 1993). The prevailing climatic conditions of the NDRC are shown in table 2.1, and indicate that this is an arid desert with a distinct winter rainfall pattern, very high temperatures in the summer, and occasional hard freezes in the winter.

The Nevada Desert FACE Facility (NDFF), which has been in operation since April 28, 1997, (Jordan et al. 1999), is the first and only FACE facility in a desert ecosystem, and thus serves as a model for desert ecosystem-based global change research around the world. The NDFF consists of nine study plots, each 25 m in diameter (490 m^2): three blower plots with elevated CO_2 concentration (550 ppm); three blower plots with ambient CO_2 concentration (370 ppm); and three nonblower control plots (fig. 2.1; see also color plate). In the elevated CO_2 plot, pure CO_2 is mixed with ambient air and is then blown by a large fan through a plenum into 32 vent pipes that surround the plot. A computer regulates the opening and closing of individual control valves at each vent pipe based on wind direction.

Table 2.1 Climate conditions at the Nevada Desert Research Center

Parameter (units)	Value	Time occurred
Precipitation (mm)		
Annual mean	146	1984–2005
Record high (1984–2005)	270	1998
Record low (1984–2005)	53	2002
Winter mean, during recharge period	81	October–February
Spring mean, during growing period	34	March–June
Summer mean	31	July–September
Temperature (°C)		
Annual mean	17.3	1984–2005
December mean daily minimum	0.1	1984–2005
December mean daily maximum	13.1	1984–2005
July mean daily minimum	21.4	1984–2005
July mean daily maximum	38.1	1984–2005
Record high (1984–2005)	47.2	July 15, 2004
Record low (1984–2005)	−9.4	December 24, 1998

Historic precipitation and temperature data are from the Western Regional Climate Center (http://www.wrcc.dri.edu/), Desert Rock WSMO, Nevada.

The air-CO_2 mixture flows through upwind vent pipes to distribute the air-CO_2 mixture across the entire plot. The computer also monitors the air and updates the flow of pure CO_2 into the fan once per second to maintain the target value of 550 ppm. The NDFF is designed to operate 365 d y^{-1} and 24 h d^{-1} for 10–20 years (commencing from 1997). Conditional shutdown of operations is programmed into the system when temperatures drop below freezing or sustained high winds occur.

The Mojave Global Change Facility (MGCF) was constructed to complement NDFF research by examining the impact of global change factors other than increased CO_2 on the Mojave Desert. The MGCF is located within 3 km of the NDFF, on the southern side of Frenchman Flat within the NTS. The MGCF treatments are based on historical climate data, as well as predictive models, and are comprised of increased summer precipitation, increased N deposition, and surface disturbance of the biological crust. The summer rain treatment is applied as three 25 mm irrigation events at three-week intervals from early July to mid-August, which increases annual rainfall by approximately 50% and triples the amount of average summer rainfall. Nitrogen deposition is applied at two levels (10 and 40 kg ha^{-1}) as calcium nitrate dissolved in 5 mm of water and delivered in October of each year via an irrigation sprinkler system. The surface disturbance treatment is performed once annually in early October (prior to the N-deposition treatment) by manual scuffing of the surface of entire plots. These treatments are applied in eight blocks with twelve 14 × 14 m treatment plots (all combinations of treatments plus a control) in each block (fig. 2.2; see also color plate). A randomized design was used to establish different combinations of treatments in each block. The MGCF treatments were phased in over a two-year period—surface disturbance was initiated in 1999, N deposition began in the late winter of 2001, and increased summer rain began in the summer of 2001.

Production Processes

We initially hypothesized that plant photosynthesis and primary production would increase with elevated CO_2, as well as in response to N deposition and added summer rain. Carbon dioxide is one of the raw materials of photosynthesis, and increasing atmospheric CO_2 concentration enhances photosynthesis because it increases the diffusion gradient of CO_2 into the leaf. Higher sustained rates of photosynthesis should equate to higher growth and plant production. Likewise, because deserts are limited in soil N and water, increasing these two key resources should increase plant photosynthesis and growth. Together, increases in all three should result in pronounced increases in photosynthesis and growth of desert plants.

Results from the NDFF support the elevated-CO_2 hypothesis, although we have observed contrasting responses in wet and dry years (table 2.2). Photosynthesis is consistently higher with elevated CO_2 during all times that stomata are open

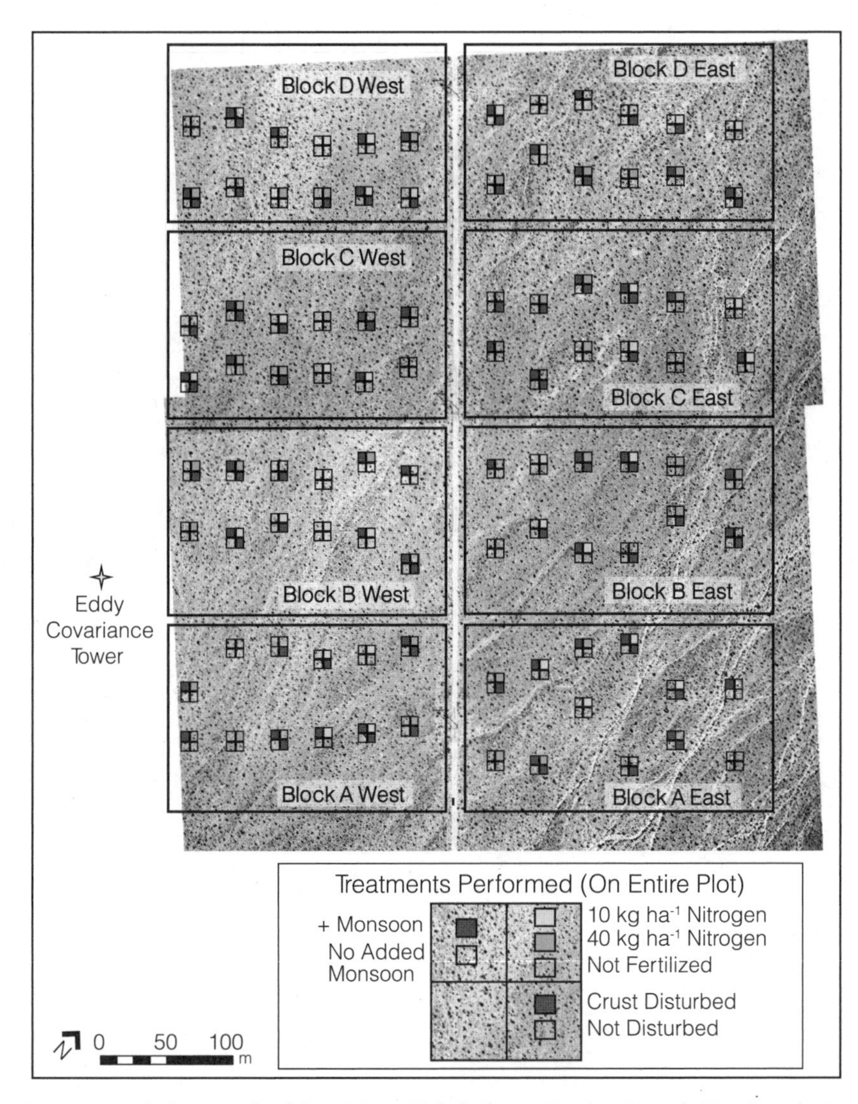

Fig. 2.2. Aerial photograph of the Mojave Global Change Facility, Nevada Test Site, depicting the distribution of ninety-six experimental plots within eight replicate blocks. For each plot, the boxes indicate the combination of treatments performed on the entire plot; *clockwise from upper left:* irrigation, fertilization, and crust disturbance. The explanation shows the treatment type on each plot. (See color plate following page 162.)

(Naumburg et al. 2003, 2004). Elevated CO_2, therefore, causes higher interannual variation in plant growth and production, with strong production enhancements in wet years but not in dry years for C_3 plants (Housman et al. 2006). Similarly, we found a pronounced positive response in photosynthesis, biomass, and seed

Table 2.2 Results from research activities at the Nevada Desert Research Center (NDRC)

Process	Parameter	Functional group	↑CO_2	↑N	↑H_2O	References
Production	Photosynthesis	Evergreen shrubs	+	0	+	Hamerlynck et al. 2002b; Naumburg et al. 2003, 2004; Barker et al. 2006
		Deciduous shrubs	+	0	+	Hamerlynck et al. 2002a; Barker et al., *unpublished*
		C_4 bunchgrass	0	0	+	Barker et al., *unpublished*
		Native annuals	+			Huxman and Smith 2001
		Exotic annual grasses	+			Huxman and Smith 2001
	Shoot growth	Evergreen shrubs	++(w) 0(d)	0	+(w) ++(d)	Housman et al. 2006; Newingham et al., *unpublished*
		Deciduous shrubs	+(w) 0(d)	+/0	0/–	Housman et al. 2006; Newingham et al., *unpublished*
		C_4 bunchgrass	0	0	++	Newingham et al., *unpublished*
		Native annuals	+	0	0	Smith et al. 2000; Irvin et al., *unpublished*
		Exotic annual grasses	+/++	0	0	Smith et al. 2000; Irvin et al., *unpublished*
	Reproduction	Evergreen shrubs	0			Newingham et al., *unpublished*
		C_4 bunchgrass	0	0	0	Newingham et al., *unpublished*
		Native annuals	+	0	0	Smith et al. 2000; Irvin et al., *unpublished*
		Exotic annual grasses	++	0	0	Smith et al. 2000; Irvin et al., *unpublished*
	Cover	Lichens		–/– –	0	Zimpfer et al., *unpublished*
Water balance	Stomatal conductance	Shrubs	–	0	0(w) ++(d)	Nowak et al. 2001, Hamerlynck et al. 2002b Naumburg et al. 2003, Barker et al. 2006
		C_4 bunchgrasses	–	0	+	Nowak et al. 2001; Barker et al., *unpublished*
		Annuals	–			Huxman and Smith 2001
	Sap flux	Evergreen shrubs	–	0	0/++	Pataki et al. 2000; Holmes and Nowak, *unpublished*
	Soil moisture		0	0	0/++	Nowak et al. 2004b; Holmes and Nowak, *unpublished*
Nutrients	N-fixation	Biological soil crusts	++	–	+	Billings et al. 2003; Zimpfer et al., *unpublished*
	Net N mineralization	Biological soil crusts	0			Billings et al. 2002a
	Available soil N	Biological soil crusts	– –			Billings et al. 2002a, 2002b
	Resorption efficiency	Shrubs	0			Housman et al., *unpublished*
	Plant C:N ratio	Perennials	+	–	+	Naumburg et al. 2003, Barker et al. 2006
		Annuals	+	–	0	Huxman et al. 1999, 2001; Irvin et al., *unpublished*
	Litter decomposition		0			Weatherly et al. 2003
C Balance	Net CO_2 exch.		–			Jasoni et al. 2005

Primary experimental manipulations at the NDRC include increasing atmospheric CO_2 (↑CO_2), nitrogen deposition (↑N), and summer rain (↑H_2O). Process studies are categorized as production, water balance, nutrients, and carbon (C) balance, with individual measurement parameters within each category.

Response categories (as percent change from controls) are as follows: "++" (> 25% increase), "+" (10–25% increase), "0" (no significant change), "–" (10–25% decrease), and "– –" (> 25% decrease). Responses followed by "(w)" are during wet years or seasons, and followed by "(d)" are during dry years or seasons. For the ↑N treatment, the symbol before each "/" is the response to the 10 kg ha^{-1} yr^{-1} treatment, and the number after each "/" is the response to the 40 kg ha^{-1} yr^{-1} treatment. For the ↑H_2O treatment, the symbol before each "/" is during the spring growing season, and the symbol after the "/" is during late summer after the added rainfall treatment. Open cells indicate that there are no data for that parameter. For multiple references, the first listed reference(s) corresponds to ↑CO_2 response, and the later reference(s) correspond to ↑N and ↑H_2O responses.

production in annual plants in a wet El Niño year but not in subsequent drier years. Additionally, we observed that the exotic *Bromus* was more responsive to elevated CO_2 than native annual grasses or forbs, with above-ground biomass 2.3 times higher (Smith et al. 2000, Nagel et al. 2004). As atmospheric CO_2 continues to increase, this trend has fairly ominous implications for Mojave Desert vegetation, given the propensity for *Bromus* to initiate, and then accelerate, fire cycles in the region (Brooks 1999a, 1999b). In contrast to C_3 plants (shrubs and annuals), elevated CO_2 did not enhance photosynthesis and growth in the C_4 bunchgrass, *Pleuraphis rigida*. This finding is consistent with many other studies that show C_4 plants to be largely unresponsive to elevated CO_2 as a result of their biochemical CO_2-concentrating mechanism. It has also led to the hypothesis that elevated CO_2 has favored the invasion of C_3 woody shrubs into C_4 grasslands over the past century (Idso 1992), resulting in today's C_3 shrub-dominated vegetation in the Mojave Desert (realizing that other important factors, such as overgrazing and drought, have also been important in this shift).

Overall, the results show that, when integrated across all growth forms, CO_2 stimulation of production is most pronounced when other key resources, such as N and water, are not limiting. When soils are dry and these resources are relatively unavailable, plants appear much less responsive to elevated CO_2. There are also key growth form differences: C_3 perennials are more responsive to elevated CO_2 than C_4 perennials, and, as stated above, the exotic annual *Bromus* appears to be the most responsive of all. If a fire cycle stimulated by exotic grasses were to indeed accelerate, then we are likely to see substantial changes in vegetation structure and biodiversity across the Mojave Desert (Smith et al. 2000), a region that has not historically experienced widespread fire as a disturbance.

Given higher photosynthesis and growth at elevated CO_2 levels, we also expected overall ecosystem CO_2 exchange to increase (i.e., for the ecosystem to become a larger CO_2 sink). However, to our surprise, in 2004 (an average rainfall year) we saw a significant decrease in net ecosystem CO_2 exchange with elevated CO_2 (table 2.2). We propose that this is because with elevated CO_2 soil respiratory losses are greater than photosynthetic gains, a hypothesis which is supported by soil respiration data (DeSoyza et al. *unpublished data*).

Water and N are soil resources that strongly limit plant growth in desert regions (Smith et al. 1997). Thus, we hypothesized that added N and summer rain would increase photosynthesis and production in all desert growth forms at the MGCF. While added summer rain stimulated production in most growth forms, there were few responses to the addition of N alone (table 2.2). Some species did respond to added N, but only in concert with added summer rain. The water response was particularly prevalent in summer-active species (evergreen shrubs and bunchgrasses) but was not observed in primarily spring-active growth forms (drought-deciduous shrubs and winter annuals), which principally rely on soil moisture recharge during winter rains. Therefore, if climate change were to result in an increase in summer monsoon activity, we might expect to see a relative increase in evergreen shrubs and bunchgrasses in the plant community.

Biological soil crusts exhibited some intriguing responses to the global change treatments. We anticipated that growth of poikilohydric organisms in the crusts would be stimulated by additional water but not necessarily by additional N due to the fact that the soil crust community has a strong biological N-fixation component (mainly free-living or mutualistic cyanobacteria). However, we observed that lichen cover did not change in response to added summer rain and declined with N addition (table 2.2). It is unclear if this is an actual negative response or simply a result of the way the N was applied—in a single event in the fall. An additional experiment showed that soil crust mosses exhibited widespread chlorosis (bleaching) during the drought of 2003–2004, which was due to a number of small precipitation events that partially hydrated the mosses but for an insufficient time to repair past desiccation damage (Barker et al. 2005).

In summary, these observations offer several clear lessons about the effects of global change on photosynthesis and production. First, specific types of desert vegetation may show significant increases in production at elevated CO_2 levels, with C_3 perennials showing a greater response than C_4 perennials, and exotic C_3 annuals showing the greatest response of all. However, the overall response of the community to elevated CO_2 will be dictated by changes in precipitation patterns and perhaps also N availability—with increased precipitation and N deposition, there could be substantial increases in primary production in response to higher CO_2, and thus an ecosystem shift to a more semiarid level of productivity. In contrast, if the climate becomes drier as some models predict, then the CO_2 response may be muted. This means that production would be higher in wet years, but due to intervening drought periods, production gain would not be maintained as higher cover or biomass.

Water Balance

Many experiments have shown elevated CO_2 to result in reduced stomatal conductance (the aperture of stomatal opening), which should decrease transpiration per unit leaf surface area. Our results have consistently shown that perennial plants have reduced leaf conductance (as well as reduced sap flux in *Larrea*) with high CO_2 levels (table 2.2). However, we have not observed a change in soil-moisture content with elevated CO_2 (Nowak et al. 2004b), which contrasts with studies in semiarid grasslands, which have shown higher soil-water contents in plots exposed to elevated CO_2 (Niklaus et al. 1998; Hungate et al. 2002). We suspect that this difference is due to two factors. First, over the course of a growing season, Mojave Desert vegetation uses all available soil water not lost via evaporation (Yoder and Nowak 1999), so there should be no net differences in soil moisture storage over time and only subtle changes in rates of water use. Second, decreases in leaf-level conductance are apparently countered by increases in total leaf area, due to higher production with elevated CO_2. This results in soil moisture at the plot scale that is the same at both ambient and elevated CO_2 concentrations.

At the MGCF, experimentally tripling summer rain has resulted in a dramatic increase in soil moisture, but there is little carryover of this extra moisture to the following spring growing season (table 2.2). This is because the high vapor pressure deficits (gradient of water vapor from the leaf to the surrounding air) in arid regions and resulting high soil-water evaporation rates tend to cause perennial vegetation to use all available soil moisture within a given growing season (Smith et al. 1995; Yoder and Nowak 1999). In contrast to increased rainfall, N deposition has not yet affected stomatal conductance and soil moisture, nor has there been a consistent effect of surface disturbance on soil-water balance at the MGCF. The former result is not surprising, at least over the short term, but the latter nonresponse is somewhat unexpected, in that we anticipated surface disturbance to significantly change surface infiltration of water. We anticipated that plot-wide surface disturbance, particularly of biological crusts (Yair 2003) and surface vesicular horizons (Hamerlynck et al. 2002b), would alter infiltration rates, and thus, would change soil-moisture contents over time.

Biogeochemistry

A majority of the biogeochemistry (nutrient cycling) work at the NDRC has been carried out by a group led by R. D. Evans (Washington State University); the following is a brief summary of their findings to date. Elevated CO_2 has had contrasting effects on biogeochemical processes—it has strongly stimulated N fixation by biological soil crusts, but has also led to a major decrease in plant-available N in the soil (table 2.2). A shift in leaf litter quality with elevated CO_2, involving an increase in the C:N ratio, is consistent with lower plant-available N in the soil. However, since there have been no significant changes in litter decomposition, net N mineralization rates, or N and P resorption efficiency by shrubs, it is not yet clear why there is a strong reduction in plant-available N in soils. Based on both fatty acid and molecular data (R. D. Evans and E. Robleto *personal communication*), there are strong shifts in the microbial communities involved in soil N metabolism in the elevated CO_2 plots, indicating that reduced soil N availability does appear to be real and already stimulating community change in the soil.

Biogeochemistry work at the MGCF is still in a preliminary stage. We have, however, conducted a number of biological soil crust studies and found that added summer rain stimulated N fixation by the crust community, while N deposition reduced N fixation and crust activity in general. Added summer rain also appeared to increase the leaf C:N ratio of the evergreen shrub *Larrea,* which in turn results in a partial down-regulation of photosynthesis (reduction in photosynthesis due to a reduction in leaf N content) during the following spring growing season (Barker et al. 2006).

Remote Sensing

The size and number of plots at the NDFF and MGCF make it virtually impossible to acquire data from all plots within phenologically appropriate time scales during the typically short spring growing season of the Mojave Desert. We have therefore been testing field-based data (leaf, canopy, and whole-plot levels) and aircraft hyperspectral remotely sensed data (ecosystem level) to assess various plant parameters at different scales. Field-based remote sensing measurements were acquired with spectral analysis systems (Analytical Spectral Devices Field-Spec Pro, PP Systems Unispec, and PP Systems Dual-Channel Unispec). These measurements have focused on biological soil crust and leaf-level performance parameters. *Larrea* has been the primary species of interest for the leaf-level measurements, but other species have also been measured. Canopy- and plot-level measurements have been acquired for all species present within the plots, with particular emphasis on *Larrea*, three deciduous shrubs (*Ambrosia*, *Lycium* spp., and *Krameria erecta*), and the bunchgrass *Pleuraphis*. Aircraft data were acquired using three different sensors: (1) the Department of Energy's Remote Sensing Laboratory's Daedalus 1268 sensor, which collects 12 bands of data, where 10 bands are from 0.40–2.38 μm wavelength and two are thermal bands at 8–10 μm wavelength, with a spatial resolution of 1.5 and 0.7 meters, respectively; (2) the Earth Search Sciences Inc. Probe 1 sensor, which collects 128 bands of data within the 0.4–2.5 μm range at a 5 m spatial resolution; and (3) the Jet Propulsion Laboratory's Airborne Visible/Infrared Imaging Spectrometer (AVIRIS) sensor, which collects 224 bands of data from 0.4–2.5 μm with 5 m spatial resolution.

For both leaf-level and canopy-level data, most of the vegetation indices calculated have revealed a significant difference between elevated and ambient CO_2 plots at the NDFF after a treatment period of 5 or more years. For example, the Normalized Difference Vegetation Index (NDVI), a "greenness" index, was significantly higher in elevated CO_2 plots for many of the perennial species, particularly after several years of treatment (table 2.3). Data from June 1998 did not show statistically significant increases in NDVI with elevated CO_2, despite higher production and higher photosynthetic rates (table 2.2). This was probably because the duration of the CO_2 treatment was not sufficient to allow leaf area to accrue. By August 1998, after the full growing season, NDVI was 1.5 times higher in *Ambrosia*. Conversely, many of the NDVI values in 2004 were not significantly different from August 1998 values, despite 7 years of elevated CO_2 exposure. This reinforces our hypothesis that the effect of elevated CO_2 will be compounded by climate variability, and that it will take time for elevated CO_2 effects to become more apparent in this arid ecosystem. At the MGCF, the NDVI and other vegetation indices were

Table 2.3 Elevated/ambient carbon dioxide (CO_2) ratios (E/A) of Normalized Difference Vegetation Index (NDVI) values for five perennial species at five measurement times at the Nevada Desert Free-air Carbon Dioxide Enrichment Facility

	Precipitation	*Larrea tridentata* (Evergreen)	*Ambrosia dumosa* (Deciduous)	*Krameria erecta* (Deciduous)	*Lycium pallidum* (Deciduous)	*Pleuraphis rigida* (C_4 bunchgrass)
June 1998	Wet	1.10	1.12	1.11	1.10	1.10
August 1998	Wet	1.09	1.50[a]	1.12	1.43[a]	1.02
June 1999	Dry	1.10	1.14	1.24[a]	0.76	0.80
July 2000	Dry	1.93[a]	1.12	1.25[a]	0.71	0.59
June 2003	Average	1.35[a]	1.26[a]	1.16	1.20[a]	1.39[a]
April 2004	Average	1.07	1.12	1.09	1.10	1.29[a]

Ambient CO_2 NDVI values ranged from 0.3–0.7 for *Larrea tridentata*, 0.1–0.5 for the deciduous shrubs, and 0.1–0.4 in *Pleuraphis rigida*.
E/A ratios followed by [a] were significantly different from 1.0 ($P < 0.05$)

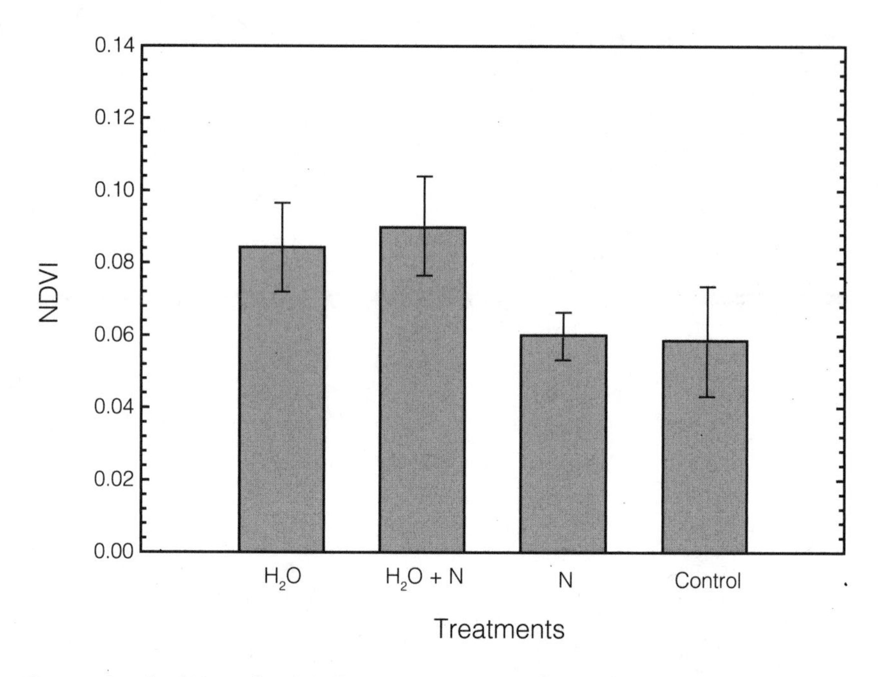

Fig. 2.3. Graph of Normalized Difference Vegetation Index (NDVI) values at the plot scale for several treatment types at the Mojave Global Change Facility. Depicted treatment types include control plots, increased summer rain (H_2O), increased nitrogen deposition (N), and both added summer rain and added N (H_2O+N). Vertical error bars are ± 1 SE.

significantly higher in the irrigated plots than in the nonirrigated plots. However, N deposition had no effect on NDVI or other indices, except in those plots that were also exposed to added summer rain (fig. 2.3). Further data collection and analyses are being performed to examine longer-term responses at the MGCF, as we anticipate NDVI to increase with N deposition over time.

The use of remote sensing has provided a means to gather synoptic information about surface conditions in response to environmental conditions and experimental treatments at spatial scales of interest, such as entire plots at the NDRC. In combination with ground-based measurements, remote sensing has enabled us to scale and better understand the potential plant community and ecosystem responses to the climate change treatments imposed at the NDRC study sites.

GLOBAL CHANGE AND MANAGEMENT ISSUES IN THE MOJAVE DESERT

Combining results from a number of global change studies with our findings at the Nevada Desert Research Center (NDRC), we can arrive at a number of broad conclusions about how the Mojave Desert ecosystem may respond to future global change (table 2.4). We know that elevated CO_2 stimulates greater plant production, but this stimulation occurs primarily in wetter years when soil resources (water and N) are more readily available. In the absence of significant precipitation change, elevated CO_2 only acts to increase the annual variability in plant production, thus exacerbating an already episodic ecosystem. Furthermore, given the differentially greater production of *Bromus* with elevated CO_2, any increase in production could be accompanied by an increased fire cycle, and thus an accelerated shift from desert shrubland into a fire-controlled annual grassland (Sage 1996; Smith et al. 2000). Although we have seen relatively few direct effects of increased N deposition at the MGCF, we predict that long-term increases in N deposition will have ecosystem-wide effects that are similar to the effects of elevated CO_2—mainly a stimulation of primary production with the greatest stimulation seen in the exotic *Bromus*. Response to changes in precipitation regimes are more difficult to predict. Higher annual rainfall will obviously increase plant production, resulting in a long-term shift from an arid to a semiarid ecosystem type, with greater abundances of deep-rooted woody species and summer-active bunchgrasses (table 2.4). Conversely, a reduction in rainfall, or even an increase in drought cycles imbedded within an overall wetter climate, may result in increased mortality of perennial plants, and thus in a more species-poor system.

It is challenging to predict how the Mojave Desert ecosystem will respond as increases in atmospheric CO_2, N deposition, and annual rainfall simultaneously occur. Responses to other combinations of change (for example, increases in CO_2 and N deposition and increased interannual precipitation variability) will be even more difficult to foresee. Hopefully, additional experimental work, particularly examining potential responses of desert ecosystems to rainfall variability, will allow the development of ecosystem models that will improve our predictive capabilities with regard to global change.

Table 2.4 Potential ecological effects of global change in the Mojave Desert

External variable	Functional response	Potential new regime
Elevated CO_2	Greater plant production	More productive desert
	Increased plant invasion	Increased fire frequency
Higher temperature	Species range shifts	Community disequilibrium
Altered precipitation		
Wetter winter	Greater production of exotics	Increased fire frequency
Wetter summer	Greater production of bunchgrasses	Semiarid ecosystem type
Drier	Increased mortality	Species-poor system
Increased N deposition	Reduced N-fixation	Loss of N-fixing species
	Greater plant production	More productive desert

Restoration ecology has become an extremely important subdiscipline within the ecological sciences. Global change presents a particularly imposing challenge for restoration ecology, because successful restoration efforts require the best information possible regarding ecosystem structure and function, which is difficult to obtain in a rapidly changing world. To allow restoration ecology to move forward, the following five issues need to be addressed. First and foremost is the fundamental uncertainty of future conditions. If we cannot predict the environmental changes that will take place in the next one or two decades, then how do we plan restoration efforts? Second, how will biodiversity reserve boundaries and corridors between reserves shift with climate change? Given the differing dispersal capabilities of species, will some (including exotic species) quickly migrate in response to climate change, while others of low vagility suffer high mortality? Third, will key mutualisms break down as climate changes and/or species differentially disperse? Fourth, will climate change potentially alleviate environmental stress, thus actually enhancing restoration efforts? And finally, how will invasive species respond to global change factors, such as continuing increases in CO_2, N, and precipitation? Invasive species already present a formidable challenge for restoration ecologists, but if global change favors the differential success of invaders, they will cause even greater complications. While this is exciting from a research perspective, it is daunting when the primary goal is to enhance the cost-effectiveness and success rate of ecosystem restoration. Clearly, significantly more research needs to be conducted on the important interface between global change and restoration ecology.

In conclusion, deserts are an extremely important environment from a global change perspective and should be used to study global change and its long-term impacts on both natural and managed ecosystems. Our research efforts at the NDRC confirm the idea that deserts are indeed quite responsive to a variety of global change processes. We suggest that elevated CO_2 and changes in precipitation, N deposition, invasive species, and land use are all critical global change phenomena that will have potentially dramatic effects on arid ecosystems, such as the Mojave Desert. In addition, there is no doubt that these potential responses to global change will have important implications for land management issues, including desertification, conservation, and restoration. None of these issues should be considered without being framed in an explicit global-change context.

ACKNOWLEDGMENTS

We gratefully acknowledge funding support from the following agencies: the Department of Energy's Terrestrial Carbon Processes program (grant DE-FG02–03ER63651), the National Science Foundation's Ecosystem Studies Program (grants DEB-98–14358 and

02–12812), the Department of Energy's Program for Ecological Research (grant DE-FG02-02ER63361), and the Andrew W. Mellon Foundation. We also thank our current and past investigators at the Nevada Desert Research Center—Bob Nowak, Dave Evans, Jay Arnone, Jim Coleman, and Jeff Seemann—as well as the many talented staff, post-docs, graduate, and undergraduate students who have assisted with the NDRC's science program.

We gratefully acknowledge the cooperation and support of the National Nuclear Security Administration's Nevada Site Office, the Brookhaven National Laboratory, and the many collaborators who have made our Mojave Desert global change research endeavors a success. We sincerely thank the reviewers and editors for the thoughtful comments that improved the clarity and content of this chapter.

REFERENCES

Aber, J. D., W. McDowell, K. Nadelhoffer, A. Magill, G. Berntson, M. Kamakea, S. McNulty, W. Currie, L. Rustad, and I. Fernandez. 1998. Nitrogen saturation in temperate forest ecosystems. *BioScience* **48**:921–934.

Barker, D. H., L. R. Stark, J. F. Zimpfer, N. D. McLetchie, and S. D. Smith. 2005. Evidence of recent drought-induced stress on biotic crust mosses of the Mojave Desert. *Plant, Cell and Environment* **28**:939–947.

Barker, D. H., C. Vanier, E. Naumburg, T. N. Charlet, K. M. Nielsen, B. A. Newingham, and S. D. Smith. 2006. Enhanced monsoon precipitation and N deposition affect leaf traits and photosynthesis differently in spring and summer in the desert shrub *Larrea tridentata. New Phytologist* **169**:799–808.

Beatley, J. C. 1974. Effects of rainfall and temperature on the distribution and behavior of *Larrea tridentata* (creosote-bush) in the Mojave Desert of Nevada. *Ecology* **55**:245–261.

Belnap, J. 2003. Factors influencing nitrogen fixation and nitrogen release in biological soil crusts. Pages 241–262 *in* J. Belnap and O. L. Lange, editors. *Biological soil crusts: structure, function, and management.* Ecological Studies 150. Second edition. Springer-Verlag, Berlin, Germany.

Belnap, J., B. Büdel, and O. L. Lange. 2003. Biological soil crusts: characteristics and distribution. Pages 3–30 *in* J. Belnap and O. L. Lange, editors. *Biological soil crusts: structure, function, and management.* Ecological Studies 150. Second edition. Springer-Verlag, Berlin, Germany.

Belnap, J., and D. Eldridge. 2003. Disturbance and recovery of biological soil crusts. Pages 363–383 *in* J. Belnap and O.L. Lange, editors. *Biological soil crusts: structure, function, and management.* Ecological Studies 150. Second edition. Springer-Verlag, Berlin, Germany.

Belnap, J., S. L. Phillips, and M. E. Miller. 2004. Response of desert biological soil crusts to alterations in precipitation frequency. *Oecologia* **141**:306–316.

Billings, S. A., S. M. Schaeffer, and R. D. Evans. 2002a. Trace N gas losses and N mineralization in Mojave Desert soils exposed to elevated CO_2. *Soil Biology and Biochemistry* **34**:1777–1784.

Billings, S. A., S. M. Schaeffer, and R. D. Evans. 2003. Nitrogen fixation by biological soil crusts and heterotrophic bacteria in an intact Mojave Desert ecosystem with elevated CO_2 and added soil carbon. *Soil Biology and Biochemistry* **35**:643–649.

Billings, S. A., S. M. Schaeffer, S. Zitzer, T. Charlet, S. D. Smith, and R. D. Evans. 2002b. Alterations of nitrogen dynamics under elevated carbon dioxide in an intact Mojave Desert ecosystem: evidence from nitrogen-15 natural abundance. *Oecologia* **131**:463–467.

Brooks, M. L. 1999a. Habitat invaisibility and dominance by alien annual plants in the western Mojave Desert. *Biological Invasions* **1**:325–337.

Brooks, M. L. 1999b. Alien annual grasses and fire in the Mojave Desert. *Madroño* **46**: 13–19.

Brooks, M. L., C. M. D'Antonio, D. M. Richardson, J. B. Grace, J. E. Keeley, J. M. Ditomaso, R. J. Hobbs, M. Pellant, and D. Pyke. 2004. Effects of invasive alien plants on fire regimes. *BioScience* **54**:677–688.

Busch, D. E., and S. D. Smith. 1995. Mechanisms associated with decline of woody species in riparian ecosystems of the southwestern U.S. *Ecological Monographs* **65**:347–370.

Clark, J. S., and C. Fastie. 1998. Reid's paradox of rapid plant migration. *BioScience* **48**: 13–24.

Crowley, T. J. 2000. Causes of climate change over the past 1000 years. *Science* **289**: 270–277.

Davidowitz, G. 2002. Does precipitation variability increase from mesic to xeric biomes? *Global Ecology and Biogeography* **11**:143–154.

Dregne, H. E. 1991. Global status of desertification. *Annals of the Arid Zone* **30**:179–185.

Easterling, D. R., B. Horton, P. D. Jones, T. C. Peterson, T. R. Karl, D. E. Parker, M. J. Salinger, V. Razuvayev, N. Plummer, P. Jamason, and C. K. Folland. 1997. Maximum and minimum temperature trends for the globe. *Science* **277**:364–367.

Ehleringer, J. R., S. L. Phillips, W. S. F. Schuster, and D. R. Sandquist. 1991. Differential utilization of summer rains by desert plants. *Oecologia* **88**:430–434.

Ellsworth, D. S., P. B. Reich, E. S. Naumburg, G. W. Koch, M. E. Kubiske, and S. D. Smith. 2004. Photosynthesis, carboxylation and leaf nitrogen responses of 16 species to elevated pCO_2 across four free-air CO_2 enrichment experiments in forest, grassland and desert. *Global Change Biology* **10**:2121–2138.

Evans, R. D., and J. R. Ehleringer. 1993. A break in the nitrogen cycle in aridlands? Evidence from $\delta^{15}N$ of soils. *Oecologia* **94**:314–317.

Evans, R. D., and J. R. Johansen. 1999. Microbiotic crusts and ecosystem processes. *Critical Reviews in Plant Science* **18**:183–225.

Fenn, M. E., R. Haeuber, G. S. Tonneson, J. S. Baron, S. Grossman-Clarke, D. Hope, D. A. Jaffe, S. Copeland, L. Geiser, H. M. Rueth, and J. O. Sickman. 2003. Nitrogen emissions, deposition, and monitoring in the western United States. *BioScience* **53**:391–403.

Galloway, J. N., F. J. Dentener, D. G. Capone, E. W. Boyer, R. W. Howarth, S. P. Seitzinger, G. P. Asner, C. C. Cleveland, P. A. Green, E. A. Holland, D. M. Karl, A. F. Michaels, J. H. Porter, A. R. Townsend, and C. J. Vörösmarty. 2004. Nitrogen cycles: past, present, and future. *Biogeochemistry* **70**:153–226.

Hamerlynck, E. P., T. E. Huxman, T. N. Charlet, and S. D. Smith. 2002b. Effects of elevated CO_2 (FACE) on the functional ecology of the drought-deciduous Mojave Desert shrub, *Lycium andersonii*. *Environmental and Experimental Botany* **48**:93–106.

Hamerlynck, E. P., J. R. McAuliffe, E. V. McDonald, and S. D. Smith. 2002a. Ecological responses of two Mojave Desert shrubs to soil horizon development and soil water dynamics. *Ecology* **83**:768–779.

Houghton, J. T., Y. Ding, D. J. Griggs, M. Noguer, P. J. van der Linden, X. Dai, K. Maskell, C. A. Johnson. 2001. *Climate change 2001: the scientific basis.* Cambridge University Press, Cambridge, England, UK.

Housman, D. C., E. Naumburg, T. E. Huxman, T. N. Charlet, R. S. Nowak, and S. D. Smith. 2006. Increases in desert shrub productivity under elevated CO_2 vary with seasonal water availability. *Ecosystems* **9**:374–385.

Hungate, B. A., M. Reichstein, P. Dijkstra, D. Johnson, G. Hymus, J. D. Tenhunen, C. R. Hinkle, and B. G. Drake. 2002. Evapotranspiration and soil-water content in a scrub-oak woodland under carbon dioxide enrichment. *Global Change Biology* **8**:289–298.

Hunter, R. B. 1991. *Bromus* invasions on the Nevada Test Site: present status of *B. rubens* and *B. tectorum* with notes on their relationship to disturbance and altitude. *Great Basin Naturalist* **51**:176–182.

Huxman, T. E., T. N. Charlet, C. Grant, and S. D. Smith. 2001. The effects of parental CO_2 and offspring nutrient environment on initial growth and photosynthesis in an annual grass. *International Journal of Plant Sciences* **162**:617–623.

Huxman, T. E., E. P. Hamerlynck, and S. D. Smith. 1999. Reproductive allocation and seed production in *Bromus madritensis* ssp. *rubens* at elevated CO_2. *Functional Ecology* **13**:769–777.

Huxman, T. E., and S. D. Smith. 2001. Photosynthesis in an invasive grass and native forb at elevated CO_2 during an El Niño year in the Mojave Desert. *Oecologia* **128**:193–201.

Huxman, T. E., K. A. Snyder, D. Tissue, A. J. Leffler, K. Ogle, W. T. Pockman, D. R. Sandquist, D. L. Potts, and S. Schwinning. 2004. Precipitation pulses and carbon fluxes in semiarid and arid ecosystems. *Oecologia* **141**:254–268.

Idso, S. B. 1992. Shrubland expansion in the American Southwest. *Climatic Change* **22**: 85–86.

Jasoni, R. L., S. D. Smith, and J. A. Arnone III. 2005. Net ecosystem CO_2 exchange in Mojave Desert shrublands after eight years of exposure to elevated CO_2. *Global Change Biology* **11**:749–756.

Jordan, D. N., S. F. Zitzer, G. R. Hendrey, K. F. Lewin, J. Nagy, R. S. Nowak, S. D. Smith, J. S. Coleman, and J. R. Seemann. 1999. Biotic, abiotic and performance aspects of the Nevada Desert Free-air CO_2 Enrichment (FACE) Facility. *Global Change Biology* **5**:659–668.

Kacholia, K., and R. A. Reck. 1997. Comparison of global climate change simulations for a $2 \times CO_2$-induced warming: an intercomparison of 108 temperature change projections published between 1980 and 1995. *Climatic Change* **35**:53–69.

Karl, T. R., R. W. Knight, and N. Plummer. 1995. Trends in high-frequency climate variability in the twentieth century. *Nature* **377**:217–220.

Keeling, C. D., and T. P. Whorf. 2002. Atmospheric CO_2 records from sites in the SIO air sampling network. *In* Trends: a compendium of data on global change. Online. Carbon Dioxide Information Analysis Center, Oak Ridge National Laboratory, U.S. Department of Energy, Oak Ridge, Tennessee. http://gcmd.nasa.gov/records/GCMD_CDIAC_CO2_SIO.html.

Kim, J. 2005. A projection of the effects of the climate change induced by increased CO_2 on extreme hydrologic events in the western U.S. *Climatic Change* **68**:153–168.

Kunkel, K. E., D. R. Easterling, K. Redmond, and K. Hubbard. 2003. Temporal variations of extreme precipitation events in the United States: 1895–2000. *Geophysical Research Letters* **30**:51–54.

Lal, R. 2004. Carbon sequestration in dryland ecosystems. *Environmental Management* **33**: 528–544.

Loik, M. E., D. D. Breshears, W. K. Lauenroth, and J. Belnap. 2004. A multi-scale perspective of water pulses in dryland ecosystems: climatology and ecohydrology of the western USA. *Oecologia* **141**:269–281.

Mack, R. N. 1981. The invasion of *Bromus tectorum* L. into western North America: an ecological chronicle. *Agro-Ecosystems* **7**:145–165.

Mann, M. E., and P. D. Jones. 2003. Global surface temperatures over the past two millennia. *Geophysical Research Letters* **30**:51–54.

Melillo, J. M., A. D. McGuire, D. W. Kicklighter, B. Moore, C. J. Vorosmarty, and A. L. Schloss. 1993. Global climate change and terrestrial net primary production. *Nature* **363**:234–240.

Nagel, J. M., K. L. Griffin, T. E. Huxman, and S. D. Smith. 2004. CO_2 enrichment reduces the energetic cost of biomass construction in an invasive desert grass. *Ecology* **85**: 100–106.

Naumburg, E., D. C. Housman, T. E. Huxman, T. N. Charlet, M. E. Loik, and S. D. Smith. 2003. Photosynthetic responses of Mojave Desert shrubs to free-air CO_2 enrichment are greatest during wet years. *Global Change Biology* **9**:276–285.

Naumburg, E., M. E. Loik, and S. D. Smith. 2004. Photosynthetic responses of *Larrea tridentata* to seasonal temperature extremes under elevated CO_2. *New Phytologist* **162**:323–330.

Nicklaus, P. A., D. Spinnier, and C. Körner. 1998. Soil moisture dynamics of calcareous grassland under elevated CO_2. *Oecologia* **117**:201–208.

Nowak, R. S., L. A. DeFalco, C. S. Wilcox, D. N. Jordan, J. S. Coleman, J. R. Seemann, and S. D. Smith. 2001. Leaf conductance decreased under free-air CO_2 enrichment (FACE) for three desert perennials in the Nevada desert. *New Phytologist* **150**:449–458.

Nowak, R. S., D. S. Ellsworth, and S. D. Smith. 2004a. Functional responses to elevated atmospheric CO_2—do photosynthetic and productivity data from FACE experiments support early predictions? *New Phytologist* **162**:253–280.

Nowak, R. S., S. F. Zitzer, D. Babcock, V. Smith-Longozo, T. N. Charlet, J. S. Coleman, J. R. Seemann, and S. D. Smith. 2004b. Elevated atmospheric CO_2 does not conserve soil water in the Mojave Desert. *Ecology* **85**:93–99.

Noy-Meir, I. 1973. Desert ecosystems: environment and producers. *Annual Review of Ecology and Systematics* **4**:25–52.

Pacala, S. W., G. C. Hurtt, D. Baker, P. Peylin, R. A. Houghton, R. A. Birdsey, L. Heath, E. T. Sundquist, R. F. Stallard, P. Ciais, P. Moorcroft, J. P. Caspersen, E. Shevliakova, B. Moore, G. Kohlmaier, E. Holland, M. Gloor, M. E. Harmon, S. M. Fan, J. L. Sarmiento, C. L. Goodale, D. Schimel, and C. B. Field. 2001. Consistent land- and atmosphere-based U.S. carbon sink estimates. *Science* **292**:2316–2320.

Pataki, D. E., T. E. Huxman, D. N. Jordan, S. F. Zitzer, J. S. Coleman, S. D. Smith, R. S. Nowak, and J. R. Seemann. 2000. Water use of two Mojave Desert shrubs under elevated CO_2. *Global Change Biology* **6**:889–898.

Peters, R. L., and T. E. Lovejoy. 1992. *Global warming and biological diversity.* Yale University Press, New Haven, Connecticut.

Pockman, W. T., and J. S. Sperry. 1997. Freezing-induced xylem cavitation and the distribution of Sonoran Desert vegetation. *American Journal of Botany* **87**:1287–1299.

Reynolds, J. F., P. R. Kemp, K. Ogle, and R. J. Fernandez. 2004. Modifying the "pulse-reserve" paradigm for deserts of North America: precipitation pulses, soil water, and plant responses. *Oecologia* **141**:194–210.

Sage, R. F. 1996. Modification of fire disturbance by elevated CO_2. Pages 231–249 *in* F. A. Bazzaz, editor. *Carbon dioxide, populations, and communities.* Academic Press, San Diego, California.

Schlesinger, W. H., J. F. Reynolds, G. L. Cunningham, L. F. Huenneke, W. M. Jarrell, R. A. Virginia, and W. G. Whitford. 1990. Biological feedbacks in global desertification. *Science* **247**:1043–1048.

Schneider, S. H. 1989. The greenhouse effect: science and policy. *Science* **243**:771–781.

Smith, S. D., C. A. Herr, K. L. Leary, and J. Piorkowski. 1995. Soil-plant water relations in a Mojave Desert mixed shrub community: a comparison of three geomorphic surfaces. *Journal of Arid Environments* **29**:339–351.

Smith, S. D., T. E. Huxman, S. F. Zitzer, T. N. Charlet, D. C. Housman, J. S. Coleman, L. K. Fenstermaker, J. R. Seemann, and R. S. Nowak. 2000. Elevated CO_2 increases productivity and invasive species success in an arid ecosystem. *Nature* **408**:79–82.

Smith, S. D., R. K. Monson, and J. E. Anderson. 1997. *Physiological ecology of North American desert plants.* Springer-Verlag, Berlin, Germany.

Smith, S. J., A. M. Thomson, N. J. Rosenberg, R. C. Izaurralde, R. A. Brown, and T. M. L. Wigley. 2005. Climate change impacts for the conterminous USA: an integrated assessment. Part 1. Scenarios and context. *Climatic Change* **69**:7–25.

Spaulding, W. G., and L. J. Graumlich. 1986. The last pluvial climatic episodes in the deserts of southwestern North America. *Nature* **320**:441–444.

Strain, B. R., and F. A. Bazzaz. 1983. Terrestrial plant communities. Pages 177–222 *in* E. Lemon, editor. *CO_2 and plants: the response of plants to rising levels of carbon dioxide.* American Association for the Advancement of Science, Symposium 84, Westview Press, Inc., Boulder, Colorado.

Tsonis, A. A., J. B. Elsner, A. G. Hunt, and T. H. Jagger. 2005. Unfolding the relation between global temperature and ENSO. *Geophysical Research Letters* **32**:L09701.

Turnage, W. V., and A. L. Hinckley. 1938. Freezing weather in relation to plant distribution in the Sonoran Desert. *Ecological Monographs* **8**:529–550.

Walther, G. R. 2003. Plants in a warmer world. *Perspectives in Plant Ecology, Evolution and Systematics* **6**:169–185.

Watson, R. T., M. C. Zinyowera, and R. H. Moss. 1996. *Climate change 1995: impacts, adaptations, and mitigation of climate change.* Cambridge University Press, Cambridge, England, UK.

Weatherly, H. E., S. F. Zitzer, J. S. Coleman, and J. A. Arnone III. 2003. *In situ* litter decomposition and litter quality in a Mojave Desert ecosystem: effects of elevated atmospheric CO_2 and interannual climate variability. *Global Change Biology* **9**:1223–1233.

Webb, R. H. 1982. Off-road motorcycle effects on a desert soil. *Environmental Conservation* **9**:197–208.

Weltzin, J. F., R. T. Belote, and N. J. Sanders. 2003. Biological invaders in a greenhouse world: will elevated CO_2 fuel plant invasions? *Frontiers in Ecology and the Environment* **1**:146–153.

Yair, A. 2003. Effects of biological soil crusts on water redistribution in the Negev Desert, Israel: a case study in longitudinal dunes. Pages 303–314 *in* J. Belnap and O.L. Lange, editors. *Biological soil crusts: structure, function, and management.* Ecological Studies 150. Second edition. Springer-Verlag, Berlin, Germany.

Yoder, C. K., and R. S. Nowak. 1999. Soil moisture extraction by evergreen and drought-deciduous shrubs in the Mojave Desert during wet and dry years. *Journal of Arid Environments* **42**:81–96.

Ziska, L. H. 2003. Evaluation of the growth response of six invasive species to past, present, and future atmospheric carbon dioxide. *Journal of Experimental Botany* **54**:395–404.

THREE

Human Population in the Mojave Desert

Resources and Sustainability

DEBRA L. HUGHSON

The concept of sustainability is regularly applied in regional ecosystem management in the United States, but few consider the broader context of the word. The American Heritage Dictionary (2000) defines *sustainable* as "the capability of being sustained" and *sustain* as: "1) to keep in existence (maintain), and 2) to supply with necessities or nourishment or to provide for." The second definition implies fluxes of resources while the first definition implies nearly continuous flux rates over an indefinitely long time period. Sustainability of current human populations, in a broad sense, is a continuation of global trade, industrial-scale agriculture, and international shipping. Sustainability, in a narrow sense, could be thought of as local resource production and consumption, independent of imports. I use the term in both a broad and narrow sense throughout this chapter. Sustainability of a supermarket is an example of the broad sense, while sustainability of a garden and a farmer's market is the corollary example of the narrow sense. Sustainability in the broader sense implies a much larger context within which ecosystems can, and should, be evaluated for long-term status.

Most people in the Mojave Desert today live in urban and suburban metropolitan areas linked by highways, utility corridors, and railroads, with necessities largely imported by aqueducts, power lines, pipelines, and by diesel-powered trucks and locomotives. The human population has dramatically increased in the Mojave Desert over the last one hundred years. U.S. Census Bureau (2001) data indicated a population of about 2.36 million in the Mojave Desert ecoregion in 2000 (fig. 3.1). Master-planned communities are being developed around Kingman, Arizona, and within the "disposal boundary" of the 1998 Southern Nevada

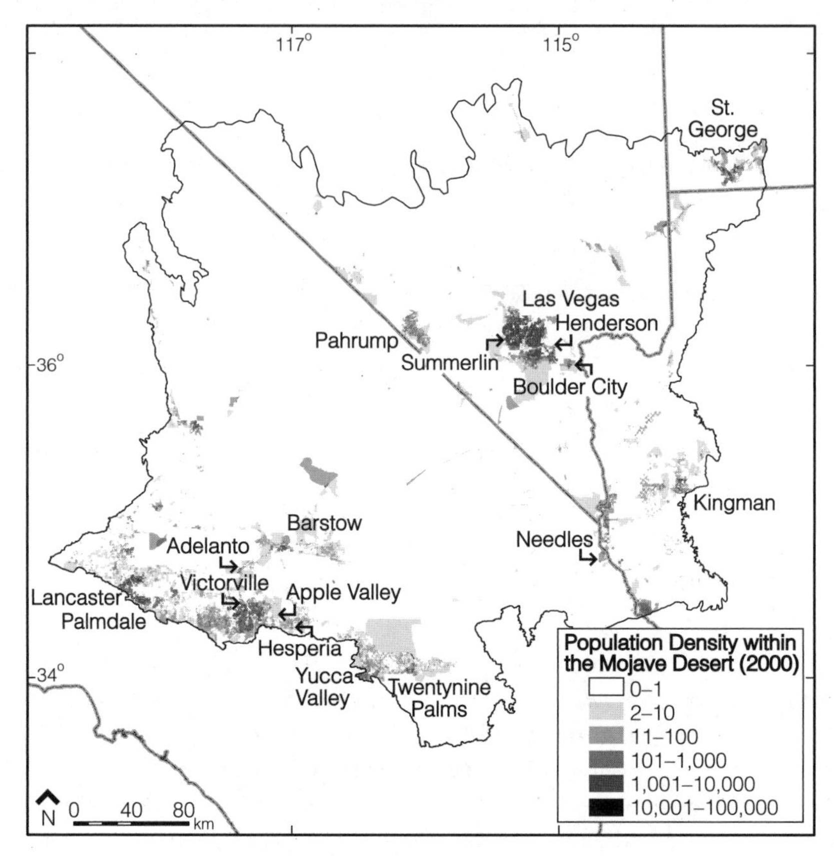

Fig. 3.1. Map of the Mojave Desert ecoregion depicting U.S. Census Bureau 2000 TIGER data of human population density at the resolution of census blocks.

Public Lands Act around Las Vegas. Continuation of this growth depends on resource inputs and/or conservation. Any loss of resources would involve considerable stress to the existing population. Although the weight of economic demand and political power draws resources from larger landscapes—perhaps from the entire world—toward this and other metropolitan clusters, the human population remains essentially constrained by these resources. Food (i.e., agriculture) depends not only on climate, but also on fertilizers, tractors, and transportation, all of which use hydrocarbon-derived energy. The current economy runs on petroleum, so any decrease in oil supply risks economic recession.

Modern suburbia is formed and enabled by the automobile. Most of the towns and cities in the Mojave Desert were founded after William Drake pioneered the first oil well in Pennsylvania, and all were small hamlets when Henry Ford made the automobile—the most ubiquitous hydrocarbon-powered vehicle—affordable. Population in the Los Angeles basin, on the southern margin of the Mojave Desert, had reached two million by 1930, associated with the discovery of oil and subsequent economic opportunities (Lockwood 1980; Yergin 1991). Construction of the interstate highway system in the 1950s and 1960s—combined with cheap gasoline and relatively inexpensive, air-conditioned housing—drew residents to new suburban tracts in the desert. Suburban expansion continues at a rapid rate, challenging municipal water purveyors to satisfy ever-increasing demands with limited supplies. Typical lifestyles in the suburban environment, which includes most residents in the Mojave Desert, involve commuting by automobile to work and dependence on gasoline stations and grocery stores.

Water usage is metered at diversions, but most water, minus evaporation, eventually returns to the flow system, albeit with degraded quality. Nevada withdraws more than its allotment of $3.70{\cdot}10^8$ m^3 y^{-1} under the Colorado River Compact, partially because of return flow credits derived from treated sewage effluent flowing down Las Vegas Wash back into Lake Mead. Similarly, fresh water diverted for thermoelectric power-generator cooling is a significant portion of the national average per capita use, which returns heated water to the environment. Evapotranspiration from outdoor landscaping, especially golf courses, forms a large part of consumptive usage in desert urban areas. The national average public and domestic per capita usage of 230 m^3 y^{-1} p^{-1} includes the eastern half of the continent, where outdoor landscaping is maintained by rainfall, whereas the Las Vegas per capita consumption of 393 m^3 y^{-1} p^{-1} in 2004 includes maintenance of outdoor landscaping by irrigation.

Complex social systems, like those in the Mojave Desert urban areas, are fueled by energy and are subject to diminishing marginal returns on the investment of additional complexity needed to sustain economic growth (Tainter et al. 1990). Diminishing returns in conventional energy development and so-far disappointing results in alternative fuels development compounds the problem. Without abundant energy to transport food, power air conditioners, and move people between urban areas, there would not be the same human density in the desert creating a demand for water. Water supply is a basin-scale issue, while energy is a global concern. At present, it is not possible to separate the complexity of metropolitan areas, such as Las Vegas, from the global petroleum supply.

HUMAN POPULATION OF THE MOJAVE DESERT

Humans arrived in the Mojave Desert at least 10,000 years ago, and perhaps earlier (Susia 1964; Fagan 1995). The climate was colder and wetter during the early Holocene, with vegetation zones, such as pinyon-juniper woodlands, about 1000 m lower than at present (Malde 1964; Mehringer 1965, 1967; Leskinen 1975; Spaulding 1990). Archeological sites dating from the early to middle Holocene are scarce in the Mojave Desert, implying that human population densities were low (Grayson 1993). Most substantial archeological sites were near dependable water sources and those that were not appeared to be used ephemerally (Fagan

1974). The beginning of the late Holocene, around 4,000 to 5,000 years ago, saw a marked increase in the number of archeological sites and in the diversity of habitat in which those sites were located (Grayson 1993). In fact, the archeological record of the late Holocene looks as though it could have been created by the native peoples encountered by the first Europeans.

Estimates of human populations in the Americas immediately before 1492 are controversial. Thorton (1987) reported estimates for the North American population (north of the Rio Grande) at first European contact ranging from less than 1 million to as many as 18 million. Ubelaker (1985) estimated that there were about 5 million people living in what is now the U.S. mainland when Columbus landed. While these may be underestimates, evidence for a large, prehistoric, human population in the Mojave Desert has not been found. The densest settlements appear to have been the agricultural communities of the Mojave tribe, or Aha Macav, from Black Canyon to the Mojave Valley and the Picacho Mountains near modern-day Parker Dam. When Francisco Garcés crossed from the Colorado River to the West Coast in 1775, the desert was occupied by Mojave, Halchidmoma, and Kohuana (Chemehuevi or Southern Paiute) tribes towards the east, and Chemehuevi (or Southern Paiute) and Desert Cahuilla tribes towards the west (Heizer and Whipple 1971).

Jedediah Smith passed through the Mojave Desert a little over 50 years after Garcés (Hafen 1982; Smith 1989). By 1844 the U.S. Army was establishing outposts in this region. European emigration increased dramatically during the final two decades of the nineteenth century, following the military defeat of the indigenous tribes and the discovery of economic mineral deposits, such as gold, silver, and borax, in Death Valley. Numerous small settlements were established, primarily associated with mineral discoveries (Lingenfelter 1988), many of which were later abandoned. The main population centers of modern times were founded near rivers, streams, or springs during this period: St. George in 1861, Las Vegas in 1905, Kingman in 1882, Needles in 1883, Barstow and nearby Daggett in about 1860, and Palmdale in 1886 (fig. 3.1).

Within a few decades of the mid-nineteenth century, human occupation in the Mojave Desert changed from widely scattered bands of hunter-gathers and primitive agriculturalists to mining settlements connected by railroads and dispersed cattle and sheep ranches. Remarkable changes have occurred in the last 100–150 years. In particular, growth in Clark County, Nevada, the only county entirely within the Mojave Desert, was approximately exponential throughout the twentieth century, doubling every decade and then recently becoming more linear with an increase of about 73,830 people per year (fig. 3.2A). High-desert cities north of the Los Angeles basin, including the Lancaster/Palmdale area and the rapidly developing San Bernardino County cities of Victorville, Apple Valley, and Hesperia, also had marked population increases in the 1980s, tripling and quadrupling in size (figs. 3.2B, 3.2C).

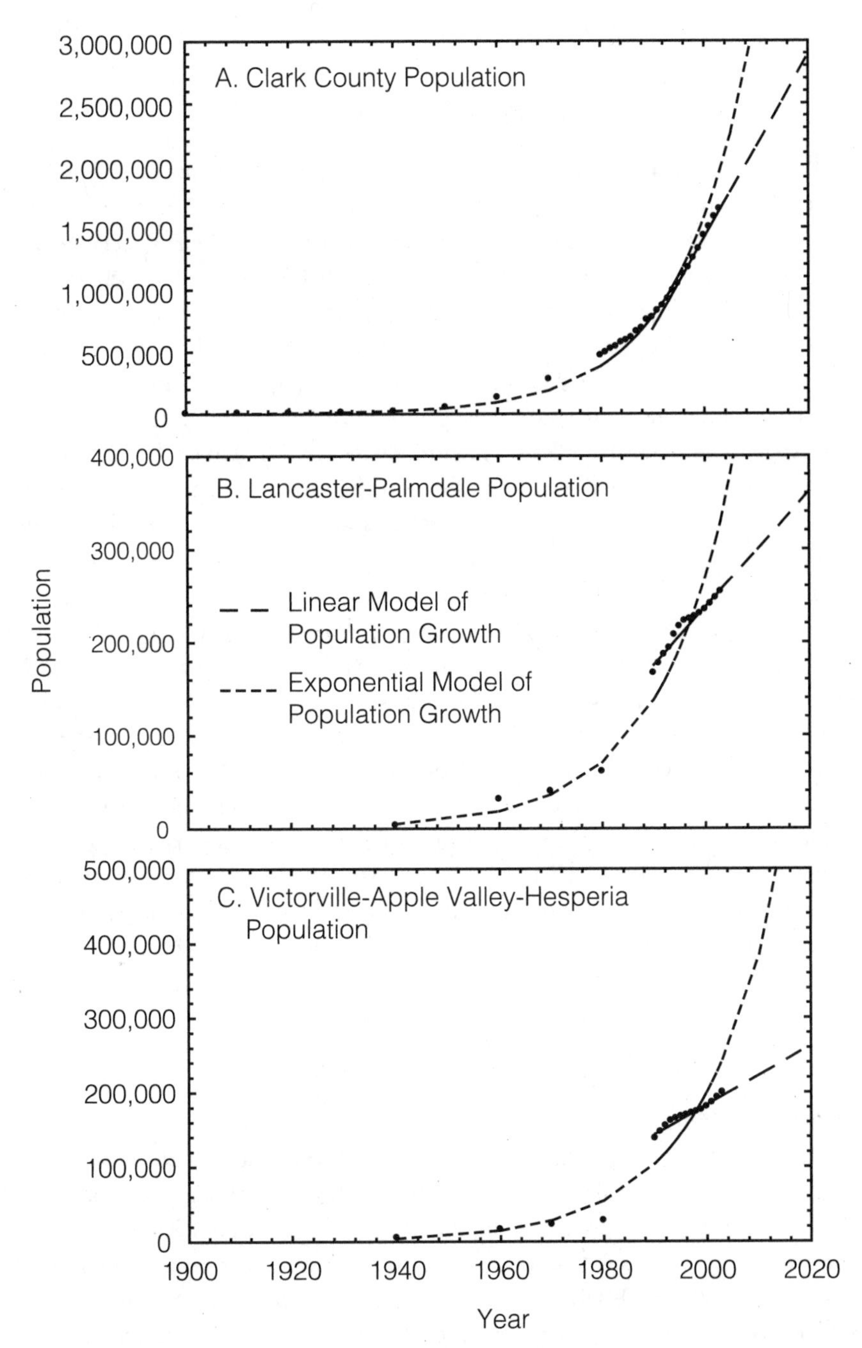

Fig. 3.2. Graphs of historical population estimates for selected cities in the Mojave Desert show a marked increase in growth after 1980, which is correlated with the expansion of suburban areas across the southwestern U.S. *A*, Clark County, Nevada; *B*, Lancaster–Palmdale, California; *C*, Victorville, Apple Valley, and Hesperia, California. Near vertical dash lines are extrapolated population from an exponential growth model; heavier dash lines are linear extrapolations from population data, 1990–2004, as shown.

Clark County (Eastern Mojave)

Population changes in the Mojave Desert are dramatically illustrated by Clark County, which was established in 1909 and is home to Las Vegas, Henderson, and the master-planned community of Summerlin. For most of the Holocene (Mehringer 1967), artesian springs fed the desert oasis that Rafael Rivera discovered in 1829 (Hopkins and Evans 1999). Mormon settlers built an adobe fort at the site in 1855 but abandoned it after two years, primarily because the indigenous population had not yet been subdued. The city of Las Vegas began as a railroad construction camp, situated to take advantage of the abundant spring water. San Pedro, Los Angeles, and Salt Lake Railroad land, owned by Montana Senator William Andrews Clark, was auctioned off in 1905 for the town site, which is now dominated by resort casinos. Easy access to artesian spring water was the reason for the location of Las Vegas, and the construction of Boulder—later Hoover—Dam was initially the reason for its rapid growth.

Work began on the nearby Boulder Canyon Project in 1931 (Stevens 1988), the same year that the Nevada legislature legalized gambling. Boulder Dam was dedicated in 1935 and renamed Hoover Dam by Congress in 1947. According to the 1940 census, the population in Clark County almost doubled to 6,414 as the unemployed came to work on the huge Depression-era project. At its peak, the Boulder Canyon Project employed 5,128 people who mostly lived in the newly formed company town of Boulder City. Population tripled by 1950, accelerated by the creation of Nellis Air Force Base to train aviators for World War II, construction of a magnesium plant near Henderson, and work on Davis Dam at Laughlin. This growth rate continued throughout the remaining decades of the twentieth century, doubling again by 1960. In 2004, the Nevada state demographer estimated population in Clark County at nearly 1.72 million.

Eastern Kern County (Western Mojave)

On a nighttime satellite image of Southern California, it appears as though an ocean of lights in Los Angeles, Orange, Riverside, and San Diego counties has washed over the San Gabriel Mountains onto the high desert beachhead of Palmdale and Lancaster. Suburban developments in this area are connected, in part, to the greater metropolis to the south, but also to Edwards Air Force Base, northeast of Palmdale and established in 1933. The beginning of Palmdale and Lancaster was marked by a Scottish settlement in 1882, the Southern Pacific Railroad in 1884, and a band of Swiss and German emigrants in 1886. Palmdale's population more than quadrupled from 12,200 in 1980 to 68,900 in 1990, while Lancaster grew from 48,000 to 97,300, largely enabled by the completion of the California Aqueduct in 1971. By 2000, 234,100 people lived in the Lancaster-Palmdale area. The Antelope Valley East Kern Water Agency (AVEK) estimated the population in its service area at 285,500 (AVEK 2005). Palmdale Water District (PMD) serves approximately 93,420 (Dudek and Associates 2004).

San Bernardino County (Central and South-Central Mojave)

Captain A. G. Lane pioneered an Anglo-American settlement near modern-day Victorville when he established a ranch at the lower Narrows of the Mojave River in 1858 (Thompson and Thompson 1995). William Holcomb struck gold near Big Bear Lake in 1860, and the boom that followed led to road construction from the Victor Valley side of Cajon Pass southward. The railroad came to Victor Valley in 1883, winding over Cajon Pass from San Bernardino. California Southern Pacific Railroad reached the Atlantic & Pacific Railroad junction at Daggett in 1885, and by 1888 the Calico Railroad was hauling ore from the Calico Mountains to the Ore Grande Milling Company at Daggett. During the 1880s, land speculators in Southern California bought and sold lots, tracts, and townships. About 1885, the community of Victor, named after railroad construction superintendent Jacob Nash Victor, was established as a railroad station approximately two kilometers northwest of the Narrows of the Mojave River. The Southern Pacific Railroad, with large holdings in the Victor Valley, promoted the township of Hesperia. Various companies such as the Apple Mesa Development Company and the California Land and Water Company organized promotions such as the giveaway of over 300,000 apple trees to land buyers during 1913 and 1914. The community of Victorville was incorporated September 21, 1962, with a population of approximately 8,100 in an area of 25 km^2. As of January 1, 1995, Victorville's population was 60,650 in an area of 175 km^2. In 2000, there were 18,200 people in Adelanto, 63,600 in Victorville, 54,000 in Apple Valley, 62,300 in Hesperia, 21,100 in Barstow, 16,800 in Yucca Valley, and 14,750 in Twentynine Palms, for a total population of 250,750 in these southern Mojave Desert towns. The Mojave Water Agency (MWA) estimated 323,550 people in its service area in 2000 and projected 373,100 people would live in the area by 2005 (MWA 2004).

WATER SUPPLY

Water for many municipalities in the Mojave Desert originates as direct and remote precipitation in the Colorado River Basin and in northern California. Because direct precipitation is insufficient for modern urban densities, agriculture, and economic growth, water is imported from outside the region. The Los Angeles Aqueduct first brought Owens Valley water to the San Fernando Valley in 1913 through the Western Mojave Desert. By 1971, the California Aqueduct East Branch was completed, bringing Feather River water from northern California through the State Water Project (SWP) to cities in the Western and South-Central Mojave Desert (fig. 3.3). The Southern Nevada Water Authority (SNWA) draws most of its water from the Colorado River at Lake Mead. Ultimately, the available

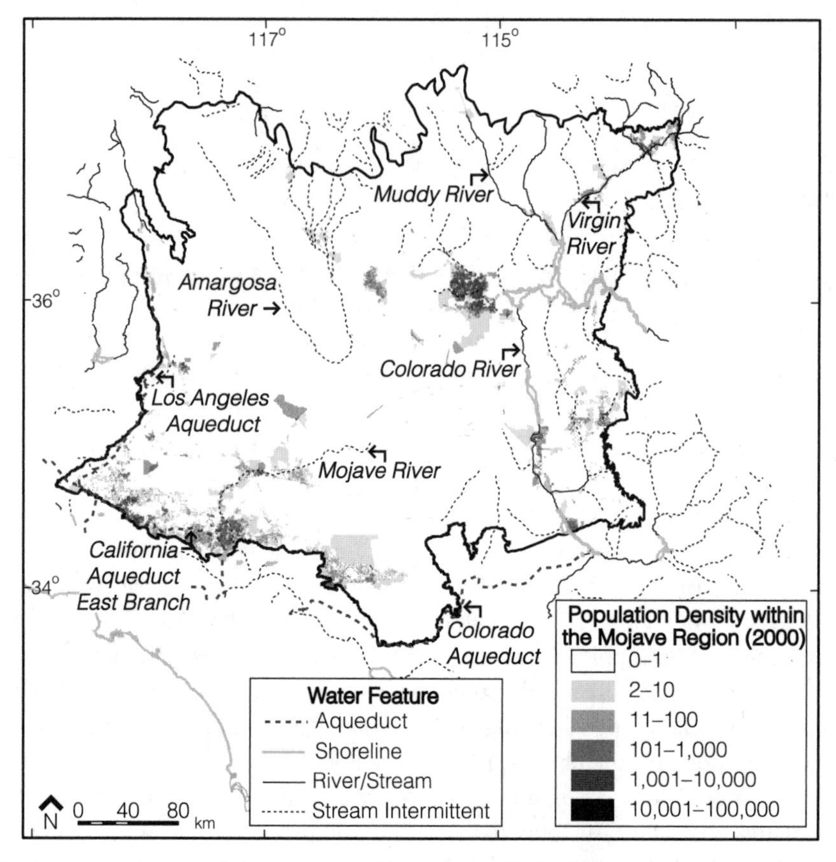

Fig. 3.3. Map showing the association of human population with rivers and aqueducts in the Mojave Desert ecoregion. Local water supply pipelines are not shown.

water supply depends entirely on precipitation in other regions, which in turn depends on climatic variability.

Water is a basic resource for industry and agriculture, as well as for public supply and personal consumption. The unit of water measurement in the U.S. is acre-feet, the depth of water in feet over the area of an acre, for the water year starting October 1 (the volume of an acre-foot is 1233.5 m^3 in SI units). Nationally, the population of the U.S. used about $5.64 \cdot 10^{11}$ m^3 y^{-1} of water in 2000, 85% of which was considered fresh (Hutson et al. 2005). This use has varied by less than 3% since 1985 (Hutson et al. 2005). The U.S. Census Bureau reported 281,421,900 people (p) in the country as of April 1, 2000, the ratio of which gives an estimated average per capita fresh water use of 1699 m^3 y^{-1} p^{-1} (cubic meters per person per year). Most of this water (45%) went toward industry, mining, and thermoelectric power-generator cooling; the next largest share (41%) went to agricultural irrigation, livestock, and aquaculture. The average per capita U.S. water consumption for public use (e.g., for pools, parks, public buildings, firefighting, flushing waterlines, and pipeline leaks) was 213 $m^3 y^{-1} p^{-1}$; domestic fresh-water use was an additional 17.6 $m^3 y^{-1} p^{-1}$ (Hutson et al. 2005). According to Gleick (1996), the personal minimum water requirement for basic human needs is at least 50 L d^{-1} (18 $m^3 y^{-1}$), which implies that domestic use nationwide in the U.S. is not excessive.

Southern Nevada Water Authority (Eastern Mojave)

Octavius Decatur Gass found the old abandoned Mormon fort at Las Vegas Springs and filed claim to the water rights in 1872 (Hopkins and Evans 1999). Discharge from the springs was probably about $6.17 \cdot 10^6$ $m^3 y^{-1}$. He gave the property to Archibald Stewart whose widow sold it to Montana Senator William Andrews Clark, the owner of the railroad that later became Union Pacific. Within a few decades the Anglo settlers had sunk almost 3,000 wells and depleted the artesian aquifer such that these famous springs stopped flowing entirely in 1962. Ironically, the historic site of Las Vegas Springs, near Alta Drive and Valley View Boulevard, is now the Springs Preserve, dedicated to sustainable living in the Mojave Desert.

Overdraft of the aquifer and declining water levels continued to accelerate until the first stage of the Southern Nevada Water System began supplying Colorado River water in 1971 (Bureau of Reclamation *unpublished data*). The state engineer's office established a policy in 1957 of issuing temporary well permits until such time as Colorado River water became available in the area. Thus, groundwater withdrawals peaked in 1970 as overdrafts supported rapid growth, and then the increasing demand brought a new source, the Colorado River, online.

The idea that this successful strategy might be repeated is a basis for the SNWA Water Resource Plan (SNWA 2005). This plan is to extract more groundwater from rural Nevada, grow the economy in Clark County, and then find new water sources in the future (SNWA 2005). The SNWA was created with a purpose of obtaining surface water and groundwater from northern Clark, Lincoln, Nye, and White Pine counties to be used in Clark County. In 1989 prior to the formation of SNWA, the Las Vegas Valley Water District filed 147 applications for groundwater rights with the state engineer in 30 different basins, and one for surface water rights on the Virgin River, although ultimately it withdrew some of these applications. The state engineer granted applications in two basins—Tikapoo Valley and Three Lakes Valley—in 2005 and a third—Spring Valley—in 2007. The state engineer is considering other SNWA applications for a Clark, Lincoln, and White Pine counties Groundwater Development Project.

Demand in the Clark County area served by SNWA is currently pushing the limit of Nevada's share of Colorado River water. Through return-flow credits (i.e., used-water recycling) and creative groundwater banking strategies, SNWA is able to withdraw about $6.17 \cdot 10^8$ $m^3 y^{-1}$ from Lake Mead, or about five-thirds of

Nevada's apportionment. The SNWA Water Resource Plan (2005) identifies "in-state resources," namely about $1.54 \cdot 10^8$ m^3 y^{-1} of groundwater withdrawals from 19 basins in Clark, Lincoln, and White Pine counties and another $1.48 \cdot 10^8$ m^3 y^{-1} of surface flow in the Muddy and Virgin rivers, as a way to grow through the middle of the twenty-first century. Some as yet undefined interbasin transfers and exchanges would then be implemented. The Colorado River water supply, however, is not guaranteed; the period from October 2000 through September 2002 saw the lowest three-year average flow in the Colorado River at Lee's Ferry, and lowest flow between 1895 and 2003 was recorded in 2002 (Webb et al. 2005). Given that demand on the Colorado River Basin already tends to exceed supply, the future of water imports to southern Nevada is an engineering challenge. One often-mentioned option for additional imports is desalination, while another possible source is agriculture retirement and transfer of irrigation water to municipal supply. Outright purchase of agricultural water rights by the more prosperous resort casinos is a possibility.

During the twentieth century, growth in Clark County was nearly exponential, doubling about every 9.8 years ($r^2 = 0.994$, $P = 0.0002$), and was approximately linear from 1993 to 2003 at a rate of about 73,830 additional people per year (fig. 3.2A). Per capita water consumption has declined from approximately 7400 m^3 y^{-1} p^{-1} in 1910 to less than 400 m^3 y^{-1} p^{-1} today, in response to the transition from a sparsely populated agricultural economy to a densely populated urban/tourist economy. Based on the Nevada state demographer's 2004 population estimate for Clark County of 1,715,337 and the SNWA water supply of about $6.73 \cdot 10^8$ m^3 y^{-1}, per capita use in 2004 was 393 m^3 y^{-1} p^{-1}. If SNWA is successful in acquiring a proposed $3.02 \cdot 10^8$ m^3 y^{-1} of in-state supply, then Clark County can accommodate about 2.5 million people without additional conservation. At the exponential growth rate of the twentieth century, this number would have been reached in 2005; at the linear growth rate of the 1990s, this limit could arrive in 2014.

The Southern Nevada Water Authority plans to get the additional water supply required to meet future growth demands primarily from in-state groundwater resources, but conservation also provides an opportunity for growth. Presently, average per capita water use in the SNWA service area exceeds the national average for public and domestic use by about 50%. Over half of the supply goes to residential use, and most of that is consumed by outdoor landscaping (SNWA 2005). Many more people could be supplied with water if conservation measures including xeric landscaping were to be implemented. By 2030, SNWA's projected supply, assuming planned in-state groundwater resources are obtained, could provide over 55 million people with the water necessary for minimum domestic needs of 18 m^3 y^{-1} p^{-1}.

Antelope Valley East Kern and Palmdale Water Districts (Western Mojave)

Other than the Sacramento-San Joaquin rivers, California's biggest waterway is the Feather River, a tributary of the Sacramento River that flows out of the northern Sierra Nevada and into a reservoir impounded by Oroville Dam. This reservoir is the principal water storage facility of the SWP. Releases from Oroville Dam flow south to the Sacramento-San Joaquin Delta, where water is pumped into the California Aqueduct for distribution to Southern California. Fourteen 59.7 MW pumping units at the A. D. Edmonston Pumping Plant near Gorman then lift the water almost 610 m to cross the Tehachapi Mountains, where it splits into east and west branches. The east branch parallels the San Andreas Fault to Palmdale, Hesperia, San Bernardino, and Palm Springs.

Along its course through the Mojave Desert, the east branch supplies water to the Littlerock Creek Irrigation District, AVEK, PWD, and the MWA, all contractors with the SWP. The maximum annual entitlement of each contractor is known as its Table A allotment. The AVEK Table A allotment is $1.74 \cdot 10^8$ m^3 y^{-1}, but this amount is delivered with a reliability of less than 1% (AVEK 2005). The agency is 90% confident that supply will approach $1.73 \cdot 10^7$ m^3 y^{-1}, but only 10% confident that delivery will exceed $1.57 \cdot 10^8$ m^3 y^{-1}. The PWD Table A allotment is $2.63 \cdot 10^7$ m^3 y^{-1}, but the average annual delivery since it started receiving SWP water in 1985 is only $1.53 \cdot 10^7$ m^3 y^{-1}. The Littlerock Creek Irrigation District, adjoining PWD, receives an average of about $6.78 \cdot 10^6$ m^3 y^{-1} from the SWP.

Groundwater pumping may have peaked in Antelope Valley, the geographic triangle in the Western Mojave bounded by the San Andreas and Garlock faults, at about $4.93 \cdot 10^8$ m^3 y^{-1} in the 1950s (AVEK 2005). Natural recharge is estimated in the range of $3.86 \cdot 10^7$ to $7.29 \cdot 10^7$ m^3 y^{-1} (Galloway et al. 1998). The Antelope Valley Groundwater Basin is not adjudicated, so landowners with existing wells have correlative water rights and may pump without restraint. All of the eight water purveyors in Antelope Valley, except for AVEK, reported groundwater pumping of a total $4.54 \cdot 10^7$ m^3 y^{-1} in 2002 (Dudek and Associates 2004). This does not account for private wells or for annual variability of groundwater pumping; for example, PWD claimed an existing groundwater supply of $1.02 \cdot 10^7$ m^3 y^{-1} in 2002 (Dudek and Associates 2004), yet reported pumping $1.49 \cdot 10^7$ m^3 y^{-1} in 2004 (PWD 2004), and projects pumping $2.19 \cdot 10^7$ m^3 y^{-1} in 2020 (Dudek and Associates 2004). Groundwater pumping caused almost 2 m of land subsidence in Antelope Valley from 1926 to 1992 (Galloway et al. 1998). Although pumping has declined substantially from midcentury peak levels, aquifer overdraft continues and additional land subsidence is expected (Galloway et al. 1998). Total annual groundwater pumping in Antelope Valley is now estimated at between $1.73 \cdot 10^8$ to $2.10 \cdot 10^8$ m^3 y^{-1} (Griset 2004).

The 1980s saw a doubling of the population in Lancaster and an increase in the size of Palmdale by a factor of 5.6. Growth up to 1980 had been relatively slow and, after 1990, became approximately linear, with about 6,200 additional people each year ($r^2 = 0.956$, $P = 0.0039$; fig. 3.2B). Overall, an exponential model fits twentieth century population growth reasonably well ($r^2 = 0.961$, $P = 0.0025$). Per capita water use in the service area of PWD was 403 $m^3\ y^{-1}\ p^{-1}$ in 2004 (PWD 2004), while in the rural agricultural area of Littlerock Creek Irrigation District, per capita use was closer to 912 $m^3\ y^{-1}\ p^{-1}$ in 2000 (Dudek and Associates 2004). The AVEK service area imported $7.34{\cdot}10^7\ m^3\ y^{-1}$ in 2000 and $6.58{\cdot}10^7\ m^3\ y^{-1}$ in 2003 from the SWP. Population is estimated between 202,750 and 350,000 (AVEK 2005), which gives per capita use in the range of 188 to 362 $m^3\ y^{-1}\ p^{-1}$. The higher estimate is probably more accurate since the reported data do not include groundwater pumping.

The largest uncertainty in AVEK and PWD water supplies is the reliability of imports from the SWP. A middle-of-the-road estimate would be that AVEK received its median of about $1.23{\cdot}10^8\ m^3\ y^{-1}$, PWD received its average of $1.53{\cdot}10^7\ m^3\ y^{-1}$, Littlerock Reservoir produced $6.78{\cdot}10^6\ m^3\ y^{-1}$, and groundwater was withdrawn at the maximum estimated safe yield of $7.29{\cdot}10^7\ m^3\ y^{-1}$. At these rates, the water supply would be adequate for about 546,000 people at a per capita rate of 400 $m^3\ y^{-1}\ p^{-1}$, while the minimum domestic needs (18 $m^3\ y^{-1}\ p^{-1}$) could be met for over 12 million people.

Mojave Water Agency (Central and South-Central Mojave)

The Mojave River flows north out of the San Bernardino Mountains into a closed basin sink near Baker. It is ephemeral over much of its lower reach, so human settlement in the region has depended on groundwater. Groundwater withdrawal, which began in the early 1900s, resulted in overdraft aquifer conditions by the mid-1940s (Stamos et al. 2001), regional and floodplain aquifer drawdown of more than 30 m in places by the early 1950s (Stamos et al. 2004), and land subsidence (Sneed et al. 2003). The MWA was created in 1959 to deal with declining groundwater levels, and groundwater resources of the Mojave Watershed were formally adjudicated in 1996. On appeal, the California Supreme Court upheld the rights of prior appropriators, the so-called Cardozo Group of farmers, in its decision of August 22, 2000.

Natural recharge to the Mojave Basin regional and floodplain aquifers is sensitive to precipitation and thus climatic variability. During the water year of 1993, for example, recharge to the floodplain aquifer between the Lower Narrows and Barstow was $2.34{\cdot}10^8\ m^3$, which is 487% above the annual average for the period from 1931 to 1994 (Stamos et al. 2003). Recharge of the Mojave Basin occurs primarily in the San Bernardino Mountain headwaters, where the long-term historical average may have been around $9.25{\cdot}10^7\ m^3\ y^{-1}$ (Lahonton Regional Water Quality Control Board 2002). Since 1991, the MWA has been importing water from the SWP California Aqueduct East Branch. Water year 2004 delivery was only 40% of the full allotment, but following the unusually wet winter of 2004, the California Department of Water Resources issued a notice to SWP contractors on January 14, 2005, increasing deliveries to 60% of full allotment. This water is artificially recharged to the aquifer through infiltration ponds fed by the Mojave River Pipeline.

The agency's 2004 Regional Water Management Plan Draft Environmental Impact Report identified the need for an additional $5.06 \cdot 10^7$ m^3 y^{-1} of artificial recharge in the Mojave regional aquifer and $2.84 \cdot 10^7$ m^3 y^{-1} of artificial recharge in the Mojave floodplain aquifer to meet increasing demands through 2020. In terms of water resources, the communities along the Mojave River Basin from Hesperia to Newberry Springs surpassed a threshold of local sustainability about a half century ago and are now dependent on interbasin transfers.

The high-desert cities boomed after 1980, increasing in population by a factor of 5 in a single decade (fig. 3.2C). Growth in these cities after 1990 was approximately linear ($r^2 = 0.956$, $P = 0.0038$), but again an exponential model fits the overall twentieth century growth pattern reasonably well ($r^2 = 0.953$, $P = 0.0031$). The combined population of Hesperia, Victorville, Apple Valley, Adelanto, Barstow, Twentynine Palms, and Yucca Valley reached 284,700 by 2003. Schlumberger Water Services estimated $8.08 \cdot 10^7$ m^3 y^{-1} net average annual aquifer recharge in 2000 (MWA 2004), but also estimated consumptive use at $1.33 \cdot 10^8$ m^3 y^{-1}, with the difference being aquifer overdraft. Total delivery to the MWA from the SWP in 2000 was $1.40 \cdot 10^7$ m^3, giving a per capita consumption of 454 m^3 y^{-1} p^{-1}, about a third of which was used for agriculture (MWA 2004). Agriculture depends on and exploits the Mojave River floodplain and regional aquifers through groundwater wells and irrigation systems.

Even without agricultural consumption, the Mojave River Basin would still be overutilized, according to the Schlumberger estimates. In 2000, urban use alone exceeded the net average annual supply by $1.40 \cdot 10^7$ m^3. State Water Project imports have been variable, and may not entirely mitigate aquifer overdraft. From 1978 to 2001, there were 11 years with no deliveries and a maximum of $3.02 \cdot 10^7$ m^3 in 1983, significantly short of the full Table A allotment. The population in the service area of the MWA is dependent on SWP imports and the aquifer continues to be in overdraft. The minimum domestic requirement of 18 m^3 y^{-1} p^{-1} could, however, be met for almost 4.5 million people from net average annual recharge alone.

ENERGY SUPPLY

Oil supplies about 40% of the energy used in the United States, while coal and natural gas make up almost half, and nuclear fission and hydroelectric the remainder. Fossil fuels together contribute over 85% of the total energy consumed in the U.S. (Energy Information Administration 2005). Contributions from the renewable sources of solar, wind, and tides are in the realm of rounding errors. Petroleum products supply nearly all of the energy used in transportation (BP 2005), and about 70% of the total petroleum liquids used in the U.S. in 2006 were imported (BP 2007, Energy Information Administration 2007). Oil imports came primarily from Canada ($1.33 \cdot 10^8$ m^3), Mexico ($9.87 \cdot 10^7$ m^3), Saudi Arabia ($8.70 \cdot 10^7$ m^3), Venezuela ($8.12 \cdot 10^7$ m^3), and Nigeria ($6.38 \cdot 10^7$ m^3), while total petroleum production in the U.S. declined again in 2006 and has been generally declining since 1970 (Energy Information Administration 2007).

Depletion of oil fields is the geologic nature of the oil business and traditionally the amount of oil used is made up for by new discoveries. However, world annual discoveries, averaged over 5-year periods, peaked in 1962 (Masters et al. 1994). The history of an oil prospect involves discovery and then production. Production increases and then stabilizes, and eventually the field matures. Finally, despite water flooding, carbon dioxide injection, horizontal drilling, formation fracturing, and other advanced recovery techniques, oil production from the deposit inevitably declines. Likewise, the history of oil-rich regions and nations evolves from initial development through peak and then decline. U.S. oil production peaked in 1970 (fig. 3.4) as predicted by Hubbert (1956). The most important Middle East producer, Saudi Arabia, might not peak until the next decade or so, but its true oil reserve data are not available (Simmons 2005). The debate focuses

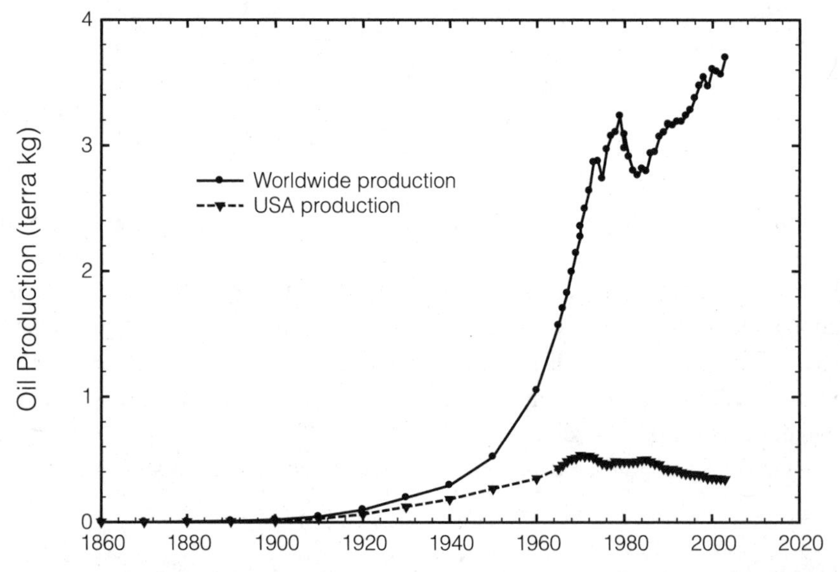

Fig. 3.4. Graph of oil production in the United States (*lower curve*) peaked in 1970 and has continued to decline despite new discoveries in Alaska during the 1980s (Etemad et al. 1991, BP 2005). Worldwide oil production (*upper curve*) continues rising to meet demand, at an annual rate of 4.5% in 2004 (BP 2005).

on when, not if, global oil production will peak—with estimates ranging from this decade to midcentury (Zucchetto 2006).

Of course we will never entirely run out of oil. There will always be some oil left in the ground simply because it will no longer be economically viable to extract. Oil that requires more energy to get out of the ground than it produces is not a source of energy, but a sink. A thermodynamic break-even point occurs when it takes more energy to discover, extract, and refine a barrel of oil than is derived from that barrel of oil. In petroleum production, the ratio of energy returned to energy invested can range from over 30 for some oil reservoirs in the Middle East to approximately 1.5 for tar sands. Extracting lower-grade energy from deposits such as the enormous tar sands in Alberta, Canada, comes with environmental costs and diminishing returns. Long before petroleum reserves are at a thermodynamic break-even point, however, demand will have outstripped supply, and it is this divergence that is of most immediate interest. Any new alternative energy supply is likely to come at an increased cost and return less on investment while population and energy demand continue to grow. Thus, the timing of a peak in world oil production is a central question in human ecology and the sustainability of our modern suburban civilization, particularly in a region such as the Mojave Desert, which is dependent upon fossil-fuel consumption for transportation and importation of basic human necessities.

DISCUSSION

People in the Mojave Desert require a greater water supply than can be generated within the region and are dependent upon imports of both food and water. Almost all of the food supply is imported by fossil-fuel-powered transport, which creates a subsidy of so-called "green water," representing the nonregional irrigation and precipitation that go into crops and food produced elsewhere. An estimated total per capita human water requirement for all needs, including food production, is about 1300 $m^3 y^{-1}$ (Falkenmark and Rockstrom 2004). Mojave Desert occupants passed this benchmark of sustainability, where nearly all food was produced locally from the available water supply, by the first half of the twentieth century. Now a new equilibrium of sustainability may develop, where urban centers draw sustenance from a much larger landscape by means of the fossil-fuel subsidy. Clearly, sustainability depends on energy supply—primarily global oil production (fig. 3.4)—and thus many other factors external to the Mojave Desert.

Water supply is a question of quality and transport. For example, Las Vegas could build desalination plants along the coast of California, sell the water to California cities, and buy it back from California's allocation of the Colorado River. Two of the most common desalination techniques—reverse osmosis and distillation—have energy requirements in the range of 2 to 10 kW h m^{-3} (Pantell 1993). At full-time operation, the city of Santa Barbara's reverse-osmosis desalination plant, for example, needs about 50 million kWh to produce $9.25 \cdot 10^6$ m^3 y^{-1} of potable water (Panfell 1993), which is in the midrange of energy usage (5.4 kW h m^{-3}). That's enough energy to drive 2,500 tractor-trailer rigs from New York to San Francisco. More modern desalination plants, such as the Tampa Bay Seawater Desalination facility, are more efficient, but some of this efficiency is related to fossil-fuel subsidies. The Tampa Bay facility is collocated with the Tampa Electric Company's coal-fired Big Bend Power Station, and takes advantage of water pumped and filtered to cool the power plant. As an added bonus, heating of the thermoelectric coolant water makes the reverse osmosis process more efficient. Since fossil fuels provide 85% of the energy consumed in the U.S., and about two-thirds of the total energy delivered by the electrical power grid, reliance on desalination for water supply implies a reliance on fossil fuels.

Most of the water allocated from the Colorado River is used for agriculture. About 65% of California's total apportionment goes to the Imperial Irrigation District, a senior appropriator and one of the largest users of Colorado River water in the lower basin, where about a third is used for irrigating alfalfa. Simply buying out the alfalfa farmers in the Salton Sink could potentially provide more than twice Nevada's legal share of the river to Las Vegas. The poorer-quality water delivered to the Imperial Valley could be traded for higher-quality water drawn from Lake Mead with a negotiated payment. Transferring irrigation water rights from agriculture to cities, however, could increase urban populations while simultaneously constraining agricultural production.

A model of population growth potential, as a function of per capita water supply, shows greater opportunity for expansion through conservation than through increased supply. As per capita use becomes more conservative, population can become quite large. Emergency personal hydration requires about 3 to 4 L p^{-1} d^{-1}, or about 0.4% of the current per capita consumption in Clark County. This minimal daily requirement could be met by importation of bottled beverages. About 6% of the current per capita consumption is enough for cooking, bathing, washing clothes, cleaning, waste disposal, and watering a small garden (World Health Organization 1998). Residential use is about 59% of the SNWA water supply, most of which goes for irrigating outdoor landscaping. Conservation in the form of landscape conversion, waste water recycling, desalination, agriculture buyouts, and water treatment are all options that hold potential for supplying the water needed for continued growth of cities in the desert. Aquifer overdrafting during dry years and artificial recharge in wet years also increase the management options available for supplying large populations.

Rather than water-supply constraints, limits to expansion of urban populations in the desert will probably manifest as quality-of-life issues on a personal basis. Energy costs, jobs, affordable housing (both in the desert and nearby cities), economic growth, tourism, and other economic factors are more likely to determine future growth patterns than water constraints. Human ecology is distinguished by its industrial metabolism whereby factories, farms, power plants, transmission lines, dams, canals, highways, railways, oil and gas fields, and all the other accoutrements of our industrial civilization are "the exosomatic equivalent of organs" (Sterrer 1993). But the "heart" that circulates resources through the "organism" of industrial civilization is the millions of years of solar energy stored in fossil fuels—a finite endowment.

As long as energy is plentiful, there is no practical limit to growth in Las Vegas or any other city in the Mojave Desert. Icebergs could be towed from the polar regions and hauled by trucks to mini-marts in plastic bottles. Fresh farm and sea produce could be delivered overnight from the tropics by jet airplane and people could continue to thrive in the desert environment. During a situation of extreme water stress in an extended drought, agriculture and outdoor landscaping could be restricted and the population could potentially survive with a comfortable margin. But there is no such cushion in the energy supply.

Fossil fuels such as oil, natural gas, and coal power the economy that sustains modern civilization. Tar sands, oil shale, and methane hydrates are other forms of fossilized solar energy that are promising but problematic and have smaller returns on investment. Hydroelectric power and harvested biomass are old standbys but produce energy at a lower level and quality than petroleum products. Investments in biodiesel and ethanol tend to barely produce profits, given the petroleum input required to grow maize on a sufficiently large scale, although some success has been achieved in Brazil using sugarcane (La Rovere 2004). Photovoltaic electricity is an expensive way to run electronics, refrigeration, and other local needs, but is inadequate to power a transportation network. Draft animals and wind are older means of utilizing solar power. Geothermal energy, another source, is derived from nuclear fission of radioactive elements within the earth. Presently, however, none of these alternatives constitute a replacement for the hydrocarbon fuels that power the transportation network that supplies the urban centers of the Mojave Desert.

A summary of alternative energy, especially for transportation, must include hydrogen fuel cell technology. Hydrogen is a means of storing energy derived from other sources and is not an energy source itself. Presently, most commercial hydrogen is produced by treating natural gas with steam, and the resulting hydrogen releases less energy than can be obtained by burning natural gas directly. Hydrogen is also obtained by electrolysis of water, but with an efficiency of less than 100%. More energy is required to break apart a water molecule by electrolysis than can be retrieved by burning hydrogen and oxygen in a fuel cell. Much spare electrical-transmission capacity comes from coal, oil, and natural-gas powered generators, while local, small-scale hydrogen electrolysis plants would need substantial photovoltaic, wind, hydroelectric, nuclear fission, or biomass inputs to compensate for any decline in global fossil fuel production. Present efforts directed towards converting electricity into hydrogen and back to electricity in fuel cells may be more profitably redirected towards photobiological, photochemical, and thin-film solar hydrogen production (National Research Council 2004). Absent new sources of hydrogen, other than natural gas and electrolysis of water, improvements in fuel cell technology would have an effect on transportation similar to improvements in battery technology. Fuel-cell technology, in its present state, is ultimately dependent upon fossil fuels. Short of harnessing nuclear fusion and, in effect, putting the sun in a bottle, the global economy remains dependent on fossil-fuel production. Uncertainties exist in how quickly conventional oil production will peak and how human institutions will react.

Ecologic considerations indicate that humans are obligate earth dwellers—dependent on the productivity and life-support services of the ecosphere for long-term sustainability (Rees 1990). The environment's maximum sustainable population is defined by resource limits, which ultimately constrain population growth rates to zero or to negative numbers (Rees 1996). The economic paradigm, in contrast, is predicated on continuous growth. Periods of no economic growth and economic contractions are experienced as recessions and depressions. This paradigm of continual economic growth will interact with resource constraints in nonlinear and unpredictable ways. Problems and resource limitations will stimulate technological innovation and resource substitutions. Concerted efforts towards conservation, innovation, and substitution could significantly postpone the inevitable depletion of conventional, nonrenewable energy resources.

ACKNOWLEDGMENTS

Robert H. Webb, David M. Miller, and other colleagues provided many helpful suggestions. This chapter benefited immensely from the astute comments of an anonymous reviewer. Mary Martin supported the Mojave Desert Science Symposium, which led to this book. The viewpoints expressed are those of the author and do not in any way reflect the position or policies of the National Park Service. The author is solely responsible for errors, omissions, and opinions stated in this chapter.

REFERENCES

American Heritage® Dictionary of the English Language, Fourth Edition. 2000. Houghton Mifflin Company, Boston, Massachusetts.

Antelope Valley-East Kern Water Agency (AVEK). 2005. *California Urban Water Management Plan, Draft December 2005.* Antelope Valley-East Kern Water Agency, Palmdale, California.

Bailey, R. G. 1995. Description of the ecoregions of the United States. U.S. Department of Agriculture, Forest Service, Ft. Collins, Colorado.

BP. 2005. Putting energy in the spotlight: statistical review of world energy, June 2005. Beacon Press, London, England.

BP. 2007. BP statistical review of world energy, June 2007. Beacon Press, London, England.

Bureau of Reclamation, U.S. Department of the Interior. Robert B. Griffith Water Project. http://www.usbr.gov/dataweb/html/griffith.html.

Dudek and Associates. 2004. *Final report municipal service review, water service—High Desert Region.* Prepared for the Local Agency Formation Commission for Los Angeles County, Glendale, California.

Energy Information Administration. 2005. *Annual Energy Review 2004, Report No. DOE/EIA-0384(2004),* U.S. Department of Energy, Washington, D.C.

Energy Information Administration. 2007. *Annual Energy Review 2006, Report No. DOE/EIA-0384(2006),* U.S. Department of Energy, Washington, D.C.

Etemad, B., J. Luciani, P. Bairoch, and J. C. Toutain. 1991. *World energy production 1800–1985.* Librarie DROZ, Geneva, Switzerland.

Fagan, J. L. 1974. Altithermal occupation of spring sites in the Northern Great Basin: University of Oregon Anthropological Papers No. 6. Department of Anthropology and the University of Oregon Museum of Natural and Cultural History, University of Oregon, Eugene, Oregon.

Fagan, B. M. 1995. *Ancient North America: the archaeology of a continent.* Thames and Hudson, London, United Kingdom.

Falkenmark, M., and J. Rockstrom. 2004. *Balancing water for humans and nature: the new approach in ecohydrology.* Earthscan Publications, Cromwell Press, Trowbridge, Wiltshire, United Kingdom.

Galloway, D. L., S. P. Phillips, and M. E. Ikehara. 1998. Land subsidence and its relation to past and future water supplies in Antelope Valley, California. Pages 529–539 *in* J. W. Borchers, editor. *Proceedings of the Dr. Joseph F. Poland symposium on land subsidence.* Association of Engineering Geologists, Denver, Colorado.

Gleick, P. 1996. Basic water requirements for human activities: meeting basic needs. *Water International* **21(2)**:83–92.

Grayson, D. K. 1993. *The desert's past: a natural prehistory of the Great Basin.* Smithsonian Institution Press, Washington D.C.

Griset, D. E. 2004. Memorandum to the Water Policy Task Force: growth and water issues in the Antelope Valley. Southern California Association of Governments, Los Angeles, California.

Hafen, L. R., editor. 1982. *Mountain men and the fur traders of the far West.* University of Nebraska Press, Lincoln, Nebraska.

Heizer, R. F., and M. A. Whipple. 1971. *The California Indians.* Second edition. University of California Press, Berkeley, California.

Hopkins, A. D., and K. J. Evans. 1999. *The first 100: portraits of the men and women who shaped Las Vegas.* Huntington Press, Las Vegas, Nevada.

Hubbert, M. K. 1956. Nuclear energy and the fossil fuels. Pages 7–25 in *American Petroleum Institute drilling and production practice proceedings (March 7–9, 1956, San Antonio, Texas, USA),* Publication No. 95. American Petroleum Institute, Houston, Texas.

Hutson, S. S., N. L. Barber, J. F. Kenny, K. S. Linsey, D. S. Lumia, and M. A. Maupin. 2005.

Estimated use of water in the United States in 2000: U.S. Geological Survey Circular No. 1268. U.S. Geological Survey, Reston, Virginia.

Lahontan Regional Water Quality Control Board. 2002. Watershed management initiative chapter. California Environmental Protection Agency, Lake Tahoe, California.

La Rovere, E. L. 2004. The Brazilian ethanol program: biofuels for transport. International Conference for Renewable Energies (June 1–4, 2004, Bonn, Germany).

Leskinen, P. H. 1975. Occurrence of oaks in late Pleistocene vegetation in the Mohave Desert of Nevada. *Madroño* **23(4)**:234–235.

Lingenfelter, R. E. 1988. *Death Valley and the Amargosa: a land of illusion.* University of California Press, Berkeley, California.

Lockwood, C. 1980. In the Los Angeles oil boom, derricks sprouted like trees. *Smithsonian* **11**:187–206.

Malde, H. E. 1964. Environment and man in arid America. *Science* **145(3628)**:123–129.

Masters, C. D., E. D. Attansi, and D. Root. 1994. World petroleum assessment and analysis. Pages 529–541 *in Proceedings of the Fourteenth World Petroleum Congress* (Stavanger, Norway). John Wiley & Sons, New York.

Mehringer, P. J., Jr. 1965. Late Pleistocene vegetation in the Mohave Desert of southern Nevada. *Journal of the Arizona Academy of Science* **3**:172–188.

Mehringer, P. J., Jr. 1967. Pollen analysis of the Tule Springs Site area, Nevada. Pages 129–200 *in* H. M. Wormington and D. Ellis, editors. *Pleistocene studies in southern Nevada.* Nevada State Museum Anthropology Papers No. 13. Carson City, Nevada.

Mojave Water Agency (MWA). 2004. Draft Mojave Water Agency 2004 regional water management plan, program environmental impact report, State Clearinghouse No. 2003101119. Mojave Water Agency, Apple Valley, California.

National Research Council. 2004. *The hydrogen economy: opportunities, costs, barriers, and R&D needs.* The National Academies Press, Washington, D.C.

Palmdale Water District (PMD). 2004. Palmdale Water District water use efficiency proposal. Palmdale Water District, Palmdale, California.

Pantell, S. E. 1993. *Seawater desalination in California.* California Coastal Commission, San Francisco, California.

Rees, W. 1990. Sustainable development and the biosphere: the ecology of sustainable development. *The Ecologist* **20(1)**:18–23.

Rees, W. 1996. Revisiting carrying capacity: area-based indicators of sustainability. *Population and Environment* **17(3)**:195–215.

Simmons, M. R. 2005. *Twilight in the desert: the coming Saudi oil shock and the world economy.* John Wiley & Sons, Hoboken, New Jersey.

Smith, J. 1989. *The southwest expedition of Jedediah S. Smith: his personal account of his journey to California 1826–1827.* G. R. Brooks, editor. University of Nebraska Press, Lincoln, Nebraska.

Sneed, M., M. E. Ikehara, S. V. Stork, F. Amelung, and D. L. Galloway. 2003. *Detection and measurement of land subsidence using interferometric synthetic aperture radar and global positioning system, San Bernardino County, Mojave Desert, California:* Water-Resources Investigations Report No. 03–4015. U.S. Department of the Interior, U.S. Geological Survey, Sacramento, California.

Southern Nevada Water Authority (SNWA). 2005. *Southern Nevada Water Authority 2005 water resource plan*. Southern Nevada Water Authority, Las Vegas, Nevada.

Spaulding, W. G. 1990. Vegetational and climatic development of the Mojave Desert: the last glacial maximum to the present. Pages 166–199 *in* J. L. Betancourt, T. R. Van Devender, and P. S. Martin, editors. *Packrat middens, the last 40,000 years of biotic change*. University of Arizona Press, Tucson, Arizona.

Stamos, C. L., B. F. Cox, J. A. Izbicki, and G. O. Mendez. 2003. *Geologic setting, geohydrology and ground-water quality near the Helendale fault in the Mojave River Basin, San Bernardino County, California*. Water-Resources Investigations Report No. 03–4069. U.S. Department of the Interior, U.S. Geological Survey, Sacramento, California.

Stamos, C. L., J. A. Huff, S. K. Predmore, and D. A. Clark. 2004. *Regional water table (2004) and water-level changes in the Mojave River and Morongo ground-water basins, southwestern Mojave Desert, California*. Scientific Investigations Report No. 2004–5187. U.S. Department of the Interior, U.S. Geological Survey, Sacramento, California.

Stamos, C. L., P. Martin, T. Nishikawa, and B. F. Cox. 2001. *Simulation of ground-water flow in the Mojave River basin, California*. Water-Resources Investigations Report No. 01–4002. U.S. Department of the Interior, U.S. Geological Survey, Sacramento, California.

Sterrer, W. 1993. Human economics: a non-human perspective. *Ecological Economics* 7:183–202.

Stevens, J. E. 1988. *Hoover Dam*. University of Oklahoma Press, Norman, Oklahoma.

Susia, M. L. 1964. Tule Springs archaeological surface survey. Nevada State Museum Anthropological Papers No. 12. Nevada State Museum, Carson City, Nevada.

Tainter, J., C. Renfrew, W. Ashmore, C. Gamble, J. O'Shea. 1990. *The collapse of complex societies*. Cambridge University Press, Cambridge, Massachusetts.

Thompson, R. D., and K. L. Thompson. 1995. *Pioneer of the Mojave: the life and times of Aaron G. Lane*. Desert Knolls Press, Apple Valley, California.

Ubelaker, D. H. 1985. North American Indian population size, AD 1500 to 1985. *American Journal of Physical Anthropology* 77:289–94.

US Census Bureau. 2001. Census 2000 TIGER/Line data. U.S. Department of Commerce, Bureau of the Census, Geography Division, Washington, D.C.

Webb, R. H., R. Hereford, and G. J. McCabe. 2005. Climatic fluctuations, drought, and flow of the Colorado River. Pages 59–69 *in* S. B. Gloss, J. E. Lovich, and T. S. Melis, editors. *The state of the Colorado River ecosystem in Grand Canyon: a report of the Grand Canyon Monitoring and Research Center 1991–2004:* USGS Circular No. 1282. U.S. Department of the Interior, U.S. Geological Survey, Reston, Virginia.

Wiken, E. 1986. *Terrestrial ecozones of Canada*. Ecological Land Classification Series 19, Environment Canada. Ottawa, Ontario, Canada.

World Health Organization. 1998. Minimum water quantity needed for domestic use in emergencies. Technical Note No. 9–7.1.05. World Health Organization, Geneva, Switzerland.

Yergin, D. 1991. *The prize*. Simon & Schuster, New York, New York.

Zucchetto, J. 2006. *Trends in oil supply and demand, the potential for peaking of conventional oil production, and possible mitigation options: a summary report of the workshop*. National Research Council, The National Academies Press, Washington, D.C.

FOUR

Impacts of Atmospheric Nitrogen Deposition on Vegetation and Soils at Joshua Tree National Park

EDITH B. ALLEN, LEELA E. RAO, ROBERT J. STEERS,
ANDRZEJ BYTNEROWICZ, AND MARK E. FENN

The Western Mojave Desert is affected by air pollution generated in the Los Angeles air basin that moves inland with the predominant westerly winds (Edinger 1972; Fenn et al. 2003b). The pollution contains both oxidized and reduced forms of nitrogen (N), which are of concern because they are deposited on soil and plant surfaces and thus fertilize plants. Nitrogen deposition may affect plant productivity differentially; some studies have found that nonnative grasses have higher rates of N uptake and production than many native species (Allen et al. 1998; Yoshida and Allen 2001, 2004; Brooks 2003), whereas others have found that nonnative grasses have similar N-uptake and production rates as natives (Padgett and Allen 1999; Salo et al. 2005).

The number of nonnative species and their abundance has increased in the desert during the last two decades (Brooks 1999a, 1999b, *this volume*), and our objective is to determine whether this increase is related to elevated N deposition. As nonnative grasses increase in productivity, native plants may become sparse (DeFalco et al. 2001; Brooks 2000, 2003, *this volume*). This is especially a concern in protected areas with rare species—areas such as Joshua Tree National Park (JTNP), which lies within both the Mojave and the Colorado deserts. The wind patterns create N-deposition gradients that have been modeled and have the highest N levels on the west side of JTNP (Tonnesen et al. 2003). We selected sites along this gradient to make finer-scale field measurements of reactive atmospheric N, as well as soil-extractable N, and to determine the response of nonnative grass cover and native species diversity to elevated soil N.

Nitrogen deposition in the shrublands and forests of the Los Angeles air basin may be as high as 30–50 kg ha^{-1} y^{-1} (Bytnerowicz et al. 1987; Fenn et al. 1998, 2003b). Most of this N arrives as dry deposition in gaseous, ionic, and particulate form during the dry summer season and is much more difficult to measure than wet deposition, which arrives in the form of rain, snow, etc. (Bytnerowicz et al. 2000). Relatively few estimates of N deposition have been done in the Mojave or Colorado deserts. At the Black Rock site in northwestern JTNP (fig. 4.1), one study calculated a value of 8 kg ha^{-1} y^{-1} (Sullivan et al. 2001), and 12 kg ha^{-1} y^{-1} in the northwestern Coachella Valley (G. S. Tonnesen et al. *unpublished data*). Total N deposition reported at the Clean Air Status and Trends Network (CASTNET) monitoring site near Black Rock ranged from 3.2 kg ha^{-1} y^{-1} in 1995 to 5.9 kg ha^{-1} y^{-1} in 2003 (CASTNET 2005). However, CASTNET underestimates dry deposition of N (Baumgardner et al. 2002, Fenn et al. 2003a), particularly in California sites where dry deposition of ammonia (NH_3) is a significant fraction of inorganic N deposition (Fenn et al. 2007). Short-term measurements at the western Salton Sea, which are recalculated on a yearly basis, ranged from 0.4 to 6.6 kg ha^{-1} y^{-1} for nitrate-N (NO_3^--N) and 2.6 to 8.7 for ammonium-N (NH_4^+-N) (Alonso et al. 2005), but such calculations are fraught with assumptions about variations in short-term deposition rates and the spatial distribution of deposition. The lack of actual measurements of N deposition in these deserts means that observed vegetation changes cannot be explained with respect to air pollution, although field observations and N fertilizer experiments suggest there may be a relationship (DeFalco et al. 2001; Brooks 2003).

Soil N gradients caused by anthropogenic deposition have been measured in western Riverside County in coastal-sage scrub vegetation (Padgett et al. 1999) and in the San Bernardino Mountains in coniferous forest (Fenn et al. 2003b), with values for extractable N increasing 5-fold across the gradients. This effect is especially pronounced when measurements are taken in seasonally dry soils, when extractable N is highest, both from dry deposition and mineralization (conversion into plant-available forms). Thus, soil-surface N measurements during the dry season can be used as another indicator of the accumulation of N from air pollution (Padgett et al. 1999).

The impacts of elevated N include changes in nutrient cycling as well as plant community composition. The rate of nutrient cycling and N leakage has increased in mesic forests of the eastern U.S. (Aber et al. 1998), as well as in seasonally dry, mixed coniferous forests in California (Fenn et al. 2003b), but the rate of N loss is expected to be lower in arid or semiarid ecosystems (Wood et al. 2006). Studies from Europe have shown a loss of diversity of native herbaceous species and an increase in native grass biomass with N deposition (Bobbink et al. 1998; Stevens et al. 2004). Nitrogen fertilizer studies in the Mojave Desert

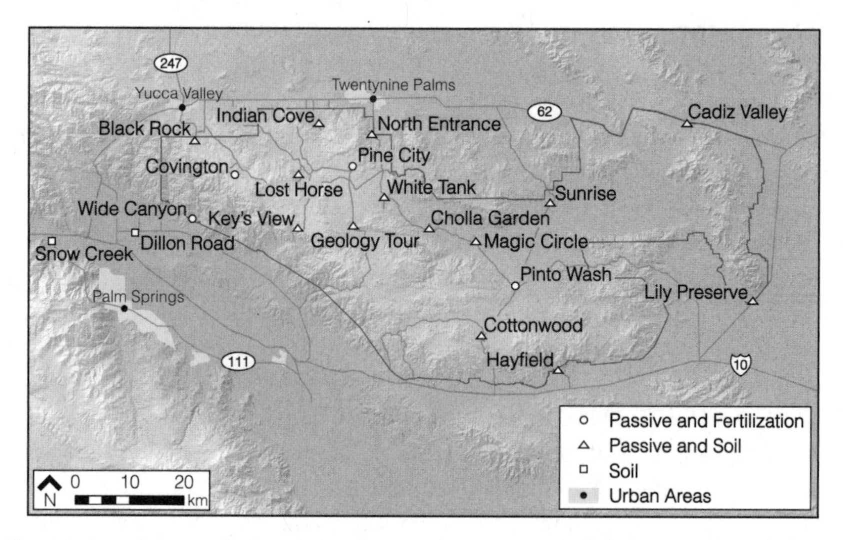

Fig. 4.1. Map showing the locations of atmospheric nitrogen (N), extractable soil N, and vegetation sample sites at Joshua Tree National Park (JTNP). The passive and nitrogen fertilization experiments are represented with circles, and the nitrogen experiments are described in table 4.1. Passive air samples and soil samples (represented with triangles) were taken from all sites in JTNP, and soils samples (represented with squares) were also taken from two sites to the west of JTNP (Snow Creek and Dillon Road).

(Brooks 2003) and in coastal sage scrub (Allen 2004) have shown increased productivity of invasive grasses and decreased productivity and diversity of native species.

To control for the natural variability of climate, soils, and vegetation inherent in any gradient, we performed N fertilization experiments to determine the impacts of N on vegetation and soils using blocked, replicated designs. This was especially critical because the Mojave and Colorado deserts have low N inputs compared with mesic climates, which would make any N response difficult to detect along a heterogeneous gradient. Air pollution measurements for the first phase of this study included ambient concentrations of ozone (O_3), nitric acid (HNO_3), and NH_3. We do not present data on N deposition (only on concentrations of atmospheric pollutants), although research on N deposition is underway. Ozone was measured because it co-varies with nitrogen oxides and has been reported at high levels in JTNP. Earlier work showed O_3 damage to native plants in the desert (Thompson et al. 1984; Bytnerowicz et al. 1988). The specific objectives of this research were to: (1) measure gaseous N pollutants and O_3 concentrations along N-deposition gradients in JTNP using passive samplers, and determine extractable soil N concentrations along the same gradients; and (2) measure nonnative grass biomass and cover, and native forb cover and species richness in N-fertilized and control plots at high and low N-pollution sites in creosote-bush scrub and pinyon-juniper woodland.

METHODS

Site Description

The research was done at Joshua Tree National Park (JTNP), with two additional locations to the west of JTNP to include areas of potentially higher N deposition (fig. 4.1). JTNP has approximately 320,000 ha that span the boundary between the Mojave and Colorado deserts. The dominant vegetation types include low-elevation *Larrea tridentata* (creosote bush) scrub, intermediate-elevation *Yucca brevifolia* (Joshua tree) woodland, and higher-elevation pinyon-juniper (*Pinus monophylla, Juniperus californica*) woodland, as well as smaller areas of riparian, grassland, and succulent vegetation types (Sawyer and Keeler-Wolfe 1995). To date, JTNP has more than 700 plant species identified. The elevation ranges from 500 to 1650 m. The geologic parent materials of JTNP consist primarily of granites with several areas of basaltic extrusions (Trent 1984). Air pollution has been increasing and visibility decreasing over the last four decades (Joshua Tree National Park 2004). Annual precipitation during the two years of study was 249 mm (2002–2003) and 180 mm (2003–2004) at the Black Rock Station (pinyon-juniper woodland), and 205 and 113 mm, respectively, at Hay Field (creosote-bush scrub, fig. 4.1). The data are reported for October 1–September 30, as the growing season is fall through spring, depending on elevation and yearly precipitation.

Air Pollution Measurements

Air samplers were deployed at 18 locations across JTNP (fig. 4.1). The locations were chosen to cover JTNP, encompass the potential west to east gradient, and be accessible to roads (although not located near any well-traveled highways that might contribute to air pollution). They were placed within the dominant vegetation types (creosote-bush scrub, Joshua tree woodland, pinyon-juniper woodland). Concentrations of ambient gaseous N pollutants [nitric oxide (NO), nitrogen dioxide (NO_2), NH_3, and O_3] were determined with passive samplers in the selected sites (Koutrakis et al. 1993). The passive samplers consisted of Teflon cartridges with pollutant-collecting filters placed in inverted PVC protective cups 2 m above ground level. Nitric acid was collected on three nylon filters placed in double rings hung inside PVC caps to protect them from wind and rain (Bytnerowicz et al. 2001). Two-week average concentrations of the pollutants were determined three times during the dry season and two times during the wet season. Results are shown for the 14 days following

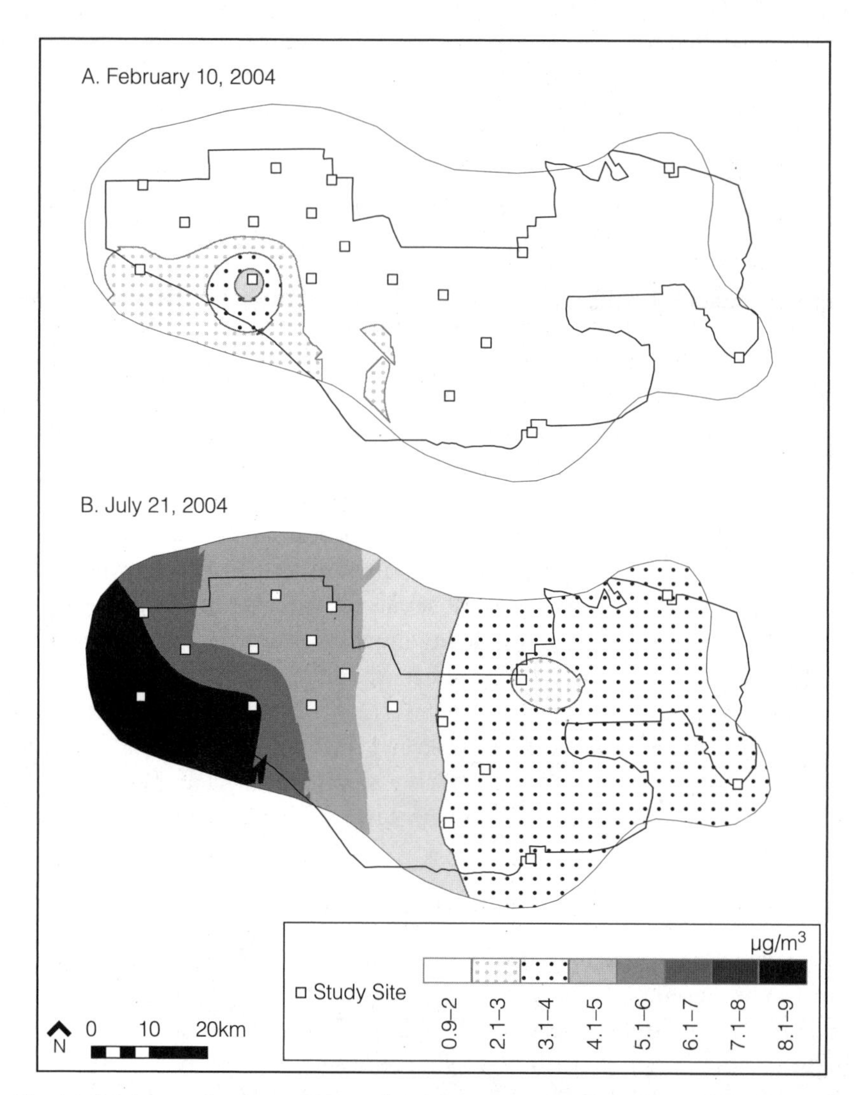

Fig. 4.2. Schematic diagram of interpolated contours depicting two-week-long average nitric acid (HNO_3) concentrations (µg m^{-3}) in the atmosphere over Joshua Tree National Park in winter (*A*) and summer (*B*) 2004 (dates on graph show start of sampling).

February 10 and July 21, 2004, which were precipitation-free periods (figs. 4.2, 4.3, and 4.4).

Soil Sampling

Soil samples for extractable N analysis were collected from the same 18 sites as the air samples during July 2004, as well as from two additional sites outside

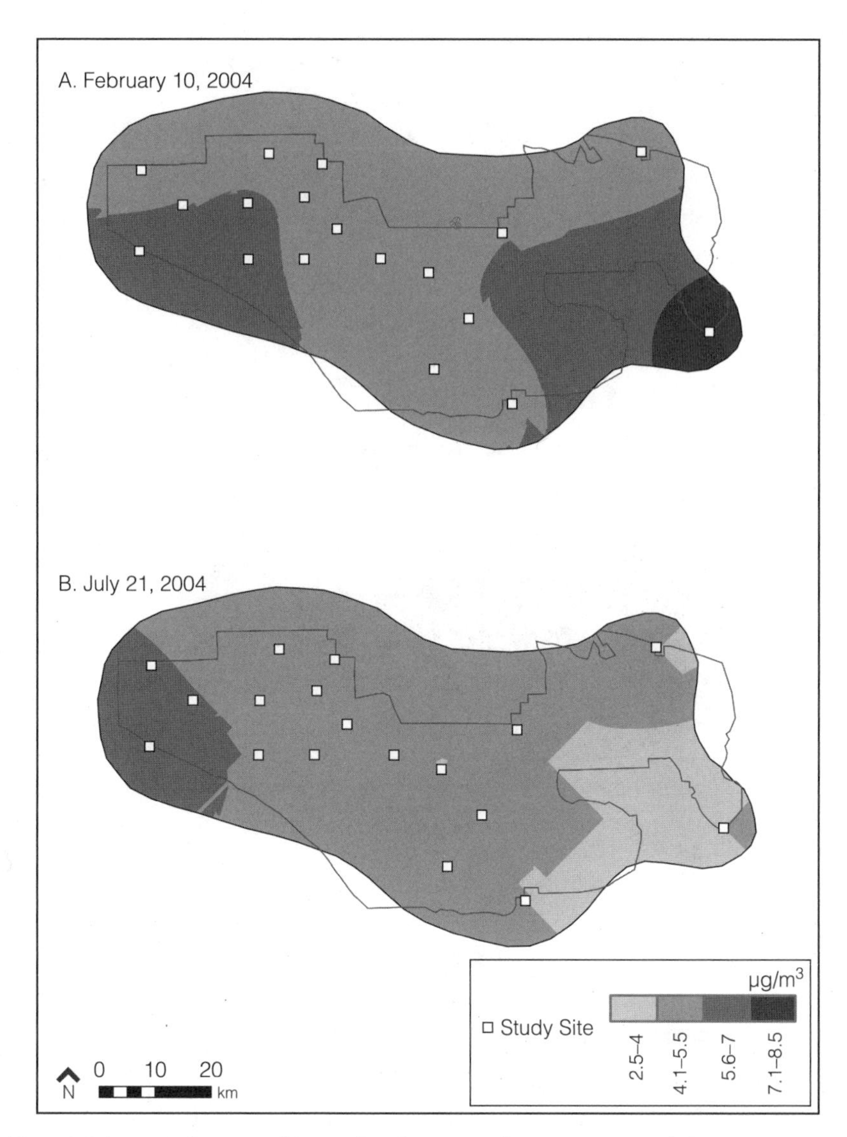

Fig. 4.3. Schematic diagram of interpolated contours depicting two-week-long average ammonia (NH_3) concentrations ($\mu g\ m^{-3}$) in the atmosphere over Joshua Tree National Park in winter (*A*) and summer (*B*) 2004 (dates on graph show start of sampling).

JTNP to the west (fig. 4.1). Dry-season samples are shown because prior analyses showed extractable N is greater in the dry season than in the winter rainy season (Padgett et al. 1999). Cores were taken 5-cm deep ($n = 10$) in the interspaces between shrubs or trees. Soils were extracted in potassium chloride (KCl), and

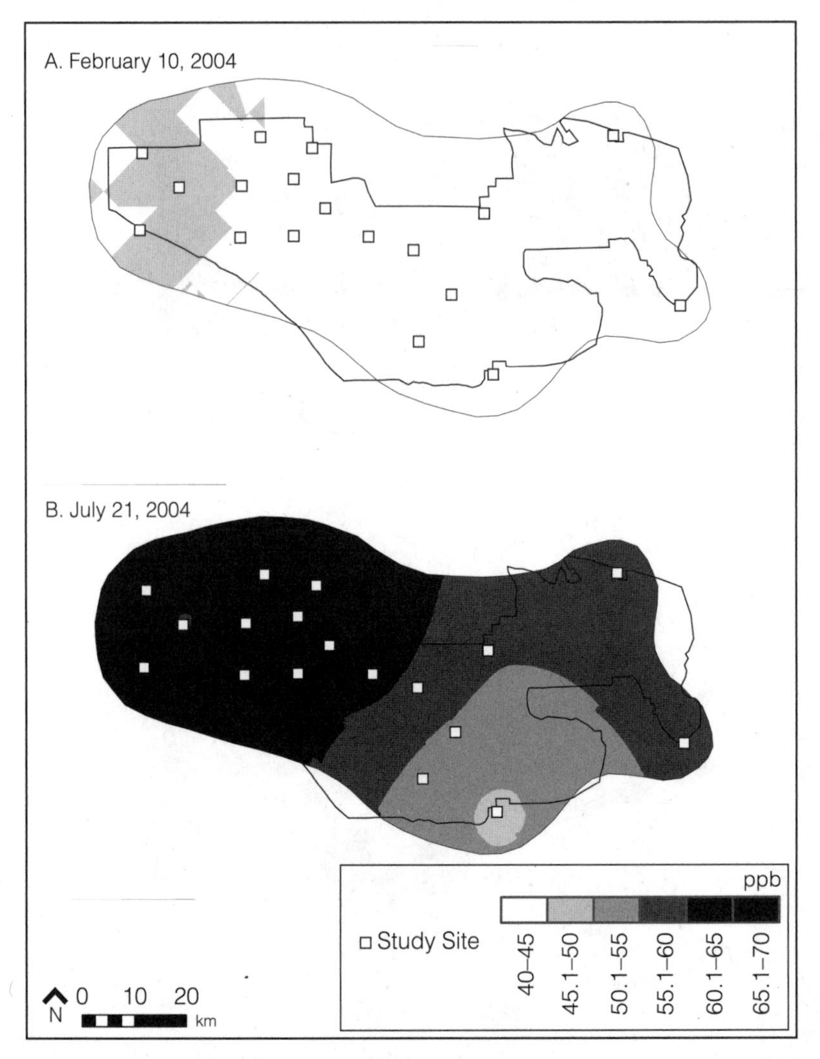

Fig. 4.4. Schematic diagram of interpolated contours depicting two-week-long average ozone (O_3) concentrations (ppb) in the atmosphere over Joshua Tree National Park in winter (*A*) and summer 2004 (*B*) (dates on graph show start of sampling).

ammonium (NH_4^+) and nitrate (NO_3^-) were measured colorimetrically using a Technicon Autoanalyzer.

Fertilization Experiment

Fertilization was done at four sites, two on the west end of JTNP and two further to the east (table 4.1 and fig. 4.1). These were in two vegetation types, creosotebush scrub in the Colorado Desert portion of JTNP, and pinyon-juniper woodland

Table 4.1 Vegetation, elevation, and hypothesized relative nitrogen (N) deposition of four sites at Joshua Tree National Park chosen for N fertilization study

Site	Vegetation type	Elevation (m)	N deposition
Pinto Wash	Creosote bush scrub	750	Low
Wide Canyon	Creosote bush scrub	550	High
Pine City	Pinyon-juniper woodland	1,400	Low
Covington Flat	Pinyon-juniper woodland	1,500	High

See fig. 4.1 for site locations.

in the Mojave Desert. The two vegetation types represent two of the most abundant vegetation types in JTNP, as well as the extremes in elevation. The relative amounts of N deposition were hypothesized based on the model of Tonnesen et al. (2003), but actual rates of N deposition are not yet known for these sites (table 4.1).

Two levels of fertilizer were used, 5 and 30 kg ha^{-1} y^{-1}, plus an unfertilized control at each site. The higher rate was chosen because 30 kg N ha^{-1} increased biomass of *Schismus* ssp. (Mediterranean split grass), *Bromus madritensis* ssp. *rubens* (red brome), and *Erodium cicutarium* (stork's bill) in another study within one growing season in the Western Mojave Desert (Brooks 2003). However, composition of low-productivity vegetation is more sensitive to N inputs, and may experience shifts even with low levels of fertilization (Theodose and Bowman 1997; Bowman and Steltzer 1998). Therefore, the treatments also included a low level of N fertilizer (5 kg N ha^{-1}).

Individual shrubs or trees were fertilized by broadcasting ammonium nitrate (NH_4NO_3) granular fertilizer into an area encompassing the tree canopy. Plot size was determined by the shrub or tree size, with 6 × 6 m for creosote bush, 8 × 8 m for juniper, and 10 × 10 m for pine. Each shrub or tree species was fertilized, and plots were selected across the landscape as 10 replicate blocks, each block containing each of the two N-fertilizer levels plus a control. Fertilizer levels were chosen because the results could be compared with the results from a previous study in the Mojave Desert, which showed increased productivity by nonnative grasses (Brooks 2003). The lower levels were chosen to determine if N would accumulate in soils in a dry climate and eventually promote a response by nonnative plants. The low level (5 kg ha^{-1} y^{-1}) was similar to the highest known level of 8 kg ha^{-1} y^{-1} at the Black Rock Station, calculated by the Environmental Protection Agency (Sullivan et al. 2001). Plots were fertilized in December 2002 and again in December 2003. The N deposition model of Tonnesen et al. (2003) indicates that more NO_3^- than NH_4^+ is deposited in Southern California, but the relative amount deposited at each of our sites is not known. Prior analyses have shown that soil NH_4^+ and NO_3^- concentrations are high in December due to accumulated dry deposition during the summer/fall dry season and remaining mineralized N from the end of the prior rainy season (Padgett et al. 1999). Soil cores were collected to 5 cm deep to determine N levels after fertilization. Growing season (March–May)

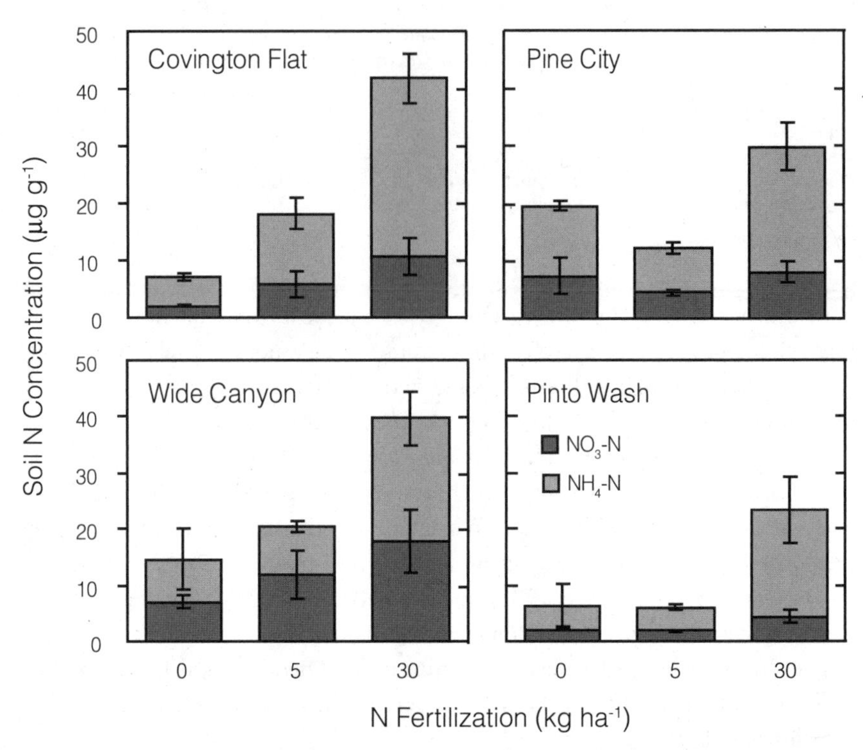

Fig. 4.5. Graphs showing the amounts of extractable soil nitrogen (N) as ammonium (NH_4^+) and nitrate (NO_3^-) in plots fertilized with ammonium nitrate (NH_4NO_3) for two years at two levels (5 and 30 kg ha^{-1}) for four sites: Covington Flat, Wide Canyon, Pine City, and Pinto Wash (see fig. 4.1 for site locations).

and dry season (July) samples were taken. July values are shown, as these were higher in extractable soil N, and represent the soil N that is available to plants for uptake at the onset of fall rains (fig. 4.5).

Vegetation Sampling

Vegetation cover was monitored at the four fertilization sites in 1.0 × 0.5 m quadrats placed just outside the drip line to the north and south of each shrub or tree. The percent cover of each species was estimated in a gridded frame with gridlines at 5, 10, and 25 cm intervals and estimates to the nearest 1% between 1%–20%, and the nearest 5% between 20%–100%; north and south sides of shrubs or trees were measured separately. Vegetation cover on the north side was on average higher than the south side, but there were no statistical interactions between the N fertilizer effect on the two sides, so the mean values of the two sides are shown (table 4.2). Nonnative grasses were clipped and measured for biomass in 10 replicate 0.25 × 0.50 m (0.125 m^2) plots within each fertilizer level to develop re-

Table 4.2 Percent cover of the most abundant native species in 0.5 m² quadrats at four sites under two nitrogen (N) fertilization levels and an unfertilized control, 2004. There were no significant differences for any of these species when analyzed individually, as there were many zero values for each species.

	Pinto Wash			Wide Canyon			Pine City			Covington Flat		
N treatment (kg/ha):	0	5	30	0	5	30	0	5	30	0	5	30
Chaenactis fremontii		1.3		3.7	2.2	1.9						
Chaenactis stevioides							2.6	4.3	3.8		1.7	1.9
Coleogyne ramosissima								2.2				
Cryptantha angustifolia				6.6	4.9	8.2						
Cryptantha pterocarya							1.5	3.2	4.8			2.9
Descurainia pinnata							2.3	3.0	4.0			
Euphorbia polycarpa				1.2								
Gilia stellata							1.5	1.1	1.6			
Malacothrix glabrata				2.5	2.0	2.0						
Mentzelia affinis										1.2	3.3	
Mirabilis californica												1.6
Pectocarya recurvata				1.5	1.5	1.4						
Phacelia distans								2.4	2.9	4.5	4.8	5.8
Poa secunda												1.1
Salvia columbariae	8.3									1.4		

All species with > 1% are annual forbs except for *C. ramosissima* (shrub) and *P. secunda* (perennial grass).

Species with < 1% cover are:

Shrubs (20 species): *Adenophyllum porophylloides, Brickellia californica, Chrysothamnus nauseosus, Echinocereus engelmannii, Ephedra nevadensis, Eriogonum fasciculatum, E. wrightii, Eriophyllum confertiflorum, Gutierrezia microcephala, G. sarothrae, Hymenoclea salsola, Juniperus californica, Lycium andersonii, Nolina parryi, Opuntia erinacea, Purshia tridentata, Quercus Cornelius-mulleri, Salazaria mexicana, Viguiera parishii, Yucca schidigera*

Native perennial grasses (4): *Achnatherum lettermanii, A. speciosum, Elymus elymoides, Erioneuron pulchellum*

Annual herbs (59): *Amsinckia tessellata, Anisocoma acaulis, Arabis pulchra, Arenaria macrademia, Aristida adscensionis, Calycoseris parryi, Calyptridium monandrum, Camissonia californica, C. campestris, C. claviformis, C. pallida, Castilleja angustifolia, Caulanthus cooperi, Centrostegia thurberi, Chaenactis macrantha, Chorizanthe brevicornu, Crassula connata, Cryptantha barbigera, C. circumscissa, C. maritima, C. micrantha, C. nevadensis, C. utahensis, Draba cuneifolia, Eriogonum davidsonii, E. maculatum, E. nidularium, E. pusillum, Eriastrum diffusum, Eriophyllum wallacei, Eschscholzia minutiflora, Eucrypta chrysanthemifolia, Euphorbia albomarginata, Filago arizonica, F. depressa, Layia glandulosa, Lepidium lasiocarpum, Linanthus aureus, L. bigelovii, L. dichotomus, L. jonesii, Loeseliastrum matthewsii, Lotus strigosus, Lupinus concinnus, Mentzelia* sp., *Nama demissum, Nemophila menziesii, Pectocarya heterocarpa, P. penicillata, P. platycarpa, P. setosa, Phacelia ciliata, P. cryptantha, Plantago ovata, P. patagonica, Rafinesquia neomexicana, Syntrichopappus fremontii, Thysanocarpus curvipes, Uropappus lindleyi*

Perennial forbs (15): *Allium parishii, Astragalus bernardanus, A. lentiginosus, A. nuttallianus, Calochortus kennedyi, Delphinium parishii, Dichelostemma capitatum, Dudleya saxosa, Eriogonum inflatum, Lomatium mohavense, Lotus argophyllus, L. rigidus, Mirablis bigelovii, Sphaeralcea ambigua, Stephanomeria exigua*

Nomenclature from Hickman (1993).

gressions of grass biomass with percent cover. Clippings were oven dried at 65C° to constant mass. Total grass biomass for the 0.5 m^2 quadrats was calculated from percent cover data based on these regressions. Biomass of native vegetation was not assessed to avoid destructive harvesting of native species, which included 77 herbaceous species (table 4.2). Vegetation was monitored in March–May in 2003 and 2004, the date depending on peak plant production according to elevation. Data are shown for 2004 only.

RESULTS

Air Pollution

Nitric acid had higher atmospheric concentrations across Joshua Tree National Park (JTNP) in July than in February (fig. 4.2), but the reverse was true for ammonia (NH_3), which had higher concentrations in February (fig. 4.3). Ozone followed a pattern similar to that of nitric acid (HNO_3) (fig. 4.4). The concentrations of HNO_3 ranged from 1.0 to 5.0 μg m^{-2} in February and from 2.0 to 9.0 μg m^{-2} in July (fig. 4.2). The concentrations fell along a gradient of high to low HNO_3 from west to east; the higher concentrations in the west are likely explained by the closer proximity to the prevailing winds that bring air pollutants from the Los Angeles basin. In winter, the highest HNO_3 value was at Key's View (fig. 4.1), a popular visitor overview on the ridge of the Little San Bernardino Mountains. This site had a higher value in the summer, although the highest summertime exposure was at Wide Canyon, one of the four experimental N fertilization sites.

Atmospheric concentrations of NH_3 ranged from 4.0 to 8.5 μg m^{-2} in February, with lower values of 2.5 to 7.0 μg m^{-2} in July (fig. 4.3). The summer concentrations of NH_3 followed the same west to east gradient as the HNO_3, but the winter pattern was different, with an area of high concentration at the east end of JTNP at the Lily Preserve site (figs. 4.1 and 4.3). Sites in the interior of JTNP had the lowest concentrations of NH_3.

Spatial and temporal patterns of ozone (O_3) concentrations were similar to HNO_3, with 39 to 57 ppb in February and 48 to 68 ppb in July (fig. 4.4). Key's View, on the western side of JTNP, was highest in O_3 in February, and Wide Canyon, also on the western side, was highest in July. However, sites on the eastern side of JTNP were also exposed to elevated levels of O_3, with intermediate values at the Lily Preserve and Cadiz Valley sites (fig. 4.1).

Soil Nitrogen

Soils collected in July 2004 on the western side of JTNP (Black Rock and Key's View) and the two sites outside the western boundary (Snow Creek and Dillon Road) had higher levels of extractable soil N than all but three sites on the eastern edge (fig. 4.6). The soils in the center of JTNP had low concentrations. The highest values were around 16–20 μg N g^{-1} soil, while the lowest values were 4 μg N g^{-1}.

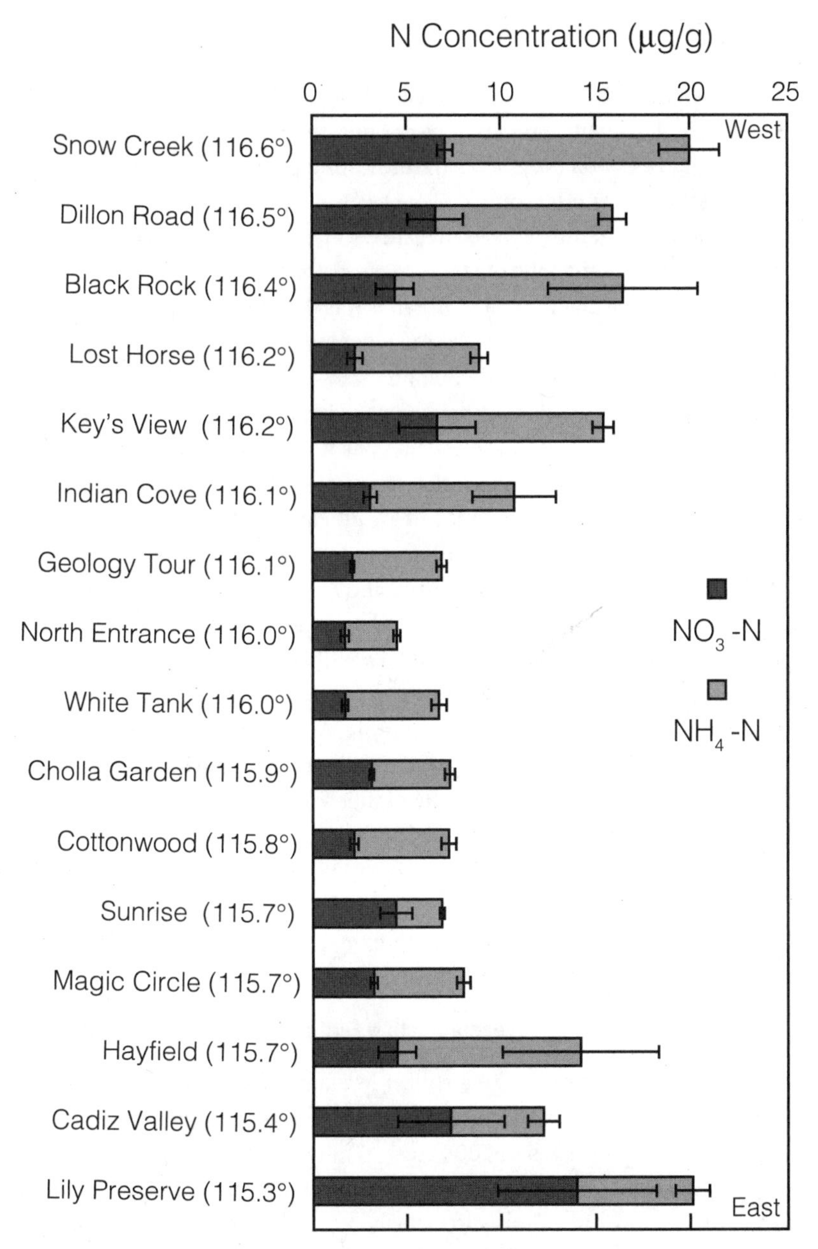

Fig. 4.6. Bar graph of extractable soil nitrogen as ammonium (NH_4^+) and nitrate (NO_3^-) for fourteen sites in Joshua Tree National Park (JTNP) and two sites west of JTNP in July 2004 (see fig. 4.1 for site locations).

There was a tendency for ammonium-N (NH_4^+-N) to be higher than nitrate-N (NO_3^--N). In general, the sites that had higher reactive atmospheric N also had high extractable soil N concentrations (Black Rock, Key's View, Wide Canyon, Lily Preserve, Cadiz Valley, and Hay Field). Soils collected during the growing season had 1–2 µg N g^{-1} (data not shown), indicating plant uptake and/or leaching of N during the growing season.

Additional edaphic factors were measured at the four N fertilizer sites. Bicarbonate-extractable phosphorus (P) ranged from 6–12 µg P g^{-1} and total P was 650–1500 µg P g^{-1} at the four sites. Total N was 0.040–0.078% and total carbon (C) was 0.22 to 0.84%. Soil pH was 6.8 at Covington Flat, 7.1 at Wide Canyon, 7.7 at Pine City and 7.9 at Pinto Wash. Soil texture was sandy loam at all the sites, with varying amounts of gravel and pebble-sized particles. Pinto Wash has the lowest amount of exposed rock on the surface, as it lies in a basin that accumulates surface sand, while the other three sites are a gravelly debris flow (Wide Canyon), rocky alluvial channel (Pine City), and an alluvial fan (Covington Flat).

Extractable N was higher with N fertilizer at the four fertilized sites, with values of 6 to 18 µg N g^{-1} with 5 kg ha^{-1} fertilizer, and 23–40 µg N g^{-1} with 30 kg ha^{-1} fertilizer in July 2004 (fig. 4.5). Pine City had unexpectedly high soil N in the control plots—as high as the fertilized plots—possibly related to small mammal activity. Covington Flat, on the western side of JTNP, had control soils with low N concentrations (6 µg N g^{-1}), which were more similar to JTNP interior sites (figs. 4.1 and 4.5). Control plots in Wide Canyon had high soil N concentration, with nearly 15 µg N g^{-1}, congruent with the high level of atmospheric reactive N. In 2003, total extractable soil N from these sites ranged from a low-control of 11.4 (SE = 2.6) to a high-fertilized (30 kg ha^{-1}) of 34.8 (12.6) µg N g^{-1}, with greater variability than in 2004 (2003 data not shown). The high and low values for 2003 both occurred at Wide Canyon, and other sites were intermediate in both their lowest and highest soil N values.

Response of Vegetation to N Fertilization

Vegetation changes were related to fertilizer level, initial soil N, and initial vegetation cover. Biomass of nonnative grasses increased significantly with N fertilization in three of the four sites in 2004 (fig. 4.7), but percent cover did not increase significantly (fig. 4.8) at any site. The lack of change in percent cover with fertilization reflects how difficult it is to visually estimate small increases in grass height (e.g., *Schismus* ssp. were only 3–6 cm); percent cover can be determined by calculating the relationships of elevated biomass to cover in fertilized vs. control plots. Percent cover of native vegetation also did not change significantly except at Pine City, where it increased in the higher fertilization treatment (30 kg N ha^{-1}) (fig. 4.8). Species richness of native vegetation decreased in the 30 kg N ha^{-1} fertil-

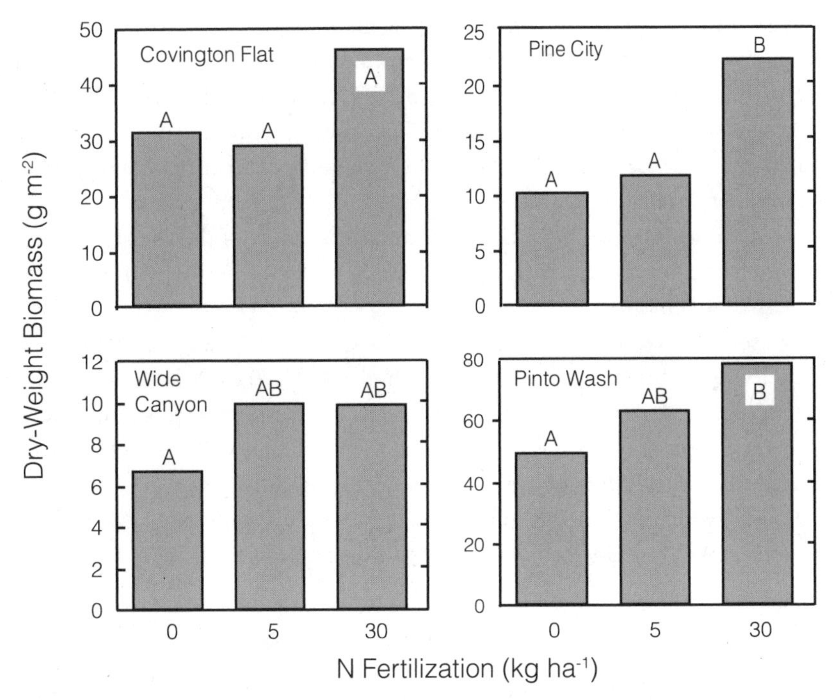

Fig. 4.7. Graphs showing nitrogen impacts on dry weight of nonnative grass in March–April 2004 following nitrogen fertilization at 5 and 30 kg ha^{-1} for four sites. Nonnative grasses were primarily *Schismus barbatus* (Mediterranean split grass) at Wide Canyon and Pinto Wash, and *Bromus madritensis* ssp. *rubens* (red brome) at Covington Flat and Pine City. Different letters above bars indicate significant differences at $P = 0.05$.

ization treatment at Pinto Wash (fig. 4.9). At Pine City, however, species richness of native vegetation was significantly higher in the 30 kg N ha^{-1} treatment (fig. 4.9, only data from 2004 are shown). In 2003, there were no significant increases in nonnative grasses with N fertilization at any of the sites, although there was an increase in native forb cover with 30 kg N ha^{-1} at Pine City as in 2004 (data not shown). As with the soil N data, variability of the plant data was also higher in 2003 than 2004.

The nonnative grass species at the two low-elevation sites, Wide Canyon and Pinto Wash, were *Schismus barbatus* and *S. arabicus* (Arabian split grass), with < 1% of the latter. *Bromus madritensis* ssp. *rubens* was the dominant nonnative grass at the two high-elevation sites, with 1%–2% of *B. tectorum* (cheatgrass) and another 1%–2% cover of *S. barbatus.* The dry mass of nonnative grass increased significantly ($P < 0.05$) with 30 kg N ha^{-1} in three of the four sites, but not at Covington Flat ($P = 0.101$). There was not a significant increase in nonnative grass biomass with 5 kg N ha^{-1} fertilizer at any of the sites (fig. 4.7). Overall, Pine City had the lowest nonnative grass biomass, and Pinto Wash had the highest even in control plots, though it had low extractable soil N and atmospheric reactive N.

Percent cover of nonnative species did not change significantly with N fertilization at any of the four sites (fig. 4.8). Total nonnative cover included grasses and *Erodium cicutarium*, but the latter contributed < 1.5% cover at each site (table 4.1). Cover of native forbs did not change significantly at any of the sites except Pine City, where it increased with elevated N. This was also the site with the lowest nonnative grass biomass.

The species richness of native herbaceous species at Pine City increased significantly following N fertilization at the highest rate, from 3.5–4.5 species per 0.5 m^2 plot (fig. 4.9). Conversely, native species richness declined significantly at Pinto Wash from 1.3–0.2 species/plot, and there was not a significant change at the other two sites. Most of the diversity of this desert vegetation is due to annual herbs, which included 59 species at the four sites combined, plus 4 perennial grasses, 15 perennial forbs, and 20 shrub species (table 4.2). Very few of these

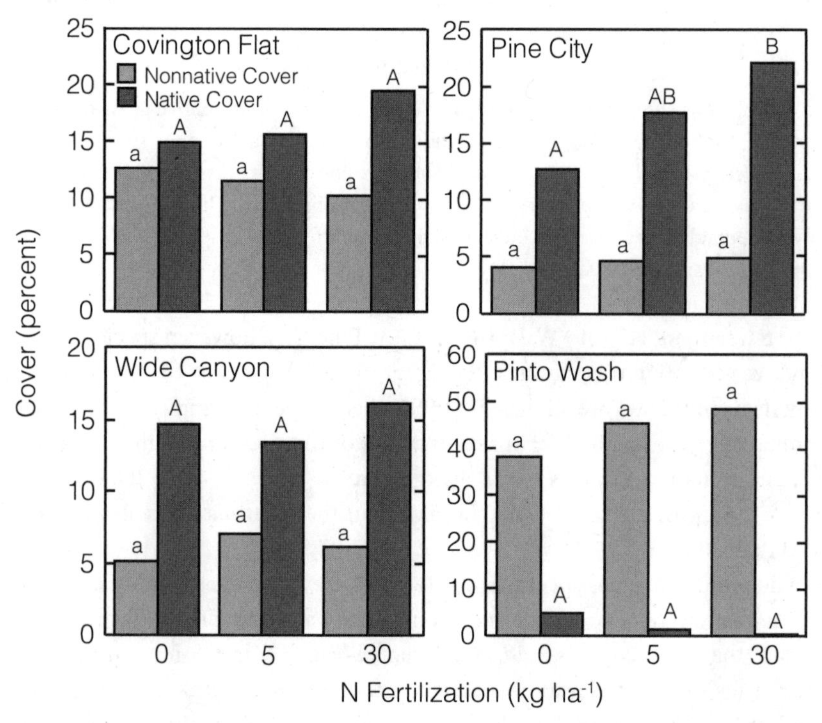

Fig. 4.8. Graphs showing nitrogen impacts on percent cover of nonnative and native herbaceous species in March–April 2004 following nitrogen fertilization at 5 and 30 kg ha^{-1} at four sites. A list of species is provided in table 4.2. Different letters above bars indicate significant differences at $P = 0.05$.

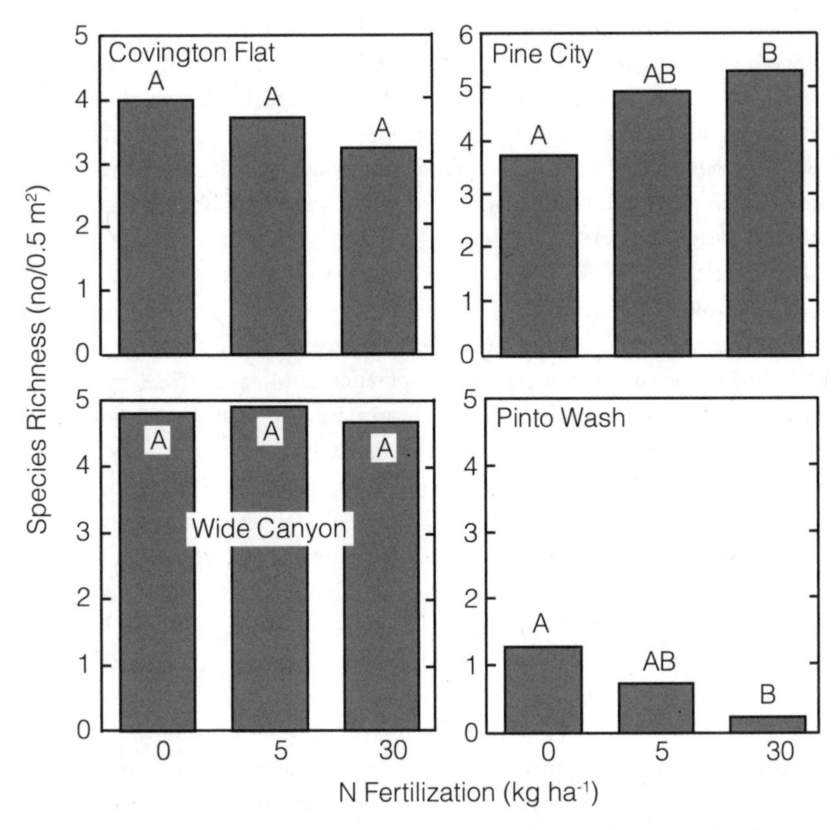

Fig. 4.9. Graphs showing nitrogen impacts on species richness (number per 0.5 m^{-2}) of native herbaceous species in March–April 2004 following nitrogen fertilization for two years at 5 and 30 kg ha^{-1} at four sites. A list of species is provided in table 4.2. Different letters above bars indicate significant differences at $P = 0.05$.

had > 1% cover, and most occurred sporadically with many not occurring at all in some treatments, so no statistical analyses could be done on individual species.

DISCUSSION

Reactive Atmospheric and Soil N

The relationship between reactive atmospheric N concentrations and soil N were consistent in most sites. The sites with highest extractable soil N (Black Rock, Key's View, Wide Canyon, Hay Field, and Lily Preserve) also had the highest atmospheric nitric acid (HNO_3) and/or ammonia (NH_3) concentrations. Cadiz Valley also had high soil N, and had higher-than-expected ozone (O_3) for an eastern site in Joshua Tree National Park (JTNP). Elevated O_3 is an indicator of poor air quality, although we did not observe elevated atmospheric N at Cadiz Valley during the time periods under study. The phenomenon of much greater eastern transport of O_3 compared with N compounds and N deposition has been observed in the adjacent San Bernardino Mountains (Alonso et al. 2003; Fenn et al. 2003b). The sites to the west of JTNP (Snow Creek) also had high soil N, and may be subject to 12 kg ha^{-1} y^{-1} of N deposition, as modeled by Tonnesen et al. (2003 *unpublished data*). Studies currently underway will determine the relationship between measured reactive atmospheric N and rates of N deposition to validate air pollution models.

High levels of O_3 in JTNP are of concern and were the subject of earlier studies on physiological responses of Mojave Desert plants (Thompson et al. 1984; Bytnerowicz et al. 1988). Concentrations of up to 100 ppb, which may occur in JTNP during the summer, affected performance of Mojave Desert plants (Bytnerowicz et al. 1988). A number of species, primarily riparian or deep-rooted species, were observed to have symptoms of O_3 damage in the summer, (Bytnerowicz et al. 1988). The visible damage was mainly in species that are physiologically active in summer.

Unlike O_3, which is easily converted to oxygen (O_2) and afterwards no longer has environmental impacts, gaseous HNO_3 and NH_3 remain active and accumulate in the soil during the dry season. Nitrogen-deposition gradients have been detected by sampling soil N in the mixed coniferous forest of the San Bernardino Mountains (Fenn et al. 1998) and in the coastal sage scrub of the Riverside-Perris Plain (Padgett et al. 1999). In both areas, soil N concentrations have been correlated with atmospheric N concentrations. Reactive atmospheric nitrogen accumulates on leaf and soil surfaces during the dry season and moves to the rooting zone via canopy throughfall, stem flow, and leaching (Fenn et al. 1998, 2003b; Padgett et al. 1999). In dry environments, soil N accumulates on the soil surface over time (Padgett et al. 1999; Wood et al. 2006). We measured higher concentrations of N in fertilized soils in 2004, the second year of fertilization, than in 2003, and we also observed significant responses by the vegetation.

In contrast, Brooks (2003) measured vegetation response after only one year of 30 kg N ha^{-1} fertilization in the Mojave Desert, with only 82 mm precipitation, but noted that there was higher-than-average precipitation in the month when germination occurred. Likewise, rainfall was greater in our study during 2004 than 2003, and significant effects on invasive grasses were not observed until the second year. Thus it is likely that, in this dry climate where opportunities for leaching are limited to infrequent wet periods, soils exposed to N pollutants and fertilized soils accumulate N over time (Walvoord et al. 2003). Along the N gradient we observed high values of 15–20 μg N g^{-1}, but at this time it is not clear whether this is an upper threshold to which N may accumulate under the current air pollution level, or whether higher soil concentrations will be observed over time. Fertilizing with 30 kg N ha^{-1} during the rainy season resulted in levels up to 45 μg N g^{-1}, so if air pollution increases, we expect to observe elevated soil N.

A drawback of the gradient approach is that it fails to detect other reasons that soils may have elevated or variable N concentrations. Soil texture, pH, parent material, moisture, and other factors control the rate of N mineralization and alter the extractable N concentration (Pastor et al. 1984). Soil texture may control growth of nonnative grasses, which have fine, shallow root systems. The two rocky/gravelly sites, Wide Canyon and Pine City, had the lowest grass biomass, and Pinto Wash, which lies in a basin that accumulates sand, had the highest grass biomass, even though it had the lowest soil N. Mineralization studies are underway at all of the sites to determine the N supply rate of these soils.

Impacts of Elevated Nitrogen on Native and Nonnative Vegetation

Although observations along the N gradient did not reveal a clear relationship between nonnative grass cover and soil N concentration, the fertilizer experiments at the four sites did show significant impacts of N on both native and nonnative plants. Pinto Wash had the highest grass cover and lowest soil N, but nonnative grass biomass was even higher following 30 kg N ha^{-1} fertilization. This level of N fertilization also caused increased *Schismus* and *Bromus* ssp. productivity in the Western Mojave Desert (Brooks 2003) in an area of low to moderate air pollution (Tonnesen et al. 2003 *unpublished data*). This suggests that if N deposition increases further at any of the sites, the nonnative grass biomass may increase with a potential for loss in productivity and species richness of native herbs. A surface-soil N concentration of 23 and 30 μg N g^{-1} in the 30 kg N ha^{-1} treatment in the two sites with lower air pollution (Pinto Wash and Pine City) resulted in nonnative grass growth response. Therefore, 23 μg N g^{-1} can be conservatively considered the low threshold for significant plant N response based on this fertilization study.

We hypothesize that sites along the gradient that have approximately 23 μg N g^{-1} soil N are already being affected by elevated N, assuming other edaphic factors are not limiting. It is not yet clear whether elevated soil N is caused by atmospheric N deposition at all of these sites, especially the sites in the eastern edge of JTNP. Furthermore, the two high-deposition N fertilizer sites (Covington Flat and Wide Canyon) had soil concentrations of 18–20 μg N g^{-1} following fertilization with 5 kg N ha^{-1}. Nonnative grass productivity is likely also elevated, but there is no longer an unpolluted control plot at these sites to test this statistically. This suggests that even small yearly N increases, such as the 5 kg N ha^{-1} over two years in this study, will eventually raise the level of soil N to values high enough to cause a significant increase in nonnative grass biomass.

The amount of initial nonnative grass biomass at each of the sites was critical to the changes that took place in native plant species richness and cover following N fertilization. At Pinto Wash, where nonnative grass biomass was the highest, the higher level of N fertilization caused a decrease in native species richness in each plot, while at Pine City, where nonnative grass biomass was lowest, the native species richness and cover increased with fertilization. This suggests that the native species are also N-limited, but that the nonnative grasses respond to N more rapidly, assuming the nonnatives have already colonized and the site is suitable to their growth. The strong competitive interaction between the nonnative and native grasses, as well as nonnative forbs, was demonstrated in a grass-removal experiment where both groups of forbs increased in density and biomass in response to nonnative grass thinning (Brooks 2000).

Another study showed that *Bromus madritensis* ssp. *rubens* has a higher rate of ^{15}N uptake than native seedlings of *Artemisia californica* in coastal sage scrub vegetation (Yoshida and Allen 2004). However, in a growth-chamber experiment, native species responded to N fertilization with the same relative percent increase as *B. madritensis* (Salo et al. 2005), a nonnative species. Since both native and nonnative species respond to N, other factors may also be involved, such as seed production and phenology of germination. The nonnative grasses germinate earlier than native species and produce seeds even in dry years when native plants do not germinate, which maintains the nonnative seed bank (Brooks 1999b, 2003, *this volume*). Thus it appears that the different responses of native species at Pinto Wash and Pine City may be due to the competitive interactions of natives with high vs. low cover of nonnative grasses.

The high grass biomass has been cited, in part, for an increase in fire frequency in the Mojave Desert (Brooks 1999a; Brooks and Esque 2002; Brooks et al. 2004; Brooks and Minnich 2006), especially at higher elevations that have higher rainfall and grass productivity. A fire of 5,500 ha burned in May 1999 at Covington Flat in *Coleogyne ramosissima* (blackbrush scrub), *Hilaria rigida* (galleta grass desert grassland), and Joshua tree and pinyon-juniper woodlands (JTNP staff *personal communication*). This is the largest fire known in JTNP, and followed the wet spring of 1998 in which there was a high production of *B. madritensis* (our N fertilization experiment at Covington Flat was in unburned vegetation). The fuel load for the fire was likely increased due to higher-than-average production of native plus nonnative species, although the grass biomass at that time is not known. The fuel threshold for nonnative grass biomass has been estimated at 0.5 to 1.0 T ha^{-1} dry matter (Fenn et al. 2003a). This level of biomass was produced in Pinto Wash in quadrats located just beyond the dripline of shrubs (50–70 g m^{-2}), but a fire would not be expected there because the grass cover is discontinuous in the interspaces.

More recent fires occurred at Snow Creek (450 ha, July 2004) and Morongo Valley (1250 ha, August 2005; fig. 4.1). Both sites lie just to the west of the park in areas of higher air pollution (Tonnesen et al. 2003), and we measured 20 μg N g^{-1} in soil at Snow Creek—enough to trigger a growth response by nonnative grasses. The nonnative species that burned at Snow Creek were *Schismus* ssp. and

Brassica tournefortii, while at the higher-elevation Morongo Valley, *B. madritensis* was the primary nonnative species that burned. Typical for burned desert vegetation (Brooks and Minnich 2006), recovery of native shrubs at Covington Flat and Snow Creek is slow, and Snow Creek remains densely covered with nonnative species (R. Steers *personal observation*).

CONCLUSIONS

This study has shown that a large pulse of 30 kg N ha^{-1}, added over two years, will increase the biomass of nonnative grass and either increase or decrease native forb species richness depending on initial nonnative grass production. However, estimated annual anthropogenic N inputs in this region are much lower than the 30 kg N ha^{-1} fertilization treatment (CASTNET 2005; Sullivan et al. 2001; Tonnesen et al. 2003). In arid environments, these small amounts may build up over time, as leaching rates are low, and N may also accumulate in lower soil horizons within and below the rooting zone (Walvoord et al. 2003; Wood et al. 2006).

The concentrations of soil N in sites along the gradient were as high as fertilized low-deposition sites that had a significant response by nonnative grasses. This indicates that long-term, low-level N inputs on the west end of Joshua Tree National Park may have already accumulated enough N in surface soils (e.g., 23 µg N g^{-1}) to affect nonnative grass productivity. Characteristics intrinsic to local sites will determine to what extent nonnative species will invade a site, but anthropogenically elevated N will cause a further imbalance if the invaders are nitrophilous and/or prolific seed producers (Brooks 2003; Yoshida and Allen 2004). Further studies on N mineralization are underway to determine the rate at which N is supplied in soils of different sites along the gradient. These may help to predict which soil types are predisposed to support greater productivity of nonnative invaders.

ACKNOWLEDGMENTS

We thank Sheila Kee, Tracy Tennant, Abby Sirulnik, Cecilia Osorio, Greg Smith, Christopher True, Robin Marushia, William Swenson, Sarah Huels, Susan Determann, Tim Blubaugh, Tony Davila, Diane Alexander, and Dave Jones for assistance in the field and lab. This research was funded by the National Park Service PMIS Number 72123, and in part by NSF DEB 04–21530 and the University of California Statewide Integrated Pest Management Exotic Pests and Diseases Program.

REFERENCES

Aber, J., W. McDowell, K. Nadelhoffer, A. Magill, G. Berntson, M. Kamakea, S. McNulty, W. Currie, L. Rustad, and I. Fernandez. 1998. Nitrogen saturation in temperate forest ecosystems: hypotheses revisited. *Bioscience* **48**:921–934.

Allen, E. B. 2004. Restoration of *Artemisia* shrublands invaded by exotic annual *Bromus:* a comparison between southern California and the Intermountain Region. Pages 9–17 *in* A. L. Hild, N. L. Shaw, S. E. Meyer, E. W. Schupp, and T. Booth, compilers. *Seed and Soil Dynamics in Shrubland Ecosystems.* Proceedings, August 12–16, 2002, Laramie, Wyoming, Proceedings No. RMRS-P-31. U.S. Department of Agriculture, Forest Service, Rocky Mountain Research Station, Ogden, Utah.

Allen, E. B., P. E. Padgett, A. Bytnerowicz, and R. A. Minnich. 1998. Nitrogen deposition effects on coastal sage vegetation of southern California. Pages 131–140 *in* A. Bytnerowicz, M. J. Arbaugh, and S. L. Schilling, editors. *Proceedings of the International Symposium on Air Pollution and Climate Change Effects on Forest Ecosystems,* February 5–9, 1996, Riverside, California. General Technical Report No. 166. U.S. Department of Agriculture, Forest Service, Pacific Southwest Research Station, Riverside, California.

Alonso, R., M. J. Arbaugh, and A. Bytnerowicz. 2003. Forest health research on a natural air pollution gradient in the San Bernardino Mountains, southern California. *Ekológia* (Bratislava) **22**:18–23.

Alonso, R., A. Bytnerowicz, J. L. Yee, and W. I. Boarman. 2005. Atmospheric dry deposition in the vicinity of the Salton Sea, California—II: measurement and effects of an enhanced evaporation system. *Atmospheric Environment* **39**:4681–4689.

Baumgardner, R. E., T. F. Lavery, C. M. Rogers, and S. S. Isil. 2002. Estimates of the atmospheric deposition of sulfur and nitrogen species: clean air status and trends network, 1990–2000. *Environmental Science and Technology* **36**:2614–2629.

Bobbink, R., M. Hornung, and J. G. M. Roelofs. 1998. The effects of air-borne nitrogen pollutants on species diversity in natural and semi-natural European vegetation. *Journal of Ecology* **86**:717–738.

Bowman, W. D., and H. Steltzer. 1998. Positive feedbacks to anthropogenic nitrogen deposition in Rocky Mountain Alpine tundra. *Ambio* **27**:514–517.

Brooks, M. L. 1999a. Alien annual grasses and fire in the Mojave Desert. *Madroño* **46**: 13–19.

Brooks, M. L. 1999b. Habitat invasibility and dominance by alien annual plants in the western Mojave Desert. *Biological Invasions* **1**:325–337.

Brooks, M. L. 2000. Competition between alien annual grasses and native annual plants in the Mojave Desert. *American Midland Naturalist* **144**:92–108.

Brooks, M. L., 2003. Effects of increased soil nitrogen on the dominance of alien annual plants in the Mojave Desert. *Journal of Applied Ecology* **40**:344–353.

Brooks, M. L., and T. C. Esque. 2002. Alien annual plants and wildfire in desert tortoise habitat: status, ecological effects, and management. *Chelonian Conservation Biology* **4**:330–340.

Brooks, M. L., C. M. D'Antonio, D. M. Richardson, J. B. Grace, J. E. Keeley, J. M. DiTomaso, R. J. Hobbs, M. Pellant, and D. Pyke. 2004. Effects of invasive alien plants on fire regimes. *Bioscience* **54**:677–688.

Brooks, M. L., and R. A. Minnich. 2006. Southeastern deserts bioregion. Pages 391–414 *in* N. G. Sugihara, J. W. van Wagtendonk, K. E. Shaffer, J. Fites-Kaufman, and A. E. Thode, editors. *Fire in California's Ecosystems.* University of California Press, Berkeley, California.

Bytnerowicz, A., J. J. Carroll, B. Takemoto, P. R. Miller, M. E. Fenn, and R. C. Musselman. 2000. Distribution and transport of air pollutants to vulnerable California ecosystems. Pages 93–118 *in* K. M. Scow, G. E. Fogg, D. E. Hinton, and M. J. Johnson, editors. *Integrated Assessment of Ecosystem Health.* Lewis Publishers, Boca Raton, Florida.

Bytnerowicz, A., P. R. Miller, and D. M. Olszyk. 1987. Dry deposition of nitrate, ammonium and sulfate to a *Ceanothus crassifolius* canopy and surrogate surfaces. *Atmospheric Environment* **21**:1749–1757.

Bytnerowicz, A., D. M. Olszyk, C. A. Fox, P. J. Dawson, G. Kats, C. L. Morrison, and J. Wolf. 1988. Responses of desert annual plants to ozone and water stress in an *in situ* experiment. *Journal of the Air Pollution Control Association* **38**:1145–1151.

Bytnerowicz, A., P. E. Padgett, M. J. Arbaugh, D. R. Parker, and D. P. Jones. 2001. Passive sampler for measurement of atmospheric nitric acid vapor (HNO_3) concentrations. *The Scientific World* **1**(S2):433–439.

Clean Air Status and Trends Network (CASTNET). 2005. JOT403 (Joshua Tree NM). Online. U.S. Environmental Protection Agency, Washington, D.C. http://www.epa.gov/castnet/sites/jot403.html

DeFalco, L. A., J. K. Detling, C. R. Tracy, and S. D. Warren. 2001. Physiological variation among native and exotic winter annual plants associated with microbiotic crusts in the Mojave Desert. *Plant and Soil* **234**:1–14.

Edinger, J. G., M. H. McCutchan, P. R. Miller, B. C. Ryan, M. Schroeder, and J. V. Behar. 1972. Penetration and duration of oxidant air pollution in the South Coast Air Basin of California. *Journal of the Air Pollution Control Association* **22**:881–886.

Fenn, M. E., J. S. Baron, E. B. Allen, H. M. Rueth, K. R. Nydick, L. Geiser, W. D. Bowman, J. O. Sickman, T. Meixner, D. W. Johnson, and P. Neitlich. 2003a. Ecological effects of nitrogen deposition in the western United States. *Bioscience* **53**:404–420.

Fenn, M. E., G. S. Haeuber, G. S. Tonnesen, J. S. Baron, S. Grossman-Clarke, D. Hope, D. A. Jaffe, S. Copeland, L. Geiser, H. M. Rueth, and J. O. Sickman. 2003b. Nitrogen emissions, deposition and monitoring in the western United States. *Bioscience* **53**:391–403.

Fenn, M. E., M. A. Poth, J. D. Aber, J. S. Baron, B. T. Bormann, D. W. Johnson, A. D. Lemly, S. G. McNulty, D. E. Ryan, and R. Stottlemyer. 1998. Nitrogen excess in North American ecosystems: predisposing factors, ecosystem responses, and management strategies. *Ecological Applications* **8**:706–733.

Fenn, M. E., J. O. Sickman, A. Bytnerowicz, D. W. Clow, N. P. Molotch, J. E. Pleim, G. S. Tonnesen, K. C. Weathers, P. E. Padgett, and D. H. Campbell. 2007. Methods for measuring atmospheric nitrogen deposition inputs in arid and montane ecosystems of western North America. *In* A. H. Legge, editor. *Developments in environmental science: relating atmospheric source apportionment to vegetation effects: establishing cause effect relationships.* Elsevier, Amsterdam, the Netherlands.

Hickman, J. C., editor. 1993. *The Jepson manual: higher plants of California.* University of California Press, Berkeley, California.

Joshua Tree National Park. 2004. Strategic plan. Online. http://www.nps.gov/jotr/manage/strategic/externals.html.

Koutrakis, P., J. M. Wolfson, A. Bunyaviroch, S. E. Froehlich, K. Hirano, and J. D. Mulik. 1993. Measurement of ambient ozone (O_3) using a nitrite-coated filter. *Analytical Chemistry* **65**:209–214.

Padgett, P. E., and E. B. Allen. 1999. Differential responses to nitrogen fertilization in native shrubs and exotic annuals common to Mediterranean coastal sage scrub of California. *Plant Ecology* **144**:93–101.

Padgett, P. E., E. B. Allen, A. Bytnerowicz, and R. A. Minich. 1999. Changes in soil inorganic nitrogen as related to atmospheric nitrogenous pollutants in southern California. *Atmospheric Environment* **33**:769–781.

Pastor, J., J. D. Aber, C. A. McClaugherty, and J. M. Melillo. 1984. Above-ground production and N and P cycling along a nitrogen mineralization gradient on Blackhawk Island, Wisconsin. *Ecology* **65**:256–268.

Salo, L. F., G. R. McPherson, and D. G. Williams. 2005. Sonoran desert winter annuals affected by density of red brome and soil nitrogen. *American Midland Naturalist* **153**: 95–109.

Sawyer, J. O., and T. Keeler-Wolf. 1995. *A manual of California vegetation.* California Native Plant Society, Sacramento, California.

Stevens, C. J., N. B. Dise, J. O. Mountford, and D. J. Gowing. 2004. Impact of nitrogen deposition on the species richness of grasslands. *Science* **303**:1876–1879.

Sullivan, T. J., C. L. Peterson, C. L. Blanchard, and S. J. Tanenbaum. 2001. Assessment of air quality and air pollutant impacts in class I national parks of California. Online. National Park Service, Air Resources Division, Denver, Colorado. http://www2.nature.nps.gov/air/Pubs/pdf/reviews/ca/CAreport.pdf.

Theodose, T. A., and W. D. Bowman. 1997. Nutrient availability, plant abundance, and species diversity in two alpine tundra communities. *Ecology* **78**:1861–1872.

Thompson, C. R., D. M. Olszyk, G. Kats, A. Bytnerowicz, P. J. Dawson, and J. W. Wolf. 1984. Effects of ozone or sulfur-dioxide on annual plants of the Mojave Desert. *Journal of the Air Pollution Control Association* **34**:1017–1022.

Tonnesen, G. S., Z. Wang, M. Omary, and C. J. Chien. 2003. Formulation and application of regional air quality modeling for integrated assessments of urban and wildland pollution. Pages 285–298 *in* A. Bytnerowicz, M. J. Arbaugh, and R. Alonso, editors. *Ozone air pollution in the Sierra Nevada: distribution and effects on forests.* Elsevier Press, Amsterdam, the Netherlands.

Trent, D. D. 1984. Geology of the Joshua Tree National Monument. *California Geology* **37**:75–86.

Walvoord, M. A., F. M. Phillips, D. A. Stonestrom, R. D. Evans, P. C. Hartsough, B. D. Newman, and R. G. Striegl. 2003. A reservoir of nitrate beneath desert soils. *Science* **302**:1021–1024.

Wood, Y. A., T. Meixner, P. J. Shouse, and E. B. Allen. 2006. Altered ecohydrologic response drives native shrub loss under conditions of elevated N-deposition. *Journal of Environmental Quality* **35**:76–92.

Yoshida, L. C., and E. B. Allen. 2001. Response to ammonium and nitrate by a mycorrhizal annual invasive grass and native shrub in southern California. *American Journal of Botany* **88**:1430–1436.

Yoshida, L. C., and E. B. Allen. 2004. ^{15}N uptake by mycorrhizal native and invasive plants from a N-eutrophied shrubland: a greenhouse experiment. *Biology and Fertility of Soils* **39**:243–248.

FIVE

Spatial and Temporal Distribution of Nonnative Plants in Upland Areas of the Mojave Desert

MATTHEW L. BROOKS

Nonnative plants have emerged during the past few decades as one of the primary concerns for land managers in the Mojave Desert. They have been implicated in the decline of native plant communities, both through direct competition (Brooks 2000; DeFalco et al. 2003) and the alteration of fire regimes (Brooks 1999a; Brooks and Pyke 2001; Brooks and Esque 2002; Brooks and Matchett 2006; Brooks and Minnich 2006). An additional concern is that nonnative plants can lead to habitat alteration and thus negatively affect wildlife communities (USFWS 1994; Brooks and Pyke 2001; Brooks and Esque 2002). Nonnative plants can also accumulate in such high amounts that they become fire hazards along roadsides and near human habitations.

The number of nonnative plant species in the Mojave Desert has recently increased, which is one of the reasons for the increase in concern. At the end of the 1970s, 9% of the Mojave Desert flora was comprised of nonnative plant species (Rowlands et al. 1982). By the end of the 1990s, this number had increased to 13% (Brooks and Esque 2002). However, these percentages are still low in comparison with most floras worldwide, which contain, on average, 16% nonnatives, and the floras of island ecosystems, which typically include 32% (Lonsdale 1997, 1999).

Although the percentage of nonnative species in the Mojave Desert flora may be relatively low compared with other ecosystems, the proportion of total annual plant productivity that nonnatives comprise can be significant. For example, the vast majority of nonnatives in the Mojave Desert are annual species, particularly winter annuals (Kemp and Brooks 1998; Brooks and Esque 2002), and these typi-

Table 5.1. *Percent dominance of the annual plant community by nonnative species in the Mojave Desert*

Nonnative Dominance	Study years	# of sites	Average value during each study year[a]	Nonnative species	Source Citation
% Biomass	1982, 1983	1	2/20, 3/15	*Schismus barbatus*	Samson 1986
			69/60, 22/55	*Erodium cicutarium*	
			1/1, 1/1	*Bromus rubens*	
	1992	1	20	*Schismus barbatus*	Jennings 1993
			11	*Erodium cicutarium*	
			1	*Bromus rubens*	
	1990–1992	1	1, 3, 10	*Schismus barbatus*	Brooks 1995
			44, 20, 21	*Erodium cicutarium*	
			0, 0, 1	*Bromus rubens*	
	1994, 1995	1	29, 17	*Schismus barbatus*	Brooks 1999b
			49, 12	*Erodium cicutarium*	
			9, 3	*Bromus rubens*	
			10, 1	*Bromus trinii*	
	1995, 1999	34	17, 31	*Schismus barbatus*	Brooks and Berry 2006
			21, 37	*Erodium cicutarium*	
			29, 18	*Bromus rubens*	
			1, 1	*Bromus trinii*	
			1, 1	*Bromus tectorum*	
			1, 1	*Descurainia sophia*	
			1, 1	*Sisymbrium irio*	
	1996, 1997	3	57/5, 34/2	*Schismus barbatus*	Brooks 2000
			34/5, 54/1	*Erodium cicutarium*	
			1/54, 1/75	*Bromus rubens*	
			1/1, 1/1	*Bromus trinii*	
			1/1, 1/1	*Bromus tectorum*	
	1998	1	86	*Schismus barbatus*	Oftedahl et al. 2002
			3	*Erodium cicutarium*	
	1996–2000	3	36/1, 36/1, 19/1, 19/1, 26/1	*Schismus barbatus*	Brooks 2002
			35/1, 33/1, 19/1, 10/1, 9/1	*Erodium cicutarium*	
			1/77, 1/79, 1/81, 1/49, 1/30	*Bromus rubens*	
% Density	1973	2	40	*Schismus barbatus*	Davidson and Fox 1976
			30	*Erodium cicutarium*	
	1978	1	6	*Schismus barbatus*	Webb and Stielstra 1979
			30	*Erodium cicutarium*	
	1981–1985	1	31/2, 8/50, 37/71, 67/59, 47/57	*Schismus barbatus*	Samson 1986
			64/96, 8/50, 22/20, 11/36, 16/14	*Erodium cicutarium*	
			0/0, 6/1, 2/1, 11/1, 2/1	*Bromus rubens*	

[a]Values reported for each study year are separated by a comma (e.g., Year 1, Year 2, etc.). Where studies included data for both interspace and beneath-canopy microhabitats (*sensu* Brooks 1999b), both were listed separated by a forward slash (e.g., 2/20 = 2% in the interspace/20% in the beneath canopy).

cally constitute the bulk of the annual plant production in this region (table 5.1). A number of studies have consistently reported the winter annuals *Schismus* ssp. (*Schismus arabicus, Schismus barbatus,* Mediterranean grass) and *Erodium cicutarium* (filaree) as the dominant nonnative species during various years of contrasting rainfall between 1973 and 2000 (Davidson and Fox 1974; Webb and Stielstra 1979; Samson 1986; Jennings 1993; Brooks 1995, 1999b, 2000; Oftedahl et al. 2002; Brooks and Berry 2006). Where another common nonnative winter annual, *Bromus madritensis* ssp. *rubens* (red brome), was also abundant, it was typically present in the relatively mesic, beneath-canopy microhabitat located under perennial shrubs (Brooks 2000, 2003). However, these studies were mostly conducted at relatively low elevations (< 800–1000 m) where *Bromus madritensis* ssp. *rubens* is least abundant at regional scales (Brooks and Berry 2006). *Bromus madritensis* ssp. *rubens* can be more abundant than *Schismus* ssp. at ecotones with more mesic regions, in addition to being more abundant at higher elevations (Brooks and Matchett 2003; Brooks and Minnich 2006). *Erodium cicutarium* often dominates landscapes over a wide range of elevations.

Unlike the human land uses that land managers regulate, ecological processes such as plant invasions cannot be as easily defined or managed. As challenging as it may be to document patterns of land use, such as off-highway vehicles (OHVs) and livestock, it is much more difficult to document the invasion patterns of nonnative plants. This is a problem because if you don't know where the populations of invading plants are, you cannot effectively manage them. In addition, a management tool such as a fence can be effective in keeping off-highway vehicles outside or livestock inside of an area. Unfortunately, fences have virtually no effect on the control and spread of nonnative plants.

Another challenging aspect of managing nonnative plants is that they do not recognize political boundaries. This means that adjacent land management units must work collaboratively to manage nonnative plants. It does little good to aggressively control a plant that is left unchecked just outside of the target management unit. In response to this fact, cooperative weed management areas have been established to promote information sharing and collaborative approaches to managing nonnative plants in the Mojave Desert.

One of the first tasks for collaborators in weed management is to compile a list of the nonnative species that are present in their geographic regions. This task involves detecting new invasions and determining the extent and rate of spread of nonnative species already present in the region. This status-and-trends information can then be used to help develop a list of species that pose the greatest threats to natural, cultural, recreational, and other resources within the region. Land managers can then prioritize and determine which species to focus on and the locations where they should implement control efforts (e.g., Warner et al. 2003).

This task requires more specific information on where species are likely to be and how widespread their current distributions are, as well as where they may spread to in the future. This is difficult enough in riparian areas, but it is even more challenging to evaluate the patterns of nonnative plants across the vast expanse of upland areas, which comprise approximately 97% of the Mojave Desert (Rowlands et al. 1982).

In this chapter I describe the factors associated with the spatial and temporal distributions of nonnative plants in upland areas within the Mojave Desert. This information is presented within a framework that can be used to evaluate the invasion potential of nonnative plant species and the invasibility of various regions and landscape features. This information is especially useful to inform the process of developing monitoring strategies for early detection and status-and-trends monitoring for nonnative species. It can also be used to help develop studies to determine the ecological effects of nonnative plants and to evaluate management tools designed to control or eradicate them. I specifically do not address riparian systems such as the Colorado, Mojave, Muddy, and Virgin rivers—plant invasions in riparian systems are discussed elsewhere in this book (Dudley *this volume*).

GENERAL THEORY OF HABITAT INVASIBILITY

Research on biological invasions indicates that many different types of environmental variables can be correlated with the establishment and persistence of nonnative plants. These factors can include disturbance, connectivity to other invaded sites, disruption of ecological processes or regimes, and fluctuating resource levels (Hobbs and Huenneke 1992; Maron and Connors 1996; Lonsdale 1999; Davis et al. 2000; Larson et al. 2001). However, two factors in particular seem to explain most of the variability in landscape invasibility by plants: (1) resource availability and (2) propagule pressure (Brooks 2007).

Resource availability is related to the amount of light, water, and mineral nutrients available for plants to use (Davis et al. 2000; Brooks 2007). It is affected by the underlying resource amounts, nutrient input rates, and rates of uptake by biological organisms. As resource input increases (e.g., through terrestrial eutrophication) and/or resource uptake decreases (e.g., through removal or thinning of existing vegetation), landscape invasibility can be enhanced. Changes in those resources that are especially limiting to plant growth in desert regions have the greatest influence on plant invasions. Such resources typically include soil moisture, nitrogen, and phosphorus (Brooks 2003). The physiological characteristics of various life forms can also influence how they respond to spatial and seasonal variations in resource availability. For example, C_3 annual plants germinate and grow during the fall through spring months and lie dormant (as seeds) during the summer. These types of plants are therefore not directly affected by summer rainfall. They might actually be more successful where summer rainfall and associated summer flora are depauperate, because they would experience less competition for mineral nutrients in the soil.

Propagule pressure is a measure of the number of propagules available to establish new populations, which can be affected by either deliberate or accidental dispersal by humans or other vectors. Propagule type is also important, because if they are not well adapted to the environments they disperse into, they are unlikely to establish populations. Species susceptible to excessive herbivory or disease in the region they are invading are similarly less likely to become established. Many invasive plant species are dispersed via human pathways; areas near roads, trails, and other human infrastructure may serve as sources of propagules (Brooks and Lair *this volume*). In addition, large urban or agricultural regions serve as major sources of propagules. In agricultural areas, farming practices and livestock feed introduce nonnative species, and in urban areas, commercial and residential landscape designs can include nonnative species.

In the sections that follow, I use the themes of resource availability and propagule pressure to evaluate the spatial and temporal patterns of nonnative plants in the Mojave Desert.

SPATIAL PATTERNS AMONG SUBREGIONS

The Mojave Desert is considered transitional between the colder Great Basin to the north and the hotter Sonoran Desert to the south (Rowlands et al. 1982). The Great Basin and Sonoran Desert climatic and floristic influences grade into each other deep into the Mojave Desert along broad ecotones extending from the north and south, respectively. Narrower ecotones also occur where influences from the California floristic province to the west and the Colorado Plateau to the east extend shorter distances into the Mojave Desert. These ecotones are comprised of continuously variable physical and biological factors, such as annual rainfall, percent summer (May–October) and winter (November–April) rainfall, temperature patterns, potential evapotranspiration rates, and floristic affinities. Collectively these variables have been used to define six major subregions within the Mojave Desert: the Northern, Eastern, South-Eastern, South-Central, Central, and Western Mojave Desert (fig. 0.1; Webb et al., Introduction, *this volume*). These regions differ in important biophysical ways that make them useful for evaluating the potential for plant invasions, especially as they relate to the spatial and temporal availability of soil moisture, land use history, and proximity to sources of new plant invaders.

Northern Mojave

The Northern Mojave subregion is the most far removed from urban influences of any subregion within the Mojave Desert. The largest town, Beatty, Nevada, has a population of only 1,200. However, it is closely associated with extensive agricultural fields in the Owens Valley to the west and the Amargosa Valley through its middle section. State Highway 95 (Hwy 95) is the major route connecting northern and southern Nevada and likely serves as an invasion pathway for nonnative plants from the Great Basin. However, it is the only major highway passing through this region. Nonnative plants associated with agricultural areas are more likely to occur in this region than species associated with urbanized areas. The presence of only one major highway also suggests that propagule pressure from adjacent regions may be relatively low.

Eastern Mojave

This subregion contains one of the two largest urban centers within the Mojave Desert. With 600,000 people in Las Vegas and 1.7 million people living in Clark County in southern Nevada, the region is heavily influenced by urbanization. Agricultural areas near Mesquite, Moapa, Overton, Pahrump, and Las Vegas are also significant. Perennial livestock grazing has been prevalent in this region in the past due to relatively large amounts of native perennial grasses supported by summer rainfall. The region is traversed by Interstate Highway 15 (I-15), Interstate Highway 40 (I-40), and Hwy 95. These are all major routes of vehicular travel that can facilitate the invasion of plants from the Colorado Plateau to the east and the Sonoran Desert to the south. Nonnative plants typical of both agricultural and urbanized areas can be prevalent in this region. Propagule pressure from adjacent regions is relatively high, due to the presence of three major highways.

Central Mojave

This subregion is the only one that does not interface with areas outside of the Mojave Desert. The largest urban center in the Central Mojave is Barstow, California, which is also the center of agricultural activity in the region, with alfalfa fields and livestock yards. The three major highways that transect the Mojave Desert—I-15, I-40, and State Highway 58 (Hwy 58)—all converge in Barstow. Barstow also serves as the major railroad hub of the Mojave Desert. High propagule pressure along these transportation corridors results in frequent opportunities for the dispersal of new nonnative plants into this subregion.

Western Mojave

Urban areas in the Western Mojave Desert are concentrated near Lancaster and Palmdale, California, in the far western part of this region and in the smaller city of Ridgecrest farther north. Propagule pressure along Hwy 58, which connects the large agricultural area of California's Central Valley to the west with the southwest Mojave, provides opportunities for nonnative species associated with agriculture to spread into the Mojave Desert. Hay straw and seed in particular can disperse off of trucks onto roadsides as they become dislodged from bales by wind.

South-Central Mojave

There are extensive rural communities which have recently become more urbanized in the vicinity of Victorville, Hesperia, Apple Valley, and Lucerne Valley in the northwest part of this subregion, and Yucca Valley, Joshua Tree, and Twentynine Palms in the southern part of the region. These areas contain many alfalfa fields and horse properties. The direct connection between the Western Mojave and the San Bernardino Valley via I-15 provides a pathway for nonnative plants to disperse from the large urbanized areas to the west.

South-Eastern Mojave

This subregion is perhaps the least populated of the six subregions, but significant population centers have recently developed along the Colorado River. These communities, while predominantly supporting recreation, contain many alfalfa fields and horse properties. The direct connection between the South-Central Mojave and this subregion via I-40 provides a pathway for nonnative plants to disperse from the large urbanized areas to the east, and Highway 95 creates a corridor linking the lower-elevation parts of the Sonoran Desert to the south with the South-Eastern Mojave and the Eastern Mojave subregions.

SPATIAL VARIATION WITHIN SUBREGIONS

Local environmental and human disturbance gradients can also affect the distribution of nonnative plants within the Mojave Desert. These gradients influence local propagule pressure, rates of resource availability, and general growing conditions.

Landscape Variation

Elevation can greatly influence landscape invasibility. At a local scale, higher elevations tend to have lower temperatures, higher rainfall, and lower potential evapotranspiration rates. Thus, growing conditions are more mesic, which is generally favorable for nonnative plants. However, high elevation soils are often more shallow and poorly developed than at lower elevations, which can hinder plant establishment.

Not all species respond the same to elevation gradients. At 34 sites in the Central, South-Central, and Western Mojave, biomass of the nonnative annual grass *B. madritensis* ssp. *rubens* increased with elevation during the high rainfall year of 1995 ($r^2 = 0.22$, $P = 0.0058$) and the low rainfall year of 1999 ($r^2 = 0.23$, $P = 0.0045$; fig. 5.1). In contrast, biomass of the nonnative annual grass *Schismus* ssp. decreased with elevation at the same sites during 1995 ($r^2 = 0.19$, $P = 0.0103$) and 1999 ($r^2 = 0.15$, $P = 0.0233$). Essentially, *B. madritensis* ssp. *rubens* was the dominant nonnative annual grass above 800–1000 m, and *Schismus* ssp. was the dominant nonnative annual grass below this elevation (fig. 5.1). This pattern may be caused

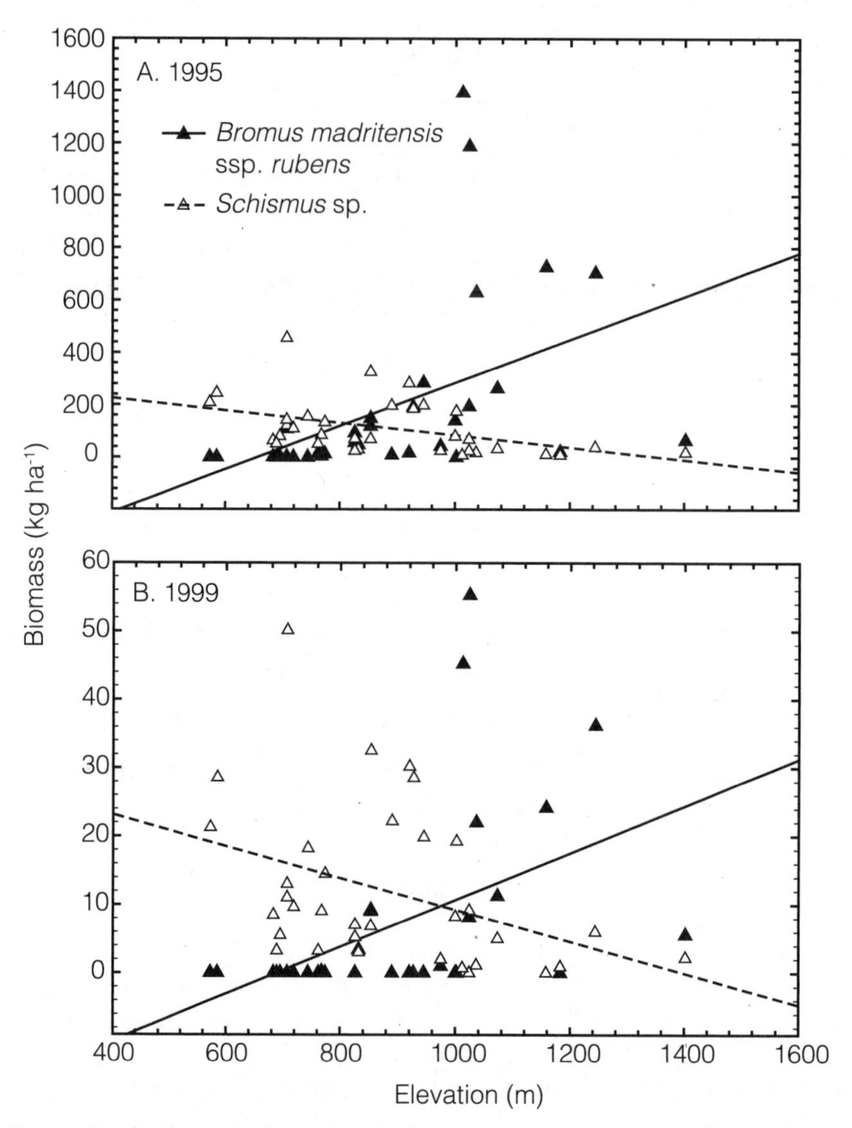

Fig. 5.1. Graphs showing relationships between elevation and biomass of two nonnative annual grasses, *Bromus madritensis* ssp. *rubens* and *Schismus* spp., in the Mojave Desert (modified from Brooks and Berry 2006).

by periodic population crashes of *B. madritensis* ssp. *rubens* during droughts at lower elevations, but not at higher elevations (Brooks and Minnich 2006). Small rainfall events during drought periods often stimulate germination of *B. madritensis* ssp. *rubens*, but subsequent soil moisture levels at lower elevations are often not high enough for them to survive to reproduce unless there is additional rain. In contrast, the more mesic conditions at higher elevations provide plants with more soil moisture for a longer period of time following rainfall events, so many individuals survive to reproduce and maintain local populations. Thus, higher elevations can serve as refugia for nonnative plants that may require relatively mesic growing conditions to survive extirpation during regional droughts.

The abundance of nonnative plants can also vary among major vegetation types, but much of this variation may be better explained by underlying soil and elevation gradients. Nonnatives that are adapted to arid conditions, such as *Schismus* ssp., can flourish in low-elevation habitats. *Schismus* ssp. reach peak biomass at sites with deep sandy loam soils and do not flourish as well where soils are shallower and especially where they are covered in desert pavement. Mustards such as *Brassica tournefortii* (Sahara mustard), which generally require more mesic conditions, often grow well in the sandy soils of low-elevation ephemeral watercourses (M. L. Brooks *personal observation*). Sites that contain siltier and more alkaline soils are particularly favorable to *Sisymbrium* ssp. mustards. These areas often occur at the margins of playas or in broad valley bottoms that are also typically the sites of livestock watering or agricultural fields. *Sisymbrium irio* (London rocket) is the dominant species of its genera across most of the Mojave Desert. Where soil moisture is more plentiful—at the margins of the Mojave Desert or at the interface with urbanized or major agricultural regions—the mustards *Sisymbrium altissimum* (tumble mustard) and *Hirschfeldia incana* (Mediterranean mustard) can also be abundant. In more harsh alkaline soils, chenopods such as *Halogeton glomeratus* (halogeton) and especially *Bassia hyssopifolia* (fivehook bassia) can be abundant, particularly in association with abandoned agricultural fields.

Regional plumes of nitrogen oxide air pollution from the burning of fossil fuels may produce depositional gradients of nitrogen in the soil that could benefit nonnative plants (Allen et al. *this volume*). These nitrogen plumes may occur in the vicinity of heavy vehicle activity, such as near Las Vegas, Nevada, or downwind from sources in Southern California, such as in the Central and South-Central Mojave. A small-scale soil nutrient enrichment study at sites in the Central, South-Central, and Western Mojave demonstrated that nonnative annual plants can benefit from enrichment, and they do so at the expense of native annuals (Brooks 2003). Specifically, soil nitrogen addition increased the density and biomass of nonnative annual plants, whereas density, biomass, and species richness of native annuals decreased.

Within-Site Environmental Variation

Microhabitat variation within sites can greatly influence the local distribution and abundance of nonnative plants. This variation can be caused by changes in topography and by the canopies of large perennial plants (Brooks 1999b; Brooks and Berry 2006). Soil resources are relatively high in washlets and in other topographic positions that accumulate water from the surrounding landscape, and also beneath the canopies of perennial shrubs. In general, shrub interspaces and run-off topographic positions are dominated by the nonnative species that can tolerate these locally arid conditions (e.g., *Schismus* ssp. and *E. cicutarium*), whereas areas beneath shrub canopies and along ephemeral washes and washlets are dominated by the nonnative species that are more successful in these relatively mesic micro-habitats (e.g., *Bromus* ssp. and *B. tournefortii*).

The abundance of nonnative plants can also be affected by which species of perennial plant they are growing under, such as was observed for *B. madritensis* ssp. *rubens* (fig. 5.2). Biomass of *B. madritensis* ssp. *rubens* was not specifically associated with a particular physiognomic type of shrub species, such as those producing large canopies and relatively mesic microsites [e.g., *Larrea tridentata* (creosote bush)], nor with a particular family of shrubs, such as leguminous species which may contribute to locally high nitrogen levels [e.g., *Psorothamnus* ssp. (indigo bush)]. The highest *B. madritensis* ssp. *rubens* biomass values generally corresponded to the presence of shrub species with affinities for the Great Basin, which tend to occur at higher elevations and latitudes within the Mojave Desert. This pattern is probably due to the fact that these perennial species are typical of vegetation types that occur in relatively mesic locations in the Mojave Desert, rather than due to any specific localized conditions created by the species itself.

Propagule pressure can also be higher under perennial shrubs when the dispersal units of plants differentially accumulate beneath them. This is often the case with *B. tournefortii;* a major mode of dispersal for this species occurs when entire plants, with seeds onboard, tumble across the landscape blown by the wind. Plants come to a stop when they reach a physical barrier, which is most often a large perennial plant, within which they lodge and eventually drop their seeds. In many cases, fencerows provide this physical barrier and serve as accumulation sites for *B. tournefortii* and *Salsola* ssp. (Russian thistle).

In some cases it seems that the dominant nonnative annual plant species beneath large perennial plants may simply be the species that established there first, preempting establishment by other species. This appeared to be the case at a site near California City in the Western Mojave, where *Bromus madritensis* ssp. *rubens, Bromus tectorum* (cheatgrass), and *Bromus trinii* (Chilean chess) were all present during the spring of 1995. However, I observed that their distributions were mutually exclusive, with individual species dominating the understories of separate *Larrea*. In many cases, adjacent *Larrea* would contain a separate monoculture of one of these three species. No obvious differences were apparent in the characteristics of the *Larrea* microhabitats at this site, and it appeared that the *Bromus* species that dominated a particular subcanopy was simply a factor of chance dispersal and establishment. This pattern appeared somewhat stable over

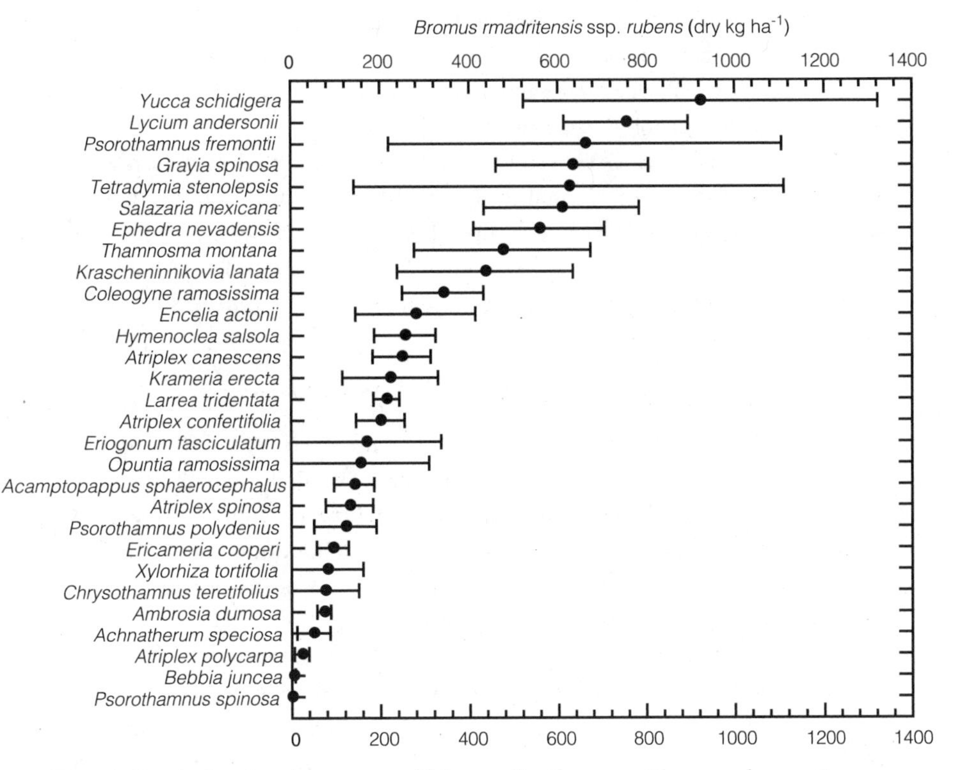

Fig. 5.2. Graph showing the amount of biomass for the nonnative annual grass *Bromus madritensis* ssp. *rubens* (red brome) measured beneath the canopies of perennial plants in the Mojave Desert. The data reported are average values (± 1 SE) collected in 1995 at thirty-four sites described in Brooks and Berry (2006).

the short term, as it was still evident ten years later when I revisited the site during the spring of 2005.

The patchy distribution of new populations of nonnative plants within a site can sometimes be apparently unrelated to landscape factors and display a relatively stochastic pattern. For example, during the winter and spring of 2005, I first observed large stands of *B. tournefortii* intermixed with *Larrea* on steep colluvial mountain slopes in the Calico Mountains of the South-Central Mojave and the River Mountains of the Eastern Mojave. This species was previously thought to only occur in great numbers on deep, sandy soils of relatively flat alluvial fans and valley bottoms. Extensive patches ranging from 1–20 ha appeared haphazardly distributed at various locations on the toe-slopes, mid-slopes, and the tops of these mountains, apparently the result of chance long-distance dispersal and establishment of nascent populations, from which the populations then spread outward. They did not fall along obvious wind dispersal pathways from the valley bottoms to the slopes. It appears that they may have been dispersed by the granivorous birds, such as horned larks and sage sparrows, which I have often observed in large flocks feeding on *B. tournefortii* seeds.

Granivorous rodents also harvest the large seeds of *B. madritensis* ssp. *rubens* and *B. tournefortii* and bury them in shallow caches in the soil where they subsequently germinate and grow in clumps of 10 to 50 plants (M.L. Brooks *unpublished data*). Ants also disperse the seeds of *Schismus* ssp. and *E. cicutarium* to their anthills where the plants can often be found growing in halos surrounding the entrance to the colony.

Within-Site Variation due to Human Disturbances

Human disturbances are perhaps the most widely recognized factors associated with plant invasions (Elton 1958; Rejmanek 1989). Disturbances can directly increase nutrient availability (e.g., atmospheric nitrogen pollution), indirectly increase nutrient availability by reducing the abundance and vigor of resident plants, and increase propagule pressure by purposefully or inadvertently providing vectors and pathways for invasion (Brooks 2007).

Initial dispersal into a region often follows major vehicular routes (Brooks and Lair *this volume*). This is particularly true along the major highways that cross the Mojave Desert, which are the primary dispersal pathways into otherwise remote and rural regions. High levels of soil moisture and mineral nutrients along roadsides facilitate survival of new populations of invaders during drought periods that may otherwise eliminate them.

Invaders can also spread along the verges of dirt roads, and their spread is often associated with the deep, loose soils of roadside berms created by periodic blading of the roadbed (Gelbard and Belnap 2003; Brooks and Lair *this volume*). However, berms do not always result in a higher abundance of nonnative plants. For example, both density and biomass of *B. tournefortii* were higher up to 10 m from dirt roads with berms, but only up to 5 m from roads without berms (fig. 5.3). However, this effect was only significant on silty/rocky/shallow soils (fig. 5.3a) and not where soils were predominantly sandy/deep (fig. 5.3b). Apparently, the deeper and looser soils created by berms were more of a novel feature in areas where soils were otherwise shallow and more compact. Where soils were already loose and deep in sandy areas, berms did not create a dramatically new soil feature that significantly benefited *B. tournefortii.*

Nonnative abundance can also be enhanced by tracks created with a single OHV pass. At a sandy loam site in the Western Mojave, within the tracks created by a single motorcycle pass, biomass of *Schismus* ssp. was 160% and 229% higher during 1994 and 1995, respectively, than in adjacent untracked soil (fig. 5.4). These tracks create local depressions 5–15 cm deep that may help increase propagule pressure by trapping seeds. They may also facilitate an increase in soil nutrient availability by accumulating organic matter and slightly increasing soil moisture levels at the bottom of the track depressions.

High densities of dirt roads and OHV trails can also lead to a higher abundance and number of nonnative plant species in the Mojave Desert (Brooks and Lair *this volume*). In the Central, South-Central, and Western Mojave Desert, the nonnative forb *E. cicutarium* had higher biomass productivity in areas that had high densities of dirt roads than in areas with low densities (Brooks and Berry 2006). At these sites the species richness of nonnative plants was highest where dirt road density was highest (fig. 5.5; $r^2 = 0.24$, $P = 0.0031$). At another site in the Western Mojave, species richness and cover of nonnatives steadily increased as the density of OHV trails increased (M. L. Brooks *unpublished data*).

Although many of the numerous livestock watering sites scattered throughout the Mojave Desert are not currently in use, they harbor many species of nonnative plants that are not particularly abundant elsewhere, except in other agricultural or urban areas (Brooks et al. 2006a). These watering sites can serve as propagule sources from which nonnatives may spread into the surrounding landscape during periods of high rainfall. Propagule pressure from years of vehicle travel to and from these sites and from the importation of contaminated hay provides numerous colonization opportunities. In addition, increased soil organic matter from the importation of hay and the deposition of livestock urine and feces coupled with reduced cover and vigor of native plants, likely lead to increased soil nutrient levels and reduced competition from native plants in the vicinity of watering sites.

An example of the response of nonnative plants to local disturbance gradients associated with livestock watering sites was recently reported from the Central Mojave (Brooks et al. 2006b). Cover of nonnative annual plants increased, whereas cover of native annuals decreased, with proximity to watering sites in this subregion. Nonnative species, which were not found anywhere else in the surrounding area, were also present immediately adjacent to the watering sites, including *Descurainia sophia* (flixweed), *Hordeum* ssp. (barley) and *Marrubium vulgare* (horehound). These effects on vegetation were only significant within 200 m of the watering sites, suggesting that efforts to manage nonnative plants and actively restore native plant communities at these watering sites should be focused within this zone.

Efforts to reduce or eliminate populations of nonnative plants should also be carefully evaluated before they are implemented. In some cases, control efforts may create disturbances that unintentionally exacerbate a nonnative plant problem. For example, rates of seed germination by *B. tournefortii* increased the year following soil disturbance where plants were killed by hoeing the soil, but not where plants were killed using herbicide (M. L. Brooks *unpublished data*).

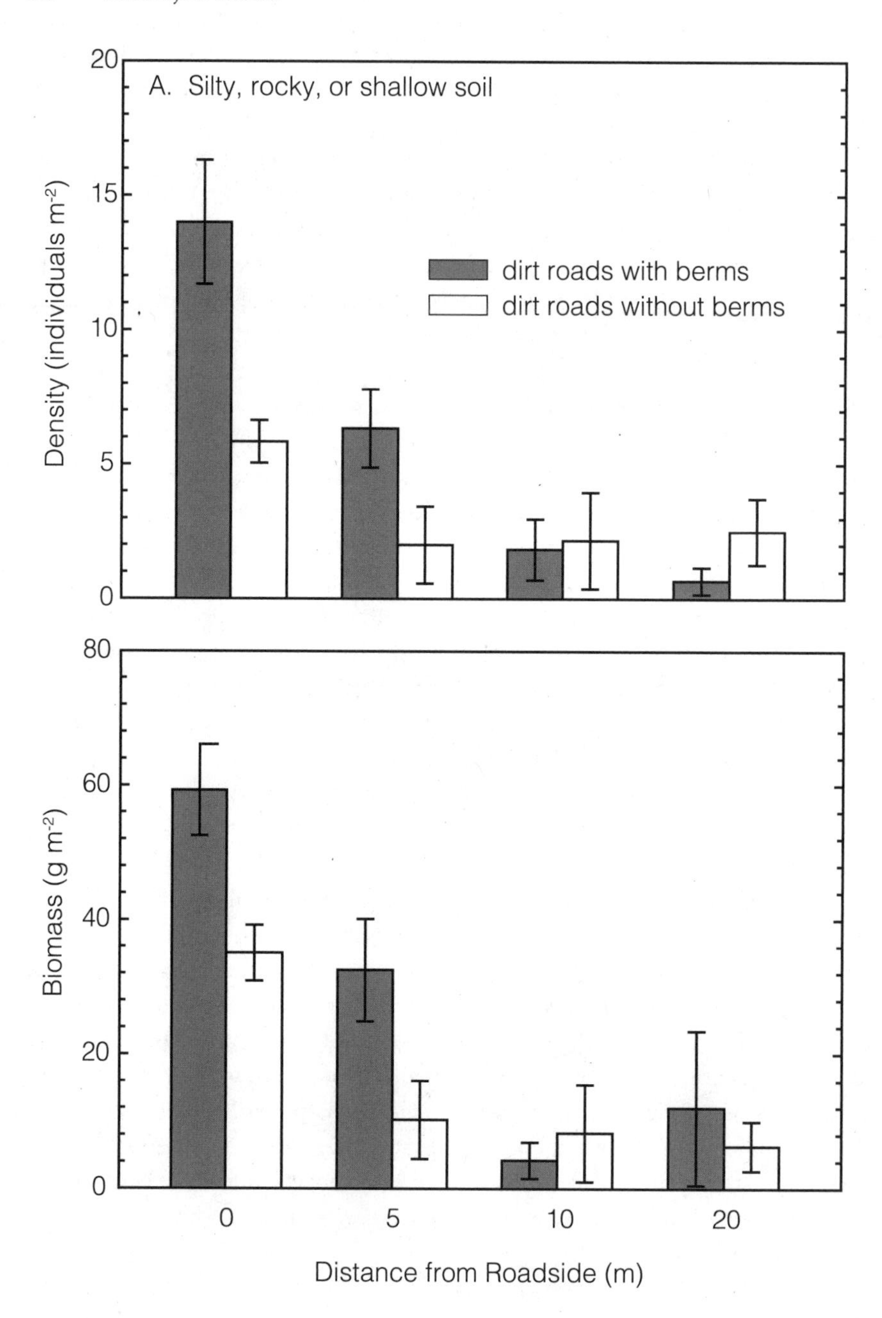
A. Silty, rocky, or shallow soil
dirt roads with berms
dirt roads without berms
Density (individuals m-2)
Biomass (g m-2)
Distance from Roadside (m)
0
5
10
15
20
40
60
80

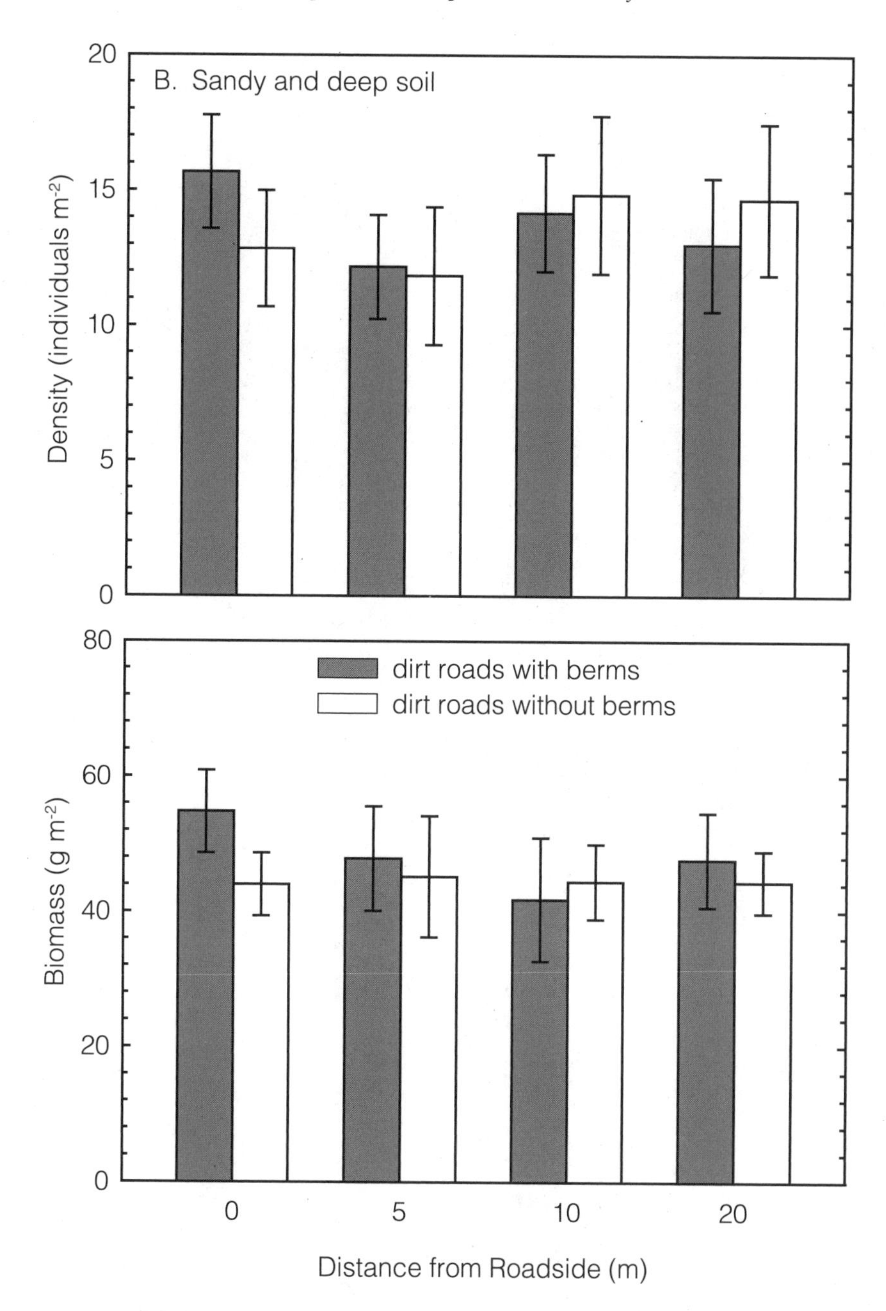

Fig. 5.3. Graphs depicting the density and biomass (±1 SE) of the nonnative *Brassica tournefortii* (Sahara mustard) at various distances from roadsides (with berms and without berms) at sites with silty/rocky/shallow (*A*) and sandy/deep soils (*B*; $n = 6$).

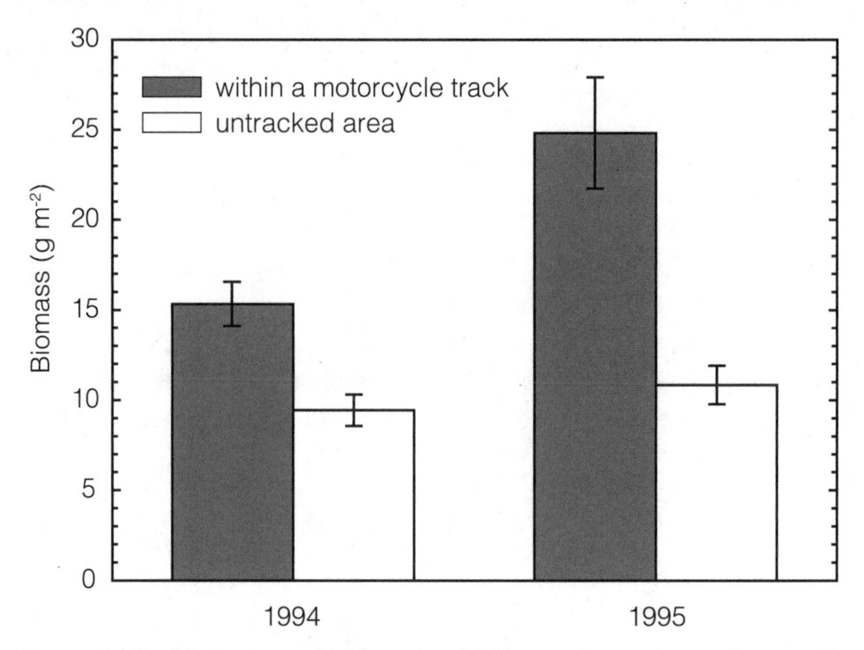

Fig. 5.4. Graph of the biomass of *Schismus* spp. (Mediterranean grass) in tracks created by a single motorcycle pass (*shaded bars*) compared to adjacent untracked areas 25 cm away (*unshaded bars*) near the Desert Tortoise Research Natural Area, California ($n = 10$ paired plots). Error bars indicate ± 1 SE.

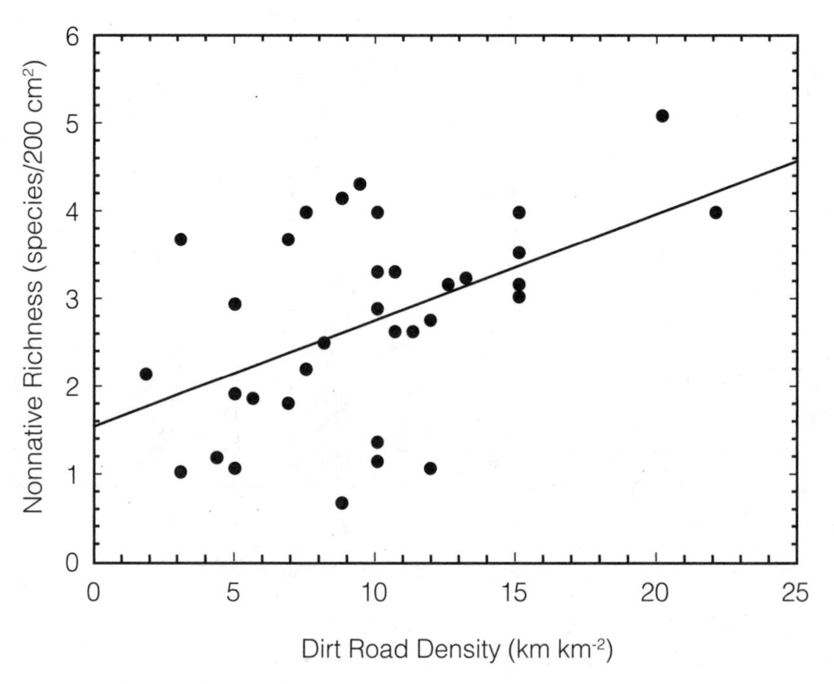

Fig. 5.5. Graph showing the relationship between species richness of nonnative plants and density of dirt roads in the Central, South-Central, and Western regions of the Mojave Desert (modified from Brooks and Berry 2006).

In addition, seed production by stands of *B. tournefortii* can be highest where densities are relatively low and biomasses of individual plants are relatively high (Trader et al. 2006). Control efforts that do not remove all individuals, or where subsequent individuals germinate and grow to maturity, may increase per capita seed production as the remaining plants achieve great size and high seed set in response to reduced intraspecific competition, resulting in no net control effect on seed production (Brooks et al. 2006b).

Similarly, targeted control of one nonnative plant species may result in an undesirable increase in dominance of other nonnatives (Klinger et al. 2006). Thinning by hand clipping *B. madritensis* ssp. *rubens* and *Schismus* ssp. resulted in higher density and biomass of the nonnative forb *E. cicutarium* compared with unthinned plots at sites in the Central, South-Central, and Western Mojave (Brooks 2000). Herbicide treatments designed to target the nonnative grass *Bromus diandrus* (ripgut brome) in Zion Canyon, on the Colorado Plateau near the Eastern Mojave, effectively reduced its cover, but also allowed for a dramatic increase in the cover of another nonnative grass, *Bromus tectorum* (M. L. Brooks, *unpublished data*).

TEMPORAL VARIATION AMONG YEARS

Interannual rainfall is highly variable in the Mojave Desert (Hereford et al. 2006; Redmond *this volume*). This variability can lead to significant differences in the species composition of the annual plant cohorts that germinate and develop into a standing annual plant community. The proportion of nonnatives among the total species in a standing community and the proportion of nonnative biomass productivity of the standing community can both vary significantly among years in response to variation of interannual rainfall amount and within-year distribution.

Many native annual plant species have stringent germination requirements that result in mass germination only during years when rainfall is likely to be sufficient to support individuals' growth to reproduction. Although this conservative approach to germination reduces the chances that germinated plants will die before reproducing, it also increases the chances that seeds will die before germinating. The longer seeds stay dormant in the soil, the greater the chance they will succumb to disease or predation. It is thought that the process of evolution will lead to an optimal bet-hedging strategy that balances these opposing selective forces in native desert annuals (Philippi 1993).

In contrast, nonnative species that did not evolve in response to the selective forces of Mojave Desert environments cannot be expected to have the same stringent germination characteristics as native desert species. In addition, the nonnative species that are most invasive tend to have high population growth rates, which enable them to attain the high densities that often lead to their undesirable ecosystem effects. These minimal germination requirements increase germination rates, which increases the chance that plants may germinate into environmental conditions that cannot sustain them through to reproduction, but decreases the chance that seeds will die before germinating.

The result of these contrasting germination strategies is that the relative species composition of the standing annual plant community can differ significantly among years. During years of low rainfall, mostly nonnative species germinate and grow, thus dominating the standing cohorts. During years of high rainfall, both native and nonnative species germinate and grow, resulting in more equitable proportions of each.

High-rainfall years that follow a series of low-rainfall years can produce standing cohorts characterized by a relatively small proportion of nonnative biomass. These are typically the years characterized as "good wildflower years" (F. C. Vasek *personal communication*). Populations of nonnatives may become depleted during dry years as many individuals germinate and die prior to reproducing. By the time a year of high rainfall returns, the numbers of nonnative propagules available to respond are reduced. Native species may key into these types of years to germinate en masse. Years of low rainfall may allow soil nutrients such as nitrogen to accumulate in the soil, making growing conditions especially good when high rainfall returns. This would have been a strong selective factor that may have shaped the germination requirements of native species.

Although the ecological theories regarding why natives and nonnatives respond differently to years of contrasting rainfall provide a useful framework, it should be realized that not all nonnative species respond the same. For example, species that evolved in more mesic ecosystems (e.g., *Bromus* ssp., most Brassicaceae) may fluctuate in dominance more interannually in response to rainfall than species that evolved in more arid ecosystems (e.g., *Schismus* ssp.).

TEMPORAL VARIATION WITHIN YEARS

The seasonal distribution of rainfall can also influence the species composition of nonnative plants. One of the most obvious examples occurs in regions with scant summer rainfall, which cannot easily support the nonnative plants that require warm-season rains (e.g., many perennial grasses). Most nonnative plants in the Mojave Desert are annual species that require winter rainfall (Brooks and Esque 2002). These species germinate in the fall, grow very slowly for much of the winter, then bolt to produce the bulk of their above-ground biomass and reproductive structures during a few months in spring. Species that can complete this cycle and reach reproductive maturity sooner than others should have a competitive advantage over later-maturing species as soil moisture wanes during middle to late spring.

Rapid growth to reproductive maturity may explain some of the competitive prowess of nonnative annuals in the Mojave Desert. Although no specific studies compare the seasonal germination, emergence, and growth patterns of nonnative versus native species, I have observed that many nonnatives reach reproductive maturity sooner than most natives. *Schismus* ssp. are perhaps the most successful at maturing early and can set seed as early as February during years of low rainfall. This is probably why they do not experience the periodic population crashes due to drought that were discussed for some species earlier in this chapter. In contrast, *B. madritensis* ssp. *rubens* typically flowers and produces seed later in the spring. If soil moisture is insufficient to keep it alive until then, it can die before reproducing. If rainfall is substantial, the early establishment and prolonged growth phase of *B. madritensis* ssp. *rubens* may result in substantial uptake of soil moisture and mineral nutrients, which may be detrimental to native plants. Native plants that were released from competition for soil moisture by experimental thinning of the nonnative annual grasses *B. madritensis* ssp. *rubens* and *Schismus* ssp. began to senesce two weeks later (i.e., had two more weeks of growth) than native plants where nonnative grasses were left in place (Brooks 2000).

MANAGEMENT RECOMMENDATIONS

The information on the spatial and temporal patterns of plant invasions presented in this chapter can be used to improve the efficiency of monitoring strategies for nonnative plants. There are four primary objectives typically associated with the monitoring of nonnative plants: (1) detect the arrival of new nonnative species; (2) document changes in the abundance and range of existing nonnatives, especially in relation to potential vectors of spread and/or land use practices; (3) determine their ecological effects; and (4) evaluate management tools designed to control or eradicate them. In all cases land managers and research scientists will never have the budgets to monitor all species, in all places, all of the time. The information in this chapter is designed to help develop cost-effective monitoring plans for evaluating the status and trends of nonnative plants in the Mojave Desert.

An early warning system to detect new invaders within a land management unit should start with a regional approach that involves collaboration with neighboring land management units. In this way, managers can know where potential new invaders may come from and can coordinate regional control efforts. Cooperative weed management areas (WMAs) are designed for this purpose, and consultation with managers of WMAs should be an integral part of the monitoring of invasive nonnative species.

Monitoring for new invaders should focus on likely pathways of invasion. Surveys should be focused specifically on areas near propagule sources and where soil water and mineral resources are relatively high (Brooks 2007). These surveys should also be focused on detecting specific suites of species that are most likely to establish within the survey region (e.g., summer annuals in summer rainfall regions), which can increase sampling efficiency.

Spatial and Temporal Recommendations

At the regional scale, it is important to identify the specific species most likely to invade from adjacent regions. For example, in the Western Mojave, a list of potential invaders should be developed that focuses on species coming from the agricultural areas of cismontane Kern County and the urban areas of cismontane Los Angeles and San Bernardino counties. Areas with the highest rainfall should also be targeted for early detection monitoring, which in the Western Mojave occurs at the ecotones with the Sierra Nevada Mountains and Transverse Ranges. At the landscape scale, surveys should focus on linear corridors along which invaders can spread, such as roadsides, railroads, and utility corridors, especially near urbanized and agricultural areas. Locally, within sites, surveys should target areas of high resource availability, such as washes and washlets, beneath large perennial shrubs, and within areas of disturbance where competition from native plants has been reduced.

Surveys should ideally be done during years of relatively high rainfall, when the full spate of nonnative species present at a site is most likely to be found. Surveys for nonnative species during low-rainfall years are easier to implement because of the low amounts of native species, which, when present, may obscure the presence of the nonnatives. However, during the early stages of invasion, when densities of a particular species are low, they may be difficult to detect except when each plant grows exceptionally large, as they do during high-rainfall years. In addition, the potential impact of nonnative species is easier to evaluate when one can observe the maximum biomass productivity they can achieve, which would be expected to occur during high-rainfall years.

Surveys should also focus on the seasonal window when a target species can be easily identified. For winter-germinating annual species, this seasonal window is typically from March through April, when annual plant biomass and flowering are at their peak. Satellite-based remote sensing has been proposed in the past as a method to detect concentrations of nonnative species that begin to develop earlier than most native winter annuals (e.g., January–February for *Schismus* ssp.), but no successful examples of this approach have yet to be reported. Summer annuals would most appropriately be sampled towards the end of the summer monsoon season (e.g., August–September). Perennial species can be sampled throughout the year, although timing surveys to correspond with peak flowering may help in detecting individual plants. All of this assumes there is a particular nonnative species or life form that is targeted for the surveys. If surveys are meant to detect all potential nonnatives at a site, then they should be done during March and April to maximize detection of both winter annuals (which are the predominant nonnative plant taxa) and perennial species.

Recommendations for Specific Monitoring Purposes

Monitoring the status and trends of known populations of invaders requires first detecting them, then developing a monitoring design centered on each localized population. Plots should be established along transects following likely vectors of spread (e.g., along roadsides, washes, or invader-specific substrate types). When testing the hypothesis that the expected vector of spread is in fact facilitating spread, then one or more additional sampling transects should be established in random directions to serve as controls.

Monitoring to determine the ecological effects of nonnative plants depends on the specific type of effect. For example, when evaluating the potential effect of an invading species on fire regimes, one must measure variables associated with fuelbed structure and potential fire behavior (Brooks et al. 2004). In the case of an herbaceous species invading a desert shrubland, the impact on fire regimes would be most significant if the species increases intershrub fuel loads (Brooks and Minnich 2006), so monitoring should focus on this microhabitat.

If monitoring is designed to evaluate the effectiveness of control treatments for a specific species, then one should consider how the effects may vary among the range of landscape and local within-site conditions in which the species occurs. For example, herbicide treatments may be most effective when applied along roadsides but less effective when applied where native shrub cover is higher and may hinder herbicide application. Also, if the ultimate reason for controlling a nonnative species is to lessen its ecological impact on a valued natural resource, then the response of the natural resource to the application of nonnative control treatments should be monitored as well. Similarly, responses of nonnative species that are not the primary target of control efforts should be monitored because they may increase in dominance after target species decline (e.g., Brooks 2000). After all, if a control treatment effectively reduces the dominance of the target nonnative but does not result in the desired response by the valued natural resource, or results in increased dominance of other nonnative species, then one must question if the treatment was the best use of limited land management resources.

ACKNOWLEDGMENTS

This chapter received helpful reviews from Tom Dudley, John Hall, and Jill Heaton.

REFERENCES

Brooks, M. L. 1995. Benefits of protective fencing to plant and rodent communities of the western Mojave Desert, California. *Environmental Management* **19**:65–74.

Brooks, M. L. 1999a. Alien annual grasses and fire in the Mojave Desert. *Madroño* **46**: 13–19.

Brooks, M. L. 1999b. Habitat invasibility and dominance by alien annual plants in the western Mojave Desert. *Biological Invasions* **1**:325–337.

Brooks, M. L. 2000. Competition between alien annual grasses and native annual plants in the Mojave Desert. *American Midland Naturalist* **144**:92–108.

Brooks, M. L. 2003. Effects of increased soil nitrogen on the dominance of alien annual plants in the Mojave Desert. *Journal of Applied Ecology* **40**:344–353.

Brooks, M. L. 2007. Effects of land management practices on plant invasions in wildland areas. Pages 147–162 *in* W. Nentwig, editor. *Biological Invasions.* Springer-Verlag, Heidelberg, Germany.

Brooks, M. L., and K. H. Berry. 2006. Dominance and environmental correlates of alien annual plants in the Mojave Desert. *Journal of Arid Environments* **67**:100–124.

Brooks, M. L., C. M. D'Antonio, D. M. Richardson, J. Grace, J. J. Keeley, J. DiTomaso, R. Hobbs, M. Pellant, and D. Pyke. 2004. Effects of invasive alien plants on fire regimes. *BioScience* **54**:677–688.

Brooks, M. L., J. V. Draper, and M. R. Trader. 2006b. Controlling Sahara mustard: evaluation of herbicide and mechanical treatments (California). *Restoration Ecology* **24**: 277–278.

Brooks, M. L., and T. C. Esque. 2002. Alien annual plants and wildfire in desert tortoise habitat: status, ecological effects, and management. *Chelonian Conservation and Biology* **4**:330–340.

Brooks, M. L., and J. R. Matchett. 2003. Plant community patterns in unburned and burned blackbrush (*Coleogyne ramosissima*) shrublands in the Mojave Desert. *Western North American Naturalist* **63**:283–298.

Brooks, M. L., and J. R. Matchett. 2006. Spatial and temporal patterns of wildfires in the Mojave Desert, 1980–2004. *Journal of Arid Environments* **67**:148–164.

Brooks, M. L, J. R. Matchett, and K. H. Berry. 2006a. Effects of livestock watering sites on plant communities in the Mojave Desert, USA. *Journal of Arid Environments* **67**: 125–147.

Brooks, M. L., and R. A. Minnich. 2006. Southeastern deserts bioregion. Pages 391–414 *in* N. G. Sugihara, J. W. van Wagtendonk, K. E. Shaffer, J. Fites-Kaufman, and A. E. Thode, editors. *Fire in California's Ecosystems.* University of California Press, Berkeley, California.

Brooks, M. L., and D. Pyke. 2001. Invasive plants and fire in the deserts of North America. Pages 1–14 *in* K. Galley and T. Wilson, editors. *Proceedings of the Invasive Species Workshop: The Role of Fire In the Control and Spread of Invasive Species, Fire Conference 2000.* The First National Congress on Fire, Ecology, Prevention and Management, November 28–December 1, 2000, San Diego, California. Miscellaneous Publication No. 11. Tall Timbers Research Station, Tallahassee, Florida.

Davis, M. A., J. P. Grime, and K. Thompson. 2000. Fluctuating resources in plant communities: a general theory of plant invasibility. *Journal of Ecology* **88**:528–534.

Davidson, E., and M. Fox. 1974. Effects of off-road motorcycle activity on Mojave Desert vegetation and soil. *Madroño* **22**:381–412.

DeFalco, L. A., D. R. Bryla, V. Smith-Longoza, and R. S. Nowak. 2003. Are Mojave Desert annual species equal? Resource acquisition and allocation for the invasive grass *Bromus madritensis* ssp. *rubens* (Poaceae) and two native species. *American Journal of Botany* **90**:1045–1053.

Elton, C. S. 1958. *The ecology of invasions by animals and plants.* Methuen, London, UK.

Gelbard, J. L., and J. Belnap. 2003. Roads as conduits for exotic plant invasions in a semi-arid landscape. *Conservation Biology* **17**:420–432.

Hereford, R., R. H. Webb, and C. I. Longpré. 2006. Precipitation history and ecosystem response to multidecadal precipitation variability in the Mojave Desert region, 1893–2001. *Journal of Arid Environments* **67**:13–34.

Hobbs, R. J., and L. F. Huenneke. 1992. Disturbance, diversity, and invasion: implications for conservation. *Conservation Biology* **6**:324–337.

Jennings, W. B. 1993. *Foraging ecology and habitat utilization of the desert tortoise (Gopherus agassizii) at the Desert Tortoise Research Natural Area, eastern Kern County, California.* Contract No. B950-C2-0014 prepared for the U.S. Bureau of Land Management. Riverside, California.

Kemp, P., and M. L. Brooks. 1998. Exotic species of California deserts. *Fremontia* **26**: 30–34.

Klinger, R. C., M .L. Brooks, and J. M. Randall. 2006. Fire and Invasive Plant Species. Pages 499–519 *in* N. G. Sugihara, J. W. van Wagtendonk, K. E. Shaffer, J. Fites-Kaufman, and A. E. Thode, editors. *Fire in California's ecosystems.* University of California Press, Berkeley, California.

Larson, D. L., P. J. Anderson, and W. Newton. 2001. Alien plant invasion in mixed-grass prairie: effects of vegetation type and anthropogenic disturbance. *Ecological Applications* **11**:128–141.

Lonsdale, W. M. 1997. Global patterns of plant invasions. *Bulletin of the Ecological Society of America* **78**:21.

Lonsdale, W. M. 1999. Global patterns of plant invasions and the concept of invasibility. *Ecology* **80**:1522–1536.

Maron J. L., and Connors P. G. 1996. A native nitrogen-fixing shrub facilitates weed invasion. *Oecologia* **105**:302–312.

Oftedal O., S. Hillard, and D. J. Morafka. 2002. Selective spring foraging by juvenile desert tortoises (*Gopherus agassizii*) in the Mojave Desert: evidence of an adaptive nutritional strategy. *Chelonian Conservation and Biology* **4**:341–352.

Philippi, T. 1993. Bet-hedging germination of desert annuals: beyond the first year. *American Naturalist* **142**:474–487.

Rejmanek, M. 1989. Invasibility of plant communities. Pages 369–388 *in* J. A. Drake, H. A. Mooney, F. di Castri, R. H. Groves, F. J. Kruger, M. Rejmanek, and M. Williamson, editors. *Biological invasions: a global perspective.* John Wiley and Sons, Brisbane, Australia.

Rowlands, P. G., H. Johnson, E. Ritter, and A. Endo. 1982. The Mojave Desert. Pages 103–162 *in* G. L. Bender, editor. *Reference handbook on the deserts of North America.* Greenwood Press, Westport, Connecticut.

Samson, D. A. 1986. Community ecology of Mojave Desert winter annuals. Ph.D. Dissertation, University of Utah, Salt Lake City, Utah.

Trader, M. R., M. L. Brooks, and J. V. Draper. 2006. Seed production by the non-native *Brassica tournefortii* (Sahara mustard) along desert roadsides. *Madroño* **53**:313–320.

U.S. Fish and Wildlife Service. 1994. *Desert tortoise (Mojave population) recovery plan.* U.S. Fish and Wildlife Service, Portland, Oregon.

Warner, P. J., C. C. Bossard, M. L. Brooks, J. M. DiTomaso, J. A. Hall, A. Howald, D. W. Johnson, J. M. Randall, C. L. Roye, M. M. Ryan, and A. E. Stanton. 2003. Criteria for categorizing invasive non-native plants that threaten wildlands. Online. California Exotic Pest Plant Council and Southwest Vegetation Management Association. http://ucce.ucdavis.edu/files/filelibrary/5319/6657.doc.

Webb, R. H., and S. S. Stielstra. 1979. Sheep grazing effects on Mojave Desert vegetation and soils. *Environmental Management* 3:517–529.

SIX

Invasive Plants in Mojave Desert Riparian Areas

TOM L. DUDLEY

Riparian areas and other wetlands are rare in desert environments, occupying less than 3% of the Mojave Desert (Rowlands et al. 1982). However, these are critical habitats with a disproportionate role in sustaining biodiversity and threatened native species (Brookshire et al. 1996; Sanders and Edge 1998). As an example, approximately 55% of the vertebrates, 50% of the invertebrates, and 40% of the plants considered sensitive by the Clark County, Nevada, Multiple Species Habitat Conservation Plan are restricted to, or closely associated with, wetland habitats (Clark County 2005). Water management and land use practices have severely limited the capacity of western rivers to support sensitive species and ecosystem processes (Horton 1977; Brookshire et al. 1996), and fewer than 10% of California's waterways retain functional riparian habitat (Riparian Habitat Joint Venture 2004). These are further degraded by invasion of nonindigenous (e.g., nonnative, introduced, exotic, alien) plants that may compete with native plants and provide poorer resources than the native vegetation they have replaced (Allan and Flecker 1993; Dudley and Collins 1995; Wilcove et al. 1998).

Deserts are among the least invaded ecosystems in North America, at least in terms of the number of nonindigenous species that have established and the proportion of the flora they represent (Rejmanek and Randall 1994; Brooks *this volume*). Roughly 5% of the flora is comprised of exotic species, presumably owing to physiological stresses caused by the harsh climate and moisture conditions. However, these stresses are apparently reduced in riparian areas, where water availability is less limiting than in the drier uplands, providing conditions that a much broader range of species can tolerate. For example, in a floral survey of the Mojave

River in the vicinity of Victorville, California (Mojave Narrows), almost 20% of the 283 documented taxa were nonindigenous, even though this survey included adjacent upland habitats (MacKay *undated*).

To gain a better understanding of which invasive species are in the Mojave region, and which may colonize other riparian habitats where they are not currently present, species are tabulated based on information gleaned from a variety of sources. These species are then further characterized according to the degree of risk they pose (or potentially pose) to biodiversity and natural resources. Invasive species, defined as any species (native or nonnative) that can spread away from the point of introduction (Richardson et al. 2000), are considered problematic from a management viewpoint when they do any of the following: (1) change the rate of nutrient and/or resource supply, (2) disrupt trophic dynamics among native species, and/or (3) alter natural disturbance regimes, including wildfire frequencies (D'Antonio and Vitousek 1992).

Maintaining native biodiversity is an obvious management concern, especially considering that invasive species rank second only to habitat destruction as detrimental to threatened or endangered species, and between 42% and 49% of federally listed species are threatened, in part, by nonnative species (Stein et al. 1996; Wilcove et al. 1998; Pimentel et al. 2001). Degraded systems still retain some value in terms of habitat, and restoration of natural riparian characteristics has been a high-priority goal for resource managers in the Southwest (Szaro and Rinne 1988).

Riparian systems are areas that carry water for a sustained period of at least several weeks (as opposed to ephemeral desert washes, which are discussed by Brooks, *this volume*), and therefore promote development of vegetation that depends on regular contact with moisture. The primary waterways within the Mojave region are the Mojave River and the Amargosa River. The Amargosa is sometimes considered to be a Great Basin system, and it may make sense biogeographically to integrate the two regions into the Basin and Range Province, with the Mojave largely representing lower elevations where *Larrea tridentata* (creosote bush) often dominates (generally < 1,200 m, Brussard et al. 1999). The lower Owens Valley is a similar transitional region generally considered to have vegetation affiliations with the Mojave. Numerous smaller waterways also traverse the area, particularly in the eastern mountain ranges of the Mojave National Preserve. Also, a portion of the Colorado River and its tributaries (particularly the Bill Williams, Virgin, and Muddy rivers) are reasonably considered ecologically allied with the Mojave ecosystem. Riparian invasive plants share affinities across most of this prescribed desert region.

After discussing the suite of nonindigenous riparian plants in this broad region, I will focus the remainder of this chapter on a single taxon (*Tamarix* ssp., a.k.a. tamarisk or saltcedar) that is considered by many to be the most destructive invader in the southwestern deserts, including in the Mojave Desert.

RIPARIAN INVADERS IN THE MOJAVE DESERT

Invasive Plant Life Forms and Representation in the Flora

Table 6.1 is a preliminary compilation of invasive wetland and riparian plants in the Mojave region. This list is certainly incomplete, considering that there are at least 315 nonnative species recorded as spontaneous in California which are not covered by the 1993 Jepson Manual (Hrusa et al. 2002). New invasives are also documented in weed manuals and state weed lists (e.g., DiTomaso and Healy 2003; Rafferty 2005; Ryan 2005). Invasions continue unabated, and there is a need to develop a more comprehensive assessment of the diversity, distribution, abundance, and impacts of riparian invaders regionally. This total includes many herbaceous species, including annuals (about 28 species) and short-lived perennials (about 29 species), woody shrubs (2 species) and trees (9 species), and aquatic plants (9 species). Vine-like invaders are conspicuously absent from this mix, perhaps because vines that invade elsewhere [e.g., *Pueraria lobata* (kudzu) in the southeastern United States and *Delairea odorata* (cape ivy) and *Hedera* ssp. (true ivies) in Pacific coastal environments] tend to require more humid atmospheric conditions to survive. Only *Datura wrightii* (jimsonweed), which may be native or a short-distance anthropogenic introduction from Mexico, displays prostrate vine-like behavior—it can be an annual and therefore does not have to tolerate conditions in all seasons.

Species diversity of both nonindigenous and native plants is related to the presence of water, and as moisture decreases, invaders tend to decline. Fifty-six nonindigenous species (about 20% of the flora) are associated with the Mojave Narrows section of the Mojave River (MacKay *undated*). These are present in the permanently wet sections of river, where channel topography and geology force groundwater near the surface, and create a productive area of high ecological significance (Lines 1996). Invasive plants tend to have the highest diversity in the same locations as native plants (biodiversity "hot spots"), and riparian areas are primary hot spots (Stohlgren et al. 1998). The Camp Cady Area of Critical Environmental Concern, farther downstream on the Mojave River, also supports a variety of riparian species (LeBlanc 2002). However, many years of water diversion (e.g., the construction of Mojave Forks Dam) and groundwater lowering have decreased the availability of water, and subsequently reduced the diversity of all species (West Mojave Planning Team 1999). Of the 37 plants in Camp Cady, six (16%) are nonnative. Under harsher arid-riparian conditions, Evens (2001) documented 13 (5%) nonnative species among the 245 identified in intermittent and ephemeral watercourses of the eastern Mojave ranges (Granite, Providence, New York, and

Table 6.1 Invasive plants known or suspected in wetlands and riparian areas of the Mojave Desert region

Species	Common name	Habit	Threat level[a]	Abundance[b]	Habitat	Notes
Acroptilon repens	Russian knapweed	Herbaceous perennial	1	o	Riparian floodplains, alkali sinks	Noxious
Agrostis stolonifera, A. semiverticillata, A. viridis	Creeping bent, water bent	Herbaceous perennial, rhizomatous	2	o	Streamsides, ditches, marshes, ponds	
Ailanthus altissima	Tree-of-heaven	Suckering tree	1	o	Rocky channels, springs	
Alhagi maurorum	Camelthorn	Shrub, rhizomatous	1	o	Ditches, streambanks	Noxious
Apium graveolens	Wild celery	Herbaceous perennial	3	o	Marshes, springs	Can be toxic
Artemisia biennis	Biennial wormwood	Herbaceous annual or biennial	3	o	Terraces, ditches, montane	
Arundo donax	Giant reed	Herbaceous perennial canes	1	o–c	Terraces, stream-sides, canal margines	Noxious
Avena fatua	Wild oat	Herbaceous annual	2	u	Floodplain terraces, understory	
Bacopa monnieri	Water-hyssop	Herbaceous perennial, sprawling	3	u	Moist soil, ponds	Recent in Riverside County
Bassia hyssopifolia	Fivehook bassia	Herbaceous annual	2	c–a	Alkali sink, dry riparian terraces	
Brassica tournefortii	Sahara mustard	Herbaceous annual	2	c	Sandy washes	Increasing rapidly
Bromus spp., esp. *B. diandrus, B. madritensis* ssp. *rubens, B. tectorum*	Bromes—rip-gut brome, red bromes, cheatgrass	Herbaceous annual	3	c	Terraces	Promote fire
Cardaria pubescens	Whitetop	Herbaceous perennial	3	u	Alkali seeps, ditches, disturbed areas	Noxious, Owens Valley
Centaurea solstitialis	Yellow starthistle	Herbaceous perennial	1	u	Floodplains, dry terraces	Toxic to horses
Chenopodium glaucum, C. chenopodioides	Pigweed, goosefoot	Herbaceous annual	3	o	Saline/alkali sinks, streambanks	Western Mojave
Cirsium arvense	Canada thistle	Herbaceous perennial	2	c	Floodplains, dry terraces, disturbed areas	Western Mojave

Cynodon dactylon	Bermuda grass	Herbaceous perennial, rhizomatous	2	c	Streamsides, sandy stream bottoms	
Datura wrightii	Jimsonweed	Prostrate annual/herbaceous perennial	3	c	Roadsides, sandy washes	Native to Mexico, Toxic
Descurainia sophia	Tansy mustard	Herbaceous annual	3	c	Dry terraces, disturbed areas	Can be toxic
Echinochloa spp. (*E. colona*)	Jungle rice	Annual	3	o, c	Streambanks, wet disturbed areas	Presence uncertain
Egeria densa	Brazilian waterweed	Aquatic perennial	2	u	Canals, ponds, slow streams	Noxious, presence uncertain
Eichhornia crassipes	Water hyacinth	Aquatic perennial	2	u	Canals, ponds, lakes	
Elaeagnus angustifolia	Russian olive	Tree	1	o	Floodplains, seeps, springs	
Eriochloa contracta	Prairie cupgrass	Annual	3	o	Ditches, stream channels	Native to central US
Erodium cicutarium	Filaree	Annual	3	c	Dry terraces, disturbed soil	
Foeniculum vulgare	Fennel	Biennial/herbaceous perennial	3	c	Seeps	
Grindelia squarrosa	Gumplant	Herbaceous biennial	3	o	Streamsides	Concentrates selenium, native to Rocky Mountains
Halogeton glomeratus	Saltlover	Annual	2	c	Alkaline sinks, disturbed soil	Toxic from oxalates
Hirschfeldia incana	Shortpod mustard	Annual/biennial	3	c	Terraces, disturbed soil, washes	
Holcus lanatus	Velvet grass	Herbaceous perennial	2	c	Floodplains, disturbed streambanks	Cultivated for forage
Hordeum marinum H. murinum	Mediterranean barley, foxtail barley	Annual	3	o	Terraces, moist disturbed soil	
Hydrilla verticillata	Hydrilla	Aquatic perennial	1	u	Canals, ponds, streams	Noxious, probably eradicated
Lactuca serriola	Prickly lettuce	Herbaceous perennial	3	c	Disturbed areas, seeps	

Table 6.1 (continued)

Species	Common name	Habit	Threat level[a]	Abundance[b]	Habitat	Notes
Lepidium latifolium	Tall whitetop, perennial pepperweed	Herbaceous perennial	1	o	Floodplains, alkali sinks, marshes	Uncommon, but likely to increase
Ludwigia peploides montevidensis	Water primrose	Aquatic perennial	3	?	Ponds, streambanks, shores	Native, but *L. p. montevidensis* is South American
Lythrum salicaria	Purple loosestrife	Herbaceous perennial, rhizomatous	3	u	Ditch margins, streamsides	Presence uncertain
Marrubium vulgare	Horehound	Herbaceous perennial or subshrub	3	u	Terraces, disturbed areas, montane	
Melilotus alba	White sweetclover	Herbaceous annual or biennial	3	c	Streamsides, ditches, roadsides	
Melilotus officinalis	Yellow sweetclover	Herbaceous annual or biennial	3	o	Streamsides	
Mentha spicata	Spearmint	Annual	3	o	Streamsides, springs, montane	
Mollugo cerviana	Carpet weed	Annual	3	u	Disturbed areas, dry washes, seasonal ponds	Presence uncertain
Nicotiana glauca	Tree tobacco	Woody shrub or tree	3	o	Terraces, disturbed areas	Presence uncertain
Paspalum dilatatum	Dallis grass	Herbaceous perennial	2	c	Floodplains, ditches, moist disturbed areas	Agricultural weed (native knotgrass, *P. distichum,* is similar and near water)
Phalaris aquatica	Harding grass	Herbaceous perennial, rhizomatous	2	?	In-channel, marshes, wet streambanks	Noxious, presence uncertain
Phalaris arundinacea	Reed canary grass	Herbaceous perennial, rhizomatous	2	u	Montane streambanks	Native, possibly European genotype / cultivar
Phragmites australis	Common reed	Herbaceous perennial	3	c	Ditches, streambanks, springs	Native, and possible nonnative genotoyopes
Pistia stratiotes	Water lettuce	Aquatic perennial	3	u	Ditches, ponds	Colorado River
Plantago major	Plantain	Herbaceous perennial	3	o	Lawns, moist streambanks	
Poa pratensis	Kentucky blue-grass	Herbaceous perennial	3	c	Terraces, alkali/ saline sinks	

Polygonum spp.	Smartweed	Submerged herb	2–3	?	Canals, ponds	Many native and nonnative species, some pests
Polypogon interruptus	Ditch polypogon	Herbaceous perennial	3	o	Canals, ditches	South American
Polypogon monspeliensis	Rabbitsfoot grass	Annual	3	c–a	Streamsides, ditches, moist seeps—often alkaline	
Potamogeton crispus	Curlyleaf pondweed	Aquatic perennial	3	u	Canals, ponds, reservoirs, streams	Similar to several native pondweeds
Ranunculus testiculatus	Bur or hornseed buttercup	Annual	2	u	Grasslands, montane, streambanks	Presence uncertain, toxic
Robinia pseudoacacia	Black locust	Tree	2	o	Streambanks, wet meadows	Native to eastern US, common in Owens Valley
Rumex obtusifolius, R. crispus	Western dock, curly dock	Herbaceous perennial	3	c	Terraces, streambanks, ditches, moist disturbed areas	Other spp., also oxalic acid, toxic
Saccharum ravennae	Ravennagrass	Herbaceous perennial bunchgrass	2	o–c	Ditch, marshes,	Horticultural, Imperial County
Salix alba, S. babylonica	White willow, weeping willow	Tree	3	u	Disturbed areas, minor escapes	Horticultural
Salsola tragus, S. paulsenii	Russian thistle, tumbleweed	Annual	2	c	Alkali sinks, ditches, terraces, especially saline areas	
Salvinia molesta	Giant salvinia	Floating annual/ perennial	2	u	Canals, ditches, slow-moving water	Reproduces by fragmentation
Saponaria officinalis	Bounding bet, soapwort	Herbaceous perennial	3	o	Streamsides	Somewhat toxic
Schismus barbatus	Schismus	Annual	2	c	Sandy washes, terraces	
Sisymbrium spp. (*S. orientale, S. irio S. altissimum*)	Hedgemustard, London rocket, tumblemustard	Annual	3	o	Disturbed uplands, terraces	

Table 6.1 (continued)

Species	Common name	Habit	Threat level[a]	Abundance[b]	Habitat	Notes
Sonchus asper, S. oleraceus	Sow thistles	Annual	3	o	Terraces, disturbed areas, seeps	
Sorghum bicolor	Sudangrass	Annual	2	c	Canal banks, disturbed areas, ditches	
Sorghum halepense	Johnsongrass	Perennial	2	c	Canal banks, disturbed areas, ditches	
Tamarix aphylla	Athel	Tree	1	c	Lake margins, planted sites	Adventive at Lake Mead
gallica, T. parviflora, T. ramosissima	Tamarisk, saltcedar	Tree	1	o–a	Floodplains, lake margins, springs	
Tribulus terrestris	Puncturevine, caltrop	Prostrate annual	2	o	Terraces, disturbed soils	Reduced by biocontrol
Ulmus pumila	Siberian elm	Tree	2	o	Floodplains, washes, disturbed areas	Common street tree
Vallisneria americana	American eelgrass	Aquatic perennial	2	u	Ponds, springs	
Veronica anagallis-aquatica	Water speedwell	Aquatic perennial	3	u	Montane, streambanks, wet meadows	
Washingtonia filifera – N	California fan palm	Tree	2	o	Seeps, streams	Many populations planted
Washingtonia robusta	Mexican fan palm	Tree	2	o	Ditches, near towns	

Dudley and Collins 1995; Dudley 1998; Evens 2001; Baldwin et al. 2002; DiTomaso and Healy 2003; Page 2004; Rafferty 2005; Ryan 2005; California Invasive Plant Council 2006; MacKay *undated.*

[a]Threat level refers to risk in natural areas; 1 = serious known impacts to biodiversity, ecosystem function or natural resources; 2 = impacts known or suspected, but do not substantially alter ecosystems; 3 = plant is present but not causing impacts, or impacts not currently known.

[b]Abundance categories: u = uncommon or rare, status unclear; o = occasional or present in scattered sites; c = common, can be found regularly; a = abundant, dominates where conditions are suitable.

N—native species rapidly expanding from original habitat.

Clark mountains). In the physiologically stressful conditions at the highly saline Ash Meadows/Carson Slough area, only 3 of the 29 plant species (*Polypogon monspeliensis* [rabbitsfoot grass], *Lactuca serriola* [prickly lettuce], and *Tamarix* ssp.) were nonnative (Soil Ecology and Research Group 2004).

Ecological Relationships of Riparian Invaders

The majority of invasive plants present in Mojave Desert riparian areas are fairly low-growing taxa with broad distributions. Weedy taxa present across the west are common in the region and include *Melilotus alba* (white sweet clover) and other sweet clovers, *Lepidium latifolium* (tall whitetop/perennial pepper weed), *Rumex* ssp. (docks), *P. monspeliensis, Sonchus* ssp. (sow thistle), *Artemisia biennis* (wormwood), *Cynodon dactylon* (Bermuda grass), and numerous others, which suggests that these plants are habitat generalists that simply require substantial moisture to colonize and establish. Other widespread roadside weeds are less dependent on moisture, although they benefit from wet conditions during establishment, and include *Lactuca, Hirschfeldia incana* (shortpod mustard), *Descurainia sophia* (tansy mustard), *Sisymbrium* ssp.(tumble mustards), *Foeniculum vulgare* (fennel) and *Erodium cicutarium* (filaree). In addition, several common riparian invaders are weeds often associated with alkaline soils, like *Salsola tragus* (Russian thistle) and *Bassia hyssopifolia* (fivehook bassia). This is unsurprising considering that desert riparian soils are typically of moderate alkalinity or salinity. Curiously, *Kochia scoparia* (forage kochia) is another water-loving alkaline associate that appears not yet to occur in the Mojave, although it is present in the nearby San Bernardino Mountains and is a common invader throughout deserts to the north.

Many of the above plants are considered ruderal species—that is, weedy species that frequently germinate on physically disturbed sites where competition is reduced, and that may be expected to decline in abundance as native or other more competitive invasive species establish and displace these early successional taxa (Grime 1979). Stream and river margins, by their natural condition of experiencing frequent physical disturbance due to flooding and desiccation (Poff et al. 1997), are colonized by such disturbance-dependent species (Hood and Naiman 2000). While these species may be both common and diverse, they do not necessarily degrade channel-margin ecosystems, owing to their tendency to drop out as the vegetation community matures. Such ruderal weeds may be largely ignored from a management perspective, so that efforts can focus on those few species that can dominate and create substantial ecological impacts.

In the California deserts, even though there may be a large number of nonindigenous taxa in riparian areas, resource managers report a lower number of problem species than in the other biogeographic provinces of California (Dudley and Collins 1995). Likewise, the Mojave Weed Management Area lists 13 serious invaders (Page 2004), of which only three are clearly riparian [*Arundo donax* (giant reed), *Tamarix ramosissima* (saltcedar/tamarisk), and *Ailanthus altissima* (tree-of-heaven)]. Three others [*Alhagi maurorum* (camelthorn), *Halogeton glomeratus* (salt lover) and *S. tragus*] are commonly associated with low-lying, seasonally wet sites but are less typical in riparian zones. One other Mojave Weed Management Area species, *Linaria dalmatica* (Dalmatian toadflax), was not included in table 6.1 because it does not seem to be riparian locally, although it is known to be a riparian invasive in other areas.

Some of the more persistent herbaceous invasives include species like *Acroptilon repens* (Russian knapweed), *Lepidium, Foeniculum, Brassica tournefortii* (Sahara mustard), *Saccharum ravennae* (ravennagrass), *Sorghum halepense* (Johnsongrass), *Bromus* ssp., *Centaurea solstitialis* (yellow starthistle), and others, which can come to dominate local habitat patches. These problematic invaders tend to form dense aggregations that offer little space for other colonizing species. They are more likely than the ruderal weeds to be perennial, and these high-density, low-diversity vegetation patches are self-sustaining over multiple seasons. Some species, like *Lepidium* and *Bromus diandrus* (rip-gut brome), are moderately tolerant of shade, and therefore can develop in association with overstory species like *Salix* ssp. (willow) and *Populus fremontii* (cottonwood), or can inhibit growth during the establishment of these later successional, native riparian species.

Dominance of riparian habitats by perennial weeds can also be a problem when invaders are fire-prone and periodically burn, impacting the fire-intolerant overstory plants (D'Antonio and Vitousek 1992), and then regenerate from roots or rhizomes to dominate post-fire environments. *Arundo* is a particular problem in this regard, and has sustained wildfires along the lower Colorado River and in other semiarid California ecosystems (Dudley 2000), as well as in the Chihuahuan Desert (T. L. Dudley *personal observation*). This bamboo-like grass is increasingly common in the Mojave region as it escapes from cultivation, especially in the Joshua Tree National Park and Barstow areas in California and the Las Vegas area in Nevada. *Arundo* is dispersed by floods that break off rhizomes and carry them into wildland environments, where it is likely to cause negative impacts to native vegetation and to reduce associated wildlife diversity (Herrera and Dudley 2003; Kisner and Kus 2003).

One of the reasons why persistent herbaceous weeds are of concern to land managers is that they can be pests outside of riparian areas as well. They may initially colonize and develop large populations in naturally disturbed riparian habitats, and then expand from these reservoirs into upland and other areas (Pysek and Prach 1993). Often they are toxic [e.g., *Lactuca, Halogeton, Centaurea* ssp., *Descurainia, Melilotus* ssp., *Datura, Nicotiana glauca* (tree tobacco), *Lepidium, Saponaria officinalis* (bouncingbet)] so they may pose some risk to human and/or livestock health (Ralphs 2002). It is surprising that *Conium maculatum* (poison hemlock) appears to be absent from the Mojave Desert, even though it is widespread in adjoining regions under moist, partly shaded conditions (Pitcher 1989; Van Devender et al. 1997). Dispersal of riparian-dependent weedy species into new areas may be restricted by the extremely dry conditions in the intervening distances, but with increasing human activity and movement among regions these barriers are being lowered (Mack et al. 2000).

Aquatic Plants. Six of the invasive plants listed in table 6.1 are fully aquatic [for example, *Eichhornia crassipes* (water hyacinth) and *Salvinia molesta* (giant salvinia)], occurring as submerged or emergent plants in perennial waters. These are a serious concern for managers of lentic or slow waters such as canals and reservoirs, and most are federally listed as noxious weeds (DiTomaso and Healy 2003). The *S. molesta* recently discovered in parts of the lower Colorado River raised great concern because of the problems that this South American floating fern has caused in the southeastern United States and around the world, and was quickly targeted for control (California Department of Food and Agriculture 1999; U.S. Fish and Wildlife Service 2000). These aquatic plants are generally not a major concern in the habitats covered in this volume, although they do have the potential for invading spring pools, such as those inhabited by desert pupfish (*Cyprinodon macularius*) and other sensitive species in Death Valley.

A native aquatic plant, *Ceratophyllum demersum* (hornwort), was noted as a management problem because of its potential to choke out spring species in Death Valley (Dudley and Collins 1995), and *Vallisneria americana* (American eelgrass) also encroaches on the habitat of sensitive species in springs of the Moapa Valley National Wildlife Refuge (Moapa Valley National Wildlife Refuge *undated*). Noxious aquatic plants typically colonize new locations as a result of disposal by aquarists (Mehrhoff 1999), which, fortunately, may be less likely to occur in isolated springs than in easily accessible riverways and reservoirs.

Woody Invaders. Even though a minority of the regional invaders are woody trees or shrubs [including *Eleaegnus angustifolia* (Russian olive), *Ulmus pumila* (Siberian elm) and *Ailanthus*], the large size and greater persistence of these plants means that their potential to affect other ecosystem elements is high. Some are not yet widespread or dominant members of the riparian assemblage, but they should be of concern, because they are well known as serious invaders in other regions (Olson and Knopf 1986; Dudley and Collins 1995; Shafroth et al. 1995a, 1995b; Moore 2003). All three are widely used as ornamentals across the West, and have escaped from cultivation. *Eleaegnus* fruits are eaten by birds like doves and jays and dispersed widely, often germinating within the riparian thickets where the birds have perched (Olson and Knopf 1986). Dense stands eventually develop and the long thorns (often 4 cm or more) make access difficult for wildlife and humans alike. They tolerate drought conditions once established, and grow vigorously, inhibiting establishment by other species. *Ailanthus* also forms dense stands, especially through suckering, and recovers quickly following wildfire. Both *Ailanthus* and *Eleaegnus* will dominate montane spring sources—for example, at Butterbredt Springs, California, an important wildlife area in the northwestern corner of the Mojave Desert, *Ailanthus* has overtaken the wetland associated with the spring. *Ulmus pumila* is not yet as widespread as the other two trees in the Mojave, but based on the similarity of habitat it shares with *Ailanthus* and *Eleagnus* in other desert regions, the presence of this wind-dispersed tree is likely to increase over time—it is already increasing in the Great Basin and arid Rocky Mountain states.

Robinia pseudoacacia (black locust) is also increasingly common in moist areas in nearby regions, particularly the Owens Valley, and this nitrogen-fixing tree will likely be able to tolerate the conditions found in Mojave riparian systems. Two willows, *Salix alba* (white willow) and *Salix babylonica* (weeping willow), are reported as minor escapes in the region.

An unusual tree-like adventive in some areas is the fan palm, both the native California *Washingtonia filifera* (fan palm) and *Washingtonia robusta* (Mexican fan palm). There is some controversy over the status of *W. filifera.* Some argue that it is a natural relict species, while others contend that it may not be native to the region and that its invasiveness has been enhanced by anthropogenic cultivation, dispersal, and land use (Cornett 1987; Spencer 1996). These palms are of moderate concern because dense growths of *W. filifera,* as well as the clearly nonnative *Phoenix dactylifera* (date palm), appear to be encroaching on spawning and foraging habitat of the endangered Moapa dace (*Moapa coriacea*), which is restricted to thermal springs in Moapa Valley National Wildlife Refuge in the Muddy River watershed (Moapa Valley National Wildlife Refuge *undated*). These and some other palms [e.g., *Brahea armata* (Mexican blue palm)] are routinely found as escapees from planted groups of palms, because seeds are carried and dropped by birds even in canyon springs far removed from the source (C. Deuser *personal communication*). Palm infestations can be fairly easily controlled, as demonstrated in western San Diego County, where their tendency to support wildfire has justified control (Kelly 2004). However, this can be costly because of the large biomass involved. Perhaps of greater concern is the fact that native palm oases are threatened by many factors, including invasion by *Tamarix* ssp.

Invasion, Ecology, and Control of *Tamarix* ssp.

Tamarix ssp. (tamarisk, saltcedar) are not only one of the most pernicious invaders in western North America (Shafroth et al. 2005), they are also the subject of innumerable control projects within the Mojave (Neill 1983; Egan 1997; Barrows 1998) and far beyond (Taylor and McDaniel 1998; Hart et al. 2005). Understanding the ecological role of *Tamarix* ssp. is important to understanding effective management prescriptions.

Origins and Establishment of Tamarix. Species of *Tamarix,* which include large shrubs and small trees, was probably introduced from Eurasia early in the 1800s for erosion control, windbreaks, and other horticultural uses. In the Mojave Desert it was commonly planted along railroad lines for erosion control because not much else would grow under such harsh conditions, and it continues to be managed for this purpose in several areas, such as near Kelso, California. In its native range, the various *Tamarix* ssp. (at least 54 are known) are often a dominant element of the vegetation in low-lying basins and river corridors where heat, salt, and irregular water availability limit the number of other riparian plants (Baum 1978). They can use saline groundwater and excrete excess salt from foliar glands on the photosynthetic surface (leaves and green stems). The leaves are cedar-like bracts, and these two traits suggest the other common name for the genus, saltcedar.

By 1877, *Tamarix* was reported as naturalized in riparian areas of the southwestern United States, and now dominates riparian vegetation along many major rivers of the region (Horton 1977). Deciduous *Tamarix* ssp. occupy an estimated 1 to 1.6 million acres or more from northern Mexico to central Montana and from central Kansas to coastal California (Robinson 1965; Brotherson and Field 1987). Infestations may incur a cost of between $127 and $291 million per year, primarily due to water loss via evapotranspiration (Zavaleta 2000). Currently, *Tamarix* is the third most common riparian woody plant in western riparian areas (Friedman et al. 2005). The most common vegetation type displaced by *Tamarix* ssp. is the classic *P. fremontii-Salix* ssp. (cottonwood-willow) gallery forest of the desert Southwest (Everitt 1998, Webb et al. 2001). Other native community types in the region that have been degraded or displaced by *Tamarix* ssp. are *Proposopis glandulosa/Prosopis pubescens* (mesquite woodlands and bosques), *Pluchea sericea* (arrowweed) and *Atriplex* ssp. (saltbush) shrublands, *Acacia greggii* (catclaw) and *Psorothamnus spinosa* (smoketree) thorn-scrub, and even *Typha latifoli*a (cattail) and *Phragmites australis* (common reed) marsh and *W. filfera* oases (D'Antonio and Dudley 1997; DeLoach et al. 2000).

There are several species involved in this invasion, but the most common form present across the Mojave region is a five-petaled type variously called *T. ramosissima, T. pentandra, T. chinensis,* and other names. A recent genetic evaluation of invasive *Tamarix* ssp. concluded that the dominant form across the West, including in the Mojave Desert, is a variable hybrid formed from two Asian species, *T. ramosissima* and *T. chinensis* (Gaskin and Schaal 2002); I will refer to this form as *T. ramosissima.* A four-petaled species, *T. parviflora,* is more common in coastal drainages of central and northern California, but is also present in scattered locations in the Mojave Desert. Other *Tamarix* species likely occur in the area, but this has not been studied in detail. An evergreen form, *T. aphylla* (athel) is a well-known shade tree used throughout the Mojave and other desert regions.

Tamarix in the Mojave Desert. *Tamarix ramosissima* is common to abundant in most drainages in the Mojave, at least where it has not been targeted for removal. Large stands remain along the Mojave River in the Barstow area, in the Narrows near Victorville, California, and at the ecologically sensitive Camp Cady and Afton Canyon reaches lower in the watershed (Egan 1997). It is also invasive in the Amargosa watershed, particularly in locations such as the Oasis Valley near Beatty, Nevada, and the Amargosa River Natural Area. Ash Meadows National Wildlife Refuge is also infested with *T. ramosissima*, and control has been attempted over several years to protect threatened aquatic species (Deuser et al. 1998; Egan 2002; Kennedy 2002; Kennedy and Hobbie 2004). In the adjacent Carson Slough at least two listed plant species, *Nitrophila mohavensis* (Amargosa niterwort) and *Grindelia fraxino-pratensis* (Ash Meadows gumplant), are at risk from infestation (Soil Ecology and Research Group 2004).

Along the Virgin River, *T. ramosissima* is the dominant riparian species (fig. 6.1) and occupies over 40.5 km^2 from upstream of the town of Mesquite, Nevada, to Lake Mead, Arizona (Busch and Smith 1995). It is common but less devastating in other tributary streams, such as the Muddy River (Fleishmann et al. 2003) and the Bill Williams River. It and various hybrid forms are also abundant on the shores of Lake Mead and in much of the lower Colorado River, occasionally codominant with the invasive *Arundo*. Stands are present in Las Vegas Wash (now a perennial stream owing to urban effluent) and in numerous smaller drainages throughout the Mojave Desert, such as Darwin Canyon and Saline Valley in Death Valley National Park, California; Last Chance Canyon in Red Rocks State Park, California; and countless remote springs.

Tamarix parviflora has also been widely planted in the region, particularly along railroad corridors such as in the Cameron/Cache Creeks drainage northwest of the town of Mojave, California. This population has expanded downstream and is now found in scattered stands as far away as California City, California. It and *T. ramosissima* are also scattered throughout the China Lake/Rosamond (California) area and in seasonally moist sinks across to the Lancaster/Palmdale (California) area. Infestations are seen along Highway 395 in the Little Lakes area (again with *Arundo*) at the junction between the Great Basin and Mojave Deserts. *Tamarix parviflora* differs from *T. ramosissima* not only in flower morphology, but also in that it tends to spread its branches more broadly, and produces pink floral displays only in the spring prior to full development of the foliage.

Tamarix aphylla, on the other hand, has been considered noninvasive because it was thought not to produce viable seed, despite its invasiveness in Australia

Fig. 6.1. Photograph of the Virgin River in Clark County, Nevada. Although cottonwoods, willows, and other native plants are present in the system, the riparian vegetation is dominated almost entirely by *Tamarix ramosissima* (saltcedar).

(Griffin et al. 1989). However, this low-quality ornamental shade tree must now be considered a potential problem, because not only is it now germinating and establishing on the shores of Lake Mead and in some Sonoran Desert locations (Walker et al. 2006), it is also hybridizing with the deciduous *T. ramosissima* (Gaskin and Shafroth 2005). The evergreen *T. aphylla* does not tolerate winter freezing, unlike deciduous *Tamarix* ssp., but we recently found that the hybrid form survives freezing to regrow from the basal crown (T. L. Dudley *unpublished data*), and therefore has the potential to invade farther north.

Why are Species of Tamarix Successful Invaders?

The success of *Tamarix* ssp. is related to its ability to take advantage of unpredictable periods of favorability to colonize sites, and its ability to tolerate harsh conditions once established. A common trait of invasive species is the frequent and abundant production of propagules (Rejmanek and Richardson 1996), at which *Tamarix* ssp. excel. It can reach sexual maturity within a year of germinating, and a single tree is reported to annually produce upwards of a half million seeds (Robinson 1965; Young et al. 2004). Although reproduction of *T. ramosissima* is concentrated in late spring, it continues to flower for six months or more, which means that seed is potentially available season-long to take advantage of suitable germination and growth conditions (Carpenter 2003). This is in contrast to native *P. fremontii* and *Salix* ssp., which are more restricted in their seed production period, and thus at a competitive disadvantage in ecosystems where precipitation is irregularly available (Johnson 1992; Busch and Smith 1995; Rood et al. 2005). *Tamarix parviflora,* however, is generally limited to a similarly brief flowering period in early to late spring, depending on climatic conditions, which likely accounts for its much less invasive behavior compared with the five-petaled forms.

Colonization typically follows high runoff events that scour substrates and remove existing vegetation. The fine silts and sand left as flooding recedes retain moisture and provide ideal seedbeds (Robinson 1965; Brotherson and Field 1987). Establishment is most successful when water levels recede slowly (e.g., along the margins of reservoirs or regulated streams), allowing the relatively slow-growing seedlings to put down roots that follow the water table (Sher et al. 2002). Seeds are wind- and water-dispersed to these sites, and it is common to see extremely abundant *Tamarix* "lawns" with densities of 5,000 or more seedlings per square meter. Seedling mortality, however, is normally extremely high because desiccation and/or repeated flooding will eliminate young plants (Cleverly et al. 1997). Juveniles are poor competitors (Sher et al. 2002), so once a stand is established, repeat infestation may be less likely until major flood events again clear the area. For these reasons, many infestations are comprised of only one or a few year classes, representing periods of suitable conditions for establishment (Friedman et al. 1995). Fortunately, seeds remain viable for only a few days to a maximum of a few weeks, so there is no persistent seed bank produced (Robinson 1965; Young et al. 2004).

Invasion by *Tamarix* ssp. is often facilitated by river management that has altered natural hydrologic and geomorphic processes (Robinson 1965; Everitt 1980, 1998; Anderson 1998; Stromberg 2001; Shafroth et al. 2002b; Shafroth et al. 2005). Its greatest expansion occurred in the early- to mid-twentieth century following the widespread construction of dams and other water projects in the western United States. Water diversions and groundwater pumping reduce availability of moisture to native obligate phreatophytes, which require nearly constant contact with free water—unlike *Tamarix* ssp., facultative phreatophytes which continue metabolic activity below saturation levels and tolerate long periods of drought (Sala et al. 1996). Water diversions and a lack of flushing flows also result in salinization of riparian soils to levels that can inhibit native plants (Busch and Smith 1995; Shafroth et al. 1995b). Land uses, such as livestock grazing and groundwater pumping, further facilitate replacement of native plants by *Tamarix* ssp. However, *Tamarix* ssp. have invaded many relatively pristine sites as well, particularly in smaller watersheds where natural flooding is infrequent, occurring at intervals of perhaps three or more years (Dudley et al. 2000).

Although flood disturbance creates conditions conducive to germination of *Tamarix* seedlings, the young plants appear to be less tolerant of repeated flooding than native *P. fremontii* and *Salix* ssp. seedlings (D'Antonio et al. 2000; Lytle and Merritt 2004), and they benefit from the more stable hydrology resulting from water management (Robinson 1965). Thus, it is often the human alteration of natural disturbance regimes, rather than simply disturbance itself, that favors invasion of *Tamarix* ssp. and other invasive plants (D'Antonio et al. 2000). In the absence of actually removing dams and other river regulation structures (Shafroth et al. 2002b; Rood et al. 2005), managed flooding may be used to recreate ecologically important hydrological elements that promote native regeneration and inhibit invasive species (Collier et al. 1997; Poff et al. 1997; Molles et al. 1998; Stromberg 2001; Graf et al. 2002).

Once established, however, *Tamarix* ssp. are resistant to infrequent disturbances such as flooding and wildfire. Even if above-ground material is removed, the plants readily regrow from below-ground basal crowns (Busch and Smith 1992). *Tamarix* ssp. can thereby maintain populations and exclude other, more desirable vegetation. Unlike the young plants, mature *Tamarix* ssp. are highly competitive, and inhibit colonization and restrict the growth of other riparian plants. These relationships are illustrated along the Bill Williams River, where *T. ramossisima* is common but does not dominate the riparian vegetation, as it does elsewhere. This is apparently related to the frequency of natural flooding, as the Bill Williams is relatively unregulated, whereas the highly modified lower Colorado River is heavily infested with *Tamarix* ssp. (Busch and Smith 1995; Shafroth et al. 2005).

A variety of physiological adaptations allow *Tamarix* ssp. to dominate such altered ecosystems, including higher sodium tolerance, osmotic salt regulation, and greater water-use efficiency (under drought conditions) than native riparian taxa. These traits shift the competitive balance in altered ecosystems and favor *Tamarix* ssp. over the native *Salix* ssp., *P. fremontii,* and other woody natives, and may be facilitating local extinctions of the *P. fremontii-Salix* ssp. forests. This decline is illustrated in the photographic time-series produced by Webb et al. (2001)—historical comparisons that indicate reductions in the abundance and extent of large native trees along parts (but not all) of the Mojave River and increases in *Tamarix* ssp., which are associated with groundwater pumping and water regulation.

Environmental Impacts of Tamarix ssp. Invasion

The displacement of native vegetation by *Tamarix* ssp. has had numerous subsequent impacts throughout the Southwest, many of which are discussed in greater detail elsewhere (Graf 1978; Everitt 1980; Hunter et al. 1988; Busch and Smith 1995; Sala et al. 1996; Smith and Devitt 1996; Lovich and DeGouvenain 1998; Dudley et al. 2000; Shafroth et al. 2005). These include reductions in the diversity and abundance of riparian-dependent wildlife, increases in soil salinity, exacerbation of over-bank flooding and channel incision, increased frequency and magnitude of wildfire, and reduction of available forage and access to water for wildlife and livestock. Channel erosion was of particular interest in the Mojave

Desert in 2005 due to catastrophic bank failures during winter flooding of the Mojave and Virgin rivers, which was likely exacerbated by the abundant *Tamarix* ssp. present. Because *Tamarix* ssp. form a dense, monotypic stand, rather than the heterogeneous vegetation structure with single tree stems which is more typical of native assemblages, flood waters are forced upward and outward and erosion is exacerbated (Graf 1978).

Probably the most important impact from a management perspective is the loss of water resources owing to excessive evapotranspiration (Smith et al. 1998; Cleverly et al. 2002; Dahm et al. 2002; Shafroth et al. 2005). Studies at the Virgin River show that many riparian plants are similar in maximum water transpiration per unit leaf area, while the key to comparing plant water use is the combination of total leaf area and plant water stress (Sala et al. 1996; Smith and Devitt 1996; Devitt et al. 1998; Dahm et al. 2002). Under productive growth conditions, *Tamarix* ssp. have roughly double the leaf area of native *Salix* ssp., thus doubling the amount of water lost to the atmosphere through evapotranspiration. Hydrological conditions can alter leaf area; for example, channel down-cutting will lower water tables, resulting in a reduction of leaf area in *Tamarix* ssp., and thus a reduction in evapotranspiration per unit land area. When water stressed, evapotranspiration is reduced in all plants, but plants that are more tolerant of water stress (i.e., *Tamarix* ssp.) will be less strongly affected. Small-scale control programs may be ineffective in saving water, because water gained at the project site is likely to promote enhanced growth and subsequent evapotranspiration downstream, thus yielding no net water gain. Control programs should therefore be designed to maximize the extent of the controlled area and reductions in total leaf area in order to achieve measurable benefits.

Biodiversity Implications. While there is widespread agreement that *Tamarix* ssp. provide relatively poor wildlife habitat, control programs invariably include the potential for the loss of even this inferior habitat (Fleishmann et al. 2003; Shafroth et al. 2005). Particularly contentious is the issue of endangered species habitat for the federally listed Southwestern Willow Flycatcher (*Empidonax traillii* ssp. *extimus*). Recent surveys indicated that it is nesting, to a limited degree, in *Tamarix* ssp., primarily in Arizona, New Mexico, and some Mojave Desert habitats (Sogge et al. 2003). It has been noted that roughly 1% of the *Tamarix* habitat in the Southwest is used by this bird, and that rehabilitating native vegetation is the long-term key to improving its status (M.K. Sogge *personal communication*). However, this remains a major issue in *Tamarix* management for southwestern deserts and has delayed or halted many control programs.

Diversity and abundance of other riparian-associated birds declines with the influx of *Tamarix* ssp. (Anderson et al. 1977; Hunter et al. 1988; Rosenberg et al. 1991; Kelly and Finch 1999; Yard et al. 2004), including along the Mojave River

(Schroeder 1993). In some cases, species richness may not differ greatly from that of native habitat types, but whole groups may be rare or absent in *Tamarix* ssp., particularly "timber drillers" (woodpeckers), cavity dwellers, and habitat specialists like Summer Tanager (*Piranga rubra*) and Yellow-billed Cuckoo (*Coccyzus americanus*) (Hunter 1988; Ellis 1995). Cohan et al. (1979) also found that fruit- and grain-feeding birds, and cavity dwellers (woodpeckers, bluebirds, owls, wrens, and others) were absent, and insectivores reduced, in stands of *Tamarix* ssp. along the lower Colorado River.

Depauperate assemblages result primarily from fewer food resources on exotic plants than on native plants (Knutson et al. 2003), as well as from the reduced structural complexity of *Tamarix* stands (Shafroth et al. 2005). The relative abundance of native plants within the vegetative mix is critical to these relationships. Yard et al. (2004) indicate that at roughly 75% dominance by *Tamarix* ssp. there is an abrupt decline in utilization by neotropical migrants. In the absence of aggressive management, many areas not infested to this degree are likely to become so, particularly in ecosystems as dynamic as desert rivers.

Other wildlife species also decline when riparian systems are invaded by *Tamarix* ssp. Herpetofauna occur in lower diversity and abundance in *Tamarix*-dominated habitats across the Southwest (Jakle and Gatz 1985; Shafroth et al. 2005). Several reptiles of special interest in the Mojave Desert are threatened, including the western pond turtle (*Clemmys marmorata*) and the endangered desert slender salamander (*Batrachoseps aridus)* (Lovich and DeGouvenain 1998). Fish habitat is also degraded by the lowered water levels and habitat alteration that results from *Tamarix* invasion (Dudley et al. 2000). In particular, desert pupfish (*Cyprinodon macularis*) and Ash Meadows speckled dace (*Rhinichthys osculus* ssp. *nevadensis*) benefit from removal of *Tamarix* in the Death Valley area (Kennedy 2002), as might the Virgin River spinedace (*Lepidomeda mollispinis* ssp. *mollispinis*), woundfin (*Plagopterus argentissimus*), Virgin River chub (*Gila seminuda*), and the Moapa dace (*Moapa coriacea*) upstream of Lake Mead. Continued removal of *Tamarix* ssp. from the lower Mojave River could help provide future habitat by enhancing water available for three sensitive species now absent from the river: Mojave tui chub (*Gila bicolor* ssp. *mohavensis*), red-legged frogs (*Rana aurora*), and arroyo toad (*Bufo microscaphus*) (West Mojave Planning Team 1999). Mammals such as the peninsular bighorn sheep (*Ovis canadensis* ssp. *cremnobates)* and Mojave River vole (*Microtus californicus* ssp. *mohavensis*) could also benefit from improved water availability and vegetation complexity (Lovich and DeGouvenain 1998; Shafroth et al. 2005).

Control of *Tamarix* ssp.

Many efforts have been undertaken in recent decades throughout the Mojave region to control infestations of *Tamarix* ssp. because of its environmental and economic impacts (Neill 1983; Kunzmann et al. 1989; Rowlands 1989; Egan 1997; Barrows 1998; Anderson et al. 2003). Conventional controls for *Tamarix* ssp.—mechanical removal and chemical treatments—have benefited native species in numerous locations, and allowed a return of surface water flow in some cases (e.g., Inglis et al. 1996; Egan 1997; Barrows 1998; West Mojave Planning Team 1999). Such methods, however, are expensive, and can damage associated aquatic resources and nontarget native riparian vegetation. In addition, these labor- and technology-intensive approaches are difficult to apply in remote or inaccessible habitats, and treated sites exhibit a high frequency of reinfestation (Shafroth et al. 2005).

Biological Control of *Tamarix* ssp. Another weed-management tool is classical biological control, in which specialist herbivores that feed on the target plant in its native environment are imported to suppress pest weed infestations. Many years of overseas and quarantine testing led to the 1994 approval for the release of two insects for control of *Tamarix* ssp.—a leaf-feeding beetle from central Asia, *Diorhabda elongata* (Chrysomelidae), and a Middle Eeastern mealy bug, *Trabutina mannipara* (DeLoach et al. 1996, 2000). However, the biocontrol program was halted when the Southwestern Willow Flycatcher was found nesting in *Tamarix* ssp. Only after several years were researchers allowed to proceed with field testing, and then only in areas where the bird did not use this plant (Dudley et al. 2000).

Field releases of the saltcedar leaf-beetle in 2001 eventually lead to excellent establishment in several states, with widespread defoliation of *Tamarix* ssp., especially in northern Nevada (DeLoach et al. 2004). Water losses due to evapotranspiration were reduced 70%–80% the first year (X. Pattison *personal communication*). In addition, there were numerous other positive responses to the introduction of biocontrol. Researchers are now working with several ecotypes of the *D. elongata* species complex from different regions and latitudes to determine which forms will be most successful in different geographic regions and with different *Tamarix* genotypes in North America (DeLoach et al. 2004; Bean et al. 2007).

At this time, however, biocontrol testing in Arizona, southern Nevada, Southern California, and New Mexico west of the Pecos River are prohibited because of fears that control of *Tamarix* ssp. will make sites unsuitable for the Southwestern Willow Flycatcher. The concern is that *Tamarix* ssp. will be eliminated too rapidly to allow time for recovery of natives, and in many sites *Tamarix* ssp. may be the only woody plant that will tolerate the degraded conditions that now exist.

These concerns are important, but they may not apply to the current situation for several reasons (DeLoach et al. 2000; Dudley et al. 2000; Dudley and DeLoach 2004). First, while recovery of native vegetation may not occur in some damaged systems and weed control efforts should certainly take into account the restoration potential of such sites, the sites where the willow flycatcher currently nests in *Tamarix* ssp. contain a substantial amount of native vegetation (Sogge et al. 2003). Therefore, it is reasonable to expect that cover of native plants will increase following control. In addition, the bird is already extirpated from most of the Mojave Desert, particularly where *Tamarix* ssp. dominate.

Second, reevaluation of the data of McKernan and Braden (2001), used by wildlife agencies to justify protecting *Tamarix* ssp. similarly to "critical habitat," suggests that reproductive output is reduced in birds nesting in *Tamarix* ssp. compared with birds nesting in native habitat (0.89 fledglings per female per year in *Tamarix* ssp. vs. 1.89 in native stands). Not controlling *Tamarix* ssp. is likely to accelerate the decline of the bird, and continued invasion is likely to result in local extirpation (Dudley et al. 2000).

Third, *Tamarix* ssp. support fewer food resources (i.e., insects) than native vegetation, which is a probable cause for the poor reproductive performance of the Southwestern Willow Flycatcher, and the introduction of herbivores (biocontrol agents) will enhance resource availability. In fact, in northern Nevada, where the beetles have been introduced, we are seeing substantial increased utilization of *Tamarix* vegetation by several migratory bird species that are feeding on the beetles (X. Hitchcock *personal communication*).

Because *Tamarix* ssp. regrow quite rapidly following defoliation, and it takes three or more years to result in substantial mortality, there will be a consistent resource base for migratory birds for several years. In addition, the slow kill rate allows time for recovery of other vegetation if the site is suitable.

Finally, invasion of *Tamarix* ssp. is considered an important threat to a whole host of threatened or sensitive species in the region, which will continue to decline as regulatory agencies practice short-sighted, single-species management and delay the application of biocontrol or other forms of control. It is almost certain that willow flycatchers will benefit from control of *Tamarix* ssp. for the reasons stated above.

In 2003 the Saltcedar Biological Control Consortium proposed to conduct research trials of the saltcedar leaf beetle to control *T. ramosissima* on the Mojave River at Camp Cady, California, and with *T. parviflora* in the Tehachapi foothills northwest of the town of Mojave, California. These trials are to be conducted using mesh sleeves within secure cages, native vegetation is present at the sites, and the sites are isolated and far from known willow flycatcher nesting habitat. In spring of 2006 the U.S. Fish and Wildlife Service gave approval for cage testing at these two sites, and the project is now moving forward under a research permit from the U.S. Department of Agriculture Animal and Plant Health Inspection Service. No other sites in the region have been approved for research on biocontrol of *Tamarix* ssp., and political considerations make it difficult to discuss the possibility of expanding this approach in the region. In the meantime, *Tamarix* invasions continue (along with increased wildfires, lowering of water tables, loss of native riparian trees, etc.) as remaining wildlife species continue to decline in riparian ecosystems of the Southwest.

SUMMARY

Invasions of Mojave Desert riparian areas and wetlands by nonindigenous plants are proceeding, to the detriment of native plant assemblages and the wildlife that they support. Fewer invasive species occur here than in less arid regions, and many of these are relatively innocuous, weedy species that do not pose severe risks to our natural resources. However, a small number of highly aggressive species are present in the region and, except where dedicated control efforts have been put into place, they appear to be expanding. Human alterations of habitats and natural disturbance regimes have important roles in fostering these invasions in many systems, in addition to providing the vectors for their transport into the region. Desert riparian areas are relatively benign habitats compared with more arid upland environments, and allow those species that have established a foothold to expand. To protect and enhance rare and critical riparian ecosystems in the Mojave Desert, it is inadequate to simply leave them alone and "let nature take its course"—instead we must actively and often intensively manage these areas in order to return them closer to their original status. This may include managing the disturbance regimes themselves by providing simulated flood scour and reducing wildfire frequency, for example, in order to create ecosystem processes more conducive to the development of native vegetation. However, all actions need to be based on knowledge of the nonnative and native communities and the ecosystem processes that maintain them.

Detailed evaluations of the status and trends in plant invasions are generally lacking, although at least some short-term floral surveys have been attempted for certain locations. There is a clear need for a more comprehensive evaluation of the invaders present and new species entering the region, and also of the Mojave riparian flora as a whole, in order to evaluate where quality habitat exists and where restoration efforts should be focused. The lack of detail is even apparent in the work of the California Invasive Plant Council, which, in its plant assessments of invasive species of concern, does not differentiate between biogeographic regions when it discusses riparian woodland and riparian scrub communities (California Invasive Plant Council 2006). Substantial discussion has taken place concerning the need to establish an invasive plants information and advocacy group for the southwestern deserts, in order to differentiate between these systems and the California floristic province, which dominates in discussions of weed management. With Weed Management Areas and statewide groups developing species lists and management guidelines for the deserts, including those in Nevada and Arizona, it seems prudent to share resources and information in order to provide a comprehensive evaluation of weeds in our deserts, both riparian and upland.

ACKNOWLEDGMENTS

Participants in the Mojave Weed Management Area provided advice in putting together the invasive species list, particularly Valerie Page. Editorial assistance was provided by Matt L. Brooks, Jill S. Heaton, and two anonymous reviewers.

REFERENCES

Allan, J. D., and A. S. Flecker. 1993. Biodiversity conservation in running waters. *BioScience* **43**:32–43.

Anderson, B. 1998. The case for saltcedar. *Restoration and Management Notes* **16**:130–134.

Anderson, B. W., A. Higgins, and R. D. Ohmart. 1977. Avian use of saltcedar communities in the lower Colorado River Valley. Pages 128–136 *in* R. R. Johnson and D. A. Jones, editors. *Importance, preservation and management of riparian habitat.* General Technical Report No. RM-43. U.S. Department of Agriculture, Forest Service, Rocky Mountain Forest and Range Experiment Station, Fort Collins, Colorado.

Anderson, B. W., R. E. Russell, and R. D. Ohmart. 2003. *Revegetation: an account of two decades of experience in the arid southwest.* Avvar Books, Blythe, California, USA.

Baldwin, B. G., S. Boyd, B. J. Ertter, R. W. Patterson, T. J. Rosatti, and D. H. Wilken. 2002. *The Jepson Desert Manual.* University of California Press, Berkeley, California.

Barrows, C. 1998. The case for wholesale removal. *Restoration and Management* Notes **16**:135–139.

Baum, B. R. 1978. The genus *Tamarix.* Israel Academy of Sciences and Humanities, Jerusalem, Israel.

Bean, D. W., T. L. Dudley, and J. C. Keller. 2007. Seasonal timing of diapause induction limits the effective range of *Diorhabda elongata deserticola* (Coleoptera: Chrysomelidae) as a biological control agent for tamarisk (*Tamarix* ssp.). *Environmental Entomology* **36**:15–25.

Brookshire, D. S., M. McKee, and C. Schmidt. 1996. Endangered species in riparian systems of the American West. Pages 238–241 *in* D. W. Shaw and D. M. Finch, editors. *Desired future conditions for southwestern riparian ecosystems: bringing interests and concerns together.* General Technical Report No. RM-272. U.S. Department of Agriculture, Forest Service, Rocky Mountain Forest and Range Experiment Station, Fort Collins, Colorado.

Brotherson, J. D., and D. Field. 1987. *Tamarix:* impacts of a successful weed. *Rangelands* **9**:110–112.

Brussard, P. F., D. A. Charlet, and D. Dobkin. 1999. The Great Basin-Mojave Desert Region. Pages 505–542 *in* M. J. Mac, P. A. Opler, C. E. Puckett-Haecker, and P. D. Doran, editors. *The status and trends of the nation's biological resources.* Volume two. U.S. Department of the Interior, U.S. Geological Survey, Reston, Virginia.

Busch, D. E., and S. D. Smith. 1992. Fire in a riparian shrub community: postburn water relations in the *Tamarix-Salix* association along the lower Colorado River. Pages 52–55 *in* W. P. Clary, M. E. Durant, D. Bedunah, and C. L. Wambolt, compilers. *Proceedings of the Symposium on Ecology and Management of Riparian Shrub Communities, May 29–31, 1991, Sun Valley, Idaho.* General Technical Report No. INT-289. U.S. Department of Agriculture, Forest Service, Intermountain Research Station, Missoula, Montana.

Busch, D. E., and S. D. Smith. 1995. Mechanisms associated with decline of woody species in riparian ecosystems of the southwestern US. *Ecological Monographs* **65**:347–370.

California Department of Food and Agriculture. 1999. Map: cities with sites of *Limnobium* or *Salvinia* species. California Department of Food and Agriculture, Integrated Pest Control Branch, Noxious Weed Information Project, Sacramento, California.

California Invasive Plant Council. 2006. *California invasive plant inventory.* Cal-IPC Publication No. 2006–02. (Online). California Invasive Plant Council, Berkeley, California. http://www.cal-ipc.org/ip/inventory/pdf/Inventory2006.pdf.

Carpenter, A. T. 2003. Element stewardship abstract for *Tamarix ramosissima* Ledebour. Online. The Nature Conservancy Wildland Weed Management and Research Program, University of California, Davis, California, USA. http://tncweeds.ucdavis.edu/esadocs/documnts/tamaram.html. Accessed July 2006.

Clark County. 2005. Clark County multiple species habitat conservation plan and environmental impact statement for issuance of a permit to allow incidental take of 79 species in Clark County, Nevada. Online. Clark County Environmental Planning Division, Las Vegas, Nevada. http://www.co.clark.nv.us/air_quality/Environmental/MultipleSpecies/chap2.pdf.

Cleverly, J. R., C. N. Dahm, J. R. Thibault, D. J. Gilroy, and J. E. Allred Coonrod. 2002. Seasonal estimates of actual evapotranspiration from *Tamarix ramosissima* stands using three-dimensional eddy covariance. *Journal of Arid Environments* **52**:181–197.

Cleverly, J. R., S. D. Smith, A. Sala, and D. A. Devitt. 1997. Invasive capacity of *Tamarix ramosissima* in a Mojave Desert floodplain: the role of drought. *Oecologia* **111**:12–18.

Cohan, D. R., B. W. Anderson, and R. D. Ohmart. 1979. Avian population responses to salt cedar along the Lower Colorado River. Pages 371–382 *in* R. R. Johnson, and J. F. McCormick, editors. *Proceedings of the Symposium on Strategies for Protection and Management of Floodplain Wetlands and other Riparian Ecosystems, December 11–13, 1978, Callaway Gardens, Georgia.* General Technical Report No. WO-12. U.S. Department of Agriculture, Forest Service, Washington, D.C.

Collier, M. P., R. H. Webb, and E. D. Andrews. 1997. Experimental flooding in Grand Canyon. *Scientific American* **276**:82–89.

Cornett, J. W. 1987. The occurrence of the desert fan palm, *Washingtonia filifera,* in southern Nevada. *Desert Plants* **8**:169–171.

Dahm, C. N., J. R. Cleverly, J. E. Allred Coonrod, J. R. Thibault, D. E. McDonnell, and D. J. Gilroy. 2002. Evapotranspiration at the land/water interface in a semi-arid drainage basin. *Freshwater Biology* **47**:831–843.

D'Antonio, C. M., and T. L. Dudley. 1997. *Saltcedar as an invasive component of the riparian vegetation of Coyote Creek, Anza-Borrego State Park.* Final Report to California Department of Parks and Recreation, San Diego, California.

D'Antonio, C. M., T. L. Dudley, and M. M. Mack. 2000. Disturbance and biological invasions: direct effects and feedbacks. Pages 413–452 *in* L. R. Walker, editor. *Ecosystems of the world.* Volume sixteen. Elsevier Press, Amsterdam, The Netherlands.

D'Antonio, C. M., and P. M. Vitousek. 1992. Biological invasions by exotic grasses, the grass/fire cycle, and global change. *Annual Review of Ecology and Systematics* **23**:63–87.

DeLoach, C. J., R. Carruthers, T. Dudley, D. Eberts, D. Kazmer, A. Knutson, D. Bean, J. Knight, P. Lewis, J. Tracy, J. Herr, G. Abbot, S. Prestwich, G. Adams, I. Mityaev, R. Jashenko, B. Li, R. Sobhian, A. Kirk, T. Robbins, and E. Delfosse. 2004. First results for control of saltcedar (*Tamarix* ssp.) in the open field in the western United States. Pages 505–513 *in* W. Cullen, editor. *Proceedings of the XI International Symposium on the Biological Control of Weeds, April 27-May 2, 2003, Canberra, Australia.* Montana State University, Bozeman, Montana.

DeLoach, C. J., R. I. Carruthers, J. Lovich, T. L. Dudley, and S. D. Smith. 2000. Ecological interactions in the biological control of saltcedar (*Tamarix* ssp.) in the US: toward a new understanding. Pages 819–874 *in* N. R. Spencer, editor. *Proceedings of the X International Symposium on Biological Control of Weeds, July 4–14, 1999, Bozeman, Montana.* Montana State University, Bozeman, Montana, USA.

DeLoach, C. J., D. Gerling, L. Fornasari, R. Sobhian, S. Myartseva, I. D. Mityaev, Q. G. Lu, J. L. Tracy, R. Wang, J. F. Wang, A. Kirk, R. W. Pemberton, V. Chikatunov, R. V. Jashenko, J. E. Johnson, H. Zeng, S. L. Jiang, M. T. Liu, A. P. Liu, and J. Cisneroz. 1996. Biological control programme against saltcedar (*Tamarix* ssp.) in the USA: progress and problems. Pages 253–260 *in* V. C. Moran and J. H. Hoffman, editors. *Proceedings of the IX International Symposium on the Biological Control of Weeds, January 19–26, 1996, Stellenbosch, South Africa.* University of Capetown, Rondebosch, South Africa.

Deuser, C., J. Haley, and I. Torrence. 1998. *Lake Mead "SWAT" team attacks tamarisk.* Natural Resource Year in Review 1997. Publication No. D-1247. U.S. Department of the Interior, National Park Service, Natural Resource Program Center, Lakewood, Colorado.

Devitt, D. A., A. Sala, S. D. Smith, J. Cleverly, L. K. Shaulis, and R. Hammett. 1998. Bowen ratio estimates of evapotranspiration for *Tamarix ramosissima* stands on the Virgin River in southern Nevada. *Water Resources Research* **34**:2407–2414.

DiTomaso, J. M., and E. A. Healy. 2003. *Aquatic and riparian weeds of the West.* California Weed Science Society, Salinas, California, USA.

Dudley, T. 1998. Exotic plant invasions in California riparian areas and wetlands. *Fremontia* **26**:24–29.

Dudley, T. L. 2000. *Arundo donax.* Pages 53–58 *in* C. C. Bossard, J. M. Randall, and M. C. Hoshovsky, editors. *Invasive plants of California's wildlands.* University of California Press, Berkeley, California.

Dudley, T., and B. Collins. 1995. *Biological invasions in California wetlands: the impacts and control of non-indigenous species in natural areas.* Pacific Institute for Studies in Development, Environment, and Security, Oakland, California.

Dudley, T. L., and C. J. DeLoach. 2004. Saltcedar (*Tamarix* ssp.), endangered species and biological weed control—can they mix? *Weed Technology* **18**:1542–1551.

Dudley, T. L., C. J. DeLoach, J. Lovich, and R. I. Carruthers. 2000. Saltcedar invasion of western riparian areas: impacts and new prospects for control. Pages 345–381 *in* *Transactions of the Sixty-fifth North American Wildlife and Natural Resources Confer-*

ence, March 24–28, 2000, Rosemont, Illinois. Wildlife Management Institute, Washington D.C.

Egan, T. B. 1997. *Afton Canyon riparian restoration project: fourth year status report.* U.S. Department of the Interior, Bureau of Land Management, Barstow, California.

Egan, T. B. 2002. Riparian restoration in the Mojave Desert. Online. Death Valley forums. http://www.death-valley.us/article244.html.

Ellis, L. M. 1995. Bird use of saltcedar and cottonwood vegetation in the middle Rio Grande Valley of New Mexico. *Journal of Arid Environments* **30**:339–349.

Evens, J. M. 2001. Vegetation in watercourses of the eastern Mojave Desert. *Fremontia* **29**:26–35.

Everitt, B. L. 1980. Ecology of saltcedar—a plea for research. *Environmental Geology* **3**:77–84.

Everitt, B. L. 1998. Chronology of the spread of saltcedar in the central Rio Grande. *Wetlands* **18**:658–668.

Fleishmann, E., N. McDonal, R. MacNally, D. D. Murphy, J. Walters, and T. Floyd. 2003. Effects of floristics, physiognomy and non-native vegetation on riparian bird communities in a Mojave Desert watershed. *Journal of Animal Ecology* **72**:484–490.

Friedman, J. M., G. T. Auble, P. B. Shafroth, M. L. Scott, M. F. Merigliano, M. D. Freehling, and E. R. Griffin. 2005. Dominance of non-native riparian trees in western USA. *Biological Invasions* **7**:47–751.

Friedman, J. M., M. L. Scott, and W. M. Lewis Jr. 1995. Restoration of riparian forest using irrigation, artificial disturbance, and natural seedfall. *Environmental Management* **19**:547–557.

Gaskin, J. F., and B. A. Schaal. 2002. *Hybrid Tamarix widespread in US invasion and undetected in native Asian range.* Proceedings of the National Academy of Sciences **99**:11256–11259.

Gaskin, J. F., and P. Shafroth. 2005. Hybridization of *Tamarix ramosissima* and *T. chinensis* (saltcedars) with *T. aphylla* (athel) (family Tamaricaceae) in the southwestern USA determined from DNA sequence data. *Madroño* **52**:1–10.

Graf, W. L. 1978. Fluvial adjustments to the spread of saltcedar in the Colorado Plateau region. *Geological Society of America Bulletin* **89**:1491–1051.

Graf, W. L., J. Stromberg, and B. Valentine. 2002. Rivers, dams, and willow flycatchers: a summary of their science and policy connections. *Geomorphology* **47**:169–188.

Griffin, G. F., D. M. Stafford Smith, S. R. Morton, G. E. Allan, and K. A. Masters. 1989. Status and implications of the invasion of tamarisk (*Tamarix aphylla*) on the Finke River, Northern Territory, Australia. *Journal of Environmental Management* **29**:297–315.

Grime, J. P. 1979. *Plant strategies: vegetation processes and ecosystem properties.* Wiley, New York, New York.

Hart, C. R., L. D. White, A. McDonald, and Z. Sheng. 2005. Saltcedar control and water salvage on the Pecos River, Texas, 1999–2003. *Journal of Environmental Management* **75**:399–409.

Herrera, A. M., and T. L. Dudley. 2003. Reduction of riparian arthropod abundance and diversity as a consequence of giant reed (*Arundo donax*) invasion. *Biological Invasions* **5**:167–177.

Hood W. G., and R. J. Naiman. 2000. Vulnerability of riparian zones to invasion by exotic vascular plants. *Plant Ecology* **148**:105–114.

Horton, J. S. 1977. The development and perpetuation of the permanent saltcedar type in the phreatophyte zone of the Southwest. Pages 123–127 *in* R. R. Johnson and D. A. Jones, editors. *Proceedings of the Symposium on the Importance, Preservation, and Management of Riparian Habitat, July 9, 1977, Tucson, Arizona.* General Technical Report No. RM-43. U.S. Department of Agriculture, Forest Service, Intermountain Forest and Range Experiment Station, Fort Collins, Colorado.

Hrusa, F., B. Ertter, A. Sanders, G. Leppig, and E. Dean. 2002. Catalogue of non-native vascular plants occurring spontaneously in California beyond those addressed in the Jepson Manual—Part I. *Madroño* **49**:61–98.

Hunter, W. C., B. W. Anderson, and R. D. Ohmart. 1988. Use of exotic saltcedar (*Tamarix chinensis*) by birds in arid riparian systems. *Condor* **90**:113–123.

Inglis, R., C. Deuser, and J. Wagner. 1996. *The effects of saltcedar removal on diurnal ground water fluctuations.* Technical Report No. NPS/NRWRD/NRTR-96/93. US Department of the Interior, National Park Service, Denver, Colorado.

Jakle, M. D., and T. A. Gatz. 1985. Herpetofaunal use of four habitats of the Middle Gila River drainage, Arizona. Pages 355–363 *in* R. R. Johnson, C. D. Ziebell, D. R. Patton, P. F. Folliott, and R. H. Hamre, editors. *Riparian ecosystems and their management: reconciling conflicting uses.* First North American Riparian Conference, April 16–18, 1985, Tucson, Arizona. General Technical Report No. RM-120. U.S. Department of the Interior, Forest Service, Fort Collins, Colorado.

Johnson, W. C. 1992. Dams and riparian forests: case study from the upper Missouri River. *Rivers* **3**:229–242.

Kelly, M. 2004. *Washingtonia robusta* (Mexican fan palm). *California Native Plant Society, San Diego Chapter Newsletter* September 2004:7.

Kelly, J. F., and D. M. Finch. 1999. Use of saltcedar vegetation by landbirds migrating through the Bosque del Apache National Wildlife Refuge. Pages 222–230 *in* D. M. Finch, J. C. Whitney, J. F. Kelly, and S. R. Loftin, editors. *Rio Grande Ecosystems: Linking Land, Water, and People, June 2–5, 1999, Albuquerque, New Mexico.* Proceedings No. RMRS-P-7. U.S. Department of Agriculture, Forest Service, Rocky Mountain Research Station, Ogden, Utah.

Kennedy, T. A. 2002. The causes and consequences of plant invasions. Ph.D. dissertation, Department of Ecology, University of Minnesota, St. Paul, Minnesota.

Kennedy, T. A., and S. E. Hobbie. 2004. Saltcedar (*Tamarix ramosissima*) invasion alters organic matter dynamics in a desert stream. *Freshwater Biology* **49**:65–75.

Kisner, D. A., and B. E. Kus. 2003. The effect of *Arundo donax* on sensitive riparian bird species in southern California. *In Invasive Plants in Natural and Managed Systems: Linking Science and Manangement, 7th International Conference on the Ecology and Management of Alien Plant Invasions, November 3–7, 2003, Fort Lauderdale, Florida.* Allen Press, Inc., Lawrence, Kansas.

Knutson, A., M. Muegge, T. Robbins, and C. J. DeLoach. 2003. Insects associated with saltcedar, *Baccharis* and willow in West Texas and their relative value as food for insectivorous birds: preliminary results. Pages 41–50 *in Proceedings of the Saltcedar and Water Resources in the West Conference, 16–17 July 2003, San Angelo, Texas.* Texas A&M Agricultural Experiment Station and Cooperative Extension, San Angelo, Texas.

Kunzmann, M. R., R. R. Johnson, and P. B. Bennett, editors. 1989. *Saltcedar control in southwestern US.* Proceedings of the Saltcedar Conference, Special Report No. 9. University of Arizona, School of Renewable Natural Resources, Cooperative National Park Resources Study Unit, Tucson, Arizona.

LeBlanc, T. 2002. Camp Cady Wildlife Area: plants. Online. California Department of Fish and Game, Sacramento, California. http://www.dfg.ca.gov/lands/wa/region6/cady_pages/cady_plants.htm.

Lines, G. C. 1996. *Ground-water and surface-water relations along the Mojave River, southern California.* U.S. Geological Survey Water-Resources Investigations Report No. 95-4189. Denver Federal Center, Denver, Colorado.

Lovich, J. E., and R. C. DeGouvenain. 1998. Saltcedar invasion in desert wetlands of the southwestern United States: ecological and political implications. Pages 44–467 *in* S.K. Majumdar, E. W. Miller, and F. J. Brenner, editors. *Ecology of Wetlands and Associated Systems, September 3–4, 1997, Grand Junction, Colorado.* Pennsylvania Academy of Science, Easton, Pennsylvania.

Lytle, D. A., and D. M. Merritt. 2004. Hydrologic regimes and riparian forests: a structured population model for cottonwood. *Ecology* **85**:2493–2503.

Mack, R. N., D. Simberloff, W. M. Lonsdale, H. Evans, M. Clout, and F. A. Bazzaz. 2000. Biotic invasions: causes, epidemiology, global consequences, and control. *Ecological Applications* **10**:689–710.

MacKay, P. J. *Undated.* Mojave River plant list. Online. Lewis Center for Educational Research, Apple Valley, California. http://kuhn.lewiscenter.org/'Emhuffine/subprojects/Instructor/Mojave%20Desert%20Collection/plantMojaveDesert.html. Accessed July 2006.

McKernan, R. L., and G. Braden. 2001. Status, distribution, and habitat affinities of the southwestern willow flycatcher along the lower Colorado River, year 5–2000. San Bernardino County Museum, Redlands, California.

Mehrhoff, L. D. H. 1999. Introduction of nonindigenous aquatic vascular plants in southern New England: a historical perspective. *Biological Invasions* **1**:281–300.

Moapa Valley National Wildlife Refuge. Undated. Refuge habitat. Online. U.S. Department of Interior, Fish and Wildlife Service, Moapa Valley National Wildlife Refuge, Las Vegas, Nevada. http://desertcomplex.fws.gov/moapavalley/habitat.htm.

Molles Jr., M. C., C. S. Crawford, L. M. Ellis, H. M. Valett, and C. N. Dahm. 1998. Managed flooding for riparian ecosystem restoration. *BioScience* **48**:749–756.

Moore, L. M. 2003. Plant fact sheet: Siberian elm (*Ulmus pumila* L.). Online. U.S. Department of Agriculture, Natural Resources Conservation Service, National Plant Data Center, Baton Rouge, Louisiana. http://plants.nrcs.usda.gov/factsheet/pdf/fs_ulpu.pdf.

Neill, W. 1983. The tamarisk invasion of desert riparian areas. Education Bulletin No. 83-4. Educational Foundation of the Desert Protective Council, Spring Valley, California.

Olson, T. E., and F. L. Knopf. 1986. Naturalization of Russian-olive in the western United States. *Western Journal of Applied Forestry* **1**:65–69.

Page, V. 2004. Mojave Weed Management Area. Online. Mojave Weed Management Area, Victorville, California. http://www.mojavewma.org/index.php.

Pimentel, D., S. McNair, J. Janecka, J. Wightman, C. Simmonds, C. O'Connell, E. Wong, L. Russel, J. Zern, T. Aquino, and T. Tsomondo. 2001. Economic and environmental threats of alien plant, animal, and microbe invasions. *Agriculture, Ecosystems and Environment* **84**:1–20.

Pitcher, D. 1989. Element stewardship abstract for *Conium maculatum*. Online. The Nature Conservancy, Wildland Weed Management and Research Program, University of California, Davis, California. http://tncweeds.ucdavis.edu/esadocs/documnts/conimac.html.

Poff, N. L., J. D. Allan, M. B. Bain, J. R. Karr, K. L. Prestegaard, B. D. Richter, R. E. Sparks, and J. C. Stromberg. 1997. The natural flow regime: a paradigm for river conservation and restoration. *BioScience* **47**:769–784.

Pysek, P., and K. Prach. 1993. Plant invasions and the role of riparian habitats: a comparison of four species alien to central Europe. *Journal of Biogeography* **20**:413–420.

Rafferty, D. 2005. Nevada Noxious Weed List. Online. Nevada Department of Agriculture, Plant Industry Division, Nevada Weed Action Committee, Reno, Nevada. http://agri.nv.gov/nwac/PLANT_NoxWeedList.htm.

Ralphs, M. H. 2002. Ecological relationships between poisonous plants and rangeland condition. *Journal of Range Management* **55**:285–290.

Rejmanek, M., and J. M Randall. 1994. Invasive alien plants in California: 1993 summary and comparison with other areas in North America. *Madroño* **41**:161–177.

Rejmanek, M., and D. M. Richardson. 1996. What attributes make some plant species more invasive? *Ecology* **77**:1655–1661.

Richardson, D. M., P. Pysek, M. Rejmanek, M. G. Barbour, F. D. Panetta, and C. J. West. 2000. Naturalization and invasion of alien plants: concepts and definitions. *Diversity and Distribution* **6**:93–107.

Riparian Habitat Joint Venture. 2004. Riparian bird conservation plan: a strategy for reversing the decline of riparian associated birds in California. Online. Point Reyes Bird Observatory Conservation Science, California Partners in Flight, Petaluma, California. http://www.prbo.org/calpif/htmldocs/riparian.html.

Robinson, T. W. 1965. Introduction, spread and areal extent of saltcedar (*Tamarix*) in the western states. U.S. Geological Survey Professional Paper No. 491-A. U.S. Government Printing Office, Washington D.C.

Rood, S. B., G. M. Samuelson, J. H. Braatne, C. R. Gourley, F. M. R. Hughes, and J. M. Mahoney. 2005. Managing river flows to restore floodplain forests. *Frontiers in Ecology and the Environment* **3**:194–201.

Rosenberg, K. V., R. D. Ohmart, W. C. Hunter, and B. W. Anderson. 1991. *Birds of the lower Colorado River Valley.* University of Arizona Press, Tucson, Arizona.

Rowlands, P. G. 1989. History and treatment of the saltcedar problem in Death Valley National Monument. Pages 46–56 *in* M. R. Kunzmann, R. R. Johnson and P. B. Bennett, technical coordinators. *Saltcedar control in Southwestern United States. Proceedings, Saltcedar Conference.* U.S. Department of the Interior—National Park Service Cooperative National Park Resources Study Unit, Special Report No. 9. University of Arizona, School of Renewable Natural Resources, Tucson, Arizona.

Rowlands, P., H. Johnson, E. Ritter, and A. Endo. 1982. The Mojave Desert. Pages 103–162 *in* G. L. Bender, editor. *Reference handbook on the deserts of North America.* Greenwood Press, Westport, Connecticut.

Ryan, M. 2005. *Checklist of the non-native plants of southern Nevada.* Publication No. SP-05-05. University of Nevada Cooperative Extension, Reno, Nevada.

Sala, A., S. D. Smith, and D. A. Devitt. 1996. Water use by *Tamarix ramosissima* and associated phreatophytes in a Mojave Desert floodplain. *Ecological Applications* **6**:888–898.

Sanders, T. A., and W. D. Edge. 1998. Breeding bird community composition in relation to riparian vegetation structure in the western United States. *Journal of Wildlife Management* 62:461–473.

Schroeder, A. M. 1993. Effect of saltcedar removal on avian distributions at Camp Cady Wildlife Area in the California Mojave Desert. M.S. thesis, Environmental Studies, California State University, Fullerton, California.

Shafroth, P. B., G. T. Auble, and M. L. Scott. 1995a. Germination and establishment of the native plains cottonwood (*Populus deltoides* Marshall subsp. *monilifera*) and the exotic Russian-olive (*Elaeagnus angustifolia* L.). *Conservation Biology* **9**:1169–1175.

Shafroth, P. B., J. Cleverly, T. L. Dudley, J. Stuart, C. Van Riper, and E. P. Weeks. 2005. Saltcedar removal, water salvage and wildlife habitat restoration along rivers in the southwestern US. *Environmental Management* **35**:231–246.

Shafroth, P. B., J. M. Friedman, G. T. Auble, M. L. Scott, and J. H. Braatne. 2002b. Potential responses of riparian vegetation to dam removal. *BioScience* **52**:703–712.

Shafroth, P. B., J. M. Friedman, and L. S. Ischinger. 1995b. Effects of salinity on establishment of *Populus fremontii* (cottonwood) and *Tamarix ramosissima* (salt cedar) in southwestern United States. *Great Basin Naturalist* **55**:58–65.

Shafroth, P. B., J. C. Stromberg, and D. T. Patten. 2002a. Riparian vegetation response to altered disturbance and stress regimes. *Ecological Applications* **12**:107–123.

Sher, A. A., D. L. Marshall, and J. P. Taylor. 2002. Establishment patterns of native *Populus* and *Salix* in the presence of invasive nonnative *Tamarix. Ecological Applications* **12**:760–772.

Smith, S. D., and D. A. Devitt. 1996. Physiological ecology of saltcedar: why is it a successful invader? Online. *In* S. Stenquist editor. *Proceedings, Saltcedar Management and Riparian Restoration Workshop, September 17–18, 1996, Las Vegas, Nevada.* U.S. Department of Interior, Fish and Wildlife Service, Portland, Oregon. http://www.invasivespeciesinfo.gov/docs/news/workshopSep96/smith.html.

Smith, S. D., D. A. Devitt, A. Sala, J. R. Cleverly, and D. E. Busch. 1998. Water relations of riparian plants from warm desert regions. *Wetlands* **18**:687–696.

Sogge, M. K., B. E. Kus, S. J. Sferra, and M. J. Whitfield, editors. 2003. Ecology and conservation of the willow flycatcher: studies in Avian Biology No. 26. Cooper Ornithological Society, Camarillo, California.

Soil Ecology and Research Group. 2004. Demographics and ecology of the Amargosa niterwort (*Nitrophila mohavensis*) and Ash Meadows gumplant (*Grendelia fraxino-pratensis*) of the Carson Slough area. Online. San Diego State University, Soil Ecology and Research Group, San Diego, California. http://www.serg.sdsu.edu/SERG/restorationproj/mojave%20desert/deathvalleyfinal.htm.

Spencer, W. A. 1996. The desert fan palm: evidence supports relict status, a rebuttal to an article by curator of Palm Springs Desert Museum. Online. http://www.xeri.com/Moapa/relict.htm.

Stein, B. A., S. R. Flack, N. B. Benton, R. M. Chipley, K. Master, D. F. Schweitzer, and

G. Timmons. 1996. *America's least wanted: alien species invasions of US ecosystems.* The Nature Conservancy, Arlington, Virginia.

Stohlgren, T. J., K. A. Bull, Y. Otsuki, C. A. Villa, and M. Lee. 1998. Riparian zones as havens for exotic plant species in the central grasslands. *Plant Ecology* **138**:113–125.

Stromberg, J. C. 2001. Restoration of riparian vegetation in the south-western United States: importance of flow regimes and fluvial dynamism. *Journal of Arid Environments* **49**:17–34.

Szaro, R. C., and J. N. Rinne. 1988. Ecosystem approach to management of southwestern riparian communities. Pages 502–511 *in* R. E. McCabe, editor. *Transactions of the Fifty-third North American Wildlife and Natural Resource Conference, March 19 1988, Washington DC.* Wildlife Management Institute, Portland, Oregon.

Taylor, J. P., and K. C. McDaniel. 1998. Restoration of saltcedar (*Tamarix* sp.)-infested floodplains on the Bosque del Apache National Wildlife Refuge. *Weed Technology* **12**:345–352.

U.S. Fish and Wildlife Service. 2000. *Environmental assessment for control of the aquatic weed, giant salvinia (Salvinia molesta) on four national wildlife refuges on the lower Colorado River (Arizona/California).* U.S. Department of Interior, Fish and Wildlife Service, Division of Refuges, Albuquerque, New Mexico.

Van Devender, T. R., R. S. Felger, and A. Burquez M. 1997. Pages 10–15 *in* M. Kelly, E. Wagner, and P. Warner, editors. Exotic plants in the Sonoran Desert region, Arizona and Sonora. *Proceedings of the California Exotic Pest Plant Council Symposium, October 2–4, 1997, Concord, California.* Volume three. California Exotic Pest Plant Council, Berkeley, California.

Walker, L. R., P. Barnes, and E. Powell. 2006. *Tamarix aphylla:* a newly invasive tree in southern Nevada. *Western North America Naturalist* **66**:191–201.

Webb, R. H., K. H. Berry, and D. E. Boyer. 2001. *Changes in riparian vegetation in the Southwestern United States: historical changes along the Mojave River, California.* U.S. Geological Survey Open File Report No. 01–245. Online. Tucson, Arizona. http://pubs.usgs.gov/of/ofr01-245/index.html. Accessed May 2006.

West Mojave Planning Team. 1999. Current management situation of special status species in the West Mojave Planning Area. Online. U.S. Department of the Interior, Bureau of Land Management, U.S. Fish and Wildlife Service, and California Department of Fish and Game. http://www.blm.gov/ca/pdfs/cdd_pdfs/pfp890295548.pdf.

Wilcove, D. S., D. Rothstein, J. Dubow, A. Phillips, and E. Losos. 1998. Quantifying threats to imperiled species in the United States. *Bioscience* **48**:607–615.

Yard, H. K., C. van Riper III, B. T. Brown, and M. J. Kearsley. 2004. Diets of insectivorous birds along the Colorado River in Grand Canyon, Arizona. *Condor* **106**:106–115.

Young, J. A., C. D. Clements, and D. Harmon. 2004. Germination of seeds of *Tamarix ramosissima. Journal of Range Management* **57**:475–481.

Zavaleta, E. 2000. The economic value of controlling an invasive shrub. *Ambio* **29**: 462–467.

PART II

Road Effects on the Mojave Desert

ROBERT H. WEBB AND JILL S. HEATON

Many forces are responsible for ecosystem changes in the Mojave Desert, including those outside the sphere of human influence. The overriding threat to this unique desert, however, is people (Hughson *this volume*), and particularly their rapid influx, either to live, work, or recreate, and their increasing need for roads. With the expansion of large cities such as Los Angeles, Las Vegas, San Diego, and Phoenix, as well as their outlying bedroom communities, comes local impacts, which include increased recreational pressures, agriculture, road and trail proliferation, depredation of plants and wildlife, and the destruction of habitat (Lovich and Bainbridge 1999).

Desertwide issues that press for continued or increasing road access include resource extraction, increased military training, air pollution, recreation, and the expansion of transportation and utility corridors. Roads, or vehicle interactions with landscapes, cause a number of direct ecosystem effects that range from soil compaction (Lei *this volume*) to road kill (Brooks and Lair *this volume*). Indirect effects include dust transport, which leads to reduced air quality; increased emissions of greenhouse gases and elevated air temperature (Smith et al., *this volume*); species habitat fragmentation (Barrows and Allen *this volume*); and vectors for introducing nonnative flora (Brooks *this volume*) and fauna that may displace native species. This section of the book highlights several of the threats posed by roads to the Mojave Desert across multiple spatial and temporal scales, and emphasizes their synergistic, interactive, and cumulative effects of roads.

A local-scale threat to the ecosystem is caused by roads, both paved and dirt, which can be severe, particularly to wildlife populations. Lei (*this volume*) discusses soil compaction associated with the creation of new paths and roads, concluding that the difference between walking, biking, and vehicular impact is one of degree. Brooks and Lair (*this volume*) review the literature on the ecological impacts of roads and present a classification scheme that ranges from two-track routes to multilane freeways. These authors also explain how vehicular routes can affect ecosystem properties immediately within route corridors, along gradients extending outward from corridors, and as the cumulative result of multiple routes across the landscape. Vogel and Hughson (*this volume*) show the proliferation of roads in the Mojave National Preserve from the late nineteenth century through the establishment of the park unit in 1994. They show that the increase in roads has been extreme, so much so that the problems of accurately determining where the first roads were established is inconsequential.

ACKNOWLEDGMENTS

We thank Jeffrey Lovich for reviewing this introductory material.

REFERENCES

Lovich, J. E., and D. Bainbridge. 1999. Anthropogenic degradation of the southern California desert ecosystem and prospects for natural recovery and restoration. *Environmental Management* **24**:309–326.

SEVEN

Rates of Soil Compaction by Multiple Land Use Practices in Southern Nevada

SIMON A. LEI

Continued exposure of desert areas to human trampling and off-road vehicles has resulted in observable damage to certain soil physical characteristics (Stebbins 1974; Luckenbach 1975). Management of public lands in the western United States has traditionally been based on the concept of multiple uses (Vollmer et al. 1976). As a result, many desert areas are used for recreational activities and some of these areas are exposed to intensive use. If usage is sufficiently intense, lasting damage may be done to the desert within a few hours (Vollmer et al. 1976). Such destruction is particularly serious in desert areas because of the fragility of these ecosystems (Vollmer et al. 1976).

Anthropogenic recreational activities in the Mojave Desert of southern Nevada and California lead to varying amounts of soil compaction. Soil compaction is a widely recognized land-degradation problem (Lovich and Bainbridge 1999; Webb 2002). Soil compaction, by definition, is the decrease in pore volume within a soil mass, resulting in an increase in bulk density and a decrease in macropores (Johnson and Sallberg 1960; Webb 1983; Webb et al. 1986). From casual observations, rut and footprint depth is a good indicator of soil compaction.

Human and livestock trampling, as well as off-road motor vehicle traffic, can adversely affect soil properties by compacting the soil (Wilshire and Nakata 1976; Scholl 1986; Webb 2002; Brooks and Lair *this volume*). In the Mojave Desert, off-road motor vehicles can cause significant compaction with as few as 1 to 10 passes (Davidson and Fox 1974; Vollmer et al. 1976; Wilshire and Nakata 1976; Webb et al. 1986; Webb 1982, 1983, 2002; Lei 2004). Motorcycle-induced compaction was studied with various numbers of passes in Southern California (Davidson and Fox 1974; Wilshire and Nakata 1976; Webb 1982, 1983) and southern Nevada (Webb et al. 1986; Lei 2004). Motorcycle racing is a common sport, with large races accommodating up to 3,000 participants, and can cause significant compaction with 10 or fewer passes (Webb 1982; Webb et al. 1986). Weaver and Dale (1978) measured soil bulk density with increasing use by horses, hikers, and motorcycles in Montana, but did not quantitatively examine the type of impact or frequency of visits.

An increase in bulk density caused by compaction changes other soil physical properties, most notably the particle-size distribution, pore size, and strength characteristics (Webb 2002). Soil strength and infiltration rates are sensitive indications of soil compaction. Soil strength typically is measured with a penetrometer and is strongly dependent on soil-water content (Greacen 1960). High soil-surface moisture is thought to have a significant effect on soil compaction in northwestern New Mexico (Scholl 1986). At low water contents, however, pore-water pressures are high, thus increasing the resistance to applied pressure to the soil surface in Southern California (Webb 2002).

The amount of soil compaction is a function of disturbance type, visit frequency, and soil-water content at the time of compaction. An understanding of soil compaction under multiple human recreational uses is necessary in order to minimize adverse effects. This paper provides a review of the soil compaction process, with an emphasis on soils in *Coleogyne ramosissima* (blackbrush) shrublands of southern Nevada (Lei 2004; Lei *unpublished data*). I considered two aspects of the compaction problem in this ecosystem type. First, I studied any variation that might exist in soil physical attributes (soil compaction and soil bulk density) at particular visit frequencies for each of the four disturbance types (human foot, bicycle, motorcycle, and motor vehicle). Then, I investigated variation in these two soil physical attributes at various human trampling frequencies under two soil moisture levels.

METHODS

Study Sites

Field studies were conducted in *Coleogyne* shrublands during fall 2002 in Kyle Canyon and during summer 2003 in Red Rock Canyon National Conservation Area (RRCNCA). Both sites are located on the eastern slope of the Spring Mountains. *Coleogyne* occurs in dry, well-drained soils that are typically sandy, gravelly, and rocky. The presence of an indurated hardpan or bedrock in subsoil makes these soil types edaphically arid. Soils are calcareous, have poorly developed soil horizons, and are composed primarily of weathered granite and limestone bedrock (Soil Survey Staff 1985). Organic decomposition and soil formation are slow due to the arid nature of the region (Soil Survey Staff 1985).

Field Design and Sampling

In order to directly investigate recreation-induced compaction, four trails—foot, bicycle, motorcycle, and four-wheel motor vehicle—were created with 1, 10, 100, and 200 passes in Kyle Canyon (table 7.1). Prior to this study, these areas showed no clear evidence of compaction or other types of soil disturbance (Lei 2004). Each trail was 100 m in length, with 50 m between trails. A human hiking trail was trampled by a 78 kg person in hiking boots, with a pass defined as one walk down the lane at a normal gait (Lei 2004). The mountain bike, motorcycle, and four-wheeled vehicle (table 7.1), with a rider weight of 78 kg, were operated at a constant speed of 32 km hr^{-1} along predetermined linear paths to avoid sudden acceleration, braking, and turning effects. Soil measurements were taken at the predisturbance level in all four trails to ensure comparability (Lei 2004). Soil measurements were also made after 1, 10, 100, and 200 passes (post-disturbance level). Within each type of disturbance, 40 soil samples were collected for a total of 160 samples; samples were excavated 100 mm in diameter to depth of 100 mm. Intervals within each trail were randomly selected to avoid biased sampling (Lei 2004).

Similarly, to experimentally examine human-induced compaction in RRCNCA, 1, 10, 100, and 200 passes (total of 8 trails) were created. Each trail was 50 m in length, with 50 m separating each trail. Prior to disturbance, four trails were moistened by sprinkling with 10 mm of water to simulate rainfall and facilitate a comparison of impacts under both moist and dry conditions. This simulated rainfall reflected a seasonal difference in precipitation between moist spring and dry summer conditions. Again, the human hiking trail was trampled by a 78 kg person in hiking boots.

Within each trail, 60 soil samples were randomly selected and were excavated 100 mm in diameter to depth of 60 mm. Intervals within each trail were also randomly chosen to avoid biased sampling. Adjacent dry, undisturbed soils were sampled as a control.

The number of passes used to study soil compaction was not adjusted to reflect the fact that trail widths were larger than tire widths (Webb 1982). Despite the carefully controlled conditions of tire impact, some of the bicycle and motor

Table 7.1 Descriptions of boots, bicycle, motorcycle, and motor vehicle used under controlled traffic study in the Coleogyne ramosissima *(blackbrush) shrubland of Kyle Canyon, southern Nevada*

Disturbance type	Brand	Boot/tire width (cm>	Gross weight (kg)
Hiking boots	Outback footwear	9	0.9
Mountain bike	Roadmaster Mountain Sport	5	17.3
Motorcycle	Yamaha YZ 250	8	88.2
Motor vehicle	Chevy Tahoe	22	3,111.4

Lei 2004

vehicle passes were not directly down the center of the trail; ruts were approximately two to three times the width of the tire. Hence, soil measurements at any point on the trail were probably underestimates (Webb 1982; Lei 2004).

Soil compaction and bulk density were measured on dry soils. The moist soil trails were moistened and measured two weeks after disturbance. In dry soil trails, samples were collected immediately after disturbance. Soil compaction was measured in the field using a portable, 30° cone penetrometer (Carter 1967), which was inserted into the soil after removing stony surface pavements. The penetrometer readings were taken at the point where the cone base reached the soil surface (point depth = 38 mm).

Laboratory Measurements

Soil samples were measured for bulk density at both sites, and gravimetric moisture content at the RRCNCA site only. Soil bulk density was measured using a 58 mm diameter coring device designed to collect intact samples (Webb 2002). Soil samples were sieved through a 2 mm mesh to remove plant litter, roots, and rocks greater than 2 mm in diameter. All tests were performed on sieved substrates that were oven-dried at 65°C for 72 h until they reached a constant mass. The relatively low temperature was used to minimize "baking" of clay minerals and a loss of structural water from clay (Webb 2002). Soil bulk density was obtained by dividing mass by volume.

Statistical Analyses

For Kyle Canyon, a two-way ANOVA (Analytical Software 1994) was conducted on soil compaction and bulk density, with disturbance type (foot, bicycle, motorcycle, and vehicle) and visit frequency (1, 10, 100, and 200 passes) as main effects. For RRCNCA, a two-way ANOVA was also performed on these three soil physical attributes, with visit frequency (1, 10, 100 and 200 passes) and moisture level (moist and dry) as main effects. Statistical significance was tested at $P \leq 0.05$.

RESULTS

The trails were noticeable immediately after the impact. The soil surface was disrupted and no longer had a uniform gravel cover (Lei 2004). The depth of the ruts and footprints was a good indication of soil compaction. Tire imprints and resulting indentations on the soil surface were evident after 10 or fewer passes, while footprints became evident after 10 passes, with a greater indentation (30–40 mm) in moist soils compared with dry (control) soils.

In Kyle Canyon, significant differences in soil compaction and bulk density were detected between one pass and 100 passes in all trails ($P \leq 0.05$, tables 7.2 and 7.3). On average, the effects of 10 human footprints were equivalent to a single motorcycle or vehicle pass, and 100 footprints were equivalent to 10 vehicle passes.

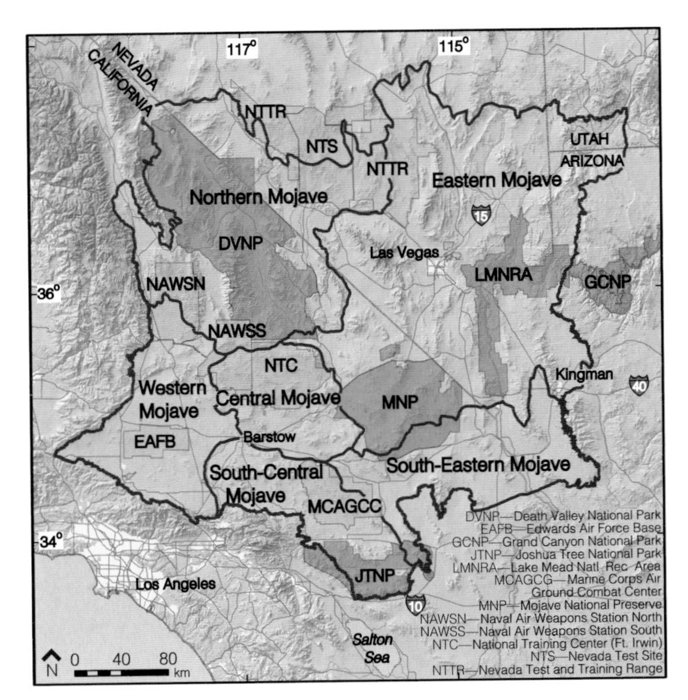

Plate 0.1. Map of the subregion boundaries within the Mojave Desert ecoregion with national park and military reserve boundaries

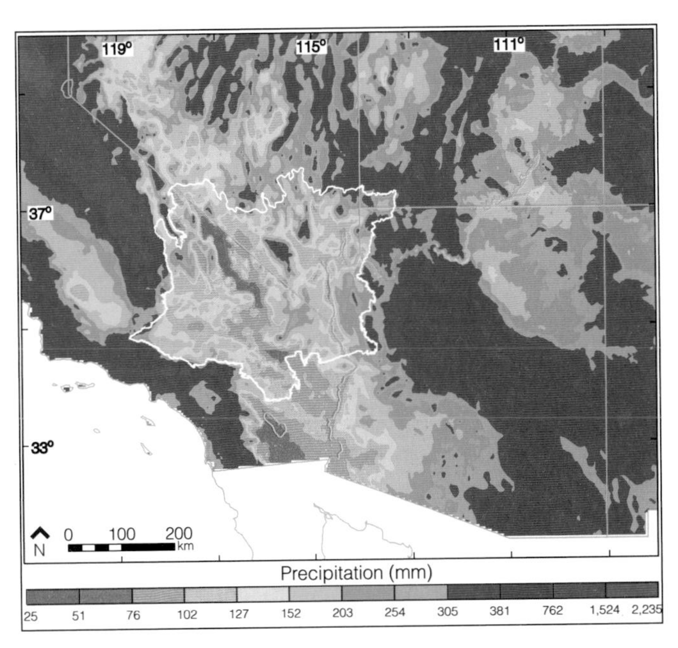

Plate 1.1. Map of mean annual precipitation, southwestern U.S., 1961–1990

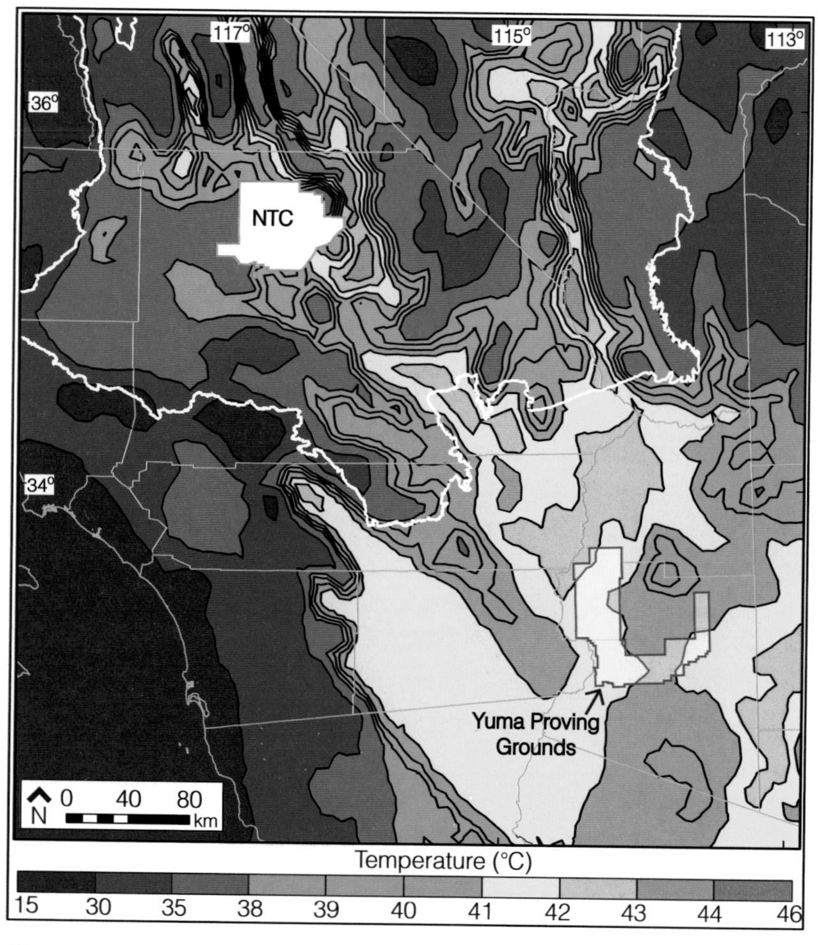

Plate 1.2. Contour map of maximum monthly mean temperature in July

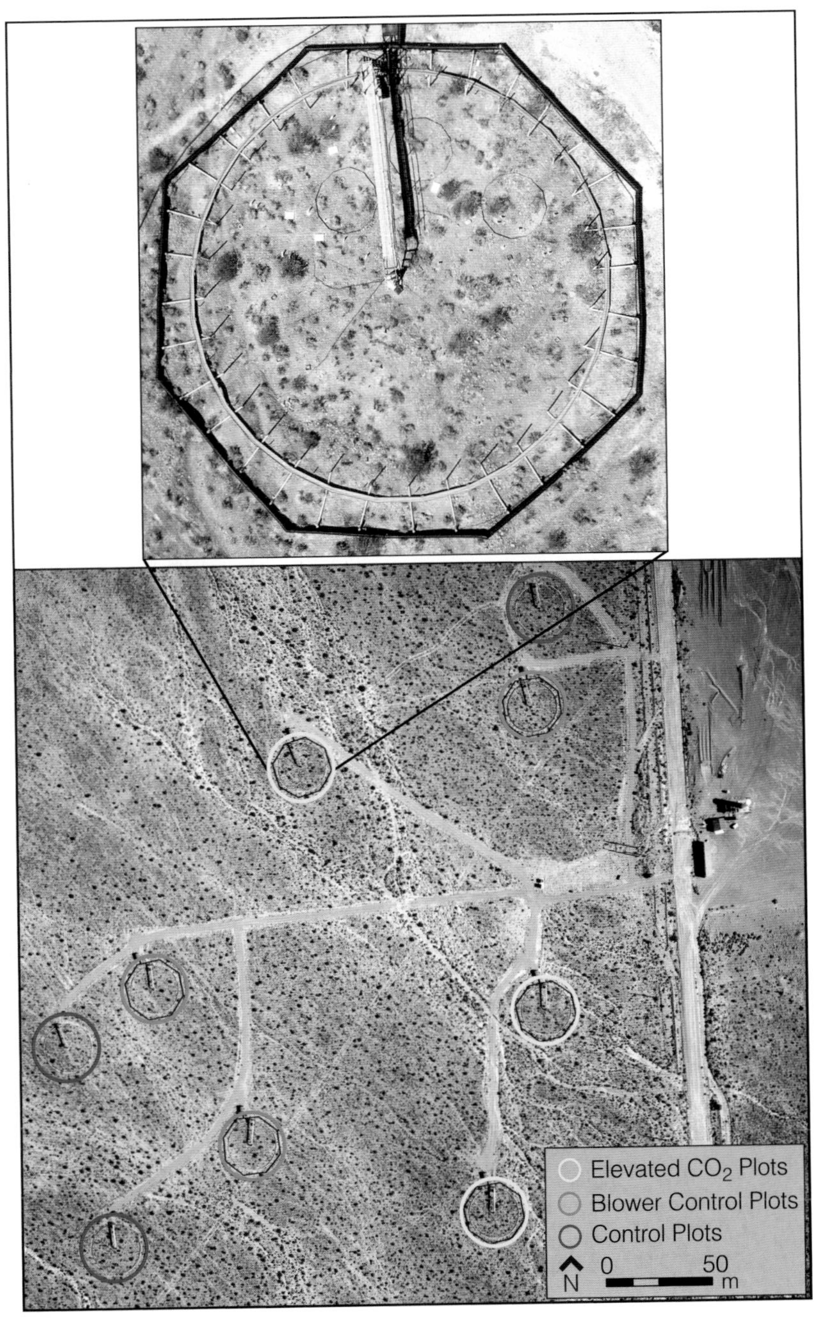

Plate 2.1. Aerial photographs of the Nevada Desert FACE Facility showing the distribution of experimental plots. A close-up aerial view (*top*) displays the infrastructure and platform used for nondestructive measurements.

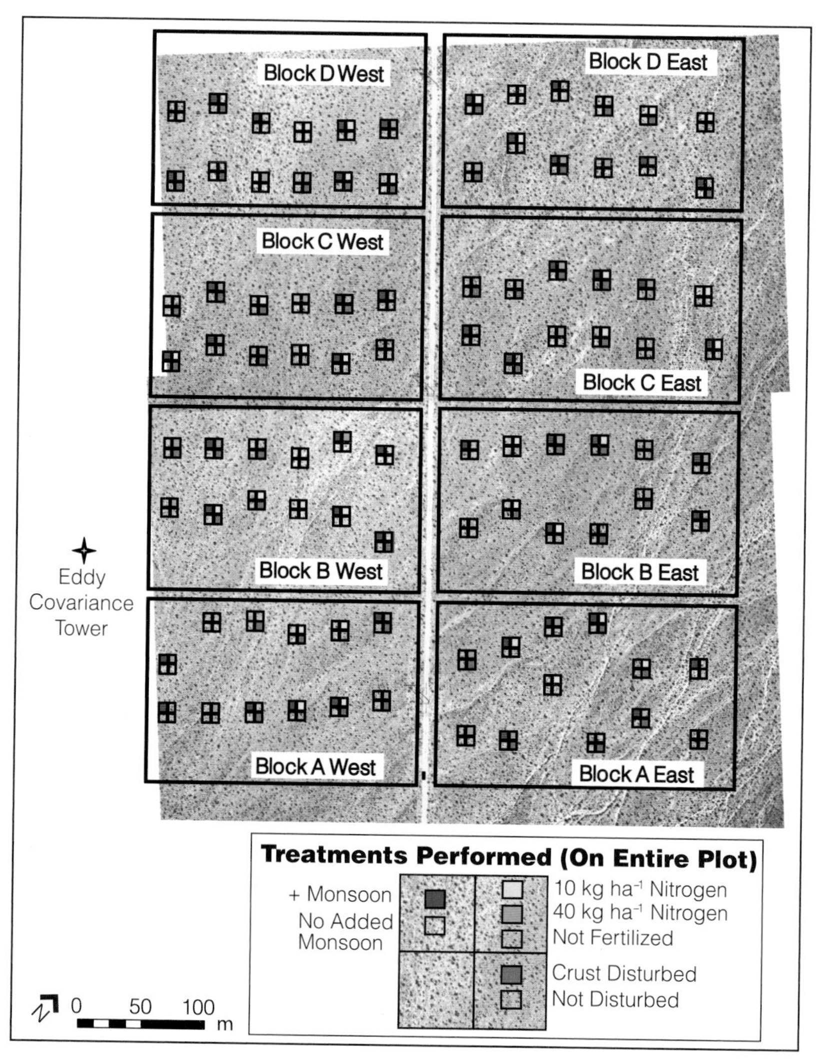

Plate 2.2. Aerial photograph of the Mojave Global Change Facility showing the distribution of experimental plots, the replicate blocks, and the treatment type on each plot.

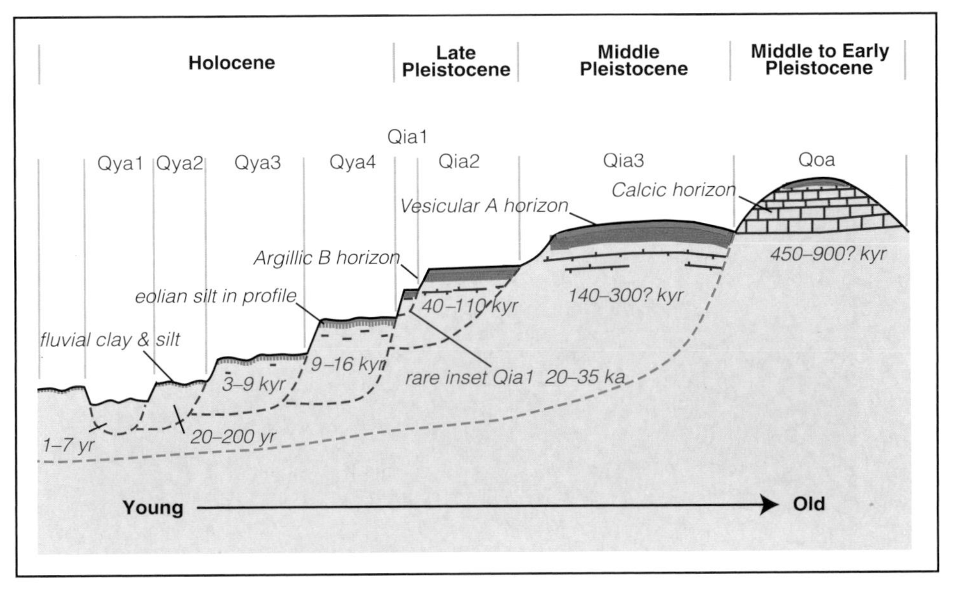

Plate 11.4. Sketch of inset piedmont deposit illustrating changes in surface morphology and soils with age

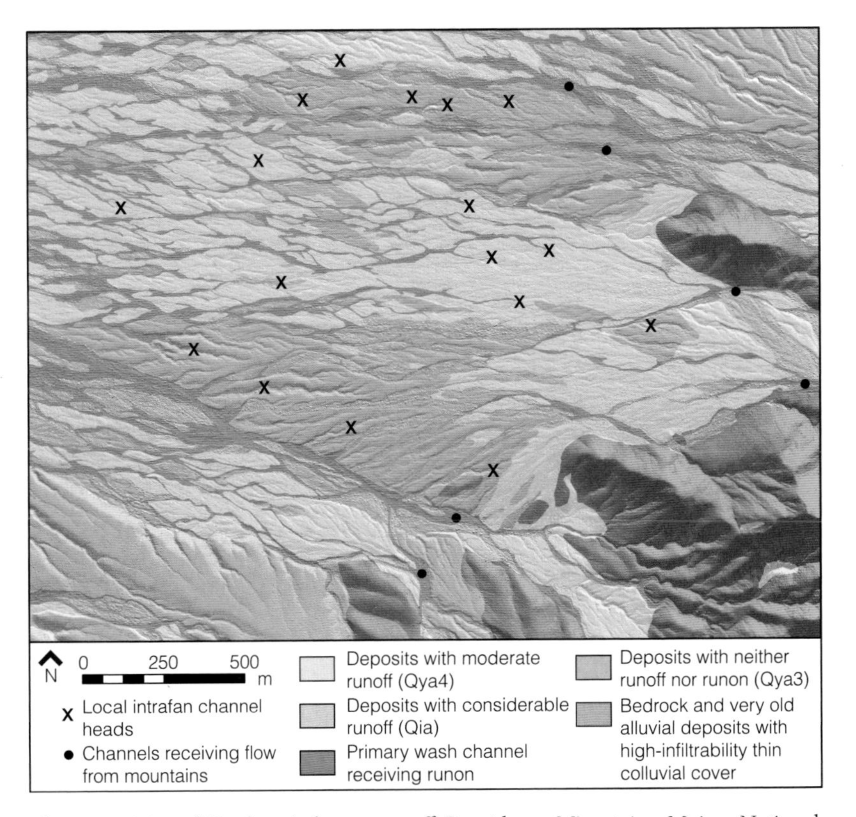

Plate 11.10. Map of Hayden piedmont runoff, Providence Mountains, Mojave National Preserve

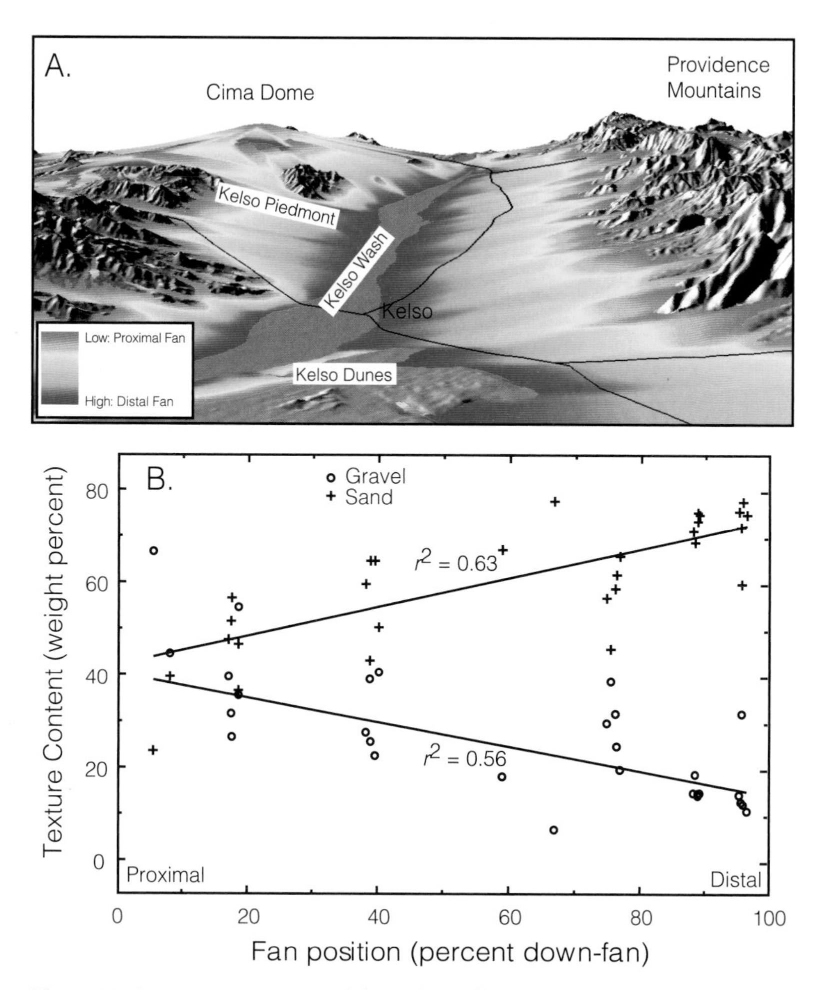

Plate 12.3. A, a perspective view of the Kelso Valley, Mojave National Preserve. *B*, graph of soil properties related to alluvial fan position.

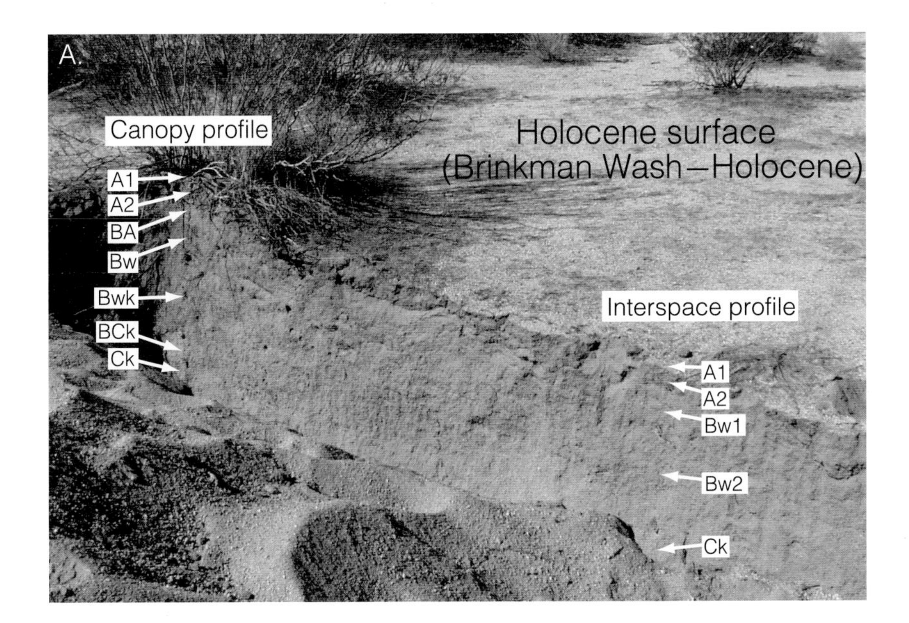

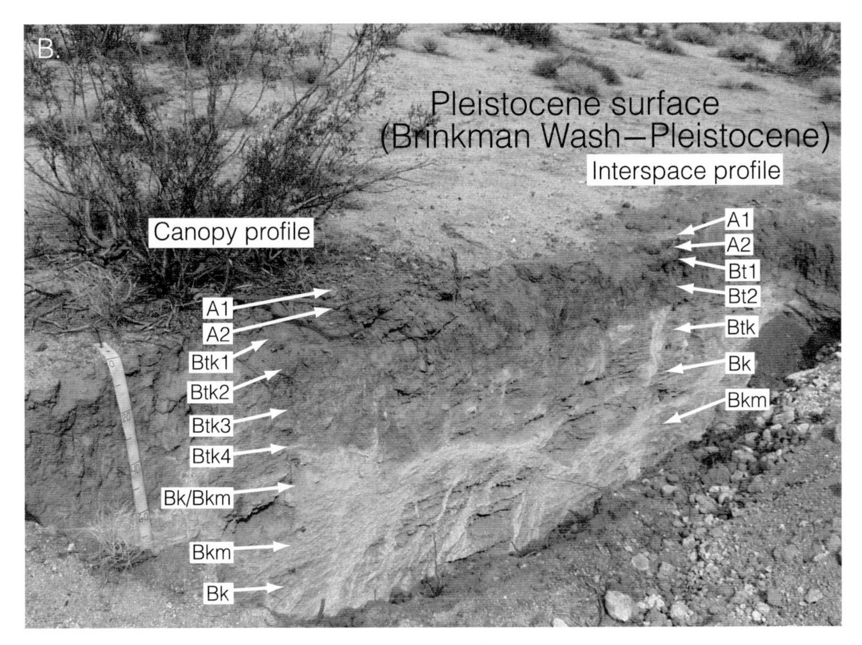

Plates 14.2 Photographs of soils in trenches at Brinkman Wash Study site, Fort Irwin

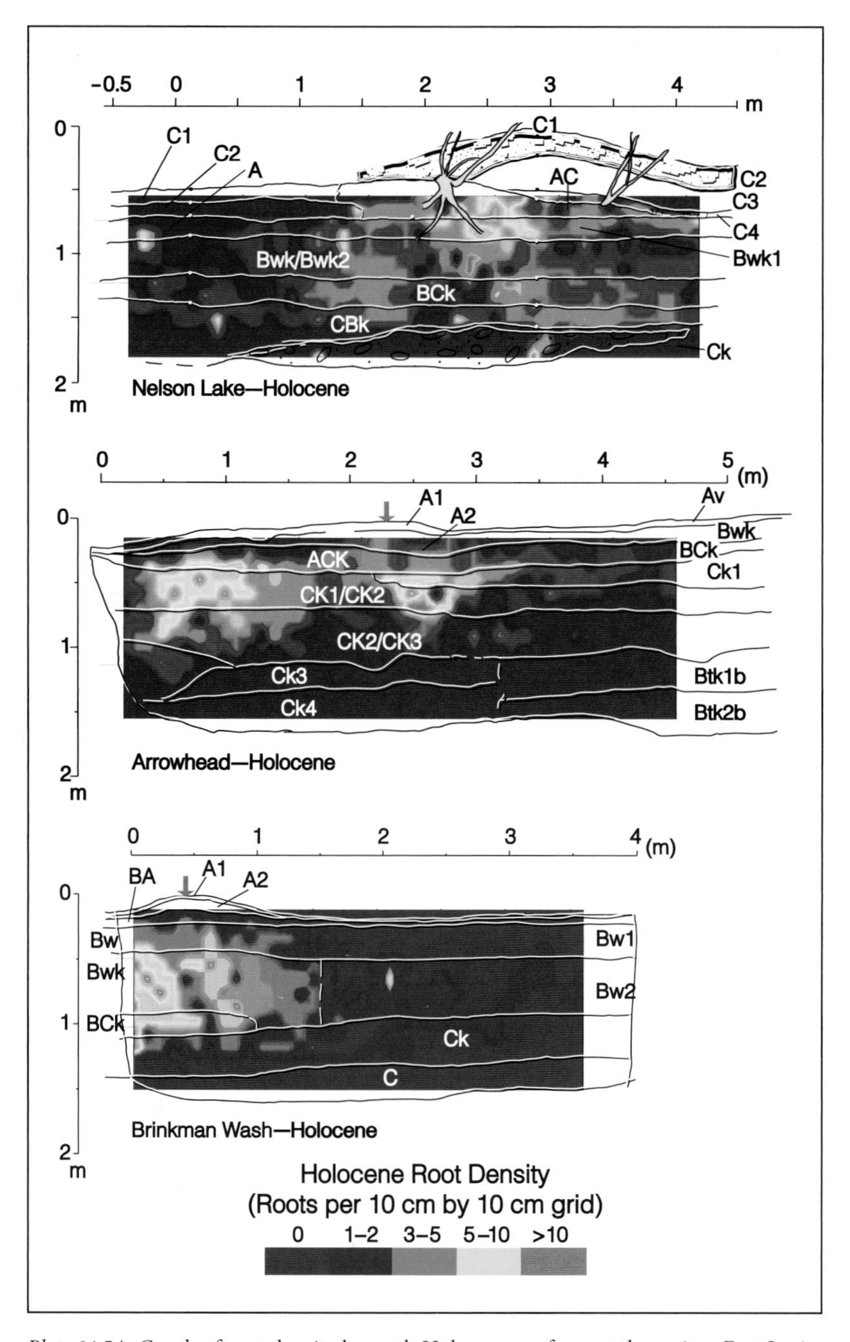

Plate 14.5A. Graph of root density beneath Holocene surfaces at three sites, Fort Irwin

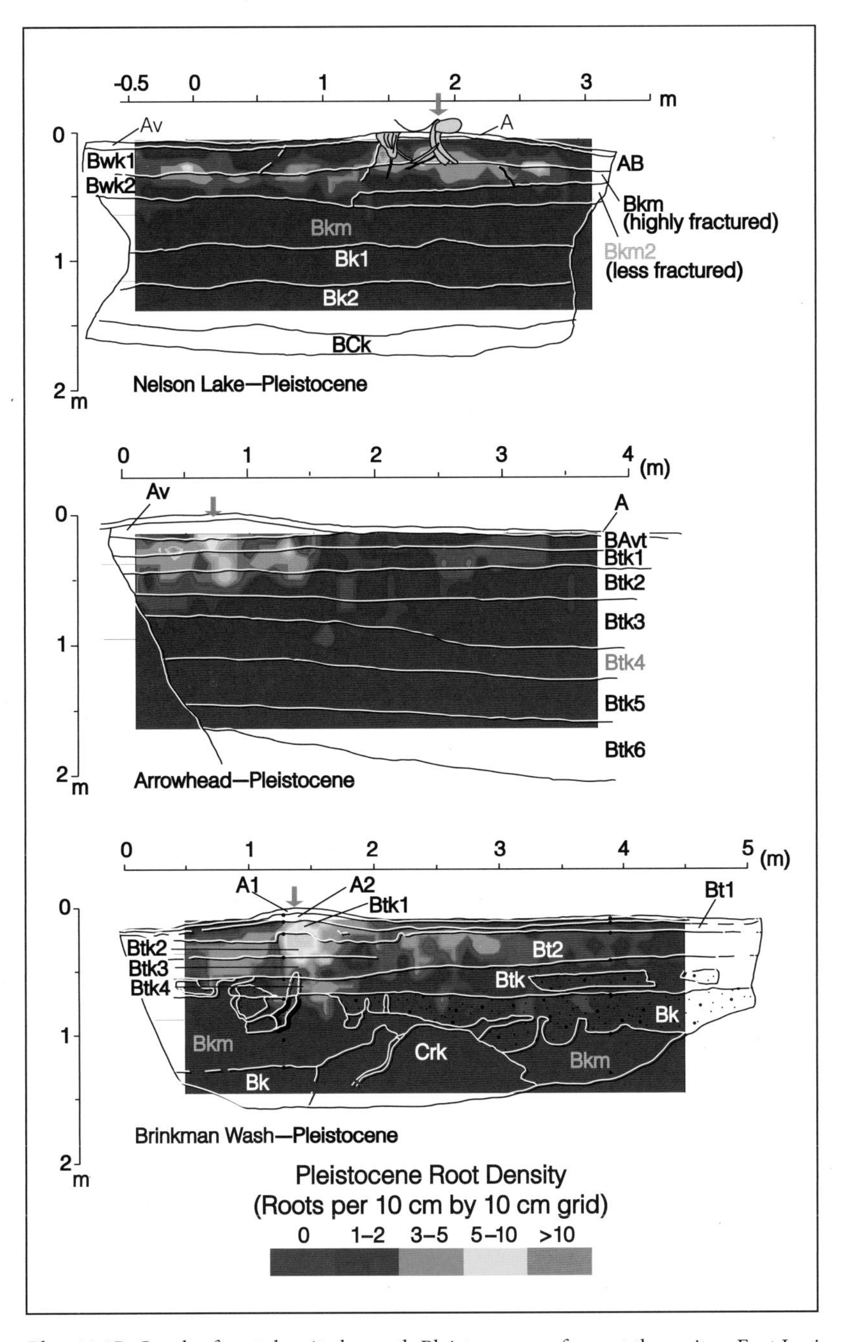

Plate 14.5B. Graph of root density beneath Pleistocene surfaces at three sites, Fort Irwin

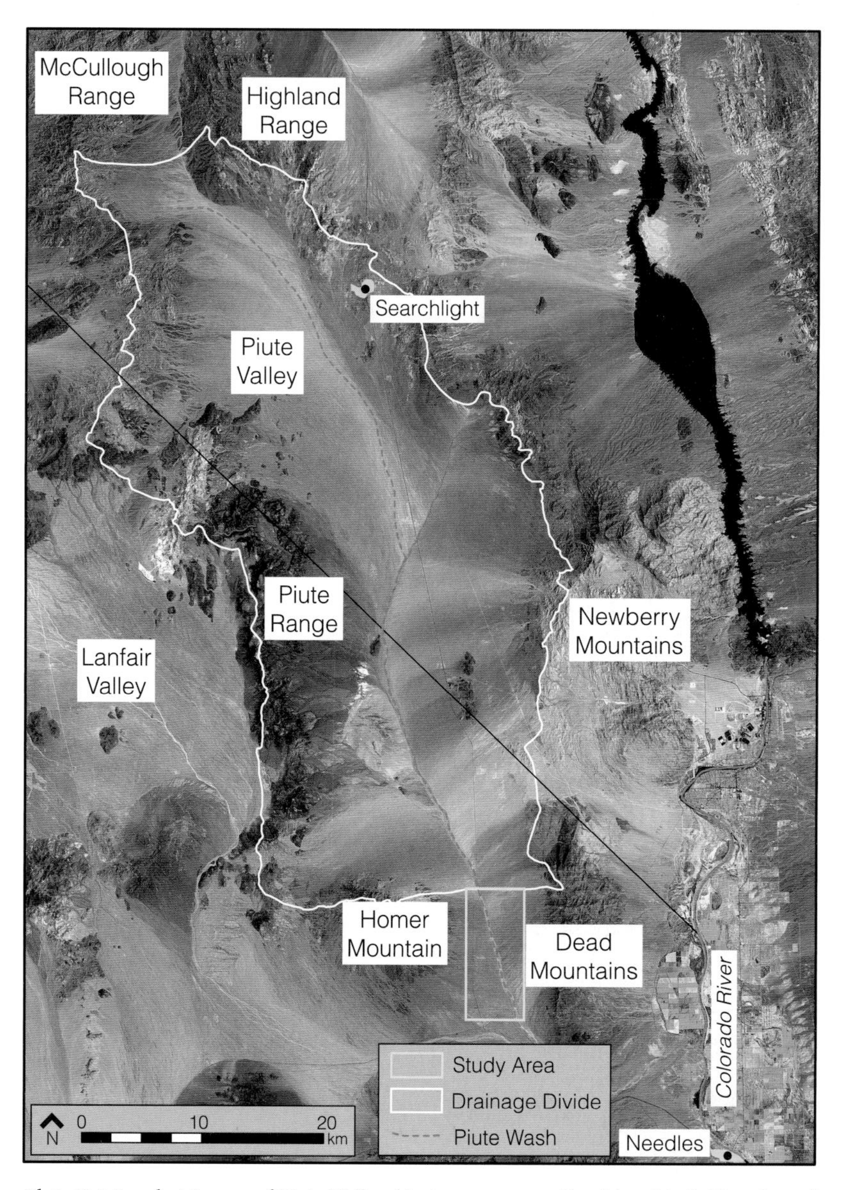

Plate 17.1. Landsat image of Piute Valley (drainage area outlined in white), Nevada and California

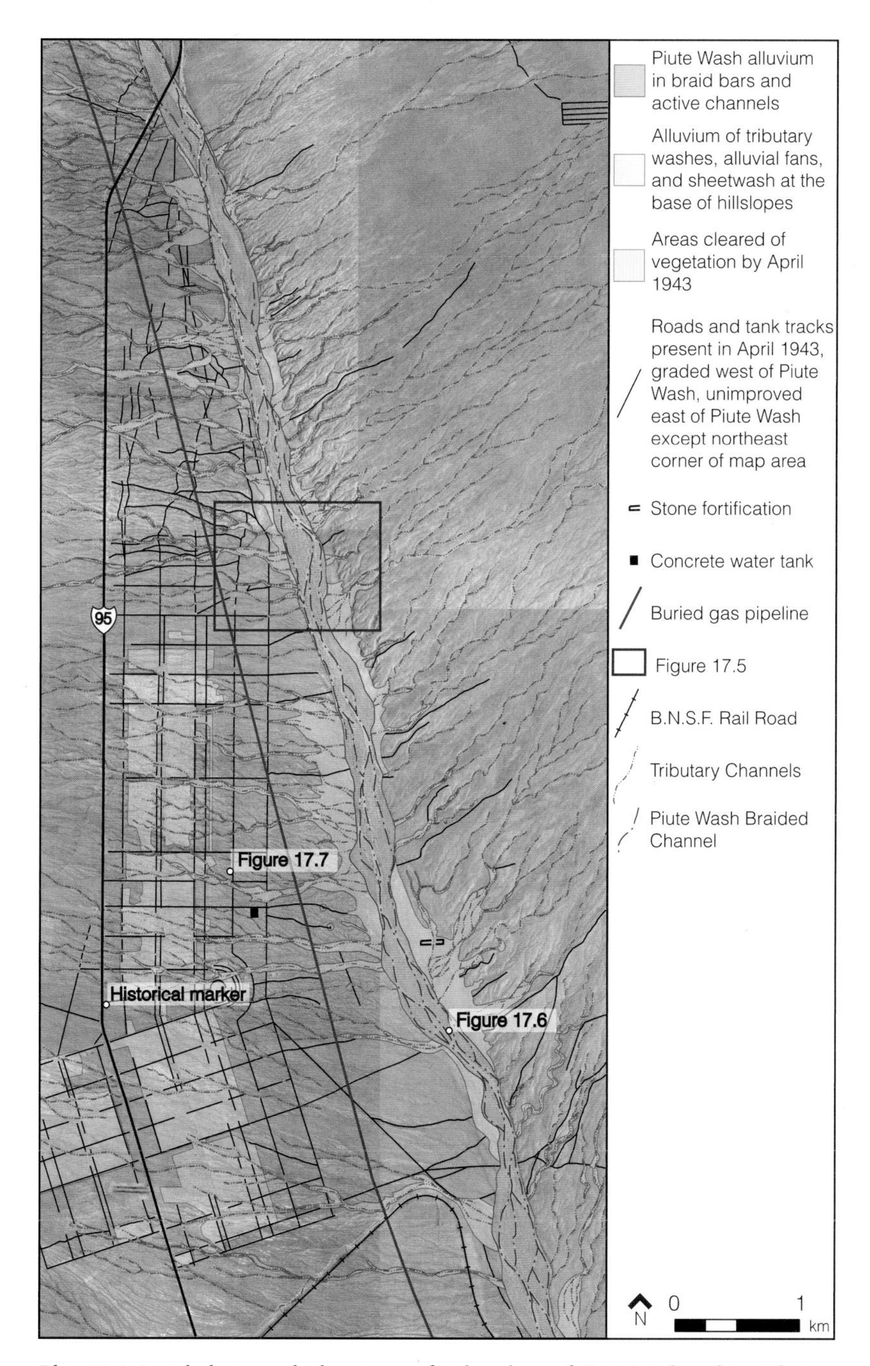

Plate 17.4. Aerial photograph showing surficial geology of Piute Wash and its tributaries and the layout of Camp Ibis

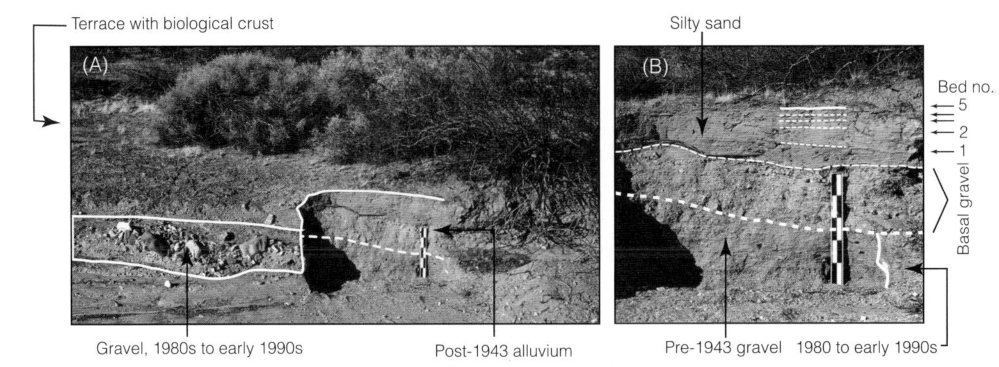

Plate 17.6. Photographs showing two exposures of braid-bar alluvium in Piute Wash at Camp Ibis

Plate 17.7. Photograph of alluvium deposited between 1943 and 2001 at Camp Ibis

Table 7.2 Changes in bulk density and compaction of the upper 10 cm of soil in response to various levels of disturbance impacts in the Coleogyne *shrubland of Kyle Canyon, southern Nevada*

Disturbance type	Number of passes	Bulk density (g cm^{-3})	Compaction (kg cm^{-2})
Hiking boots	Undisturbed	1.30 ± 0.07	6.0 ± 0.3
	1	1.33 ± 0.06	6.1 ± 0.2
	10	1.39 ± 0.06	6.4 ± 0.2
	100	1.51 ± 0.05	7.0 ± 0.2
	200	1.55 ± 0.05	7.3 ± 0.1
Mountain bike	Undisturbed	1.31 ± 0.07	6.0 ± 0.3
	1	1.35 ± 0.08	6.2 ± 0.2
	10	1.43 ± 0.06	6.6 ± 0.2
	100	1.55 ± 0.06	7.2 ± 0.1
	200	1.58 ± 0.06	7.3 ± 0.1
Motorcycle	Undisturbed	1.33 ± 0.07	6.1 ± 0.2
	1	1.38 ± 0.07	6.4 ± 0.3
	10	1.46 ± 0.06	6.8 ± 0.2
	100	1.57 ± 0.05	7.4 ± 0.1
	200	1.59 ± 0.06	7.4 ± 0.1
Motor vehicle	Undisturbed	1.30 ± 0.09	6.0 ± 0.3
	1	1.39 ± 0.07	6.4 ± 0.2
	10	1.51 ± 0.06	7.0 ± 0.2
	100	1.62 ± 0.06	7.5 ± 0.1
	200	1.63 ± 0.06	7.5 ± 0.1

Lei 2004
Mean ± SE; n = 40 per treatment, per attribute.

Table 7.3 Summary of two-way ANOVA showing effects of disturbance type, visit frequency, and their interactions on soil bulk density and compaction in Kyle Canyon, southern Nevada

	Bulk density		Compaction	
Source of variation	*F*	*P*	*F*	*P*
Disturbance type	14.41	< 0.0001	3.93	0.0244
Visit frequency	163.95	< 0.0001	117.30	< 0.0001
Disturbance by frequency	1.51	0.2036	1.61	0.1696

df = 3 for disturbance type, df = 4 for visit frequency, df = 12 for disturbance type by frequency, $P \leq 0.05$

Soils became significantly compacted after a single pass on the vehicle trail and after 10 passes on the foot, bicycle, and motorcycle trails. Conversely, differences between bicycle and motorcycle trails, and between 100 and 200 passes on all four trails, were not statistically significant ($P > 0.05$, tables 7.2 and 7.3). Similarly, the disturbance type and frequency interactions were not statistically significant for soil compaction and bulk density ($P > 0.05$, tables 7.2 and 7.3).

In RRCNCA, soil compaction and bulk density were significantly greater in human-trampled soils than in adjacent undisturbed soils ($P \leq 0.05$, tables 7.4 and 7.5). The moist soil had significantly greater compaction and bulk density than the dry soil ($P \leq 0.01$, tables 7.4 and 7.5). The average soil moisture in the upper 60 mm at the time of human trampling was 5.9% and 1.1% for the moist and dry soils, respectively.

Table 7.4 Changes in bulk density and compaction of the upper 6 cm on human hiking trails, with various levels of trampling under moist and dry soils in the Coleogyne *shrubland of Red Rock Canyon National Conservation Area, southern Nevada*

Soil moisture level	Number of passes	Bulk density (g cm^{-3})	Compaction (kg cm^{-2})
Control	Untrampled	1.28 ± 0.06	5.9 ± 0.3
Dry	1	1.31 ± 0.05	6.0 ± 0.4
	10	1.38 ± 0.06	6.4 ± 0.3
	100	1.54 ± 0.06	7.0 ± 0.2
	200	1.57 ± 0.05	7.3 ± 0.2
Moist	1	1.36 ± 0.05	6.2 ± 0.3
	10	1.49 ± 0.06	6.6 ± 0.3
	100	1.65 ± 0.07	7.3 ± 0.2
	200	1.68 ± 0.06	7.4 ± 0.1

Mean ± SE; $n = 60$ per treatment, per attribute

Table 7.5 Summary of two-way ANOVA showing effects of moisture level, human trampling frequency, and their interactions on soil bulk density and compaction in Red Rock Canyon National Conservation Area, southern Nevada

	Bulk density		Compaction	
Source of variation	*F*	*P*	*F*	*P*
Moisture level	39.89	0.0004	5.89	0.0456
Trampling frequency	67.48	<0.0001	49.25	<0.0001
Moisture by frequency	1.35	0.3329	0.25	0.862

df = 1 for moisture level, df = 3 for human trampling frequency and for moisture by trampling frequency combination, $P \leq 0.05$

Regardless of soil moisture conditions in RRCNCA, human disturbance at a slight level (10 passes) resulted in a significant increase in compaction and bulk density compared with undisturbed soils ($P \leq 0.05$, tables 7.4 and 7.5). A moderate or severe human disturbance (100 and 200 passes, respectively) caused additional significant changes in these two soil properties, especially in moist soils. However, between moderate and severe disturbance, soil compaction and bulk density were not significantly different, regardless of soil moisture level ($P > 0.05$, tables 7.4 and 7.5). Likewise, no significant interactions were detected between moisture level and disturbance frequency for soil compaction and bulk density ($P > 0.05$, tables 7.4 and 7.5).

DISCUSSION

Anthropogenic activities significantly increased soil compaction and soil bulk density under conditions of controlled anthropogenic impacts. As human trampling and off-road vehicle use increases, soil particles are pressed closer together and the bulk density increases (Davidson and Fox 1974). Soil compaction and bulk density can be expressed as logarithmic functions of the number of passes in all experimental trails; the greatest effects occurred during the first few passes, with changes per pass decreasing as the number of passes increased (Lei 2004). My results are similar to those of Webb (1982, 1983), who only studied the effects of motorcycle traffic. Moist desert soils had a significant increase in these two soil physical attributes compared with dry soils.

At both the Kyle Canyon and RRCNCA sites, the proportional extent of impact and the statistical variability of soil compaction and bulk density decreased with increasing numbers of passes. On all trails, significant differences were detected between 1 pass and 100 passes, but differences between 100 and 200 passes were not statistically significant (Lei 2004). Again, my results agree with those of Webb (1982), indicating that soil bulk density increases in the upper 60 mm of soil and increases in proportion to the number of passes. The greatest bulk density increases and related soil property changes per pass will occur during the first few passes (Webb 1982). The amount of soil compaction is a function of disturbance type, visit frequency, and moisture level, when examining these three factors independently. However, the disturbance type and visit frequency interaction, as well as the visit frequency and moisture level interaction, were not statistically significant for soil compaction and bulk density. Additional passes caused no significant further compression because the soils were already severely compacted. Thus, the impacts of hiking and biking slowly increased over time relative to the impacts of motor vehicle traffic (Lei 2004). In RRCNCA, I observed that water falling on severely compacted, moist soils was trapped at the surface, subject to evaporation rather than infiltration. The increase in bulk density of compacted soils probably decreases the amount of water soils can hold and the rate at which water can flow through to subsoils. This study is in agreement with previous studies, indicating that moist soils are often much more susceptible to compaction than dry soils (Lull 1959; Edmund 1962), especially at the surface where a thin, impermeable layer may form (Warren et al. 1986). If the amount of water applied to the moist trails had been increased, even greater disparity may have developed between moist and dry soils in terms of compaction and associated physical properties at the time of trampling (Warren et al. 1986).

Scholl (1986) also proposed that high soil-surface moisture, especially following snow melt, is a major seasonal influence on soil compaction in the semiarid lands of northwestern New Mexico. Spring trampling had a tendency toward greater compaction than fall trampling in loamy sands and sandy loams (Scholl 1986). Such poorly sorted soils are typical of the Mojave Desert (Webb 1982, 1983, 2002).

Soil texture is a major factor determining the magnitude of density increases under applied loads (Webb 1982, 1983). Sandy loam and loamy sand are highly susceptible to compaction from human trampling and off-road motor vehicles (Webb 1983). Soils in this arid region often have abundant rocks in the profile, leading to very poorly sorted soils, and usually have a gravel, cobble, or stone cover. These soils are present on bajadas and alluvial fans, which occupy a great portion of the Mojave Desert (Webb 1983). Nevertheless, mixtures with equal-size particles, such as dune sand and clay, will not compact as much as mixtures of different-size particles (Webb 1982, 1983). Soil compaction from anthropogenic activities will be minimal whenever dry dune sand or clay is used (Webb 1982). In deserts, playas or sand dunes are best for recreational use when the objective is to minimize compaction, although biological systems may be seriously disrupted by such use (Webb 1983).

ECOLOGICAL IMPLICATIONS

Although Kyle Canyon and RRCNCA are open to the public year-round, there are considerably more visitors during spring and summer. Typically, Mojave Desert soils are considered dry most of the year, with wetting occurring infrequently and only to shallow depths, although soils may remain somewhat moist during winter and spring months (Webb 1983). When compared with the relatively dry summer and fall seasons, winter and spring are periods in which soil is highly susceptible to compaction, especially during and shortly after rainfall and snow melt.

Over time, recreational activities can lead to increased soil compaction and bulk density. Ecologists and park rangers nationwide must educate visitors that human trampling on moist soils during spring seasons can cause significantly greater compaction than trampling on dry soils during summer and fall seasons. As expected, humans and motor vehicles go off-trail periodically (Lei 2003), dispersing compaction effects, particularly when soils are moist. Visitors must stay within established trails at all times when viewing nature reserves. Although anthropogenic degradation of the southern Nevada mid-elevation desert ecosystem is recognized, the public continues to use Kyle Canyon, RRCNCA, and other areas of the Mojave Desert for year-round recreational activities (Lei 2004).

ACKNOWLEDGMENTS

The author gratefully acknowledges Steven Lei, David Valenzuela, and Shevaun Valenzuela for valuable field and laboratory assistance. Steven Lei assisted with statistical analyses. Critical review provided by Robert Webb, Jill Heaton, and Durant McArthur greatly improved this chapter. Departments of Biology at the College of Southern Nevada (CSN) and the Nevada State College (NSC) provided logistical support.

REFERENCES

Analytical Software 1994. Statistix 4.1, an interactive statistical program for microcomputers. Analytical Software, St. Paul, Minnesota.

Carter, L. M. 1967. Portable recording penetrometer measures soil strength profiles. *Agricultural Engineering* **48**:348–349.

Davidson, E., and M. Fox. 1974. Effects of off-road motorcycle activity on Mojave Desert vegetation and soil. *Madroño* **22**:381–390.

Edmund, D. B. 1962. Effects of treading pasture in summer under different soil moisture levels. *New Zealand Journal of Agricultural Resources* **5**:389–395.

Greacen, E. L. 1960. Water content and soil strength. *Journal of Soil Science* **11**:313–333.

Johnson, A. W., and J. R. Sallberg. 1960. Factors that influence field compaction of soils. Highway Research Board Bulletin No. 272. Department of Transportation, Washington, D.C.

Lei, S. A. 2003. Soil compaction and moisture status from large mammal trampling in *Coleogyne*(blackbrush) shrublands of southern Nevada. Southern California Academy of Sciences Bulletin **102**:119–129.

Lei, S. A. 2004. Soil compaction from human trampling, biking and off-road motor vehicle activity in a blackbrush (*Coleogyne ramosissima*) shrubland. *Western North American Naturalist* **64**:125–129.

Lovich, J. E., and D. Bainbridge. 1999. Anthropogenic degradation of the Southern California desert ecosystem and prospects for natural recovery and restoration. *Environmental Management* **24**:309–326.

Luckenbach, R. A. 1975. What the ORVs are doing to the desert. *Fremontia* **2**:3–11.

Lull, H. W. 1959. *Soil compaction on forest and rangelands.* Miscellaneous Publication No. 768. U.S. Department of Agriculture, Forest Service, Washington, D.C.

Scholl, D. G. 1986. Soil compaction from cattle trampling on a semiarid watershed in northwest New Mexico. *New Mexico Journal of Science* **29**:105–112.

Soil Survey Staff. 1985. *Soil survey of Las Vegas Valley area, Nevada: part of Clark County.* Soil Conservation Service, U.S. Department of Agriculture, Washington, D.C.

Stebbins, R. C. 1974. Off-road vehicles and the fragile desert. *American Biology Teacher* **36**:203–208.

Vollmer, A. T., B. G. Maza, P. A. Medica, F. B. Turner, and S. A. Bamberg. 1976. The impact of off-road vehicles on a desert ecosystem. *Environmental Management* **1**:1–13.

Warren, S. D., T. L. Thurow, W. H. Blackburn, and N. E. Garza. 1986. The influence of livestock trampling under intensive rotation grazing on soil hydrologic characteristics. *Journal of Range Management* **39**:491–495.

Weaver, T., and D. Dale. 1978. Trampling effects of hikers, motorcycles, and horses in meadows and forests. *Journal of Applied Ecology* **15**:451–457.

Webb, R. H. 1982. Off-road motorcycle effects on a desert soil. *Environmental Conservation* **9**:197–208.

Webb, R. H. 1983. Compaction of desert soils by off-road vehicles. Pages 51–79 *in* R. H. Webb and H. G. Wilshire, editors. *Environmental effects of off-road vehicles.* Springer-Verlag, New York, New York.

Webb. R. H. 2002. Recovery of severely compacted soils in the Mojave Desert, California, USA. *Arid Land Research and Management* **16**:291–305.

Webb, R. H., J. W. Steiger, and H. G. Wilshire. 1986. Recovery of compacted soils in Mojave Desert ghost towns. *Soil Science Society of America Journal* **50**:1341–1344.

Wilshire, H. G., and J. K. Nakata. 1976. Off-road vehicle effects on California's Mojave Desert. *California Geology* **29**:123–132.

EIGHT

Ecological Effects of Vehicular Routes in a Desert Ecosystem

MATTHEW L. BROOKS AND BRIDGET M. LAIR

Several recent papers have summarized the ecological effects of roads (Forman 1995; Forman and Alexander 1998; Spellerberg and Morrison 1998; Spellerberg 1998, 2002; Forman et al. 2003), including a series of papers in the journal *Conservation Biology* in 2000. These reviews concur that construction of roads, the presence of roads in the landscape, and the vehicles that travel upon roads can have a wide range of ecological effects, from changes in the physical and chemical properties of ecosystems to alterations in the population and community structure of organisms. These papers, while important sources of information on road effects, understate or do not fully address some very important issues, which limits their applicability to specific situations, such as the development of local land management plans. In addition, there are currently no reviews of road effects in the Mojave Desert, an important region where roads have a large ecological impact.

One issue that has not been specifically addressed is an integrated understanding of the ecological effects due to both roads, generally defined as routes which have been intentionally created to accommodate vehicular travel, and off-highway vehicle (OHV) trails, defined as routes created through the process of repeated off-road vehicular travel. Previous road ecology reviews, by definition, focus on the effects of roads and typically do not address the wide range of potential effects associated with OHVs. The effects of OHVs have been studied much less worldwide, although some of the most significant work has been done in the Mojave Desert (Webb and Wilshire 1978, 1983; Rowlands 1980). Planning decisions for land management often focus on the nexus between roads and OHV recreation because they are inextricably interrelated in rural landscapes. This focus is emphasized in desert regions where sparse vegetation and minimal fencing facilitate relatively easy access by OHVs to the landscape, much more so than in less arid ecosystems where greater amounts of vegetation may hinder OHV travel.

Repeated OHV trail use often leads to the creation of routes that typically are not included in road databases. In fact, the 6.3 million km of public roads reported in the United States (National Research Council 1997) may be a significant underestimation due to unrecognized roads created by OHVs (Forman et al. 2003). These types of roads comprise the majority of the 608% increase in the total length of roads between 1885 and 2000 in the Mojave National Preserve in the Eastern Mojave Desert (Vogel and Hughson *this volume*), and the 1,180% increase in the total length of dirt roads and OHV trails between 1965 and 2001 at the Dove Springs Open Area in the Western Mojave Desert (Matchett et al. 2004). Thus, there is a need for more inclusive evaluations of the relative effects of both roads and OHV trails, which we refer to collectively as vehicular routes.

Another poorly understood topic is the relative effects of different types of vehicular routes. Much of what is known in the general literature is derived from studies of paved roads, whereas public land managers in the Mojave Desert primarily manage dirt roads and OHV trails. The characteristics of various types of vehicular routes can vary widely in ways that likely lead to varied ecological effects.

The ecological effects of vehicular routes can also vary among spatial scales, and land managers need to understand these relationships to reliably link their land management actions, which generally occur at local scales, to their land management objectives and goals, which are typically defined at landscape scales. Unfortunately, when managers turn to the technical literature upon which to base their management decisions, they often cannot find studies linking local actions to landscape effects. Scientific studies typically take place at only one spatial scale. On the rare occasion when scientists evaluate both local and landscape processes, the links between scales are often only vaguely described. A conceptual framework is needed to compare and contrast the potential ecological effects of different types of vehicular routes at different spatial scales. Managers can then use this framework to more accurately infer the potential effects of their management actions from the results of past studies.

By presenting the full range of possible vehicular route effects, past reviews typically lack the details necessary to evaluate specific effects within a given ecosystem. For example, habitat fragmentation is often cited as an ecological effect of vehicular routes (Barrows and Allen *this volume*), but this effect may be more pronounced where routes create major structural gaps in forests than where the contrast between vehicular route corridors and the surrounding landscape is more subtle, such as in shrublands. To be most relevant to land managers, summaries should describe the effects of vehicular routes tailored for specific ecosystems.

In this chapter, we provide a conceptual framework describing the ecological

Table 8.1 Distinguishing features of major types of vehicular routes in the Mojave Desert

Route type	Route corridors surface	Shoulder Characteristics	Berms	Width	Frequency of travel	Density and area on the landscape	Other factors
OHV trails	Single or 2-track dirt	None	None	< 1 m–3 m	Low–intermittently moderate	High	Pervasive in wildlands, source of dust, most created since the 1960s, some topsoil may be present
Unimproved local roads	1-lane dirt	None	Low	3–4 m	Low	High	Pervasive in wildlands, source of dust, some topsoil may be present, perennial plants may be growing in the roadbed
Improved local roads	1- or 2-lane, graded dirt or gravel	None or narrow	Med	5–7 m	Low–moderate	Moderate–high	Source of dust
Collector roads	2-lane, dirt, gravel, or paved	Narrow	High	7–10 m	Moderate	Moderate	
Arterial roads	2-lane, paved	Wide	High	10–50 m	High	Low	Fencing, culverts, artificial lighting
Limited-access highways	Multi-lane, paved	Very wide	High	< 50 m	Very high	Very low	Fencing, culverts, median, over-pass and interchange structures, artificial lighting

characteristics of various types of vehicular routes, from OHV trails to limited-access highways. We discuss some of the major processes that operate across spatial scales, providing specific examples from the Mojave Desert. Although this review is most relevant to desert ecosystems, and the Mojave Desert in particular, it provides an example of how other ecosystems might be affected by vehicular routes as well.

STUDY REGION

The Mojave Desert (fig. 0.1, Webb et al., Introduction, *this volume*) is transitional between the Great Basin Desert to the north and the Sonoran Desert to the south (Rowlands et al. 1982; Rundel and Gibson 1996). It is a semiarid to arid region with highly variable rainfall which can range from virtually zero to more than 250 mm during any given year. The landscape is characterized by a basin-and-range topography with elevations that are typically between 600 and 900 m. Vegetation is comprised primarily of shrublands and shrub-steppe on deep soils at low and middle elevations, and scattered xeric conifer woodlands on shallow soils at high elevations.

Human disturbances in the Mojave Desert are primarily caused by livestock grazing, mining, military training, and other land uses associated with proximity to large populations in Los Angeles to the southwest and Las Vegas to the northeast (Lovich and Bainbridge 1999). Increasing regional human populations in Southern California and Nevada have led to greater visitation to the Mojave Desert, especially since the 1970s, and associated increases in the frequency and dominance of nonnative plant species (Brooks et al. 2006; Brooks *this volume*), the frequency of wildfires (Brooks and Minnich 2006; Brooks and Matchett 2006), and the density of vehicular routes (Matchett et al. 2004; Vogel and Hughson *this volume*). Because vehicular routes facilitate access by humans to the landscape, the presence of routes exacerbates all human-mediated disturbances. In fact, the intensity of disturbance within and adjacent to vehicular routes, coupled with recurrent disturbance along routes that have high rates of vehicular travel and regular route maintenance, make vehicular routes one of the most intense and pervasive forms of anthropogenic disturbance in the Mojave Desert. Accordingly, managing vehicular routes is a current focus of land management planning efforts in this region.

TYPES OF VEHICULAR ROUTES

Different types of vehicular routes are distinguished by fundamental characteristics that influence their effects on ecosystems. These characteristics include surface type, presence of shoulders or berms, width of the route corridor, frequency of vehicular travel, density and total area of routes on the landscape, and other factors such as the presence of infrastructure, including medians, fences, culverts, and artificial lighting (table 8.1). In the following sections, we describe the fundamental ways that vehicular routes differ and discuss their implications for land management.

OHV Trails and Unimproved Local Roads

When a vehicle passes across the landscape it leaves a track. Many times these tracks are not driven over again, but sometimes they are, especially if they traverse an obvious path of entry into the landscape (e.g., a wash) (Matchett et al. 2004). Repeated tracking eventually creates an enduring trail, which is the most basic form of vehicular route (fig. 8.1a). Trails created by two-wheeled motorcycles consist of a single, narrow footprint less than 1 m wide. If four-wheeled vehicles also use these trails, they may widen the footprint to 2–3 m, creating a two-track jeep trail. In areas of very frequent OHV use, such as OHV "pits" or staging areas, or in areas of intensive military ground training, multiple routes may merge into a very broad area devoid of perennial vegetation 10–100 m or more across. Although these areas represent highly intense and focused surface disturbance, they only comprise a fraction of the total area encompassed by the less intensively disturbed but more extensive networks of OHV trails in the Mojave Desert.

Once a highly visible trail is created, it becomes more susceptible to regular use, and at some point may widen even further and become recognized as an unimproved local road (fig. 8.1b). Rather than try to define the specific point at which an OHV trail becomes an unimproved road, we consider these two types of routes as components of overall OHV route networks. Land managers commonly view their travel management programs in this way, and we think they should be presented in a similar integrated fashion in literature reviews and other decision-support tools.

Off-highway vehicle trails and unimproved local roads are typically less than 4 m wide with a dirt surface (fig. 8.1a, b; table 8.1). By definition, they are typically not bladed, filled, or otherwise improved, so they do not possess many of the ecologically significant characteristics of more developed roads, such as large widths, high berms, or broad shoulders (table 8.1). In limited cases single passes by a grading blade may occur along some sections of unimproved local roads. Berms along the midline of unimproved local roads may develop over time, especially on roads that have evolved from two-track jeep trails. Some topsoil may still be in place and emergent perennial shrubs and grasses may grow up within the roadbed, especially along midline berms. Frequency of travel is typically low on OHV trails and unimproved local roads, except on holiday weekends or during OHV races when use can increase dramatically, and in the vicinity of designated OHV Open Riding Areas where vehicle travel is not restricted to specific routes. Individually, OHV trails and unimproved local roads may lack intense ecological impacts, but collectively they represent a significant threat when trails are dense and comprise a large portion of the landscape (Matchett et al. 2004).

Land managers typically do not plan or direct the establishment of OHV trails or unimproved local roads. These routes may develop in areas that cannot sustain their long-term functioning as vehicular routes. For example, routes straight up hillsides (e.g., hill climbs) facilitate the downslope flow of water and promote rills and gullies that ultimately impede vehicular travel. This process in turn leads to multiple redundant routes parallel to each other that characterize heavily used hillslopes (Wilshire 1978).

OHV trails may be poorly marked by signs in some places, causing OHV riders to inadvertently leave a designated trail and create a new one. Thus, the presence of OHV trails can lead to the development of new routes (Goodlett and Goodlett 1993) that result in trail networks with densities that increase until individual routes become indistinguishable from one another (Matchett et al. 2004). In contrast to OHV trails, unimproved local roads are clearly distinguishable when traveling across the landscape in a vehicle, and there is less chance that vehicle operators will inadvertently lose their way and travel off these routes. However, these routes often lack the benefits of civil engineering and can become eroded or otherwise degraded over time. Route degradation can then promote detours as people drive around the degraded stretches, and these detours may eventually become parallel redundant routes.

Most OHV trails in the Mojave Desert are probably less than 40 years old, since OHV recreation first became popular during the late 1960s (Bureau of Land Management 1999). Old trails may become abandoned over time, but the number of new trails created generally exceeds the number of abandoned trails, resulting in a net increase over time in some parts of the Mojave Desert (Matchett et al. 2004). Unimproved local roads are often older, since they typically developed in response to historical needs for access to the landscape for mining, livestock operations, and maintenance of wildlife guzzlers.

Improved Local Roads and Collector Roads

Improved local roads and collector roads represent a significant step up in terms of ecological effects. They are wider than unimproved local roads and OHV trails, and typically bladed, which removes topsoil and creates berms and shoulders (fig. 8.1c, d; table 8.1). They may also have fill, gravel, or asphalt added to create a more stable road surface. These features cause physical and chemical changes in soil properties. Although frequency of travel on improved local roads and collector roads is higher, the occurrence of these improved roads is less frequent on the landscape, and the total area covered by them is lower than that of unimproved local roads and OHV trails in the Mojave Desert.

Arterial Roads and Limited-Access Highways

Arterial roads and limited-access highways facilitate long-distance travel between regions, in contrast to the previously discussed types of vehicular routes, which primarily provide access to landscapes within a region. The most extreme examples of roadbed, shoulder width, and engineered surfaces characterize these routes (fig. 8.1e, f; table 8.1). They also possess features rarely found in other types

A.

B.

C.

D.

E.

F.

Fig. 8.1. Photographs of the six types of vehicular routes listed in table 8.1. *A,* an OHV trail; *B,* an unimproved local road; *C,* an improved local road; *D,* a collector road; E, an arterial road; and *F,* a limited-access highway.

of routes, including fenced right-of-ways, medians, broad shoulders, culverts, artificial lighting, and overpass and interchange structures. Frequency of travel and vehicle speed are the highest among all route types, but their frequency of occurrence on the landscape and area covered are the lowest of all types of vehicular routes.

SPATIAL SCALES OF VEHICULAR ROUTE EFFECTS

The ecological effects of vehicular routes can be characterized at three spatial scales: (1) direct local effects within route corridors, (2) indirect local effects distributed along gradients radiating outward from route corridors, and (3) dispersed landscape effects resulting from the cumulative effects of multiple routes across landscapes (fig. 8.2). Ecological effects at each spatial scale are not mutually exclusive, as the cumulative influence of smaller-scale local effects associated with individual routes translates into larger-scale landscape effects resulting from the net influence of multiple routes. To be most useful to land managers, information on the effects of vehicular routes should be presented in the spatial context at which ecosystem processes or human-use patterns occur, and the links between spatial scales should be explicitly described.

Direct local effects occur within the footprint of vehicular routes, which includes features that may be created through continued maintenance (e.g., medians, shoulders, or berms). We call this area the vehicular-route corridor (sensu "road corridor" of Forman and Alexander 1998). Initial effects associated with the creation of routes are obvious and dramatic, including alteration of soils and direct mortality of vegetation and wildlife. Ongoing, repeated effects are associated with patterns of vehicular use (e.g., vehicle types, rates of speed, frequency of use) and the continued maintenance of the route (e.g., blading or spraying herbicides along shoulders). Direct effects vary among different types of vehicular routes, with the most severe occurring along paved highways and the least severe occurring along OHV trails.

Although direct local effects are relatively consistent across ecosystems, their interactions with the unique characteristics of individual ecosystems dictate how they translate into the indirect and dispersed landscape effects which are of primary concern for land managers. For example, direct effects of vehicular routes on soil moisture will likely have greater ecological effects in arid ecosystems than in more mesic ecosystems, because water is more limiting to primary productivity in the former than in the latter.

Indirect local effects influence areas immediately adjacent to vehicular routes, otherwise known as the vehicular-route effect zone (sensu "road-effect zone" of Forman and Alexander 1998). The width of this zone varies greatly among the different types of routes. Characteristics of this zone may also be influenced by ecological gradients along the length of vehicular-route corridors (Forman and

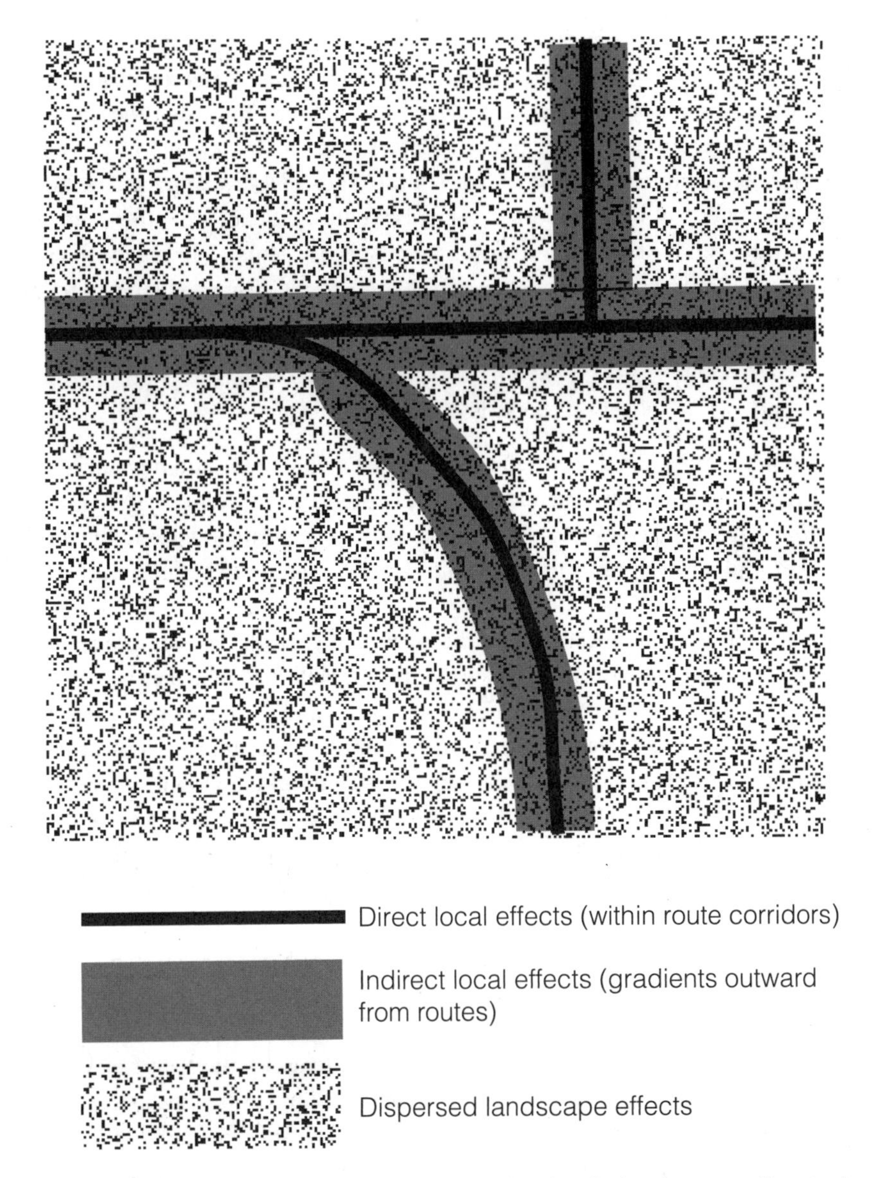

Fig. 8.2. Schematic diagram showing the three spatial scales of vehicular-route effects in the Mojave Desert.

Deblinger 2000), the variable responses of different ecological factors to vehicular routes (Forman et al. 2003), or the unique properties of different types of vehicular routes. Thus, definitions of vehicular route-effect zones should be tailored for specific types of ecosystems, ecological response factors, and vehicular routes, and may not be accurate beyond the context in which they were developed.

Dispersed landscape effects of vehicular routes can be very difficult to determine in a landscape of multiple land uses, such as the Mojave Desert. In addition, even when a significant correlation is established between route densities of various types and environmental response variables, the primary reasons for this relationship can be difficult to identify (M. L. Brooks *unpublished data*). Interactions among the effects of various types of vehicular routes, the effects of other land use disturbances, and the characteristics of specific ecosystems all influence the net effect on environmental response variables. As a result, dispersed landscape effects are also context-specific and should be generalized very cautiously.

EFFECTS OF VEHICULAR ROUTES

Effects on Soils in the Mojave Desert

Vehicular routes can directly affect soils by removing them, adding material to them, changing their physical and chemical composition, or covering them with gravel or asphalt. Many of these changes have effects that extend beyond the route corridor and contribute to indirect and dispersed landscape effects on plants and animals.

One of the most significant ecological effects that vehicular routes have on soils in desert regions involves changes in water runoff patterns. Vehicular routes that run straight up hillslopes can promote soil erosion and the development of rills and gullies. This most often occurs with OHV trails and unimproved dirt roads. Sediment yield during rainfall events can be 10–20 times higher on Mojave Desert hillslopes following OHV use (Iverson 1980). In Dove Springs Canyon in the Western Mojave Desert, 0.3 m of soil eroded downslope along OHV "hill-climb" trails between 1973 and 1975 (Snyder et al. 1976). Off-highway vehicle use accelerates water erosion on hillslopes by decreasing infiltration rates, loosening soil surfaces, channelizing runoff in vehicular tracks, and reducing microtopographic roughness oriented perpendicularly to the slope (Iverson 1980).

Vehicular routes that run parallel to elevation contours can also alter runoff patterns by redirecting water along roadside ditches to low points along the road, at which point water crosses the road and continues downslope in a more concentrated stream than otherwise would have occurred. This process concentrates channels at higher slope positions (Montgomery 1994), resulting in more elongated first-order drainage basins and accelerated erosion rates (Forman and Alexander 1998). These effects become more pronounced as the route corridor becomes more impervious to surface flow, such as along raised roadbeds or where diversion berms, levees, or "chevrons" have been constructed upslope of paved highways or other types of linear human infrastructure (Schlesinger and Jones 1984; Schlesinger et al. 1989). These effects may also increase as the impermeability of the soil and the size of the watersheds feeding each culvert increase. The result is a significant redistribution of landscape water, increasing soil moisture on the upslope side of vehicular routes and along the channels that flow from culverts on the downslope side, and decreasing soil moisture on the upland areas between these downslope channels. This can have significant repercussions on patterns of plant productivity and species composition.

Heavily traveled routes can produce significant amounts of air pollution that create gradients of heavy metals in the soil and plants within 20–200 m from route corridors (Trombulak and Frissell 2000). Bioaccumulation of heavy metals in animals that eat affected plants is a significant concern, especially when increased levels reduce their life spans and reproductive rates. The desert tortoise (*Gopherus agassizii*) is a federally threatened species that has declined in numbers during recent years due to increased incidence of respiratory tract and shell diseases. Increased levels of heavy metals along roadsides may facilitate the contraction of these diseases (K. Berry *personal communication*).

High rates of vehicular travel may also be positively correlated with nitrogen oxide pollution and increased levels of nitrogen in the soil. Increased soil nitrogen affected plant communities up to 200 m from a highway in Britain (Angold 1997). Experiments in the Mojave Desert suggest that increased soil nitrogen can promote the growth of nonnative annual plants, and reduce growth and diversity of native annual plants (Brooks 2003; Allen et al. *this volume*). Vehicular routes with dirt surfaces can also be a significant source of dust. OHV recreation in particular has been identified as the cause of dust plumes covering areas as large as 1700 km^2 (Nakata et al. 1976).

Effects on Vegetation in the Mojave Desert

Vegetation cover and productivity can significantly increase along vehicular routes with paved (Johnson et al. 1975; Vasek et al. 1975; Lightfoot and Whitford 1991) and dirt (Johnson et al. 1975; Vasek et al. 1975; Hessing and Johnson 1982; Starr and Mefford 2002) surfaces. This effect has been attributed to either release from competition from nearby plants removed along the vehicular-route corridor, enhancement of soil moisture from rainfall flowing off the route surface to the base of the berm facing the route, or enhancement of rainfall flowing off of the upslope landscape to the base of the berm facing the surrounding desert (Johnson et al. 1975; Vasek et al. 1975; Schlesinger et al. 1989). The latter two hypotheses are supported by observations of effects on vegetation where berms were present and where berms were absent along improved dirt roads (Starr and Mefford 2002). However, Vasek et al. (1975) observed that enhancement of plant productivity along dirt roads can also occur where obvious drainage factors do not apply. Johnson et al. (1975) suggested that water running off road surfaces affects plant productivity because the roots from shrubs on adjacent berms tend to grow towards the roadside, and that upslope runoff from the surrounding desert is important since productivity can be much higher on the upslope than on the downslope side of paved roads.

We have generally observed that plant productivity does not increase significantly along the margins of OHV trails and is greater along paved than dirt roads. It also appears that the enhancement of vegetation does not scale up proportionately to road width from smaller paved roads to limited-access highways (Johnson et al. 1975). This latter observation may be due to the broader shoulders along major highways compared with smaller paved roads, which puts more distance between water flowing from the edges of impermeable road surface and the closest stands of vegetation.

Vehicular routes are also a primary pathway for plant invasions into arid and semiarid ecosystems (Johnson et al. 1975; Amor and Stevens 1976; Greenberg et al. 1997; Brooks and Pyke 2001; Gelbard and Belnap 2003). Resource availability and propagule pressure (density of propagules) are the primary factors associated with plant invasions (Brooks 2007), and vehicular routes can lead to increased levels of both. Resource availability is often enhanced along the verges of vehicular routes, as discussed above, and vehicles are a major vector for the dispersal of nonnative plant propagules (Clifford 1959; Schmidt 1989; Lonsdale and Lane 1994).

In a study conducted in the deserts of Kuwait, single passes by OHVs create tracks that can provide favorable microsites for annual species (Brown and Schoknecht 2001), while in the Mojave Desert, they can provide favorable microsites for the nonnative annuals *Schismus barbatus* and *Erodium cicutarium* (Davidson and Fox 1974). Biomass of *Schismus* ssp. in the Mojave Desert can be 200%–300% higher within tracks created by single passes of OHVs than in adjacent areas during years of average to low rainfall in the Mojave Desert (Brooks *this volume*). During years of above-average rainfall when soil moisture is more plentiful, this microsite enhancement becomes less prominent.

We have repeatedly observed linear concentrations of nonnative plants along Mojave Desert roadsides, especially of nonnative mustards such as *Brassica tournefortii, Sisymbrium irio, Descurania sophia,* and *Hirschfeldia incana.* On the Colorado Plateau, northeast of the Mojave Desert, cover of the invasive grass *Bromus tectorum* was three times higher along verges of paved roads compared with two-track jeep trails, and compared with cover of five common nonnative forbs on verges of paved roads as well (Gelbard and Belnap 2003). Total nonnative plant cover and species richness were both over 50% higher in the vehicular route-effect zone 50 m from paved routes compared with two-track routes. In the Mojave Desert, species richness of annual plants was higher along roadsides, especially along paved roads, and most of this difference was attributed to the nonnative annuals *E. cicutarium, S. barbatus,* and *Bromus madritensis* ssp. *rubens* and the ruderal natives *Amsinckia tessellata* and *Descurainia pinnata* (Johnson et al. 1975). Density and biomass of *B. tournefortii* were higher along dirt roads compared with areas 10 m away where soils were shallow, silty, and rocky, but not where soils were deeper and sandier in the Central Mojave Desert (Brooks *this volume*). There is also evidence that these indirect effects of vehicular routes may translate into dispersed landscape effects, since nonnative species richness and biomass of the nonnative forb *E. cicutarium* were positively correlated with density of dirt roads in the Mojave Desert (Brooks 1998; Brooks et al. 2006).

The typical pattern of plant invasions into the Mojave Desert traces the following course. In the first phase, nonnative invaders appear along roadsides near their adjacent regions of origin. For example, the nonnative mustard *Brassica tournefortii* spread northward along paved highways into the South-Eastern and South-Central Mojave Desert from its initial point of colonization in the Sonoran Desert (Minnich and Sanders 2000; Brooks *this volume*), then on through to the Northern Mojave Desert and into the Colorado Plateau (Brooks *this volume*). In some cases invaders may "island hop" into the region by establishing first in urbanized or agricultural regions within the Mojave Desert, then move outward along roadsides into less developed areas. Once within the region, invaders are prepositioned to begin the second phase of invasion, the spread away from roadsides into wildland areas. We have often observed that the initial stages of spread away from vehicular routes occurs within landscape features (e.g., washes or north-facing hillslopes) or microsites (e.g., beneath perennial shrubs) where soil moisture levels are locally high. Disturbed areas, such as utility corridors, areas with high levels of OHV use (Davidson and Fox 1974), or burned areas adjacent to roadsides (Milberg and Lamont 1995), are also more readily invaded (M. L. Brooks *personal observation*). The third and final stage of invasion, which is achieved by relatively few species in the Mojave Desert (Brooks and Esque 2002; Brooks et al. 2006), is the naturalization of self-sustaining invader populations in wildland areas away from roads.

Effects on Animals in the Mojave Desert

Animals are directly affected by the habitat loss associated with the development of vehicular routes, and by mortalities caused by collisions with vehicles traveling on these routes. Studies of *Gopherus agassizii* indicate that population densities are lower near vehicular routes (Nicholson 1978; Berry and Turner 1987; Boarman et al. 1997). Fenced exclusion of ground-dwelling vertebrates from a limited-access highway in the Western Mojave Desert reduced road kills of *Gopherus agassizii* by 93% and of vertebrates in general by 88% (Boarman and Sazaki 1996). However, another study from a limited-access highway in the Northern Mojave Desert suggests that rodents rarely crossed the highway (Garland and Bradley 1984). Thus, generalizations about the direct effects of vehicular routes on rates of animal mortality are difficult to make because of limited, and somewhat contradictory, information. It is also probable that effects on animals vary among route types, vegetation types, and taxa with differing behavioral characteristics and habitat preferences.

Enhanced productivity of vegetation along improved roads, especially those that are paved, can lead to increased abundances of insects (Lightfoot and Whitford 1991) and rodents (Garland and Bradley 1984). However, these two studies do not span more than one year of sampling, and road effects can vary among years of contrasting rainfall. For example, densities of Antelope ground squirrels (*Ammospermophilus leucurus*) in the Western Mojave Desert decreased with the increasing distance intervals of 0–100 m, 100–200 m, and greater than 200 m from improved dirt roads during two years of low rainfall, whereas the trend reversed during an intervening year of high rainfall (Starr 2001). The explanation was that animals were drawn to areas near roads during the years of low rainfall because their annual plant forage was more abundant there compared with areas further from roads. During the year of high rainfall, forage was more abundant across the landscape and may have not been the limiting factor it was during the low-rainfall years. The negative relationship between ground squirrels and roads when rainfall was high indicated there may have been other negative factors associated with roads that are either only manifested during years of high rainfall, or the effects are masked by the positive influence of greater forage availability close to roads during years of low rainfall.

Increased vegetation structure along improved roads may also increase the diversity of bird communities. However, one study that evaluated the general relationships between vegetation structure and bird community diversity in the Mojave Desert did not find significant correlations (Brooks 1999). The apparent increase in habitat quality along road edges may have a net negative effect as animals are drawn from the surrounding desert towards roadsides where they are more likely to be killed by passing vehicles or where they may bioaccumulate heavy metals concentrated there.

A basic question relates to how the direct and indirect local effects of vehicular routes translate into dispersed effects on animal populations and communities across the landscape, and how these effects vary as rainfall fluctuates from year to year. In particular, studies are needed to determine the characteristics of vehicular routes that create barriers or filters to animal movement patterns and lead to habitat fragmentation for animals. Although habitat fragmentation is often cited as a major concern related to vehicular routes in the Mojave Desert, no studies have directly evaluated the role of vehicular routes in fragmenting wildlife habitat in this region.

SUMMARY OF EXISTING RESEARCH

Most of what is known regarding the ecological effects of vehicular routes in the Mojave Desert is focused on OHV networks of trails and unimproved local roads (39 of 47 studies) (table 8.2). These studies provide important insights for inferring the potential ecological effects of other types of vehicular routes. All of the studies on vehicular routes addressed some aspect of direct local effects: five studies addressed indirect effects and four addressed dispersed effects. Very few addressed multiple scales: five studies addressed direct plus indirect effects, four studies addressed direct plus dispersed effects, and none addressed all three scales of effects.

Most studies have quantified direct effects of vehicular routes by comparing conditions within the road corridor with reference conditions at a single distance outside of the corridor. These studies can produce misleading information if the reference site lies within the indirect vehicular route-effect zone, thus not serving as a true control. Studies focused on the local effects of individual routes should be designed to evaluate both direct and indirect effects and incorporate a gradient of sites at various distances from the route. Gradient study designs offer an effective way to evaluate the local effects of vehicular routes, because they can identify inflection points and asymptotes of ecological responses to routes. True controls can then be defined as the area beyond the distance at which the gradient effect reaches its asymptote. Gradient data can also be used to develop transfer functions for modeling the dispersed landscape effects of multiple routes, an effect essential to include in land management plans.

Most studies (41 of 47) evaluated creosote-bush scrub habitat. Only 9 studies evaluated effects on shadscale scrub, 3 evaluated Joshua tree woodland, 4 evaluated saltbush scrub, 4 evaluated desert psammophytie scrub, 2 evaluated blackbrush scrub, 1 evaluated microphyll woodland, and 4 were not specific to a particular vegetation type (table 8.2). Creosote-bush scrub is the dominant vegetation type in the Mojave Desert (Rundel and Gibson 1996; Rowlands et al. 1982), but saltbush scrub, blackbrush scrub, and Joshua tree woodland also cover a considerable amount of the region and probably need to be included in vehicular-route studies proportionally more than they have been. Pinyon-juniper woodland has not been studied at all, although it is of concern to land managers because it occurs in relatively small and disjunct stands that may be especially vulnerable to landscape disturbances such as those stemming from vehicular routes.

Soils were included as a response variable in half of the studies (24 of 47), but all but two of these studies focused exclusively on OHV trails or unimproved local roads (table 8.2). Similarly, of the 12 annual plant studies, all but three were focused on OHVs and unimproved local roads. Thus, what is known about soil and annual plant responses to vehicular routes in the Mojave Desert is mostly derived from studies that are focused on OHV effects and at local scales. In contrast, the 29 studies that addressed perennial plants and the 10 that addressed animals were more equally distributed among vehicular-route types. Few studies addressed multiple combinations of soil, annual plant, perennial plant, and animal response

Table 8.2 Published studies evaluating ecological effects of vehicular routes in the Mojave Desert

Route type[a]	Spatial scale[b]	Plant communities[c]	Soils	Response variables: Annual plants	Perennial plants	Animals	Source Citation
OHV	Direct	Creosote bush scrub	Compaction, moisture soil strength	Cover, density	NE[d]	NE	Adams et al. 1982a, b
OHV	Direct, indirect	Creosote bush scrub	NE	NE	Cover, density	NE	Artz 1989
OHV	Direct	Joshua tree woodland, creosote brush scrub	Crust cover, nitrogenase activity, texture	NE	NE	NE	Belnap 2002
OHV	Direct, indirect	Creosote bush scrub	NE	NE	NE	Mortality of small, terrestrial vertebrates and desert tortoise	Brattstrom and Bondello 1983
OHV	Direct	Creosote bush scrub	NE	NE	Germination and survival	NE	Brum et al. 1983
OHV	Direct	Creosote bush scrub	NE	NE	Density	Density, live weight, and relative abundance of desert tortoise	Bury 1987
OHV	Direct	Creosote bush scrub, desert psammophytic scrub, shadscale scrub	NE	Density, species composition, volume	Cover, density, species composition, volume	Biomass, density, and species composition of invertebrates, reptiles and rodents	Bury and Luckenbach 1983
OHV	Direct, dispersed	Creosote bush scrub	NE	NE	Cover, density	Abundance, biomass, density, and habitat of desert tortoise	Bury and Luckenbach 2002
OHV	Direct, dispersed	Creosote bush scrub	NE	NE	NE	Biomass and density of lizards	Busack and Bury 1974
OHV	Direct	Creosote bush scrub	Bulk density, compaction	Density, species composition	Cover, species composition	NE	Davidson and Fox 1974
OHV	Direct	Creosote bush scrub	Moisture, texture	NE	NE	NE	Dyck and Stukel 1979
OHV	Direct	Not specific to a vegetation type	Desert pavement	NE	NE	NE	Elvidge and Iverson 1983

OHV	Direct	Creosote bush scrub	Bulk density, compaction	NE	NE	NE	Fox 1973
OHV	Direct	Not specific to a vegetation type	Erosion, runoff, sediment yield	NE	NE	NE	Hinckley et al. 1983
OHV	Direct	Creosote bush scrub	Erosion, sediment yield	NE	NE	NE	Iverson 1980
OHV	Direct	Creosote bush scrub	NE	NE	Cover, density	NE	Lathrop 1978
OHV	Direct	Creosote bush scrub, desert psammophytic scrub, shadscale scrub	NE	Cover	Cover, density, diversity	NE	Lathrop 1983a
OHV	Direct	Creosote bush scrub	NE	NE	Cover, density, diversity, productivity	NE	Lathrop 1983b
OHV	Direct	Blackbrush scrub	Bulk density, compaction, pore space	NE	NE	NE	Lei 2004
OHV	Direct	Creosote bush scrub, desert psammophytic scrub, shadscale scrub, desert microphyll woodland	NE	Density	Cover, density, volume	Biomass, density, and species richness of reptiles and mammals	Luckenbach and Bury 1983
OHV	Direct	Creosote bush scrub	Bulk density, infiltration rate, moisture, particle size, penetration resistance, texture	NE	Cover, germination of transplanted seedlings	NE	Marble 1985
OHV	Direct	Creosote bush scrub	Compaction, infiltration rate, moisture, penetration resistance, texture	NE	Cover, density	NE	McCarthy 1996

Table 8.2 (continued)

Route type[a]	Spatial scale[b]	Plant communities[c]	Soils	Response variables: Annual plants	Response variables: Perennial plants	Animals	Source Citation
OHV	Direct	Creosote bush scrub, saltbush scrub, desert psammophytic scrub	NE	Reproduction, species richness	Reproduction, species richness	NE	Pavlik 1979
OHV	Direct	Creosote bush scrub	Bulk density, compaction	NE	Cover, density (species categorized as long- or short-lived)	NE	Prose et al. 1987
OHV	Direct	Creosote bush scrub	Bulk density, compaction, erosion, moisture, runoff, sediment yield	Cover	Cover	NE	Snyder et al. 1976
OHV	Direct	Creosote bush scrub	Compaction, moisture, texture	NE	NE	NE	Tullock 1983
OHV	Direct	Shadscale scrub	NE	Density	Density	Abundance, biomass and density of rodents and lizards	Vollmer et al. 1976
OHV	Direct	Creosote bush scrub	Compaction, infiltration	NE	NE	NE	Webb 1980
OHV	Direct	Creosote bush scrub, shadscale scrub	Bulk density, compaction, penetration resistance, shear stress	NE	NE	NE	Webb 1983
OHV	Direct	Not specific to a vegetation type	Compaction	NE	Cover, density	NE	Webb et al. 1983
OHV	Direct	Not specific to a vegetation type	Inorganic elements microfloral elements	NE	NE	NE	Wilshire 1983

OHV	Direct	Creosote bush scrub	Compaction	NE	NE	NE	Wilshire and Nakata 1976
OHV, ILR	Direct	Creosote bush scrub	Bulk density, compaction, moisture, nutrient levels, pH, total N, N pool, N mineralization, available P	NE	Cover, density, species composition, volume	NE	Bolling and Walker 2000
OHV, ULR, ILR	Direct, dispersed	Creosote bush scrub, blackbrush scrub	NE	Cover, diversity, seedbank density	Cover, diversity	NE	Brooks, *unpublished data*
ULR	Direct	Creosote bush scrub, shadscale scrub	Bulk density, compaction, moisture, texture	NE	Cover, density, species composition	NE	Webb and Wilshire 1980
ULR	Direct	Creosote bush scrub, shadscale scrub	NE	NE	Density, frequency	NE	Wells 1961
ULR, ILR	Direct	Creosote bush scrub	Bulk density, compaction, moisture, nutrient levels, pH, total N, N pool, N mineralization, available P	NE	Cover, density, species composition, volume	NE	Bolling 1996
ULR, ILR	Direct, dispersed	Creosote bush scrub	Total N	Biomass, density, species richness	Cover, species richness	NE	Brooks and Berry 2006
ULR, ILR	Direct	Creosote bush scrub, saltbush scrub	NE	NE	Abundance, cover, density	NE	Vasek et al. 1975
ILR	Direct	Creosote bush scrub	NE	NE	Biomass, cover, density, richness	NE	Lathrop and Archibold 1980a
ILR	Direct	Creosote bush scrub	NE	NE	Biomass, cover, density, richness	NE	Lathrop and Archibold 1980b

Table 8.2 (continued)

Route type[a]	Spatial scale[b]	Plant communities[c]	Response variables				Source Citation
			Soils	Annual plants	Perennial plants	Animals	
ILR	Direct, Indirect	Creosote bush scrub, Saltbush scrub, Joshua tree woodland	NE	NE	NE	Cover, density, diversity of antelope ground squirrel	Starr 2001
ILR	Direct, Indirect	Creosote bush scrub, Saltbush scrub, Joshua tree woodland	NE	NE	Abundance cover density, species richness	NE	Starr and Mefford 2002
ILR	Direct	Creosote bush scrub, shadscale scrub	Bulk density, pH, texture, total N, water retention	NE	NE	NE	Walker and Powell 2001
ILR, AR	Direct	Creosote scrub	NE	Density species composition	Biomass cover density	NE	Johnson et al. 1975
LAH	Direct	Creosote bush scrub, shadscale scrub	NE	NE	NE	Mortality of small, terrestrial vertebrates and desert tortoises	Boarman and Sazaki 1996 Boarman et al. 1997
LAH	Direct, indirect	Creosote bush scrub	NE	NE	cover	Abundance, species richness of rodents	Garland and Bradley 1984

[a]OHV = OHV trails, ULR = unimproved local roads, ILR = improved local roads, CR = collector roads, AR = arterial roads, and LAH = limited-access highways.
[b]Direct local, indirect local, dispersed landscape.
[c]Munz (1974).
[d]NE = not evaluated.

variable categories: 15 addressed some combination of two categories, 6 addressed combinations of three, and none simultaneously addressed all four.

Future research should evaluate multiple scales, including direct, indirect, and dispersed effects of vehicular routes. Understudied vegetation types should be evaluated to improve the breadth of knowledge across different ecological conditions within the Mojave Desert. The generality of responses of soils and annual plants to OHV trails and unimproved local roads needs to be tested in response to other types of vehicular routes. Studies that integrate multiple combinations of soils, annual plants, perennial plants, and animals would also help address management questions regarding the effects of vehicular routes on higher order ecosystem responses such as wildlife populations and communities.

Management Implications

The current decision process of route designation in the Mojave Desert is site-specific and relies, to various degrees, on biological, cultural, and recreational information. For example, if a route passes through high priority habitat for sensitive species or provides access to sensitive cultural sites, then the route may be considered undesirable and targeted for closure. However, if the route provides access to recreation areas, then it may be deemed desirable and targeted for possible designation as an open route. Effects of routes on physical processes (e.g., dust production or soil erosion) are rarely considered. Final decisions must balance different aspects of resource protection with other land uses, and the decision process needs to be supported by as much objective science as possible for decisions to withstand intense scrutiny.

The biggest challenge to public land management is developing objective criteria upon which route designation decisions can be made and later justified. Another challenge is selecting and defining indicators for successful management; that is, being able to determine when management actions produce desired results. High-priority information needs often expressed by land managers include the need to understand the effects of vehicular routes on plant invasions, native animal populations, and local biotic communities (M. L. Brooks *personal observations*).

Key questions also include: What characteristics of route networks most effectively promote plant invasions? What features of route networks result in significant habitat fragmentation for animals? Which affects animals more, the type of vehicular activity that occurs on a route or the characteristics of the route itself? How do indirect and dispersed landscape effects differ among types of vehicular routes? Are there signals to indicate when effects at smaller scales will lead to effects at larger scales? How do the effects of vehicular routes compare to the effects of other land uses and landscape disturbances?

In conclusion, future studies should address thresholds of ecological responses to roads, and thresholds of ecological recovery from past road effects following restoration efforts, and translate directly into criteria for route designation. Ideally, these criteria should provide land managers with an early-warning system to determine when and where the effects of vehicular routes will cause the biggest ecological problems. This information could be used to prioritize management actions related to vehicular routes among the multitude of land use issues that public land management agencies must balance.

ACKNOWLEDGMENTS

This chapter received helpful reviews from Karen Phillips and James Weigand.

REFERENCES

Adams, J. A., A. S. Endo, L. H. Stolzy, P. G. Rowlands, and H. B. Johnson. 1982a. Controlled experiments on soil compaction produced by off-road vehicles in the Mojave Desert, California. *Journal of Applied Ecology* **19**:167–175.

Adams, J. A., L. H. Stolzy, A. S. Endo, P. G. Rowlands, and H. B. Johnson. 1982b. Desert soil compaction reduces annual plant cover. *California Agriculture* **36**:6–7.

Amor, R. L., and P. L. Stevens. 1976. Spread of weeds from a roadside into sclerophyll forests at Dartmouth, Australia. *Weed Research* **16**:111–118.

Angold, P. G. 1997. The impact of roads on adjacent heathland vegetation: effects on plant species composition. *Journal of Applied Ecology* **34**:409–417.

Artz, M. C. 1989. Impacts of linear corridors on perennial vegetation in the east Mojave Desert: implications for environmental management and planning. *Natural Areas Journal* **9**:117–129.

Belnap, J. 2002. Impacts of off-road vehicles on nitrogen cycles in biological soil crusts: resistance in different US deserts. *Journal of Arid Environments* **52**:155–165.

Berry, K. H., and F. B. Turner. 1987. Notes on the behavior and habitat preferences of juvenile desert tortoises (*Gopherus agassizii*). Pages 111–130 *in* M.W. Trotter, editor. *Proceedings of the Eighth Annual Desert Tortoise Council Symposium, March 31–April 2, 1984, Lake Havasu City, Arizona.* Desert Tortoise Council, Beaumont, California.

Boarman, W. I., and M. Sazaki. 1996. Highway mortality in desert tortoises and small vertebrates: success of barrier fences and culverts. Pages 169–173 *in* G. J. Evink, P. Garrett, D. Zeigler, and J. Berry, editors. *Trends in addressing transportation related wildlife mortality.* Proceedings of the Transportation Related Wildlife Mortality Seminar. Florida State Department of Transportation Environmental Management Office, Tallahassee, Florida.

Boarman, W. I., M. Sazaki, and W. B. Jennings. 1997. The effects of roads, barrier fences, and culverts on desert tortoise populations in California, USA. Pages 54–58 *in* J. V. Abema, editor. *Proceedings of the 1993 International Conference on Conservation, Restoration, and Management of Tortoises and Turtles, July 11–16, 1993, State University of New York at Purchase, New York.* New York Turtle and Tortoise Society, New York.

Bolling, J. D. 1996. Ecological succession on abandoned roads in the Lake Mead National Recreation Area. Masters thesis, University of Nevada, Las Vegas, Nevada.

Bolling, J. D., and L. R. Walker. 2000. Plant and soil recovery along a series of abandoned desert roads. *Journal of Arid Environments* **46**:1–24.

Brattstrom, B. H., and M. C. Bondello. 1983. Effects of off-road vehicle noise on desert vertebrates. Pages 167–204 *in* R. H. Webb and H. G. Wilshire, editors. *Environmental effects of off-road vehicles: impacts and management in arid regions.* Springer-Verlag, New York, New York.

Brooks, M. L. 1998. Ecology of a biological invasion: alien annual plants in the Mojave Desert. Ph.D. Dissertation, University of California, Riverside, California.

Brooks, M. L. 1999. Effects of protective fencing on birds, lizards, and black-tailed hares in the western Mojave Desert. *Environmental Management* **23**:387–400.

Brooks, M. L. 2003. Effects of increased soil nitrogen on the dominance of alien annual plants in the Mojave Desert. *Journal of Applied Ecology* **40**:344–353.

Brooks, M. L. 2007. Effects of land management practices on plant invasions in wildland areas. Pages 147–162 *in* W. Nentwig, editor. *Biological invasions.* Springer-Verlag, Heidelberg, Germany.

Brooks, M. L., and T. C. Esque. 2002. Alien plants and fire in desert tortoise (*Gopherus agassizii*) habitat of the Mojave and Colorado Deserts. *Chelonian Conservation and Biology* **4**:330–340.

Brooks, M. L., and J. R. Matchett. 2006. Spatial and temporal patterns of wildfires in the Mojave Desert. 1980–2004. *Journal of Arid Environments* **67**:148–164.

Brooks, M. L, J. R. Matchett, and K. H. Berry. 2006. Effects of livestock watering sites on plant communities in the Mojave Desert, USA. *Journal of Arid Environments* **67**: 125–147.

Brooks, M. L., and R. A. Minnich. 2006. Southeastern deserts bioregion. Pages 391–414 *in* N. G. Sugihara, J. W. van Wagtendonk, K. E. Shaffer, J. Fites-Kaufman, and A. E. Thode, editors. *Fire in California's ecosystems.* University of California Press, Berkeley, California.

Brooks, M. L., and D. Pyke. 2001. Invasive plants and fire in the deserts of North America. Pages 1–14 *in* K. Galley and T. Wilson, editors. *Proceedings of the Invasive Species Workshop: The Role of Fire in the Control and Spread of Invasive Species, Fire Conference 2000.* The First National Congress on Fire, Ecology, Prevention and Management (November 28–December 1, 2000, San Diego, California) Miscellaneous Publication No. 11, Tall Timbers Research Station, Tallahassee, Florida.

Brown, G., and N. Schoknecht. 2001. Off-road vehicles and vegetation patterning in a degraded desert ecosystem in Kuwait. *Journal of Arid Environments* **49**:413–427.

Brum, G. D., R. S. Boyd, and S. M. Carter. 1983. Recovery rates and rehabilitation of powerline corridors. Pages 303–311 *in* R. H. Webb and H. G. Wilshire, editors. *Environmental effects of off-road vehicles: impacts and management in arid regions.* Springer-Verlag, New York, New York.

Bureau of Land Management. 1999. *The California Desert Conservation Area Plan 1980, as amended.* U.S. Department of the Interior, Bureau of Land Management, Riverside, California.

Bury, R. B. 1987. Off-road vehicles reduce tortoise numbers and well-being. Research Information Bulletin No. 87–6. U.S. Department of the Interior, Fish and Wildlife Service, National Ecology Research Center, Fort Collins, Colorado.

Bury, R. B., and R. A. Luckenbach. 1983. Vehicular recreation in arid land dunes: biotic responses and management alternatives. Pages 207–221 *in* R. H. Webb and H. G. Wilshire, editors. *Environmental effects of off-road vehicles: impacts and management in arid regions.* Springer-Verlag, New York, New York.

Bury, R. B., and R. A. Luckenbach, 2002. Comparison of desert tortoise (*Gopherus agassizii*) populations in an unused and off-road vehicle area in the Mojave Desert. *Chelonian Conservation and Biology* **4**:457–463.

Busack, S. D., and R. B. Bury. 1974. Some effects of off-road vehicles and sheep grazing on lizard populations in the Mojave Desert. *Biological Conservation* **6**:179–183.

Clifford, H. T. 1959. Seed dispersal by motor vehicles. *Journal of Ecology* **47**:311–315.

Davidson, E., and M. Fox. 1974. Effects of off-road motorcycle activity on Mojave Desert vegetation and soil. *Madroño* **22**:381–412.

Dyck, R. I. J., and J. J. Stukel. 1979. Fugitive dust impacts during off-road vehicle (ORV) events in the California desert. *Environmental Science and Technology* **10**:1046–1048.

Elvidge, C. D., and R. A. Iverson. 1983. Regeneration of desert pavement and varnish. Pages 225–241 *in* R. H. Webb and H. G. Wilshire, editors. *Environmental effects of off-road vehicles: impacts and management in arid regions.* Springer-Verlag, New York, New York.

Forman, R. T. T. 1995. *Land mosaics: the ecology of landscapes and regions.* Cambridge University Press, Cambridge, UK.

Forman, R. T. T., and L. E. Alexander. 1998. Roads and their major ecological effects. *Annual Review of Ecology and Systematics* **29**:207–231.

Forman, R. T. T., and R. D. Deblinger. 2000. The ecological road-effect zone of a Massachusetts (USA) suburban highway. *Conservation Biology* **14**:36–46.

Forman, R. T. T., D. Sperling, J. A. Bissonette, A. P. Clevenger, C. D. Cutshall, V. H. Dale, L. Fahrig, R. France, C. R. Goldman, K. Heanue, J. A. Jones, F. J. Swanson, T. Turrentine, and T. C. Winter. 2003. *Road ecology: science and solutions.* Island Press, Washington, D.C.

Fox, M. 1973. Compaction of soil by off-road vehicles at three sites in the Mojave Desert. Pages 1–13 *in* K. H. Berry, editor. *Preliminary studies on the effects of off-road vehicles in the northwestern Mojave Desert: a collection of papers.* U.S. Department of the Interior, National Biological Service, Ridgecrest, California.

Garland, T., and W. G. Bradley. 1984. Effects of a highway on Mojave Desert rodent populations. *American Midland Naturalist* **111**:47–56.

Gelbard, J. L., and J. Belnap. 2003. Roads as conduits for exotic plant invasions in a semiarid landscape. *Conservation Biology* **17**:420–432.

Goodlett, G. O., and G. C. Goodlett. 1993. Studies of unauthorized off-highway vehicle activity in the Rand Mountains and Fremont Valley, Kern County, California. Pages 163–187 *in Proceedings of the Seventeenth Desert Tortoise Council Symposium, March 5–9, 1992, Las Vegas, Nevada.* Desert Tortoise Council, Beaumont, California.

Greenberg, C. H., S. H. Crownover, and D. R. Gordon. 1997. Roadside soil: a corridor for invasion of xeric scrub by nonindigenous plants. *Natural Areas Journal* **17**:99–109.

Hessing, M. B., and C. D. Johnson. 1982. Disturbance and revegetation of Sonoran Desert vegetation in an Arizona powerline corridor. *Journal of Range Management* **35**: 254–258.

Hinckley, B. S., R. M. Iverson, and B. Hallet. 1983. Accelerated water erosion in ORV-use areas. Pages 81–94 *in* R. H. Webb and H. G. Wilshire, editors. *Environmental effects of off-road vehicles: impacts and management in arid regions.* Springer-Verlag, New York, New York.

Iverson, R. M. 1980. Processes of accelerated pluvial erosion on desert hillslopes modified by vehicular traffic. *Earth Surface Processes* **5**:369–388.

Johnson, H. B., F. C. Vasek, and T. Yonkers. 1975. Productivity, diversity and stability relationships in Mojave Desert roadside vegetation. *Bulletin of the Torrey Botanical Club* **102**:106–115.

Lathrop, E. W. 1978. *Plant response parameters to recreational vehicles in the California Desert Conservation Area (CDCA).* U.S. Department of the Interior, Bureau of Land Management, Riverside, California.

Lathrop, E. W. 1983a. The effect of vehicle use on desert vegetation. Pages 153–165 *in* R. H. Webb and H. G. Wilshire, editors. *Environmental effects of off-road vehicles: impacts and management in arid regions.* Springer-Verlag, New York, New York.

Lathrop, E. W. 1983b. Recovery of perennial vegetation in military maneuver areas. Pages 264–276 *in* R. H. Webb and H. G. Wilshire, editors. *Environmental effects of off-road vehicles: impacts and management in arid regions.* Springer-Verlag, New York, New York.

Lathrop, E. W., and E. F. Archbold. 1980a. Plant response to utility right of way construction in the Mojave Desert. *Environmental Management* **4**:215–226.

Lathrop, E. W., and E. F. Archibold. 1980b. Plant response to Los Angeles aqueduct construction in the Mojave Desert. *Environmental Management* **4**:137–148.

Lei, S. A. 2004. Soil compaction from human trampling, biking, and off-road motor vehicle activity in a blackbrush (*Coleogyne ramosissima*) shrubland. *Western North American Naturalist* **64**:125–130.

Lightfoot, D. C., and W. G. Whitford. 1991. Productivity of creosote foliage and associated canopy arthropods along a desert roadside. *American Midland Naturalist* **125**:310–322.

Lonsdale, W. M., and A. M. Lane. 1994. Tourist vehicles as vectors of weed seeds in Kakadu National Park, Northern Australia. *Biological Conservation* **69**:277–283.

Lovich, J. E., and D. Bainbridge. 1999. Anthropogenic degradation of the southern California desert ecosystem and prospects for natural recovery and restoration. *Environmental Management* **249**:309–326.

Luckenbach, R. A., and R. B. Bury. 1983. Effects of off-road vehicles on the biota of the Algodones Dunes, Imperial County, California. *Journal of Applied Ecology* **20**:265–286.

Marble, J. R. 1985. *Techniques of revegetation and reclamation of land damaged by off-road vehicles in the Lake Mead National Recreation Area.* Report No. 027/03. Cooperative National Park Resources Studies Unit, University of Nevada, Las Vegas, Nevada.

Matchett, J. R., L. Gass, M. L. Brooks, A. M. Mathie, R. D. Vitales, M. W. Campagna, D. M. Miller, and J. F. Weigand. 2004. *Spatial and temporal patterns of off-highway vehicle use at the Dove Springs OHV Open Area, California.* Report prepared for the Bureau of Land Management, California State Office, Sacramento, California.

McCarthy, L. E. 1996. Impact of military maneuvers on Mojave Desert surfaces: a multiscale analysis. Ph.D. Dissertation, University of Arizona, Tucson, Arizona.

Milberg, P., and B. B. Lamont. 1995. Fire enhances weed invasion of roadside vegetation in southwestern Australia. *Biological Conservation* **73**:45–49.

Minnich, R. A., and A. C. Sanders. 2000. *Brassica tournefortii* (Gouan.) Sahara mustard.

Pages 68–72 *in* C. Bossard, M. Hoshovsky, and J. Randall, editors. *Noxious wildland weeds of California.* University of California Press, Berkeley, California.

Munz, P. A. 1974. *A flora of southern California.* University of California Press, Berkeley, California.

Montgomery, D. R. 1994. Road surface drainage, channel initiation, and slope stability. *Water Resources Research* **30**:1025–1932.

Nakata, J. K., H. G. Wilshire, and G. C. Barnes. 1976. Origin of Mojave Desert dust plume photographed from space. *Geology* **4**:644–648.

National Research Council. 1997. *Toward a sustainable future: addressing the long-term* effects of motor vehicle transportation on climate and ecology. National Academy Press, Washington, D.C.

Nicholson, L. 1978. The effects of roads on desert tortoise populations. Pages 127–129 *in Proceedings of the Third Desert Tortoise Council Symposium, April 1–3,1978, Las Vegas, Nevada.* Desert Tortoise Council, Beaumont, California.

Pavlik, B. M. 1979. *The biology of endemic Psammophytes, Eureka Valley, California and its relation to off-road vehicle impact.* U.S. Department of the Interior, Bureau of Land Management, Riverside, California.

Prose, D. V., S. K. Metzger, and H. G. Wilshire. 1987. Effects of substrate disturbance on secondary plant succession: Mojave Desert, California. *Journal of Applied Ecology* **24**:305–313.

Rowlands, P. G. 1980. *Effects of disturbance on desert soils, vegetation, and community processes with emphasis on off-road vehicles: a critical review.* U.S. Department of the Interior, Bureau of Land Management, Riverside, California.

Rowlands, P. G., H. Johnson, E. Ritter, and A. Endo. 1982. The Mojave Desert. Pages 103–162 *in* G. L. Bender, editor. *Reference handbook on the deserts of North America.* Greenwood Press, Westport, Connecticut.

Rundel, P. W., and A. C. Gibson. 1996. *Ecological communities and processes in a Mojave Desert ecosystem: Rock Valley, Nevada.* Cambridge University Press, New York, New York.

Schmidt, W. 1989. Plant dispersal by motor cars. *Vegetatio* **80**:147–152.

Schlesinger, W. H., and C. S. Jones. 1984. The comparative importance of overland runoff and mean annual rainfall to shrub communities of the Mojave Desert. *Botany Gazette* **145**:116–124.

Schlesinger, W. H, P. J. Fonteyn and W. A. Reiners. 1989. Effects of overland flow on plant water relations, erosion, and soil water percolation on a Mojave Desert landscape. *Soil Science Society of America Journal* **53**:1567–1572.

Snyder, C. T., D. G. Frickel, R. F. Hadley, and R. F. Miller. 1976. *Effect of off-road vehicle use on the hydrology and landscape of arid environments in central and southern California.* U.S. Geological Survey Water Resources Investigations Report No. 76–99. Denver, Colorado.

Spellerberg, I. F. 1998. Ecological effects of roads and traffic: a literature review. *Global Ecology and Biogeography Letters* **7**:317–333.

Spellerberg, I. F. 2002. *Ecological effects of roads.* Science Publishers, Inc., Ehnfield, New Hampshire.

Spellerberg, I. F., and T. Morrison. 1998. *The ecological effects of new roads—a literature review.* New Zealand Department of Conservation, Wellington, New Zealand.

Starr, M. J. 2001. Assessing the effects of roads on desert ground squirrels. *Papers of the Applied Geography Conferences* **24**:35–40.

Starr, M. J., and J. N. Mefford. 2002. The effects of roads on perennial shrubs in the Mojave Desert, California. *Papers of the Applied Geography Conference* **25**:253–260.

Trombulak, S. C., and C. A. Frissell. 2000. Review of ecological effects of roads on terrestrial and aquatic communities. *Conservation Biology* **14**:18–30.

Tullock, R. J. 1983. *Study relating the impacts of ORVs on soils that may result from proposed motorcycle race.* U.S. Department of the Interior, Bureau of Land Management, Sacramento, California.

Vasek, F. C., H. B. Johnson, and G. D. Brum. 1975. Effects of power transmission lines on vegetation of the Mojave Desert. *Madroño* **23**:114–130.

Vollmer, A. T., B. G. Maza, P. A. Medica, F. B. Turner, and S. A. Bamberg. 1976. The impact of off-road vehicles on a desert ecosystem. *Environmental Management* **1**:115–129.

Walker, L. R., and E. A. Powell. 2001. Soil water retention on gold mine surfaces in the Mojave Desert. *Restoration Ecology* **9**:95–103.

Webb, R. H. 1980. The effects of controlled motorcycle traffic on a Mojave Desert soil. Masters Thesis, Stanford University, Palo Alto, California.

Webb, R. H. 1983. Compaction of desert soils by off-road vehicles. Pages 51–76 *in* R. H. Webb and H. G. Wilshire, editors. *Environmental effects of off-road vehicles: impacts and management in arid regions.* Springer-Verlag, New York, New York.

Webb, R. H., and H. G. Wilshire. 1978. *A bibliography of the effects of off-road vehicles.* U.S. Geological Survey Open-File Report No. 78–149. U.S. Department of the Interior, Menlo Park, California.

Webb, R. H., and H. G. Wilshire. 1980. Recovery of soils and vegetation in a Mojave Desert ghost town, Nevada, USA. *Journal of Arid Environments* **3**:291–303.

Webb, R. H., and H. G. Wilshire. 1983. *Environmental effects of off-road vehicles: impacts and management in arid regions.* Springer-Verlag, New York, New York.

Webb, R. H., H. G. Wilshire, and M. A. Henry. 1983. Natural recovery of soils and vegetation following human disturbance. Pages 279–300 *in* R. H. Webb and H. G. Wilshire, editors. *Environmental effects of off-road vehicles: impacts and management in arid regions.* Springer-Verlag, New York, New York.

Wells, P. V. 1961. Succession in desert vegetation on streets of a Nevada ghost town. *Science* **134**:670–671.

Wilshire, H. G. 1978. *Study results of 9 sites used by off-road vehicles that illustrate land modifications.* U.S. Geological Survey Open File Report 77–601. U.S. Department of the Interior, Palm Springs, California.

Wilshire, H. G. 1983. The impact of vehicles on desert soil stabilizers. Pages 31–47 *in* R. H. Webb and H. G. Wilshire, editors. *Environmental effects of off-road vehicles: impacts and management in arid regions.* Springer-Verlag, New York, New York.

Wilshire, H. G., and J. K. Nakata. 1976. Off-road vehicle effects on California's Mojave Desert. *California Geology* **29**:123–132.

NINE

Historical Patterns of Road Networks in Mojave National Preserve

JOHN VOGEL AND DEBRA L. HUGHSON

The presence of roads has large effects on the desert landscape (Brooks and Lair *this volume*). Roads divert overland flow, changing the patterns of washes (Nichols and Bierman 2001; Hereford *this volume*), reducing biomass in newly water-deficient regions (Schlesinger and Jones 1984), and enhancing road-edge vegetation (Johnson et al. 1975). Roads may fragment the habitat of some vertebrate species (Barrows and Allen *this volume*). Human presence associated with roads can also affect desert ecology through increased hunting pressure, collecting of plants and animals, frequency of fire ignitions, and land disturbance. Roads themselves have consequences through increased eolian dust transport and as avenues for species invasions (Nakata et al. 1976; Epps et al. 2005; Brooks *this volume*). Finally, abandoned roads require considerable time, typically more than a century, for total recovery (Webb et al., Natural Recovery, *this volume*), and create visual scars on the landscape that both attract continued use and detract from wilderness aesthetics. In this chapter we show an example of road proliferation in the area of Mojave National Preserve (MNP), California, from 1885 to the creation of MNP in 1994.

The most primitive type of road in the Mojave Desert simply consists of the tracks left by vehicles traversing the landscape. With repeated use, these tracks can become well-worn ruts (fig. 9.1). The first two-track trails in the Mojave Desert were cut by the wagon wheels of European emigrants. Prior to the arrival of the Europeans, trails had been created by humans, but these were traversed solely by foot and pack animals (Mann 2005). Wagons in the nineteenth century, and automobiles and military vehicles in the early twentieth century, left distinctive two-track roads in the vulnerable desert soils, some of which can still be seen today (Casebier 1986; Hereford *this volume*). As each new road was forged, it remained visible for long periods and attracted subsequent travelers. As more and more newly arrived emigrants began to explore the desert, two-track roads proliferated and widened.

Simple road construction techniques—road cuts to make hillsides passable, fill to span channels, culverts to allow drainage, and diversions to protect road beds—were applied to improve access. Frequently used roads were subject to routine maintenance and were eventually widened to multiple lanes. Bridges, borrow pits, concrete drainage structures, and imported gravel surfaces added the next layer of improvements. A small asphalt refinery began operation in California in 1864 and, by the 1920s, asphalt paving was in widespread use for road surfacing (Barth 1962). Within slightly more than a decade of its official designation in 1926, the famous Route 66 was paved all the way from Chicago to Santa Monica, crossing the Mojave Desert from Kingman to San Bernardino through Needles and Barstow (Powell 2001). In 1972, Interstate 40 bypassed Route 66 and Interstate 15 replaced U.S. Highway 91 (Cooper 1996). Vehicle routes, from trails to multilane interstate highways, are classified here by type or usage (trail road, unimproved road, light-duty road, paved road, or highway), which differs from the classification system used by Brooks and Lair (*this volume*) because of data limitations.

Road proliferation may be an indicator of the intensity of human presence on the landscape. Some positive feedbacks likely exist, whereby proliferation of roads tends to increase the density, and thereby intensity, of land use by making remote places more easily accessible to more people. Impacts of changing land use practices in the Mojave Desert ecoregion (Bailey 2005) are illustrated by historical maps (Carrico and Norris 1978). The rapid proliferation of roads in MNP may be representative of changes in the intensity of human presence over the larger Mojave Desert ecoregion. Coincident changes include a decline in desert tortoise populations (Berry 1990; U.S. Fish and Wildlife Service 1994), the spread of exotic vegetation (Baker 1986; Brooks 1998, *this volume*), increased fire frequency (Brooks 1999), and consequent ecosystem degradation. The causality of human effects on these coincident changes is often unclear, but their concurrence is not.

In this chapter, we have assembled historic data to create maps representing the status of roads in what is now MNP (fig. 0.1) for approximately the years 1885, 1929, 1955, 1980 (not shown), and 1994. The increasing network of roads indicated in this sequence represents the cumulative presence of vehicular traffic and correlated anthropogenic influence in this part of the Mojave Desert.

METHODS

Maps compiled by Westec Services (Carrico and Norris 1978) provide data for the earliest time period (circa 1885). These maps cover about 101,000 km^2 of the

Fig. 9.1. Photograph, taken December 6, 1917, showing the rutted condition of a road across a silt plain between Palo Verde and Winters Wells in the Mojave National Preserve, California (C. P. Ross 63, Thompson 1921, courtesy of the U.S. Geological Survey Photographic Library).

Mojave Desert but lack detail and appear somewhat schematic in places. The external boundary of Mojave National Preserve encompasses an area of approximately 6475 km^2. For the period from 1910 to 1950, we relied on U.S. Geological Survey (USGS) 1:62,500 topographic maps created in the 1950s, which provide accurate and detailed historical land use data and are the first in the region to have been created by modern surveying techniques. Automobile Club of Southern California road maps for the 1929 period are very similar to the USGS maps. For the period from the late 1970s through the early 1980s, we used 1:100,000 digital line graphic data, derived from USGS 1:62,500 and 1:24,000 topographic maps and augmented by aerial photographs. Hardcopy historical maps were scanned at 250 to 400 dots per inch. The resulting images were georeferenced with Leica Geosystems IMAGINE® 8.5 software.[1] Images were displayed in ESRI's ArcMap® software, roads were digitized, and attributes interpreted based upon information obtained from historical paper maps. Algorithms in ArcMap® software were used to compute cumulative road lengths for each time frame. The last data set was compiled with excellent accuracy (according to one author's field observations) by San Bernardino County during the public debate over the passage of the California Desert Protection Act (D. Moore *personal communication*). Data sets were visually compared with USGS digital orthophoto quadrangles taken between 1992 and 1999 to assess accuracy.

RESULTS

Trails and wagon roads crossed the region in the late nineteenth century (fig. 9.2). At that time, the Mojave Road was a major east-west route through the region (Casebier 1986). Competition from the railroad, which paralleled the Mojave Road to the south, led to its gradual abandonment after 1883, about the time that cattle ranching and mining were becoming significant factors in the Mojave Desert. The total length of these early (circa 1885) wagon roads in MNP was about 605 km. Their character was that of two-track ruts cutting across a sparsely populated landscape. A path from Camp Rock Spring, an official Army post established in 1866 on the Mojave Road, followed the course of Wheaton Wash south towards the Atchison, Topeka and Santa Fe Railway (now Burlington Northern Santa Fe) and another wagon road led to Providence in the Providence Mountains near the Bonanza King mine (Vredenburgh 1995). The Nevada Southern Railway, which connected the town of Goffs to Manvel (later renamed Barnwell), was operated by Isaac Blake as the primary distribution point for a large area, following the discovery of silver in the Rock Springs Mining District in 1863 (Vredenburgh 1995).

By the early twentieth century, the road network had developed considerably due to ranching, mining, homesteading, and the beginnings of automobile tourism (fig. 9.3). The Automobile Club of Southern California encouraged the populace to embark on motor tours of the Mojave Desert in the still-novel automobile, and they provided roadmaps to the region's scenic attractions. The road network by this time (circa 1929) had expanded to approximately 1659 km. Maintained gravel roads added a new character to the more heavily traveled corridors. The Union Pacific Railroad corridor, passing west to east through MNP with Kelso as a refueling and water station, was completed in 1905. While no road was indicated by the data where the Union Pacific Railroad crosses the Devils Playground south and east of Soda Lake, we added this important and well-known transportation corridor to figures 9.3 and 9.4. The wagon trail following Wheaton Wash and the section of the Mojave Road that crossed the Soda Lake playa are indicated as abandoned and not shown on the 1929 map, although they must have persisted visually, and were likely used at least occasionally. Route 66 between Goffs and Fenner appears as a highway.

By circa 1955, the road network had grown to a length of 2724 km (fig. 9.4). Electrical transmission lines were constructed to bring power from Hoover Dam to Los Angeles in the 1940s (Vastek 1975). Sections of a single-lane power line road were paved with asphalt. Parts of U.S. Highway 91 (now Interstate 15) are indicated along the northeast boundary of MNP. Much detail is missing in these data, however—Lanfair Valley, for example, was settled, with post offices at Dunbar and Lanfair, from 1912 to 1927, and minor roads likely proliferated between homesteads. Much of the detail seen in the 1994 data was almost certainly created prior to 1955.

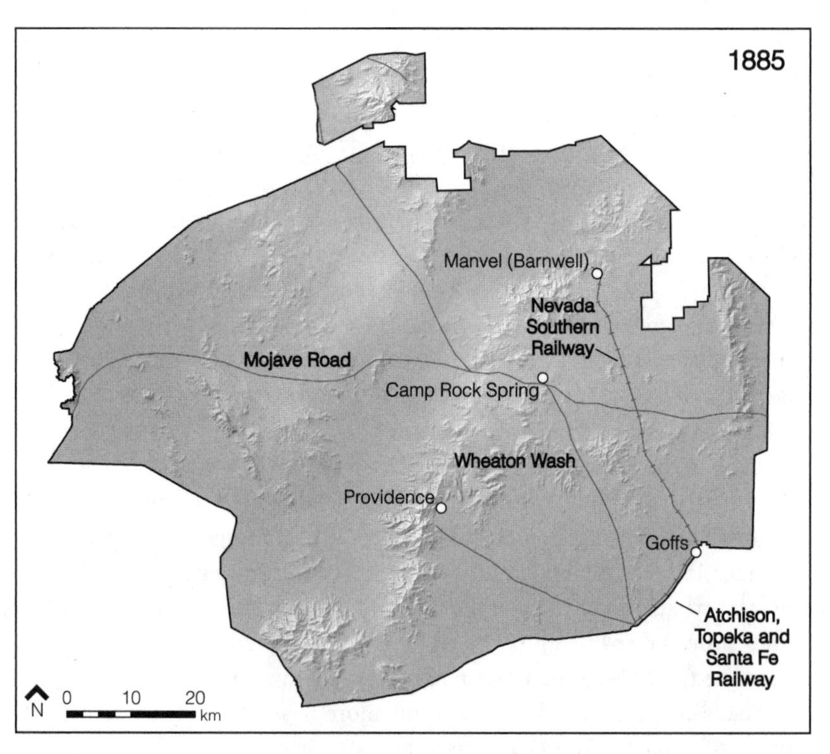

Fig. 9.2. Map of roads in the Mojave National Preserve, ca 1885, in area that became the Mojave National Preserve. The small detached area is part of the reserve north of Interstate 15. Early wagon roads had 605 km of difficult desert terrain in 1885. The Mojave Road was established by late 1859; Camp Rock Spring was established in 1866; Providence was a community of miners at the Bonanza King mine from 1882 to 1924; and Manvel was a small town that became Barnwell, a terminal on the Nevada Southern Railway, in 1893 (Vredenburgh 1995).

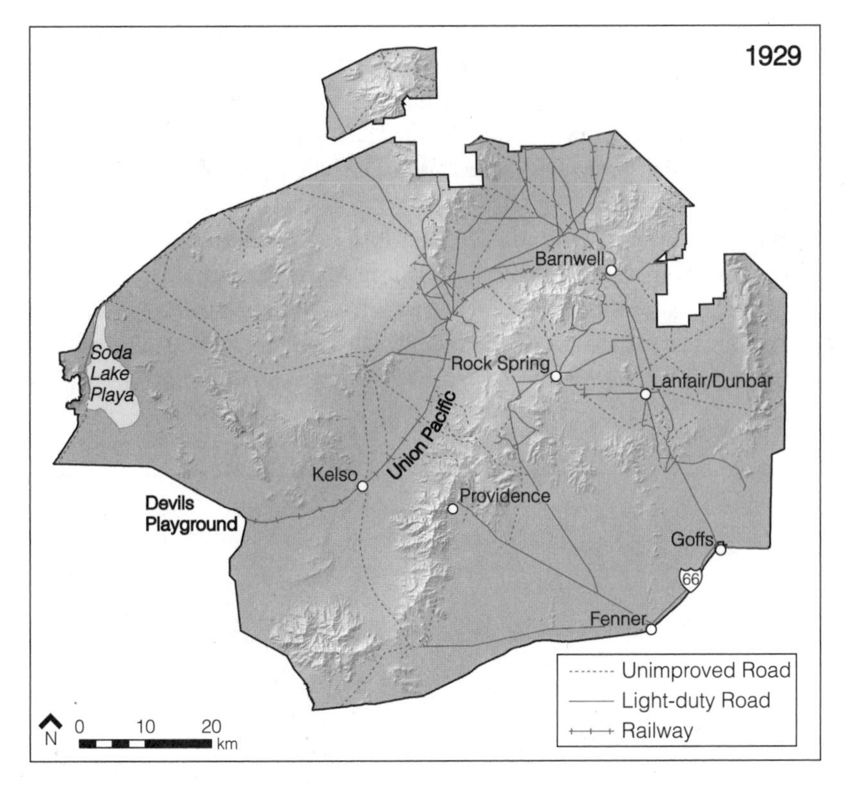

Fig. 9.3. Map of roads in the Mojave National Preserve, ca 1929. In the early 1900s, roads had expanded to approximately 1659 km in length. A post office was established at Dunbar in 1912 (D. Casebier *undated*). The Union Pacific Railroad was completed across the Mojave Desert in 1905, and Route 66, a dirt road in 1929, was a paved highway by 1938 (Powell 2001).

The USGS 1:100,000 digital line graphs for MNP circa 1980 include more detail. As a result, the total length of roads is greater, increasing to 3577 km, but we do not show the resulting map, since attributes for classifying road types were unavailable. The final road dataset shows the maximum extent of the road network (3701 km) at the time MNP was established in 1994 (fig. 9.5). Wilderness designations derived from the California Desert Protection Act simultaneously closed 726 km of two-track roads. Interstate 40 and Interstate 15 are shown in figure 9.4 for reference but are not included in the total length calculation since they are outside the boundary of MNP.

DISCUSSION

Changes in the spatial data for the Mojave National Preserve (MNP) road network represent the evolution of cartography as well as the proliferation of roads.

Error in the earliest maps is so large that the road paths are merely schematic. Roads in the 1885 data cross erosional features at odd angles and take impossible paths across mountains (fig. 9.2). The different data sets also show biased variation. For example, the Union Pacific Railroad, constructed in 1905, was not shown through Devils Playground in the 1929, 1955, or 1980 data and was only included in the 1994 data.

Road cartographic data from 1929 (fig. 9.3) show an improvement in accuracy, but occasionally miss the visual tracks, as seen in the orthophotos. For example, consider the data for the Nevada Southern Railway, which branched off the Atchison, Topeka and Santa Fe Railway at Goffs (Vredenburgh 1995). The 1885 data show sections of straight-line track awkwardly crossing hills and vales (fig. 9.6).

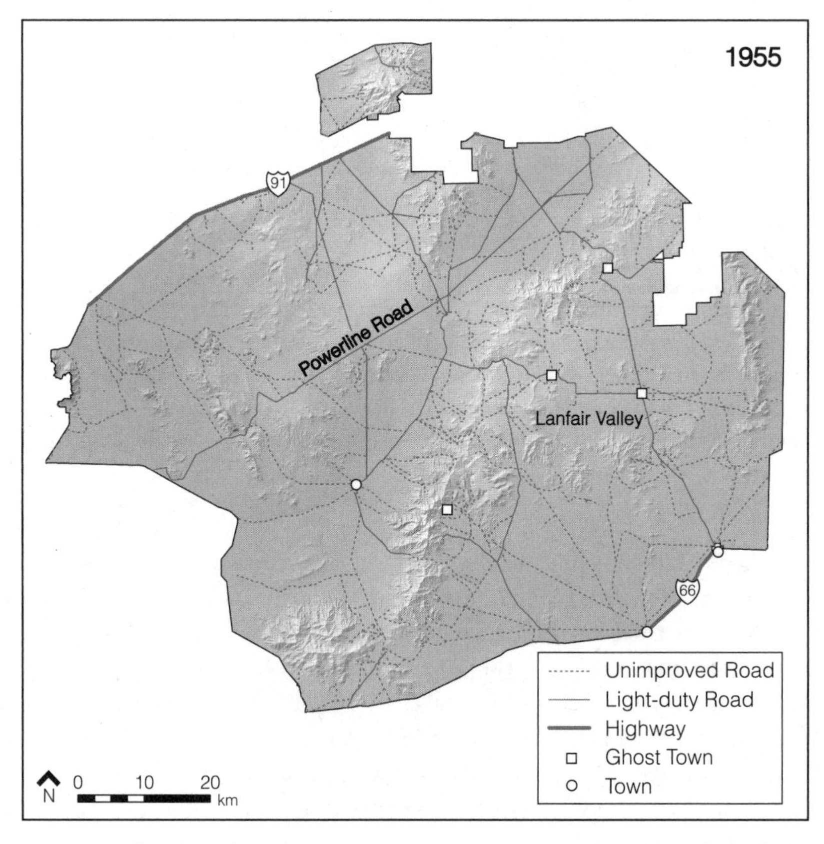

Fig. 9.4. Map of roads in the Mojave National Preserve, ca 1955. The U.S. Geological Survey 1:62,500 topographic maps for the Mojave National Preserve indicate 2724 km of roads by 1955, which is probably an underestimate for reasons discussed in the text. Powerline Road was paved in places for the construction and maintenance of electrical power lines from Hoover Dam to Los Angeles in the 1940s (Vasek et al. 1975).

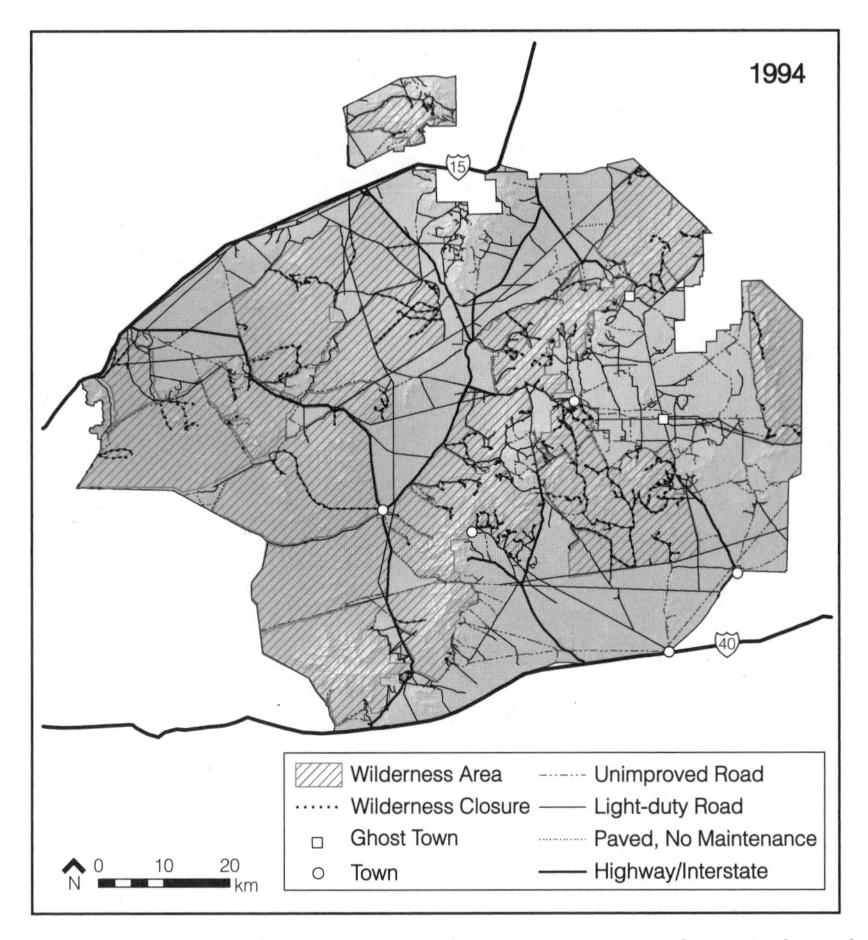

Fig. 9.5. Map of roads in the Mojave National Preserve, ca 1994. By the 1990s, the total length of the road network peaked at 3683 km. Wilderness designation closed 726 km of backcountry trail roads.

The 1929 data show roads that nearly follow the correct track, but deviate by 200 m or more in places. While both the 1955 and 1994 road data are similar, only the 1994 data show the original track of the Nevada Southern Railway, now impassable by four-wheel drive vehicles in places. Despite these cartographic errors and omissions, the road data still clearly indicate a history of road proliferation prior to changes in land management.

First carts and wagons, and then cars, trucks, and military vehicles left indelible marks on the landscape. Historical pathways of indigenous people evolved into wagon roads, followed by railroads and regularly maintained gravel roads, and finally asphalt paved roads and multilane interstate freeways. Consequent changes to the landscape were dramatic. Obviously, the important interstate corridors

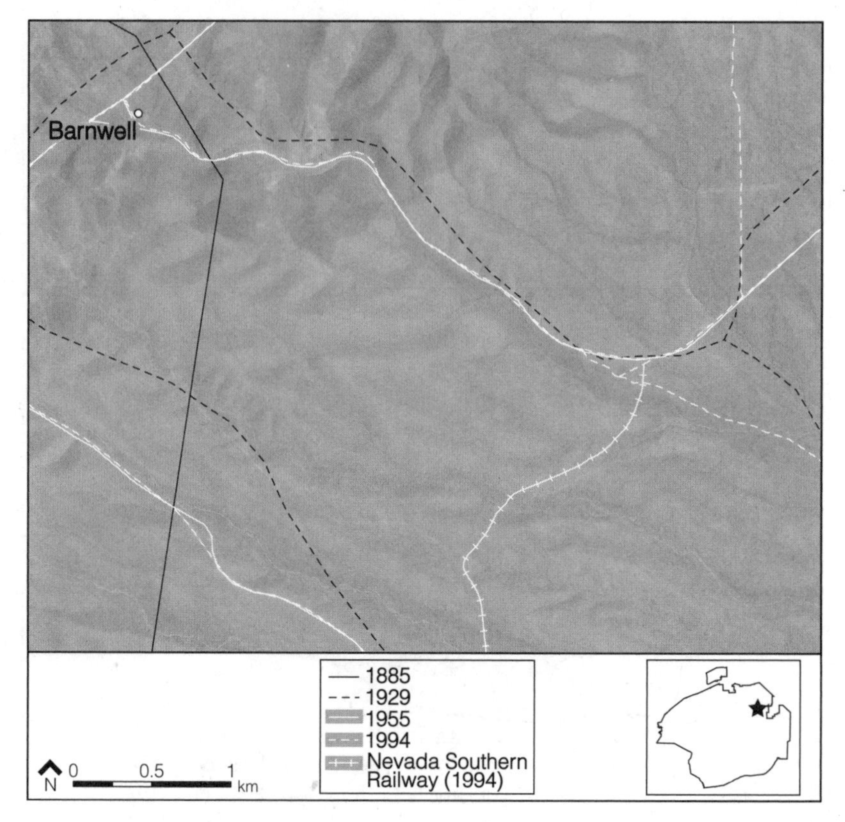

Fig. 9.6. Aerial photograph showing a comparison of road data from different years of the Barnwell, California, area within Mojave National Preserve. The 1885 Westec Services (Carrico and Norris 1978) data show a road crossing the landscape in an unnaturally straight line while schematically representing the Nevada Southern Railway, which was in operation by 1893 (Vredenburgh 1995). The Automobile Club of Southern California maps, ca 1929, show errors of at least 200 m in this projection, while 1955 and 1994 data compare fairly well. The 1994 data are the most accurate and have the most detail, including the accurate path of the Nevada Southern Railway.

of I-40 and I-15—with weekend traffic jams and the roar of long-haul transport trucks audible for several kilometers—were dramatic changes to the landscape. Less obvious, but still significant, are the increasing flocks of ravens (*Corvus corax*), which are subsidized predators (Boarman et al. 1995; Boarman and Coe 2002) drawn to human dumps, road kill, and litter lining the freeways. In addition, there is a strong spatial correlation of fire ignitions with main roads (Brooks and Esque 2002). All of these are dramatic changes from how the landscape must have appeared when bands of Indians camped along the Mojave River.

Accepting that humans are a natural part of the ecosystem, and in fact are a dominant component, leads to questions of how anthropogenic activities affect the nonhuman ecosystem. If all parts of the ecosystem are interconnected (Leopold 1968), but one species dominates, then how does this dominant species affect the rest of the ecosystem? Could roads be an indicator of the intensity of anthropogenic landscape-scale effects on the ecosystem? As one simplified analysis shows, the population of Clark County, Nevada, is correlated [$r \approx 0.79$, 95% confidence interval (-0.31, 0.99)] with the total length of roads in MNP for the five time periods included here.

Reversal of the general trend of road proliferation in MNP in 1994, with the closure of some unimproved roads, indicates an increasing general appreciation of the Mojave Desert's unique value. Land management legislation, such as the Federal Land Policy and Management Act of 1976 and the California Desert Protection Act of 1994, resulted in a concentration of some of the road-proliferation pressures, including restriction of all-terrain vehicles and dune buggy recreation to designated off-highway vehicle areas. In other areas, designated as wilderness, all motorized activity was prohibited. This designation of legally bounded areas for different types and intensities of uses has resulted in a patchwork of land use pressures on the larger Mojave Desert. Human population pressure, which precipitated the Southern Nevada Public Land Management Act of 1997, has resulted in a mosaic of suburbs and golf courses in the Las Vegas Valley. This new pressure occurs primarily on the northern boundary of MNP, but given the policies of the National Park Service, this should be represented in MNP, over the past decade and into the future, by increased frequency of traffic along existing road corridors, rather than proliferation of new roads.

Although road networks have existed at least since Roman times, the science of road ecology is relatively new. Roads can be barriers to and facilitators of plant and animal movements, barriers to and conduits of sediment transport, sources of introduced contaminants, death sinks for some wildlife populations, and a food resource bonanza for other wildlife populations. Roads affect, and are strongly affected by, runoff and sediment transport, as demonstrated by the aftermath of torrential rainfall that followed the Hackberry Complex fire in 2005. The main north-south corridor through MNP was closed for several weeks in July and August 2005 while San Bernardino County maintenance crews, using heavy equipment, moved tons of sedimentary deposits. Many road corridors still bear visual reminders of this event. It may be that some obscure backcountry trail roads will not be passable by four-wheel drive vehicles again.

Light maintenance is occasionally done to make backcountry roads passable by vehicle. One consequence of this maintenance is that often new road segments are created to detour around an obstacle. Longer-term effects of major lineaments on land forms can be easily seen in aerial and satellite photographs. One obvious effect is the diversion of runoff channels, as is apparent along the Union Pacific

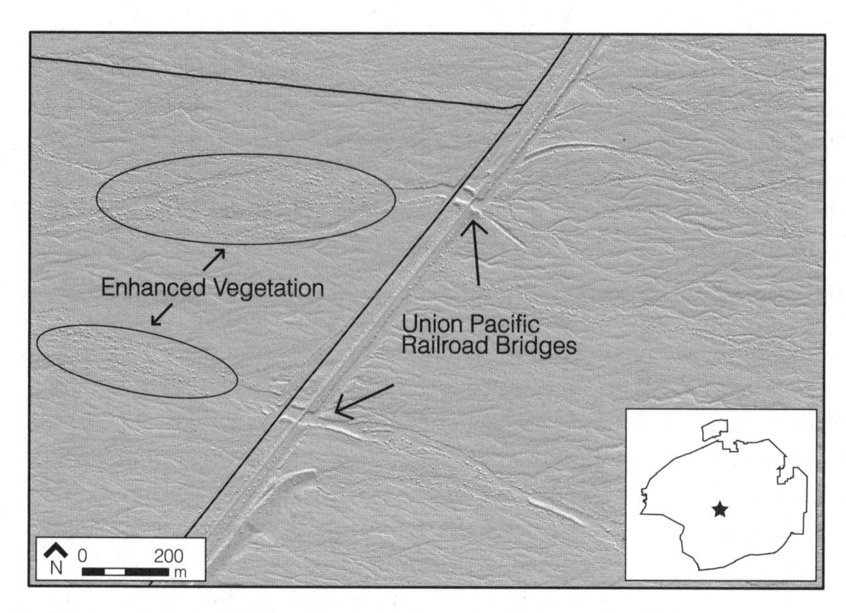

Fig. 9.7. Figure of a digital hillshade model of topography for an alluvial fan segment east of Kelso, California. The model was derived from Lidar data taken in 2005, and the topography slopes downward to the upper left. Overland flow from precipitation runoff is deflected through Union Pacific Railroad bridges. Enhanced vegetation, primarily *Larrea tridentata* (creosote bush), can be seen down-gradient in the outflow channels, while channels extant prior to construction of the railroad tend to be enhanced up-gradient and subdued down-gradient of the tracks.

Railroad north of Kelso, where runoff on the alluvial fan is focused through concrete underpasses (fig. 9.7). Channels are more incised up-gradient of the tracks, which cut across the natural flow direction, and are less incised on the down-gradient side (D. Miller *unpublished data*). *Larrea tridentata* (creosote bush) tends to favor the channels down-gradient from the railroad tracks, growing larger and denser in the washes with focused outflow (Johnson et al. 1975; Schlesinger and Jones 1984; Schlesinger et al. 1989).

Weeds and other exotic plants tend to follow railroad tracks and paved, heavily traveled roads (Brooks *this volume;* Brooks and Lair *this volume*). Weed invasion may be facilitated by the frequent disturbances that result from routine road maintenance and exacerbated by intentional and unintentional human transport of seeds. Ravens tend to congregate along roads and power line corridors, where they find perches and scavenge road kill. Cottontail and jackrabbits are killed on paved highways in great numbers, providing food for ravens and vultures. Cars also kill desert tortoises, especially during spring rains (Nicholson 1978; Berry and Turner 1984; Boarman et al. 1997). Given their long lives and slow reproduction, even an occasional tortoise road kill can drive small populations to extinction if too many mature reproductive females are lost (Congdin et al. 1994). Along the main paved highways in MNP and its two bounding interstates, this road mortality may significantly affect wildlife populations.

On the network of gravel roads, maintained and unmaintained dirt roads, and four-wheel drive trails, dust is a more noticeable effect than wildlife mortality. Rooster-tail dust plumes can follow fast-moving vehicles for a kilometer or more, leaving a low-hanging haze on the horizon. Dust mobilization caused by roads must be judged against a background of extensive, naturally occurring eolian transport. Dust storms are common events in the Mojave Desert—prevailing winds sweep small soil particles into large dune systems, often in association with alluvial transport (conveyor belt) systems that provide fine-grained sediment. Two examples—Dumont Dunes along the Amargosa River, a designated Off-Highway Vehicle Area, and Kelso Dunes on the Mojave River, a designated Wilderness Area—have opposite land management policies in terms of vehicle use. Eolian transport is important in soil and crust formation (Reheis et al. 1989; Miller et al. *this volume*) and the formation of habitat for rare, endemic species (Rahn and Rust 2000; Saul-Gershenz 2000; Saul-Gershenz and Gershenz *this volume*). The effects of increased dust mobilization from roads may be significant in some cases; however the effects of both natural and anthropogenic dust mobilization are the subject of ongoing research (Brooks and Lair *this volume*).

CONCLUSIONS

Roads are a consequence of human presence on the landscape, and the density of road networks is correlated with population. Brooks and Lair (*this volume*) show ecosystems, particularly wildlife populations, are affected by road density. Highway overpasses and underpasses and strategically placed fences have been shown to be effective in reducing wildlife road mortality (Clevenger et al. 2001) and may be necessary along some highways in Mojave National Preserve. Other ways to reduce impacts on desert tortoises could include reducing raven subsidies by routinely removing carcasses along highways and removing all legally hunted wildlife carcasses, as well as targeted raven control, highway tortoise fences, and culvert underpasses.

Invasive plants are also exacerbated by road density (Brooks and Lair *this volume*), and weed control may become even more difficult with the possible advent of climatically induced shifts in vegetation cover. Almost overnight *Brassica tournifortii* (Sahara mustard) swept across and carpeted entirely new areas of the Mojave Desert in the spring of 2005.

Regardless of what other landscape changes may occur, changes will also apply to the humans and road networks on the landscape. For example, the most recent widening of Interstate 15 is ongoing at the time of this writing. Decades hence it may become a corridor for high-speed trains and include separate paths for cars, bicycles, and animals—all of which could induce additional changes to the landscape of Mojave National Preserve.

ACKNOWLEDGMENTS

We thank University of Arizona students Joseph Riffe and Jeremy Pitts for many volunteer hours spent digitizing features from the scanned maps. Doug Schenk provided the expertly scanned and documented 1:62,500 topographic maps, as well as documentation for map projections. Kristin Berry supplied copies of the Westec Services (Carrico and Norris 1978) report and maps. David Miller provided photographs and high-resolution digital elevation data. Fran Evanisko graciously provided his GIS expertise and data. The staff of Mojave National Preserve, including Andy Leszcykowski, Robert Bryson, and Ted Weasma, generously provided their data and suggestions. Dennis Casebier of the Mojave Desert Heritage and Cultural Association shared his knowledge of the Mojave Desert region.

NOTE

1. Use of trade names does not imply endorsement by the US Geological Survey.

REFERENCES

Bailey, R. G. 2005. Identifying ecoregion boundaries. *Environmental Management* **34**: S14-S26.

Baker, H. G. 1986. Patterns of plant invasion in North America. Pages 44–57 *in* H. A. Mooney and J. A. Drake, editors. *Ecology of biological invasions in North America and Hawaii.* Springer-Verlag, New York, New York.

Barth, E. J. 1962. *Asphalt: science and technology.* Gordon and Breach, New York, New York.

Berry, K. H. 1990. *The status of the desert tortoise in California in 1989* (with amendments to include 1990–1992 data sets). U.S. Department of the Interior, Bureau of Land Management, Portland, Oregon.

Berry, K. H., and F. B. Turner. 1984. Notes on the behavior and habitat preferences of juvenile desert tortoises (*Gopherus agassizii*). Pages 111–130 in *Proceedings of the Eighth Annual Desert Tortoise Council Symposium, March 31–April 2, 1984, Lake Havasu City, Arizona.* Desert Tortoise Council, Beaumont, California.

Boarman, W. I., R. J. Camp, M. Hagan, and W. Deal. 1995. *Raven abundance at anthropogenic resources in the western Mojave Desert, California.* National Biological Service, Riverside, California.

Boarman, W. I., and S. J. Coe. 2002. An evaluation of the distribution and abundance of common ravens at Joshua Tree National Monument. *Bulletin of the Southern California Academy of Sciences* **101**:86–102.

Boarman, W. I., M. Sazaki, and W. B. Jennings. 1997. The effects of roads, barrier fences, and culverts on desert tortoise populations in California, USA. Pages 54–58 *in* J. Van Abbema, editor. *Proceedings: Conservation, Restoration, and Management of Tortoises and Turtles: An international conference, July 11–16, 1993, Purchase, New York.* New York Turtle and Tortoise Society, New York, New York.

Brooks, M. L. 1998. Ecology of a biological invasion: alien annual plants in the Mojave Desert. Ph.D. dissertation. University of California, Riverside, California.

Brooks, M. L. 1999. Alien annual grasses and fire in the Mojave Desert. *Madroño* **46**:13–19.

Brooks, M. L., and T. C. Esque. 2002. Alien plants and fire in desert tortoise (*Gopherus agassizii*) habitat of the Mojave and Colorado deserts. *Chelonian Conservation Biology* **4**:330–340.

Carrico, R. L, and F. Norris. 1978. *A history of land use in the California Desert Conservation Area.* Westec Services, Inc., Richland, Washington.

Casebier, D. G. *Undated.* Mojave National Preserve history: Lanfair Valley, a black homesteading experience. Online. Mojave Desert Heritage and Cultural Association, Essex, California. http://www.nps.gov/moja/mojahtdu.htm.

Casebier, D. G. 1986. *Mojave road guide.* Tales of the Mojave Road Publishing Company, Essex, California.

Clevenger, A. P., B. Chruszcz, and K. E. Gunson. 2001. Highway mitigation fencing reduces wildlife-vehicle collisions. *Wildlife Society Bulletin* **29**:646–653.

Congdon, J. D., A. E. Dunham, and R. C. van Loben Sels. 1994. Demographics of common snapping turtles (*Chelydra serpentina*): implications for conservation and management of long-lived organisms. *American Zoologist* **34**:397–408.

Cooper, C. 1996. History of the US highway system: from dirt paths to superhighways. Online. http://www.gbcnet.com/ushighways/history.html. Accessed 2006.

Epps, C. W., P. J. Palsboll, J. D. Wehausen, G. K. Roderick, R. R. Ramey II, and D. R. McCullough. 2005. Highways block gene flow and cause a rapid decline in genetic diversity of desert bighorn sheep. *Ecology Letters* **8**:1029–1038.

Johnson, H. B., F. C. Vasek, and T. Yonkers. 1975. Productivity, diversity and stability relationships in Mojave Desert roadside vegetation. *Bulletin of the Torrey Botanical Club* **102**:106–115.

Leopold, A. 1968. *A Sand County almanac: with other essays on conservation from Round River.* Oxford University Press, Oxford, UK.

Mann, C. C. 2005. *1491: new revelations of the Americas before Columbus.* Alfred A. Knopf, New York, New York.

Nakata, J. K., H. G. Wilshire, and G. C. Barnes. 1976. Origin of Mojave Desert dust plumes photographed from space. *Geology* **4**:644–648.

Nichols, K. K., and P. R. Bierman. 2001. Fifty-four years of ephemeral channel response to two years of intense World War II military activity, Camp Iron Mountain, Mojave Desert, California. *Reviews in Engineering Geology* **XIV**:123–136.

Nicholson, L. 1978. The effects of roads on desert tortoise populations. Pages 127–192 *in Proceedings of the Desert Tortoise Council Symposium, Las Vegas, Nevada.* Desert Tortoise Council, Beaumont, California.

Powell, J. R. 2001. A brief history of US Highway 66 and the Route 66 Association of Missouri. Online. The Route 66 Association of Missouri, Springfield, Missouri. http://www.missouri66.org/history.html. Accessed 2006.

Rahn, M. E., and R. W. Rust. 2000. Nested Coleoptera and Orthoptera on sand dunes in the Basin and Range Province of western North America. *Journal of Insect Conservation* **4**:33–43.

Reheis, M. C., J. W. Harden, L. D. McFadden, and R. R. Shroba. 1989. Development rates of late Quaternary soils, Silver Lake Playa, California. *Soil Science Society of America Journal* **53**:1127–1140.

Saul-Gershenz, L., and J. Hafernik. 2000. Beetle larvae cooperate to mimic bees. *Nature* **405**:35–36.

Schlesinger, W. H., P. J. Fonteyn, and W. A. Reiners. 1989. Effects of overland flow on plant water relations, erosion, and soil water percolation on a Mojave Desert landscape. *Soil Science Society of America Journal* **53**:1567–1572.

Schlesinger, W. H., and C. S. Jones. 1984. The comparative importance of overland runoff and mean annual rainfall to shrub communities of the Mojave Desert. *Botanical Gazette* **145**:116–124.

Thompson, D. G. 1921. *Routes to Desert Watering Places in the Mohave Desert Region, California.* US Geological Survey Water Supply Paper No. 490-B. U.S. Government Printing Office, Washington D.C.

U.S. Fish and Wildlife Service. 1994. *Desert tortoise (Mojave population) recovery plan.* U.S. Fish and Wildlife Service, Portland, Oregon.

Vasek, F. C., H. B. Johnson, and G. D. Brum. 1975. Effects of power transmission lines on vegetation of the Mojave Desert. *Madroño* **23**:114–131.

Vredenburgh, L. M. 1995. A brief summary of the history of mining in the east Mojave Desert, 1863–1947. *San Bernardino County Museum Association Quarterly* **42**:83–84.

TEN

Conservation Implications of Fragmentation in Deserts

CAMERON W. BARROWS AND MICHAEL F. ALLEN

The effects of fragmentation on natural areas continue to be a focus of conservation biology research, and fragmentation is considered one of the most serious threats to biological diversity worldwide (Wilcove et al. 1998). Most fragmentation research has focused on temperate and tropical forests (Janzen 1983; Wiens et al. 1985; Wilcove 1985; Laurence 1991; Murcia 1995; Laurence et al. 2002), and semiarid shrublands (Bolger et al. 1991, 1997; Kristan et al. 2003); little of this research has focused on desert habitats, even in the face of increasing land use and urban sprawl in the arid lands of the southwestern United States. Over the past 4 years we have been studying patterns of biodiversity within the Coachella Valley, California. This rapidly growing region is home to a large number of endemic species-of-concern that have adapted to the valley's highly variable desert environment. This valley provides an ideal opportunity to study the development, successes, and failures of habitat conservation planning.

Historically, aeolian sand habitats dominated the floor of the Coachella Valley, which covers roughly 25,900 ha. Today as much as 95% of that habitat has been lost to human development (Barrows 1996). Active aeolian sand landscapes are dynamic, with temporal changes in dune positions, vegetation cover, and substrate character. Winds in the western portion of the valley have the greatest velocity, and that is where the aeolian sand landscape exhibits the greatest amount of change. The aeolian sands provide habitat for numerous endemic taxa, including *Uma inornata,* the Coachella Valley fringe-toed lizard (federally listed as threatened), and *Astragalus lentiginosus* var. *coachellae* (Coachella Valley milkvetch, federally listed as endangered). Many of the endemic species are found primarily in the active dunes, a habitat inhospitable to species that have not evolved to survive the extreme drought, high winds, sand abrasion, and low soil nutrients that typify the habitat. Active dunes can become stabilized when vegetation density increases (due, for example, to persistent higher-than-average rainfall), wind velocity is reduced, or new sand sources are restricted by roads and development (Lancaster 1995). When this occurs, the dunes may no longer provide suitable habitat for endemic species.

Conservation networks must encompass the entirety of population and ecosystem processes in order to maintain population viability. Ecosystem processes maintain the dynamic character of habitat patches, and prevent a trajectory toward a more homogeneous landscape, which may fail to provide the diversity of habitat characteristics necessary to sustain species richness levels. The frequency of processes such as sediment flows, flooding, and fire can be essential to maintaining habitat heterogeneity. To the extent that fragmentation modifies these processes, habitat quality is affected, as well as the ability to sustain viable populations.

From the standpoint of organisms, a natural system consists of a mosaic of habitat patches of varying suitability (Wiens 1997). The porosity of a landscape matrix, which enables species to move between patches of suitable habitat, can be a critical landscape feature for sustaining viable populations. To the extent that the movement between patches is impaired, populations can lose genetic fitness (i.e., reduced heterozygosity) and can be restricted from access to critical resources that vary spatially and temporally.

Within ecological reserves, fragmentation research often focuses on habitat area and the viability of populations, many of which function as subsets of a larger population connected by periodic immigration and emigration, known as metapopulations. Whether the habitat is composed of a single unit or of a series of smaller patches, is there sufficient habitat and interpatch connectivity to sustain those populations? As the number of accessible habitat patches is reduced due to fragmentation, there is an increased risk of local extinction. With the incremental loss of metapopulations, the local populations may sink below critical-size thresholds.

A closer analysis reveals that the boundary between the habitat patches and the anthropogenic landscape may change the suitability of the habitat itself. Generally referred to as "edge effects," the primary characteristics that distinguish processes at habitat boundaries include: (1) abiotic gradients unique to those boundaries, (2) access to spatially separated resources, and (3) species interactions (Wiens et al. 1985; Murcia 1995; Laurence et al. 2002; Ries et al. 2004). Collectively, these characteristics create the basis for a conceptual framework for understanding ecological boundary responses. Population responses can be positive, negative, or neutral.

An analysis of fragmentation effects needs to be species-specific. Landscape constraints for a beetle or bee may be very different than those for a coyote (*Canis latrans*) (Wiens and Milne 1989) or for the desert bighorn (*Ovis canadensis)* (Cochran and Smith 1983; Bleich et al. 1990). Below we provide species-specific examples in the Coachella Valley to illustrate how fragmentation can impact the ability to maintain biodiversity in desert habitats.

LANDSCAPE FRAGMENTATION

Flat-tailed horned lizards (*Phrynosoma mcalli*) reach the northwestern edge of their distribution in the Coachella Valley (fig. 10.1). Within the aeolian sand landscape, flat-tails occupy habitat patches at the edges of active dunes that have moderate perennial vegetation cover and fine sands that are less mobile than the sparsely vegetated active dunes, and where harvester ants, their primary food, are most abundant (Norris 1949; Rorabaugh et al. 1987; Beauchamp et al. 1998). At different times, flat-tail lizards will also go on long treks (walk-a-bouts) looking for suitable egg-laying sites or new habitats (M. Fisher *personal observation*). They may or may not return, presumably depending on predation and on the resources encountered.

In the past, habitats conducive to flat-tail survival existed in patches throughout the Coachella Valley floor; historic flat-tail occurrences, derived from museum records, verify that distribution. Especially within the windier and more dynamic western valley, flat-tail populations likely underwent local extinctions when habitat characteristics shifted away from the preferred type. The lizards could later recolonize sites by tracking habitat suitability across an unfragmented sand landscape, perhaps assessed during their periodic walk-a-bouts. Repeated surveys between 2003 and 2005 have failed to locate any flat-tails within their historic range in the western portion of the valley floor; they have been replaced by desert horned lizards (*Phrynosoma platyrhinos*), which prefer the coarser sand and gravel conditions that currently dominate those areas (Allen et al. 2005; M. Fisher *personal communication*). Flat-tails are now confined to the one remaining, relatively large habitat patch found in the less dynamic, east-central portion of the valley floor (Allen et al. 2005; Barrows et al. 2006). The lizards are restricted from reoccupying their former range because of the numerous barriers created by suburban development and roadways (fig. 10.1).

FRAGMENTING ECOSYSTEM PROCESSES

Another species affected by habitat fragmentation in the Coachella Valley is the fringe-toed lizard. One of the key predictive variables for fringe-toed lizard

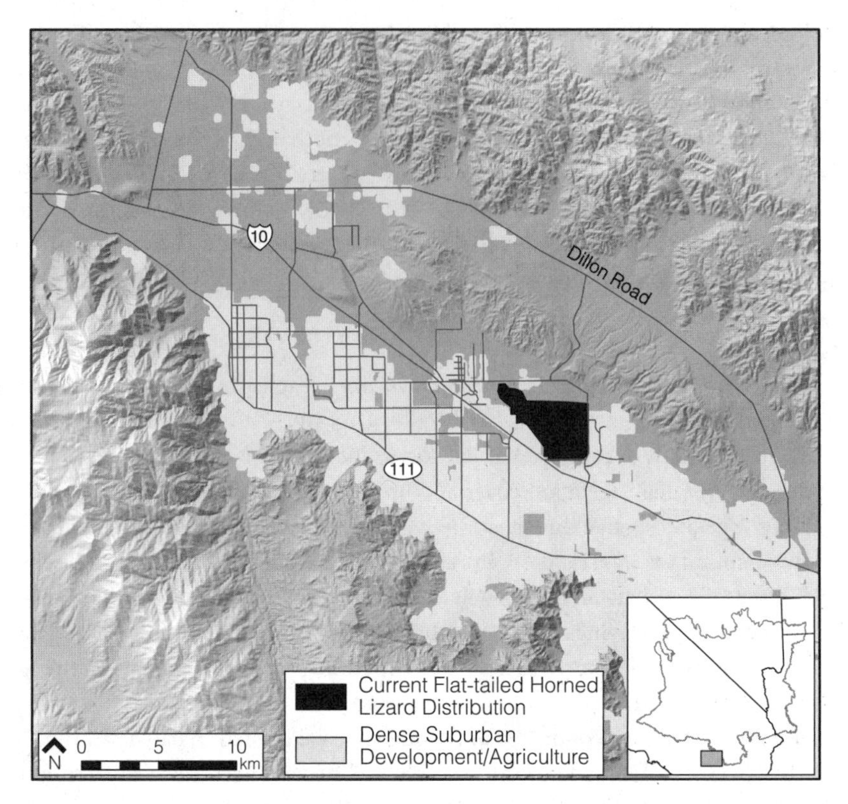

Fig. 10.1. Map of the Coachella Valley, California. Major roads are shown as solid lines, and areas impacted by dense suburban development or agriculture are shown in light gray shade. Flat-tailed horned lizards (*Phrynosoma mcalli*) once occupied most of the southeastern two-thirds of the valley; their current known distribution is shown by the solid dark polygon.

occupancy is loose, unstabilized, eolian sand (Barrows 1997). Any structures (such as a tree rows, walls, fences, buildings, or introduced invasive plants) that reduce or constrain the periodic floods that deliver new sand to the dune fields, or that block wind from sorting and transporting sand into dune habitats, will result in dune stabilization (Turner et al. 1984; Barrows 1996). As the Coachella Valley has been developed, flood control has been implemented and numerous tree rows have been planted to protect homes and golf courses from wind and blowing sand. These structures have fragmented the habitat and blocked the key processes that maintain the dunes. Relatively large contiguous dune habitats (up to ± 260 ha) on the valley floor, where the sand transport processes have been blocked, are now stabilized, and the endemic species that once thrived there, including the fringe-toed lizard, are gone (Allen et al. 2005). The importance of maintaining unfragmented, key ecosystem processes cannot be overstated, and has become a central focus in the development of a multiple-species habitat conservation plan for the Coachella Valley.

HABITAT FRAGMENTATION

How much habitat is sufficient to maintain a viable population has long been a central question in conservation biology. Unfortunately, it is a question that has been answered for only a few desert species. Chen et al. (2005) began work to address this question for the Coachella Valley fringe-toed lizard by modeling patch size and population persistence. They used a 20-year data set that tracked population dynamics at multiple individual habitat patches. Their model assumed that these habitat patches were physically isolated; no interpatch movement occurred, and the surrounding habitat matrix was inhospitable for short- or long-term fringe-toed lizard occupancy. According to the model, they determined that only habitat patches greater than 150–200 ha would be sufficient to maintain a viable population.

Answers to the question of minimum habitat size are complicated by landscape and ecosystem considerations. The long-term data set that Chen et al. (2005) used was from sites in the central-eastern portion of the valley, where rainfall is more variable and averages roughly half that of the western valley. Droughts in the eastern valley are more severe and food resources more variable; lizard populations decline dramatically during severe droughts to nearly undetectable levels (Barrows 2006). The data also revealed an annual rainfall threshold for positive population growth of roughly 45–50 mm; rainfall levels below this amount invariably resulted in population decline. Over the past 20 years, annual rainfall totaled less than 50 mm 45% of the time on these eastern valley sites (Western Regional Climate Center, Palm Springs, and Indio reporting stations). Lizards living in the western valley exist in a less variable, more benign environment. Here, annual rainfall totaled less than 50 mm just 25% of the time over those same 20 years. According to the Chen et al. (2005) model, a less variable environment would allow a population to be sustained by a smaller habitat area, since there is less risk of weather-related, catastrophic population declines.

The model, however, does not account for the fact that the configuration of the habitat patches may affect estimates of required minimum habitat area. Habitat patches less than 150–200 ha could still sustain a lizard population, as long as the patches were aggregated with other patches, and the connecting habitat between the patches was conducive to lizard movement. Under these conditions, lizards in separate habitat patches function as a metapopulation.

Our observations support these qualifications to the Chen et al. (2005) model. Six isolated habitat patches (each less than 100 ha) in the drier, central-eastern portion of the valley were occupied by fringe-toed lizards in 1993, but none were found during repeated surveys in 2004 and 2005. However, the lizards were found continuously throughout this same period on all sites where the aggregate area of connected patches was greater than 150–200 ha, even though some of the individual patches were less than 100 ha. The wetter western valley's habitat patches, which in aggregate were less than 100 ha, were continuously occupied by fringe-toed lizards as well (Barrows *personal observation*).

EDGE EFFECTS

Anthropogenic habitat fragmentation creates ecological boundaries with unique characteristics and processes. In contrast to fragmented forested biomes, where the boundary often is characterized by a moisture gradient from drier grass and shrublands into the more mesic forest core, in arid lands the opposite moisture gradient can occur. Irrigated landscapes in suburban areas can create a forest- or savanna-like landscape within an often sparsely vegetated desert.

The added vegetation in the irrigated area can provide nest sites for native birds (Germaine et al. 1998; Germaine and Wakeling 2001; Boal et al. 2003). In an analysis of edge effects at the boundary between an aeolian sand landscape and a suburban landscape in the Coachella Valley, we also found that American kestrels (*Falco sparverius*) and loggerhead shrikes (*Lanius ludovicianus*) were more abundant along habitat edges (Barrows et al. 2006). This was due, in part, to newly available nest sites, but also due to the power lines along the native habitat margins, which the birds use to launch their hunting sorties. Planted palm trees and power lines provide otherwise limiting resources to the kestrels, enabling them to occupy areas previously unavailable to them.

Analyzing a larger community assemblage in the Coachella Valley, we found no consistent responses to habitat boundary proximity for round-tailed ground squirrels (*Spermophilus tereticaudus chlorus*), sidewinders (*Crotalus cerastes*), western shovel-nosed snakes (*Chionactis occipitalis*), or harvester ants (*Pogonomyrmex* ssp., including *P. californicus* and *P. magnacanthus*). There appeared to be narrow edge-related reductions in abundance for fringe-toed lizards and desert kangaroo rats (*Dipodomys deserti*), which may have been related to the more compacted sands along the habitat edge due to road grading activities (Barrows et al. 2006).

The only species that demonstrated a significant reduced population along the patch edge, extending up to 150 m from the habitat boundary, was the flat-tailed horned lizard (fig. 10.2). This edge effect was traced to the shrikes and kestrels foraging from the power lines; the lizards suffered heavy predation rates from these birds within that boundary (Barrows 2006). Because of the sensitivity of flat-tails to these edge effects, the effective amount of land set aside to protect this species

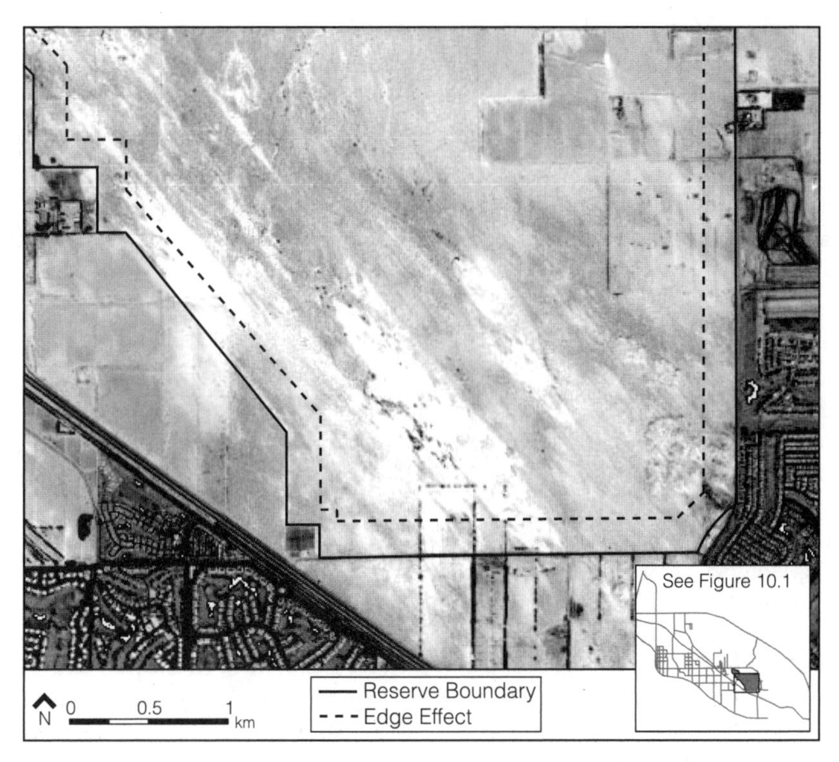

Fig. 10.2. Satellite image illustrating the juxtaposition of habitat reserve lands and surrounding suburban development in the Coachella Valley, California. The solid black line represents the reserve boundary, while the dash line represents the spatial extent to which there is a measurable edge effect for the flat-tailed horned lizard. Between the reserve boundary and the dash line, flat-tails are either absent or their numbers are greatly reduced, largely due to predation.

decreased dramatically (fig. 10.1), and individual habitat patches along the edge suffered local extinctions.

CONCLUSION

Compared with more mesic biomes, the effects of fragmentation in deserts have received little research attention. Nevertheless, desert species can be sensitive to fragmentation. However, that sensitivity is species-specific, and therefore generalized predictions are elusive. Minimizing fragmentation should be a key goal for desert conservation efforts, especially with respect to maintaining processes such as flooding, sediment transport, and the ability of species to move between habitat patches that may otherwise be too small to sustain viable populations.

Conservation areas also need to be large enough to maintain the natural dynamics that create and maintain viable habitats. Edge effects can significantly reduce the area of available habitat within conservation boundaries. Adjacent human development needs to be designed to minimize those effects, or the conservation areas need to be large enough to provide a buffer zone to minimize those impacts. In some areas, existing suburban sprawl limits the ability to employ these conservation design tenets. However, in most desert areas there are still opportunities to create well-designed, sustainable conservation land networks.

ACKNOWLEDGMENTS

The Coachella Valley Association of Governments (CVAG), and the California Department of Fish and Game provided the financial support to conduct this research. Special thanks go to Kim Nicol, Brenda Johnson, Eric Loft, and James Sullivan. Donna Thomas, Marisa Sripracha, Thomas Prentice, Brandon Mutrux, Darrell Hutchinson, and Monica Swartz provided essential support in collecting and summarizing data. Invaluable editorial suggestions were provided by Katherine Barrows.

REFERENCES

Allen, M. F., J. T. Rotenberry, C. W. Barrows, V. M. Rorive, R. D. Cox, L. Hargrove, D. Hutchinson, and K. Fleming. 2005. *Coachella Valley Multiple Species Habitat Conservation Plan Monitoring Program: 2002–2005 progress report.* Center for Conservation Biology, University of California, Riverside, California.

Barrows, C. W. 1996. An ecological model for the protection of a dune ecosystem. *Conservation Biology* **10**:888–891.

Barrows, C. W. 1997. Habitat relationships of the Coachella Valley fringe-toed lizard (*Uma inornata*). *Southwestern Naturalist* **42**:218–223.

Barrows, C. W. 2006. Population dynamics of a threatened sand dune lizard. *Southwestern Naturalist* **51**:514–523.

Barrows, C. W., M. F. Allen and J. T. Rotenberry. 2006. Boundary processes between a desert sand dune community and an encroaching suburban landscape. *Biological Conservation* **131**:486–494.

Beauchamp, B., B. Wone, S. Bros, and M. Kutilek. 1998. Habitat use of the flat-tailed horned lizard (*Phrynosoma mcallii*) in a disturbed environment. *Journal of Herpetology* **32**: 210–216.

Bleich, V. C., J. D. Wehausen, and S. A. Holl. 1990. Desert-dwelling mountain sheep: conservation implications of a naturally fragmented distribution. *Conservation Biology* **4**:383–390.

Boal, C. W., T. S. Estabrook, and A. E. Duerr. 2003. Productivity and breeding habitat of loggerhead shrikes in a southwestern urban environment. *Southwestern Naturalist* **48**:557–562.

Bolger, D. T., A. C. Alberts, and M. Soulé. 1991. Occurrence patterns of bird species in habitat fragments: sampling, extinction, and nested species subsets. *The American Naturalist* **137**:155–166.

Bolger, D. T., T. A. Scott, and J. T. Rotenberry. 1997. Breeding bird abundance in an urbanizing landscape in coastal southern California. *Conservation Biology* **11**:406–421.

Chen, X., C. W. Barrows, and B. Li. 2005. Is the Coachella Valley fringe-toed lizard (*Uma inornata*) on the edge of extinction at Thousand Palms Preserve? *Southwestern Naturalist* 51:28–34.

Cochran, M. H., and E. L. Smith. 1983. Intermountain movements by desert bighorn sheep (*Ovis canadensis*), the genetic implications. *Transactions of the Desert Bighorn Council* **27**:1–2.

Germaine, S. S., S. S. Rosenstock, R. E. Schwweinsburg, and W. S. Richardson. 1998. Relationships among breeding birds, habitat, and residential development in greater Tucson, Arizona. *Ecological Applications* **8**:680–691.

Germaine, S. S., and B. F. Wakeling. 2001. Lizard species distributions and habitat occupation along an urban gradient in Tucson, Arizona, USA. *Biological Conservation* **97**: 229–237.

Janzen, D. H. 1983. No park is an island: increase in interference from outside as park size decreases. *Oikos* **41**:402–410.

Kristan, W. B., A. J. Lynam, M. V. Price, and J. T. Rotenberry. 2003. Alternative causes of edge abundance relationships in birds and small mammals of California coastal sage scrub. *Ecography* **26**:29–44.

Lancaster, N. 1995. Pages 10–43 *in Geomorphology of desert dunes.* Routledge, New York, New York.

Laurance, W. F. 1991. Edge effects in tropical forest fragments: application of a model for the design of nature reserves. *Biological Conservation* **57**:205–219.

Laurance, W. F., T. E. Lovejoy, H. L. Vasconcelos, E. M. Bruna, R. K. Didham, P. C. Stouffer, C. Gascon, R. O. Bierregaard, S. G. Laurance, and E. Sampaio. 2002. Ecosystem decay of Amazonian forest fragments: a 22-year investigation. *Conservation Biology* **16**: 605–618.

Murcia, C. 1995. Edge effects in fragmented forests: implications for conservation. *Trends in Ecology and Evolution* **10**:58–62.

Norris, K. S. 1949. Observation on the habits of the horned lizard, *Phrynosoma mcallii. Copeia* **1949**:176–180.

Ries, L., R. J. Fletcher Jr., J. Battin, and T. D. Sisk. 2004. Ecological responses to habitat edges: mechanisms, models and variability explained. *Annual review of Ecology, Evolution and Systematics* **35**:491–522.

Rorabaugh, J. C., C. L. Palermo, and S. C. Dunn. 1987. Distribution and relative abundance of the flat-tailed horned lizard (*Phrynosoma mcallii*) in Arizona. *Southwest Naturalist* **32**:103–109.

Turner, F. B., D. C. Weaver, and J. C. Rorabaugh. 1984. Effect of reduction in windblown sand on the abundance of the fringe-toed lizard (*Uma inornata*) in the Coachella Valley, California. *Copeia* **1984**:370–378.

Wiens, J. A. 1997. The emerging role of patchiness in conservation biology. Pages 93–107 *in* S. T. A. Picket, R. S. Ostfeld, M. Shachak, and G. E. Likens, editors. *The Ecological Basis for Conservation.* Chapman and Hall, New York, New York.

Wiens, J. A., C. S. Crawford, and J. R. Grosz. 1985. Boundary dynamics: a conceptual framework for studying landscape ecosystems. *Oikos* **45**:421–427.

Wiens, J. A., and B. T. Milne. 1989. Scaling of 'landscapes' in landscape ecology, or, landscape ecology from a beetle's perspective. *Landscape Ecology* **3**:87–96.

Wilcove, D. S. 1985. Nest predation in forest tracts and the decline of migratory songbirds. *Ecology* **66**:1211–1214.

Wilcove, D. S., D. Rothstein, J. Dubow, A. Philips, and E. Losos. 1998. Quantifying threats to imperiled species in the United States. *BioScience* **48**:607–615.

PART III

The Role of Soils for Plant Communities

LYNN F. FENSTERMAKER, ERIC V. MCDONALD,
AND ROBERT H. WEBB

One of the difficulties facing resource managers today is the need to better understand how ecosystems work without adequate financial and human resources to perform the necessary studies. Managers typically lack information on specific ecosystem attributes and how they are related to ecosystem processes and function. This section provides some good examples of how basic research is an important interface between science and management. From a management perspective, applied research is more likely to provide information for immediate management of a known problem. For example, knowing rates of natural recovery (Webb et al., Natural Recovery, *this volume;* Hereford *this volume*) can aid in developing a restoration plan, or in deciding on an appropriate location for a specific land use activity. Basic research, on the other hand, is more likely to identify problems that otherwise might go unnoticed, particularly ones that may loom large in the future (e.g., the impact of climate change on invasive species). Basic research also helps to guide applied research, and thus both approaches are equally important in the quest to better understand Mojave Desert processes, function, and sustainability, from both scientific and management perspectives.

Whether research projects are categorized as basic or applied, the results will always be impacted or compounded by the soil substrate. When McDonald (1994) first mapped the age of soils on the bajadas of the Providence Mountains as part of his Ph.D. dissertation, it is doubtful that the Bureau of Land Management viewed this effort as immediately relevant to its land management practices. However, his findings inspired new research that links soils, hydrology, and vegetation and has direct land management applications (McAuliffe and McDonald 1995; Hamerlynck et al. 2002; McDonald 2002; Young et al. 2004). This body of research, which was initiated as a basic science study, had no clear relevance to managers, but the knowledge gained is now critical for understanding a wide range of ecosystem processes.

The age of geomorphic surfaces, and the amount of soil formation that geomorphic surfaces have undergone, has large effects on physical and biological properties (Miller et al. *this volume;* Bedford et al. *this volume*). Soil hydrology is strongly affected by soil development, particularly the thin surface horizon known as the vesicular A horizon, which represents the accrual of dustfall (Miller et al. *this volume*). Changes in soil hydrology affect everything from rainfall-runoff relations (Miller et al. *this volume*) to the patterns and biomass of common perennial species (Bedford et al. *this volume*). The idea that the age of a geomorphic surface is the most important factor in soil hydrology and ecology is somewhat simplistic, and modeling of, for example, soil particle-size distributions in relationship to the source distance (Bedford et al. *this volume*) begins to reveal the controls on these properties.

Root distributions are strongly influenced by soil structure (Schwinning and Hooten *this volume*), particularly restrictions placed on root elongation by high-strength soil horizons, such as argillic (clay-rich) or calcic (caliche) horizons (Stevenson et al. *this volume*). Therefore, the above-ground plant size/biomass and hence productivity, given the typical relationship between above- and below-ground biomass, are strongly affected by soil structure. Two papers in this volume (Bedford et al. *this volume;* Stevenson et al. *this volume*) provide a fresh and contrasting perspective on the relation between geomorphic surfaces and the biomass of common desert plants [(notably *Larrea tridentata* (creosote bush)], thus providing new information concerning abiotic controls of ecosystem productivity in the Mojave Desert.

Basic research into ecosystem processes and function is fundamental to public education about the value of the Mojave Desert. Our public lands, especially in a desert environment that is home to, and in close proximity to, millions of human beings, represent the perfect classrooms and laboratories for educating future generations. Basic research is fundamental in raising public interest and awareness through the unveiling of ecological stories, large and small, that help raise awareness of the need for land management practices designed to sustain resources for future generations. Sharing the wealth of information gained from basic science is the foundation of the Mojave Desert Science Symposium, and it forms a baseline of knowledge that helps to direct applied research and the quality of stewardship of the Mojave Desert.

REFERENCES

Hamerlynck, E. P., J. R. McAuliffe, E. V. McDonald, and S. D. Smith. 2002. Ecological responses of two Mojave Desert Shrubs to soil horizon development and soil water dynamics. *Ecology* **83:**768–779.

McAuliffe, J. R., and E. V. McDonald. 1995. A piedmont landscape in the eastern Mojave Desert: examples of linkages between biotic and physical components. *San Bernardino County Museum Association Quarterly* **42**:55–63.

McDonald, E. V. 1994. The relative influences of climatic change, desert dust, and lithologic control on soil-geomorphic processes and soil hydrology of calcic soils formed on Quaternary alluvial-fan deposits in the Mojave Desert, California. Ph.D. dissertation. University of New Mexico, Albuquerque, New Mexico.

McDonald, E. V. 2002. Numerical simulations of soil water balance in support of revegetation of damaged military lands in arid regions. *Arid Land Research and Management* **16**:277–291.

Young, M. H., E. V. McDonald, T. C. Caldwell, S. G., Benner, and D. Meadows. 2004. Hydraulic properties of desert pavements in the Mojave Desert, USA. *Vadose Zone Journal* **3**:956–963.

ELEVEN

Mapping Mojave Desert Ecosystem Properties with Surficial Geology

DAVID M. MILLER, DAVID R. BEDFORD, DEBRA L. HUGHSON, ERIC V. MCDONALD, SARAH E. ROBINSON, AND KEVIN M. SCHMIDT

The Mojave Desert is a water-limited system (Smith and Nowak 1990; Whitford 2002), and water availability and quality, along with other ecosystem properties, such as soil stability, nutrient availability, and interspecies competition, contribute a great deal to the structure of the landscape mosaic of plants. Most of the landscape is xeric; apart from riparian areas, the moisture available to plants is primarily that which has been stored in the soil. Near-surface soil moisture dynamics are driven by spatial and temporal interactions among climate, vegetation, and soil (Noy-Meir 1973; Rodriquez-Iturbe 2000). Recent studies exploring the linkages between geology and desert ecology have demonstrated that soil development and the geology of surficial deposits in the Eastern Mojave Desert (fig. 11.1) provide a useful template for understanding soil moisture and plant dynamics (McAuliffe 1994; McAuliffe and McDonald 1995; Hamerlynck et al. 2002).

In this chapter, we examine the mapping of potential soil moisture and examine geologic controls on lateral and vertical heterogeneity of soil moisture. We also develop examples of the ecological uses of surficial geologic maps. Within the desert landscape, piedmonts form mosaic landscapes amenable to study and also comprise the areas of greatest human impact; piedmonts are therefore the main focus of this chapter.

SOIL MOISTURE

"How much water is contained in a location within the soil and for how long?" is a driving question for desert ecosystems, because desert ecosystems are primarily governed by availability of soil moisture (Whitford 2002). Soil moisture

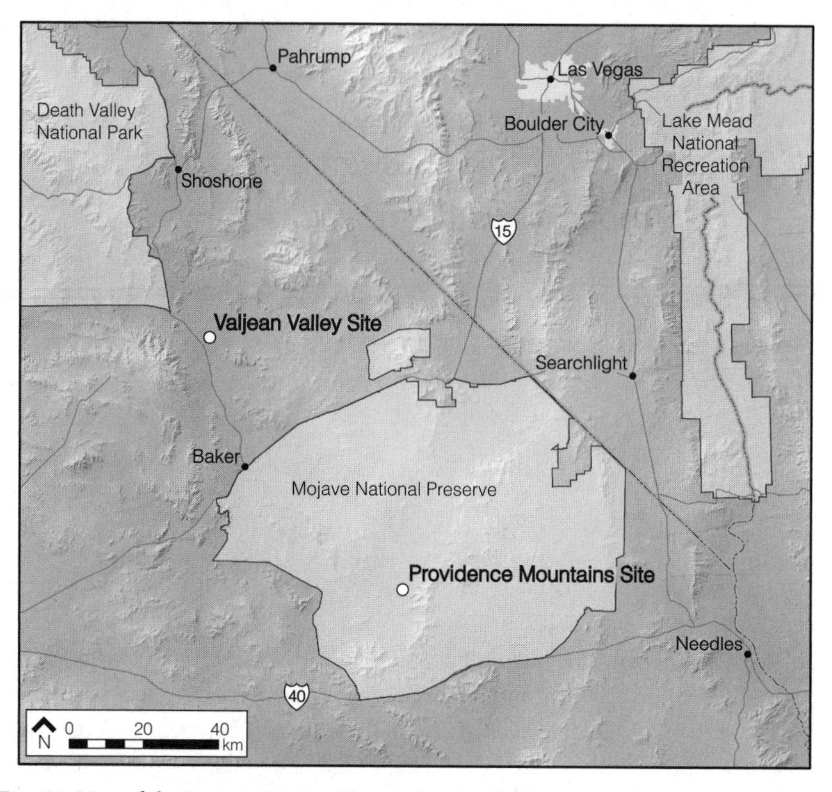

Fig. 11.1. Map of the Eastern Mojave Desert showing the locations of national park lands, major cities, and study sites (indicated as circles).

is difficult to measure directly because of the low moisture values in the desert, the vertical and lateral variation, and changes over time. Soil moisture is currently measured using invasive techniques or through remotely sensed data. Soil moisture probes can provide good site-specific information but are expensive and time consuming to install and maintain, and also involve disturbing the soil. Remote sensing techniques that either measure soil moisture directly or measure a proxy for soil moisture (such as plant cover) have been used in desert regions. Radar has been used to monitor soil moisture because radar backscatter signal is related to the dielectric constant, which depends on soil moisture. However, radar backscatter is also a function of surface roughness, soil type, vegetation, and sensor parameters (Sabins 1997). Thus, to use radar for soil moisture analysis, models must be developed for specific terrain types, making the method difficult to apply.

While plants can be mapped with remotely sensed data, plant patterns are an integrated record of soil moisture over time; without detailed understandings of plant-soil relations this integrated signal cannot be unraveled to produce quantitative values of soil moisture. Remotely sensed data of plants and other soil moisture proxies are appealing because they provide spatially continuous depictions of moisture properties and can be presented at multiple scales. However, these data are most useful after attaining a process-level understanding of soil moisture and mapping the variables most directly responsible for soil moisture distribution.

SURFICIAL GEOLOGY

The landscape of the Eastern Mojave Desert consists of rugged mountains separated by broad valleys, which generally have *Larrea tridentata* (creosote bush) scrub on alluvial fan piedmonts, denser vegetation bordering axial washes, and unvegetated playas at terminal drainage points. All parts of the physical landscape, from regional features such as mountains and valleys, to small features such as the degree of wash incision and surface-material cementation, can be understood effectively by the study of surficial geology and geomorphology.

Surficial geology describes the geologic characteristics of the landscape and surface materials in terms of the fundamental processes that have formed those characteristics. Arid land geomorphic processes that form sedimentary deposits include both transportation and deposition, and erosional processes further modify the deposits and landscape. These transportation and deposition processes can be grouped fundamentally as: (1) eolian (wind-transported); (2) fluvial, alluvial, and lacustrine (water-transported); and (3) mass-wasting (gravity-transported). Each of these transportation and deposition processes has unique characteristics in terms of the physical properties of the ensuing deposits, as well as the morphology of the deposit surface. Processes that act on these deposits and surfaces include erosion, weathering, and pedogenic soil development. Of these, pedogenic soil development generally alters the deposit more than the other processes—weathering occurs at a very slow rate in the desert environment.

Surficial geologic mapping is conducted by field methods and remote sensing interpretations. Field methods include examining landforms, surface characteristics, and soil profiles in shallow hand-dug pits and natural wash-bank exposures. Remote sensing methods primarily involve interpretation of landforms and surface characteristics on stereoscopically viewed aerial photographs. The aerial photographs are supplemented with multispectral (and hyperspectral) remote sensing data, such as Landsat 7 Thematic Mapper. A broad suite of field data, including location, geomorphic and pedogenic parameters, and vegetation attributes, is collected and referenced in a digital database. Maps are created using geographic information systems software by heads-up digitizing on digital orthophoto quadrangles that either have 1 m spatial accuracy or are custom rectified images at higher resolution. Digitization is done at much higher resolution than the ultimate publication scale, leading to a product with acceptable resolution. Examples of intermediate and detailed-scale surficial geologic maps in

Table 11.1 Characteristics of surficial geologic deposits in the Mojave Desert

Deposit name	Depositional process	Landscape position	Texture			Lateral and vertical variability
			Grain size	Sorting	Silt and clay content	
Alluvial fan	Fluvial, debris flow in places	Piedmont	Coarse gravel to sand	Low	Low	High
Eolian dune and sheet	Eolian	Commonly valley bottom	Fine to medium sand	High	Low	Low
Playa	Fluvial, lacustrine, eolian	Valley bottom	Sand, silt, clay	Moderate	High	Moderate
Axial valley wash	Fluvial, eolian	Valley axis	Fine gravel, sand, silt	Low to moderate	Moderate	Moderate
Hillslope materials	Mass-wasting	Mountain	Coarse gravel	Low	Low	High
Wetland deposit	Fluvial, eolian	Distal piedmont, valley axis	Sand, silt	Moderate	High	Low

the Eastern Mojave Desert include Schmidt and McMackin (2006) and Bedford (2003), respectively.

Soil maps, a standard product of the USDA Natural Resources Conservation Service (NRCS), provide detailed soil attributes that can be used to infer moisture retention properties. However, these maps have two drawbacks: they are generally unavailable for the Mojave Desert, and they do not typically incorporate depositional process information. Surficial geology provides a robust alternative to NRCS maps as a means for presenting and extrapolating soil data.

Surficial deposits vary in particle size, cohesiveness, bulk density, lateral and vertical homogeneity, and degree of sorting, based on the depositional processes (table 11.1 and appendix 11.1). As a result, the deposit types also vary considerably in moisture retention. The deposit types harbor distinct plant communities in many cases, so much so that vegetation maps of the desert utilize geomorphic criteria to describe plant communities (e.g., Thomas et al. 2004).

The physical substrate elements of the landscape are important because they govern factors such as the availability of soil moisture (Ehlinger 1985; Smith et al. 1995; Hamerlynck et al. 2002) and nutrients. Physical characteristics are also key to predicting properties such as susceptibility to compaction and erosion (Webb 1982), as well as recovery of vegetation from disturbance (Webb et al. 1988). Geomorphology also provides information on surface hydrology, which can be used to predict flooding and soil moisture. Natural surface roughness, for example, is a geomorphic trait that can be useful in understanding susceptibility to wind erosion (Gillette 1978; Logie 1982) and predicting habitat for organisms dependent on physical niches.

SURFICIAL DEPOSITS OF DESERT ALLUVIAL PIEDMONTS

Piedmont surficial deposits can be generalized with respect to properties such as infiltration rate and field capacity (the amount of water that soil can hold against the force of gravity) if the texture of the deposits is known in three dimensions. However, piedmont deposits tend to be rocky, poorly sorted, and irregularly bedded, with highly variable properties. The original texture of piedmont deposits depends primarily on the source bedrock, depositional position, process of deposition, and, to a lesser degree, topographic and climatic factors. Many bedrock types lead to rocky fans, but granitic rocks that weather to grus yield piedmont deposits that are finer textured and more homogenous than rocky fans (appendix 11.2). Grus-generated piedmont deposits are common in much of the Mojave Desert. In general, the mean grain size of deposits decreases with distance from the mountain front, as do topographic features such as channel incision and slope (Blair and McPherson 1994). Proximal piedmonts result from debris-flow deposition more commonly than distal deposits, which results in much coarser texture in proximal positions.

Fig. 11.2. Photograph of the Panamint Range piedmont in Death Valley showing the different ages of alluvial fan deposits, which are discernable primarily by variations in desert varnish color. Darker areas correspond to older deposits. Similar mosaics are apparent at more detailed scales.

Superimposed on these textural and topographic patterns are characteristics from the history of the piedmont's deposition, which range from remnant fragments of very old deposits to segments of active deposition (fig. 11.2). The resulting geomorphic patterns drive many landscape-scale ecosystem patterns, because different landforms have characteristic moisture infiltration and storage properties (McFadden and Knuepfer 1990; McAuliffe 1994; McDonald 1994), as described in the following section. The soils, surface characteristics, and geomorphology associated with various ages of piedmont deposits are illustrated in figure 11.3, described in appendix 11.2, and summarized in table 11.2. Figure 11.4 (see also color plate) depicts typical features of piedmont deposits ranging in age from Holocene to Pleistocene, and shows the common "terrace" inset relation of deposits in the medial to proximal piedmont position, which provides relative age information.

SOIL DEVELOPMENT AND EFFECT ON SOIL MOISTURE

Soil development in Mojave Desert surficial deposits affects many ecological processes because it dramatically changes the permeability and storage capacity of water in sediments (McAuliffe 1994; McDonald et al. 1996; Hamerlynck et al. 2002). Soil development begins when the surface of a deposit is stabilized, is no longer episodically buried by fluvial or eolian depositional processes, and is undergoing little erosion. The development of soil can be described by a series of progressive changes to the sediment that proceed in conjunction with changes to the deposit's surface morphology (described as microrelief) and other indicators, such as pavement development and varnish accumulation on surface clasts. The

Table 11.2 Summary of principal characteristics for piedmont deposits of the eastern Mojave Desert

Unit label	Surface topography	Pavement	Varnish	Av	B	Calcic morphology	Plants
Qya1	Active wash or fan	None	None	None	None	None	Few; *Hymenoclea salsola*
Qya2	Bar and swale strong	None	None–very weak	Generally none	None	None	*Hymenoclea salsola, Larrea tridentata, Ambrosia dumosa*
Qya3	Remnant bar and swale, somewhat flat	Incipient	Weak	Weak, sandy silt	Weak (Bw) reddish	Stage I	*Larrea tridentata, Ambrosia dumosa*
Qya4	Weak remnant bar and swale, fairly flat	Weak, some leveling of pebbles	Weak–moderate	Weak-sandy silt, loose	Bw	Stage I	*Larrea tridentata, Ambrosia dumosa*
Qia1	Flat, faint bar and swale	Weak–moderate	Moderate	Structured silt, 2–6 cm	Strong red, weak clay	Stage II	Sparse plants
Qia2	Flat	Moderate–strong even pebble size, pebble tops level	Moderate–strong	Structured silt, 4–8 cm	Bt usually, moderate clay	Stage II–III	Sparse; *Yucca schidigera*
Qia3	Crowned	Strong–degraded, exposed Av	Strong, purplish casts	Structured silt, 4–15 cm	Bt, strong clay	Stage III–IV	Sparse
Qoa	Whaleback	Calcic chips in pavement	Secondary or none	Secondary or none	Secondary or none	Stage IV	Common plants, many species

Yount et al. 1994; Miller et al. 2007. Correlations with other geology classifications are provided by Menges et al. (2001)

progressive changes correlate strongly with the age of the deposit (Harden and Taylor 1983; McFadden et al. 1987; Reheis et al. 1989; McDonald 1994), but soil processes are also influenced by factors such as climate variability, parent material, topography, and the biotic components of the landscape. Through detailed study, the effects of time can be disentangled from other variables, allowing us to describe the soil development in a chronosequence (cf. Birkeland 1990) related to the age of deposit.

Pedogenesis proceeds by progressive infiltration of fine-grained eolian materials (dust), chemical deposition, and weathering within sediment deposits (McFadden et al. 1987; Pavich and Chadwick 2004). Downward translocation of physical and chemical materials results in a layered system, with layers (horizons) varying in hydrologic and other properties. Pedogenesis is a continuous process, and its effects become more accentuated with time—older deposits generally exhibit more intense soil horizonation, with correspondingly greater influence on soil moisture properties, than younger soils. Because piedmont environments are dynamic, with episodic changes in fan deposition and channel incision, alluvial

Fig. 11.3. Photographs showing alluvial fan deposits that range in age from modern (active) to older than 100,000 years. Map units that correspond to table 11.2 are labeled in the photographs. *A,* an example of young fan deposits Qyal, Qya2, and Qya3. Unit Qya2 is unusually coarse-grained in this example. Note the inset of an active channel into older deposits. An approximately 40 cm tall rebar stake in the photograph provides a measure of scale. *B,* an example of a Qya4 deposit with weakly developed pavement. A rebar stake about 40 cm tall is visible in the photograph. *C,* an example of a coarse-grained proximal fan with a Qyal channel set into a Qya3 deposit by about 60 cm. *D,* an example of a proximal fan with Qyal and Qya2 inset into Qya3 deposits, which, in turn, lie 1.6 m below a Qia2 deposit. The pavement on the Qia2 surface is disturbed in places. Note the tonal differences among the deposits.

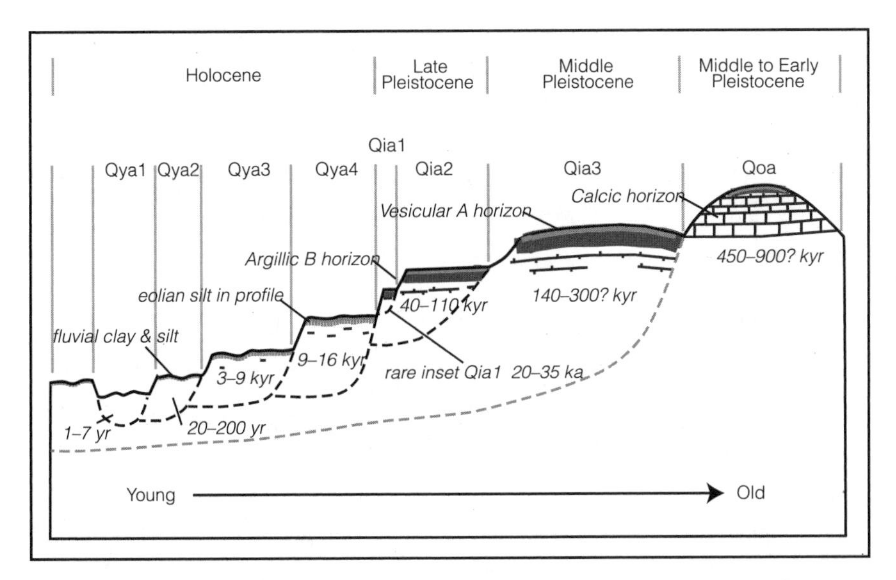

Fig. 11.4. Idealized sketch of inset piedmont deposits that depicts changes in surface morphology and soils with age. Deposit labels correspond to table 11.2. (See color plate following page 162.)

fan deposits sometimes become stranded (isolated) above more geomorphically active (and younger) areas, and are allowed to develop soil horizons. The piedmont environment evolves over time into a complex mosaic of varying soil properties due to the progressive development of soils and the presence of deposits of varying ages in the piedmonts.

The formation of desert pavement and the underlying fine-textured vesicular A horizon (Av) are two additional important features that coincide with progressive development of soils on desert piedmonts. Eolian silt and sand, common across the Mojave Desert, collect immediately below a gravel pavement, forming the Av horizon and promoting the formation of desert pavements (McFadden et al. 1987, 1998; McDonald et al. 1995). Varnish accumulates on surface clasts of desert pavement, producing the dark surface coloration common to Mojave Desert piedmonts (fig. 11.2). The degree of varnish accumulation and soil development are the two most diagnostic parameters for deposit age (McFadden et al. 1987, 1998; McDonald et al. 1995; Bull 1991). Changes in surface microrelief also coincide with soil development and the formation of desert pavement. Surface relief tends to decrease with time from the original hummocky bar and swale to a nearly flat surface as a result of eolian additions in swales and local redistribution of materials by sheetflow.

Recent studies indicate that the development of Av horizons, and, to some degree, the underlying clay-rich Bt horizons, has a profound impact on infiltration

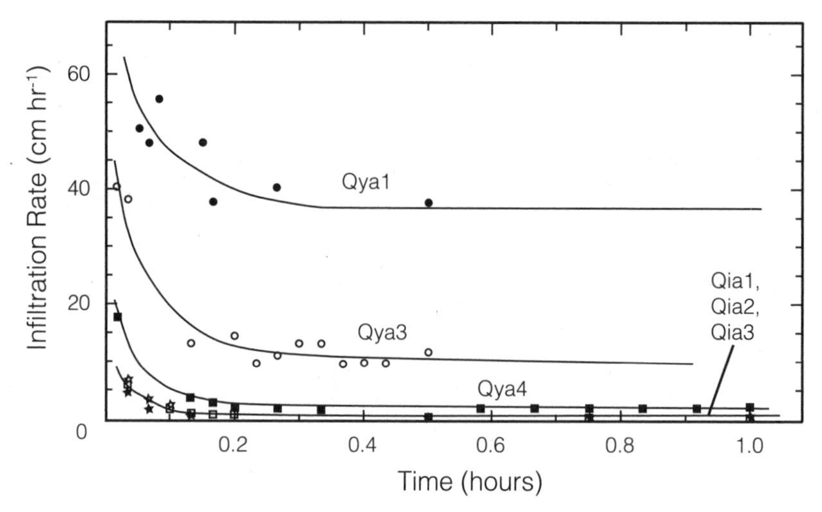

Fig. 11.5. Graph of infiltration rates for soils formed on alluvial fans along the western piedmont of the Providence Mountains, Mojave National Preserve. Infiltration was measured by double-ring infiltrometry (inner ring diameter = 60 cm, outer ring diameter = 100 cm) and 3 cm of hydraulic head. A decrease in infiltration corresponds with the development of Av horizons due to accumulation of eolian dust (modified from McDonald 1994).

rates and plant-available moisture (McDonald 1994; McDonald et al. 1996; McDonald 2002; Young et al. 2004). Measured infiltration rates for piedmont soils are reduced to as little as 1% of the infiltration rate for coarse-textured deposits lacking Av horizons (fig. 11.6). Results indicate that even small additions of eolian silt to the upper soil provide sharp reductions in infiltration. Progressive accumulation of eolian dust into soil B horizons also has a strong effect on soil-water balance by increasing the soil's water-holding capacity (McDonald et al. 1996; McDonald 2002).

The increases in surface runoff that coincide with decreases in infiltration promote surface instability. The runoff causes erosion of soils and desert pavements, eventually leading to the establishment of surface drainages and the ultimate widespread dissection of the deposit.

Although not generally considered a large factor in soil development, biologic activity, such as plant rooting and burrowing by insects, reptiles, and mammals, is important in producing some soil characteristics. In addition, stabilization of surfaces by plants probably plays a key role in reducing erosion and allowing soil development to proceed.

In the Eastern Mojave Desert, piedmont deposits show systematic relations between age and plant species composition (table 11.2), and perennial plant cover systematically decreases with the age of the deposit (McAuliffe and McDonald 1995; Bedford et al. *this volume*) with two exceptions: (1) areas undergoing active deposition, such as alluvial fan channels and washes, exhibit little or no soil development and are subject to unstable rooting conditions and abrasion by floodwaters; and (2) very old alluvial fan deposits have many or all soil horizons stripped by erosion, and typically exhibit plant assemblages similar to younger Holocene deposits.

INVESTIGATIONS OF SURFICIAL GEOLOGY AND DESERT ECOLOGY

Our studies in the Eastern Mojave Desert have explored relations between surficial geology and other ecological attributes, such as biological soil crusts, susceptibility to compaction, rates of infiltration and runoff, and perennial plant species composition and pattern. In this part of the chapter we provide examples from some of these studies to illustrate the relations between surficial deposits and ecosystem attributes. One of the strongest relations between surficial geology and ecology is that of plant species composition and pattern (Bedford et al. *this volume*). Changes in plant cover and composition vary consistently with age of deposit.

Near-Surface Texture

The Av horizon and other near-surface fine particles provide a major limit on infiltration in rocky piedmonts. We examined near-surface texture and infiltration rates at Valjean Valley (fig. 11.1) to explore the effects of fine materials in the rocky piedmont deposits. Specifically, we examined the particle-size distribution (PSD) of the upper 10 cm of a number of deposits in Valjean Valley where deposits of several ages lie within about 200 m of one another and display inset relations that indicate relative age. All deposits received metamorphic rock detritus from the nearby Silurian Hills. Figure 11.6 shows the PSD results as a function of age for samples taken at 1–5 cm depth. This depth interval tracks the Av horizon and any other shallow accumulation of fines.

The results are compared with multiple samples of active wash (Qyal) sediment, with the assumption that active wash sediments approximate the initial texture of all deposits, and departures from that texture represent pedogenic and other effects. An increase in clay and silt is evident in the Qya2 deposits, although they lack an Av horizon. With the accumulation of weakly (Qya3), moderately (Qya4), and strongly developed (Qia2, Qia3) Av horizons (table 11.3), the PSD displays a progressive increase in clay to fine sand. Not represented in the PSD is the change in the structure of the Av horizon from loose to platy (table 11.3), which apparently affects infiltration.

As described above, the thickness and maturity of the Av horizon in older deposits vary systematically with the deposit age. Young deposits that lack Av horizons also show a decrease in infiltration rate with age, and this relation also appears to be driven by the addition of silt and fine sand over time. With devel-

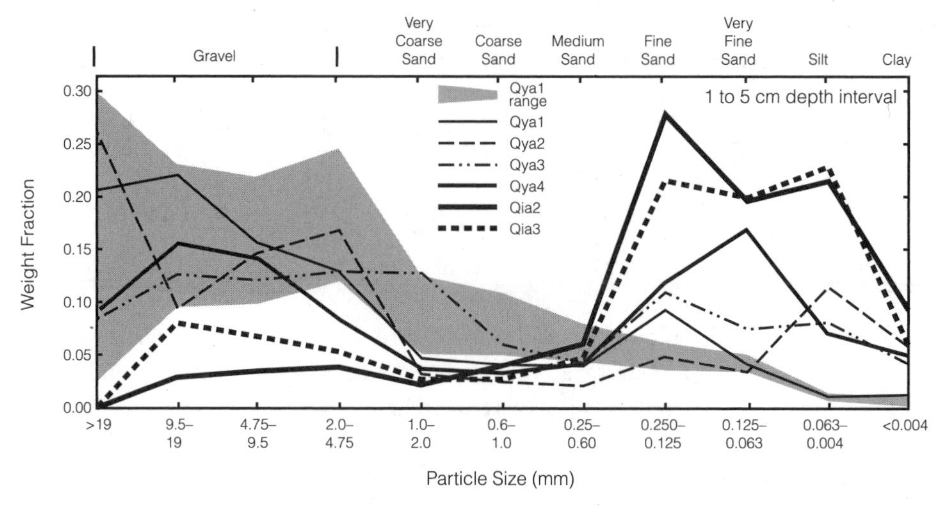

Fig. 11.6. Graph of particle-size distribution as a function of age for piedmont deposits at Valjean Valley illustrating changes in texture with age. Samples from 1–5 cm depth are compared to a reference envelope (in gray) of four bulk samples of the active wash (Qya1); this envelope represents one standard deviation from the weighted mean value and was calculated by weighting the mean by the sample size. The increase in clay, silt, very fine sand, and fine sand in older deposits is interpreted chiefly as a consequence of progressive soil development. Coarser particle size fractions decrease in relative proportion with age, because all fractions sum to one. Sizes larger than sand display large fluctuations in percentage weight fraction, because the sample size was inadequate for valid representation.

opment of the Av horizon, infiltration rates decline an order of magnitude. The results support a model in which fluvial overbank deposition of clay and silt in the Qya2 deposit reduces infiltration and enhances accumulation of eolian fines, manifested as Av horizons in older deposits.

Infiltration Rates and Moisture Retention

Measurements for piedmont deposits in the Eastern Mojave Desert, using double-ring and tension-disk infiltrometers, reinforce the idea that infiltration decreases systematically with increasing age and the degree of soil development (McDonald et al. 1996; Young et al. 2004). Results indicate infiltration for soils with well-structured Av horizons (at least 3 cm thick) is commonly less than 5 mm hr^{-1} (fig. 11.5). Measured infiltration indicates that even small additions of eolian fines to the soil surface provide sharp reductions in infiltration. At Valjean Valley, tension-disk infiltrometer tests on three deposits exhibit a similar reduction of hydraulic conductivity with age of deposit (table 11.3).

Progressive accumulation of eolian fines into soil B horizons also has a strong effect on soil-water balance by increasing the soil's water-holding capacity

Table 11.3 Av horizon characteristics and measured saturated hydraulic conductivity (cm hr^{-1}), measured with a tension-disc infiltrometer) for Valjean Valley piedmont deposits

				Saturated hydraulic conductivity		
Deposit	Av thickness	Field texture	Structure	At surface	Under pavement	Under Av horizon
Qya1	None			8.9		
Qya2	None					
Qya3	1.5 cm	Fine sand	Loose	4.6		7.7
Qya4	3 cm	Sand and silt	Moderately compact			
Qia2	8 cm	Silt	Platy			
Qia3	3 cm	Silt	Platy	1.3	1.3	8.4

(McDonald et al. 1996; McDonald 2002; Young et al. 2004). Well-developed Bt and Btk horizons in the intermediate age Pleistocene deposits commonly have loam, sandy clay loam, and clay loam textures rich in accumulated silt and clay. By comparison, the weakly developed B horizons in younger Holocene deposits commonly have sand, loamy sand, and sandy loam textures (McDonald 1994). The addition of the silt and clay enhances the formation of soil structure and increases the number of micropores. This combination increases retention of soil moisture, commonly limiting infiltration and redistribution to the uppermost 50 cm of the soils (McDonald et al. 1996; McDonald 2002; Young et al. 2004).

Modeling Soil Moisture Using Surficial Geology Data

Ecological models link the physical, biogeochemical, and meteorological processes with empirical data to provide a quantitative understanding of soil moisture dynamics. Equally important, models provide the means to extrapolate beyond measurement sites to predict system dynamics on broader temporal and spatial scales. The fundamental soil moisture model we used combines conservation of water mass with the simplified idea that flow is proportional to pressure head differential. For partially saturated soils an additional simplification ignores the gas phase and considers water conductivity through the soil pores as a function of the geometry of soil particles. A key concept in this relationship is that finer textured soil not only has lower permeability but also binds water more tightly to soil particles by capillary force due to the smaller sizes of the pores, and thus retains higher moisture content than coarse-textured soils. The resulting field equation, known as Richards' equation (Richards 1931), is solved with one of the several different functional forms to approximate moisture retention (e.g., Gardner 1958; Brooks and Corey 1964; van Genuchten 1980). Given the soil's properties and boundary conditions, the field equation can be solved for pressure and moisture content in space and time.

Because hydraulic properties of soils are a function of texture, the soil-water relationship is the result of the configuration of water-filled pores of varying sizes, and curves of moisture content versus hydraulic pressure are used to characterize pore size distribution (e.g., Childs and Collis-George 1948; Purcell 1949). Conversely, soil texture is used to approximate the soil moisture retention function and relative hydraulic conductivity (e.g., Miller and Miller 1956; Nimmo et al. 2002). Soil structure also influences soil hydraulic properties, but is not generally included in the models.

The model's boundary condition at the soil surface is the mathematical statement of how much precipitation infiltrates into the soil column. Hillel (1971) defined the term "infiltrability" as the maximum rate at which water can infiltrate a given soil under atmospheric conditions. Infiltrability is a function of vegetative cover, surface roughness, soil crusts, texture, structure, and initial wetness. In general, initially dry desert soils exhibit high infiltrability. As soil is moistened, infiltrability decreases and approaches a steady-state rate approximately equal to the saturated hydraulic conductivity of the soil (fig. 11.5). For precipitation rates less than the infiltrability, the model boundary condition equals the rainfall intensity. To estimate the infiltration boundary condition when rainfall exceeds infiltrability we used a simplified physical approach developed by Green and Ampt (1911), which assumes that the wetting front in the soil column is a sharp boundary between the initial moisture content and saturated soil above. Precipitation that does not infiltrate is presumed to run off instead of ponding on the surface.

Figure 11.7 shows modeled rainfall redistribution in a coarse-textured active alluvial wash and an older Pleistocene deposit following a 30-minute summer thunderstorm on the afternoon of August 24, 2003, and a 24-hour winter rain on November 12, 2003. According to the Green-Ampt model (a one-dimensional model based on Richard's equation) all of the 15 mm that fell during the summer thunderstorm infiltrated the coarse soil in the active wash, but more than half ran off of the Pleistocene surface. During the winter, all of the rain again infiltrated the active wash. However, four-fifths (22 of 26 mm) infiltrated the fine-textured Pleistocene soil. Soils behaved differently after wetting as well—while the alluvial wash soil dried out in a few days, the Pleistocene soil held water near the level of plant roots. These results are in accord with modeling of soil water balance (McDonald et al. 1996; McDonald 2002).

The actual measurements, using dielectric soil moisture probes buried in the alluvial wash, compare moderately well with the model results (fig. 11.7) but indicate some differences and higher dimensional processes not included in this one-dimensional model—namely a lower permeability soil layer about 40 cm deep and precipitation runoff from the adjacent Pleistocene surfaces into the active wash.

Runoff

The modeled and field-determined infiltration rates (figs. 11.5 and 11.7) can be used to estimate thresholds at which rainfall will exceed infiltration and generate runoff (fig. 11.8). Rainfall events that are longer than 10–15 min will cause the soil to reach a quasi steady state of infiltration (fig. 11.5), after which it is predicted that the age of the deposit will vary inversely with the rainfall rate required to create runoff.

We have investigated field relations between the age of the deposit and runoff during rainfall events. We instrumented piedmonts of the Providence Mountains (fig. 11.1) to acquire rainfall rates over a 3-year time span (February 2003–May 2006) and repeatedly examined evidence for runoff in several places. We used data logging, tipping-bucket rain gauges that allow temporal resolution of a few seconds to measure rainfall intensity. The daily rain record (fig. 11.9) shows that several summer and winter precipitation events exceeded 10 mm d^{-1}, and potentially delivered sufficient rain to initiate runoff. However, high-total rainfall events exhibit a seasonal variation, with summer intensities as great as 20–40 mm hr^{-1}, and winter intensities typically less than 2 mm hr^{-1}. Summer rainfall intensities for the three most intense storms were greater than 60 mm hr^{-1}.

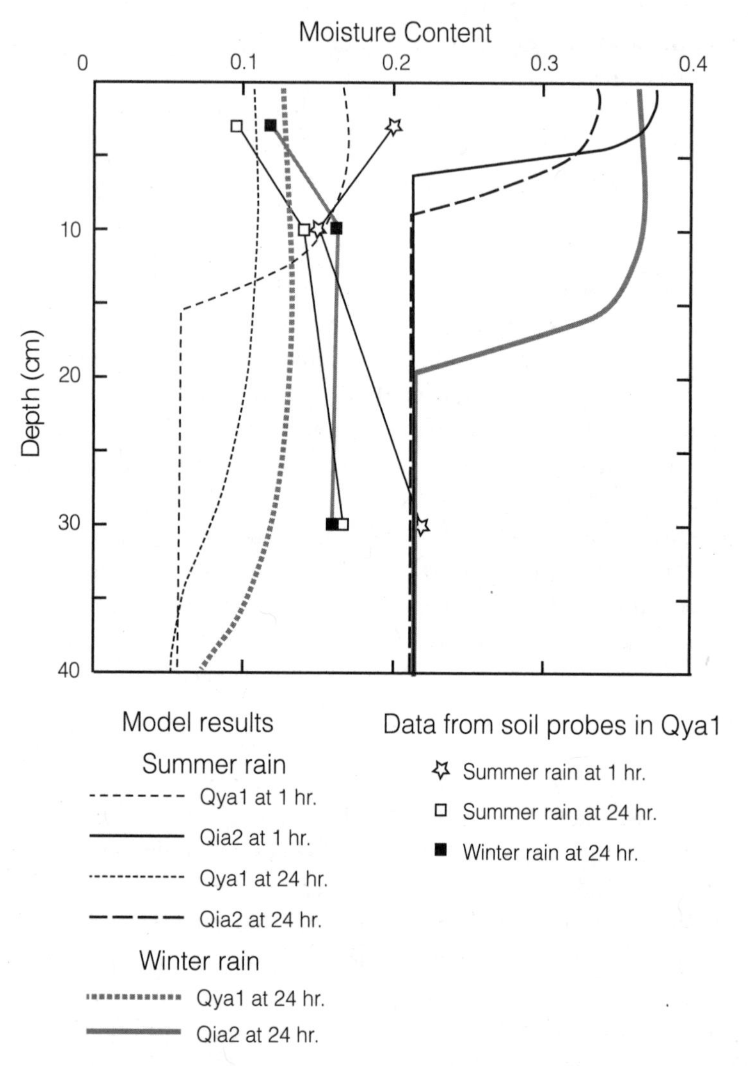

Fig. 11.7. Graph comparing infiltration rates from a numerical model to field data recorded using dielectric constant probes on the Providence Mountains piedmont, Mojave National Preserve. The model does not incorporate evapotranspiration. The altitude is the same as rain data for fig. 11.9 (~ 850 m). The summer rain occurred August 24, 2003; winter rain occurred November 12, 2003.

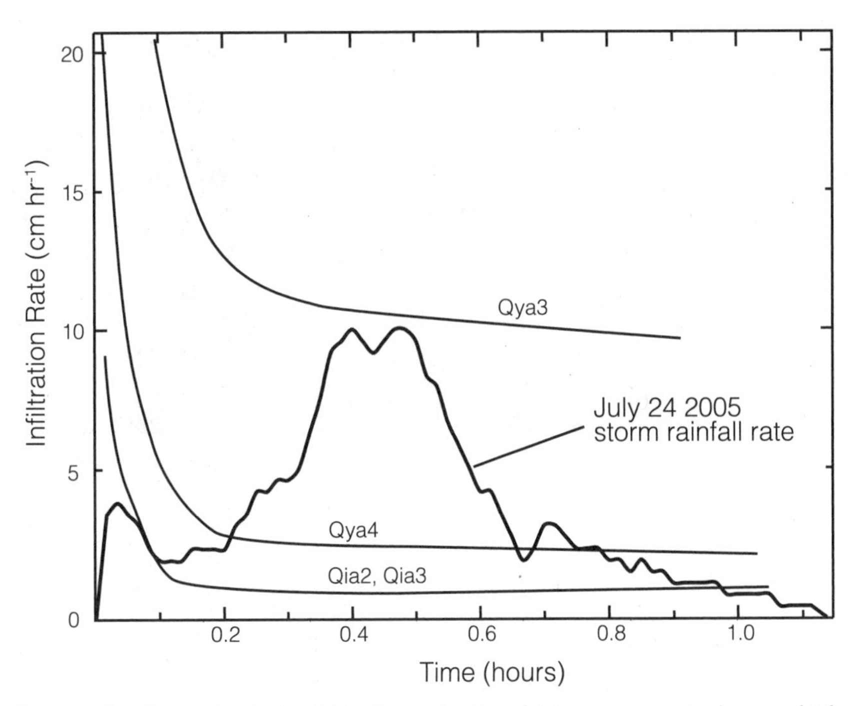

Fig. 11.8. Graph showing infiltrability determined by field measurements during a high-intensity summer rain event. Infiltration curves were measured in the field (fig. 11.6) for several ages of deposits, and rainfall data are from a measured storm in 2005 (fig. 11.9). The area above the curves and below the rain intensity line approximately represents runoff.

We studied the piedmont following rain events to establish empirical runoff results to compare with predictions from models. Field observations after storms that we attributed to runoff included: (1) redistributed sand and gravel, newly deposited sheets of overbank fines, and redistributed organic litter in large channels; (2) fine-scale depositional features in tiny upland depressions; and (3) upland areas of uni-directionally bent grasses in which litter had been removed. These indications of runoff provide a means for discriminating between local runoff and cases of channel flow caused by runoff farther up the piedmont or in mountains.

Our field observations confirm that winter storms did not create piedmont runoff during the 3-year time of study. Winter storms exhibited low maximum intensities, and the peak-intensity intervals were short compared with the overall rain event. In contrast, several summer rain events generated runoff from older deposits due to high-intensity rainfall lasting 30–60 min. Infiltration rates for Qia and Qya4 deposits were exceeded by several intense summer storms and rates for the Qya3 deposits were exceeded by two of the summer storms we recorded.

The field results support the model's prediction that older deposits exhibit

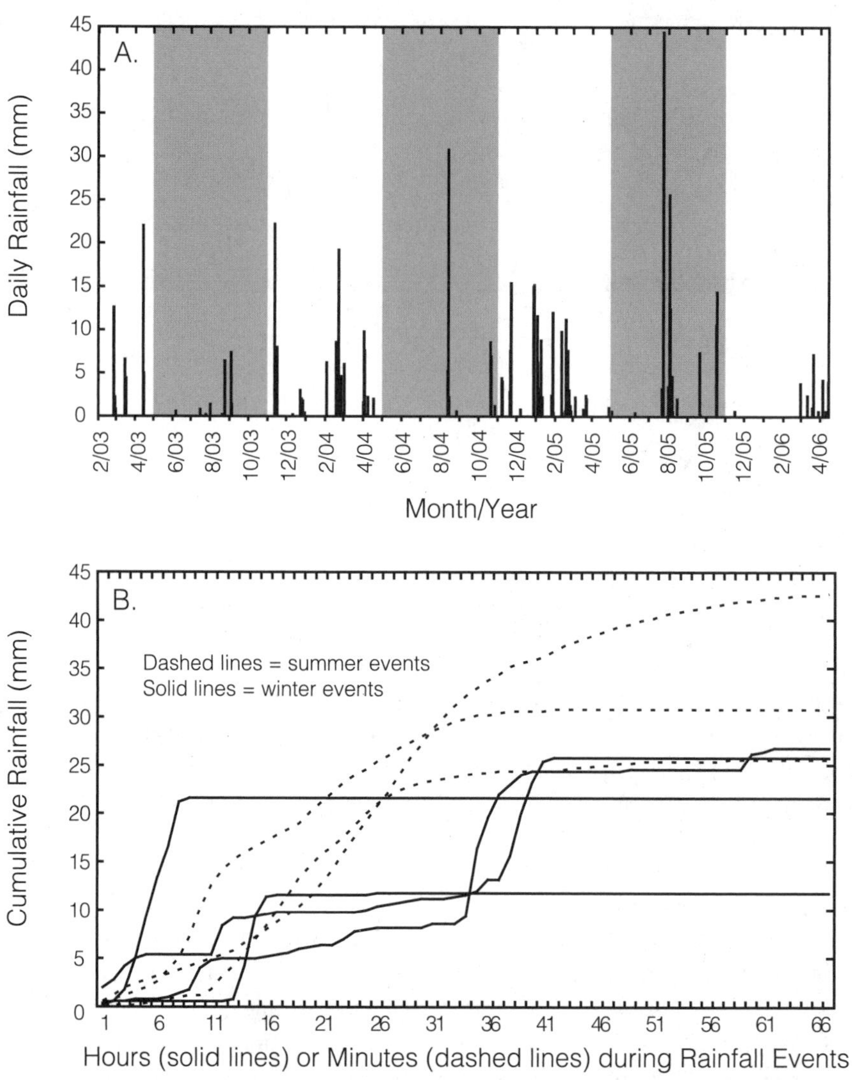

Fig. 11.9. Graphs of rainfall intensity for the Providence Mountains piedmont, Mojave National Preserve. The measurement station was located ~ 850 m altitude. *A,* a three-year rainfall record presented as daily totals. Both winter and summer events exhibit daily totals exceeding 15 mm. *B,* examples of intense rainfall events presented as cumulative rainfall. Thin solid lines represent winter events presented in hourly increments, and wide-dash lines represent summer events in minute increments. Different time increments were used, because winter rain events require one or more days to accumulate 15 mm, whereas summer events require as little as 10–15 min.

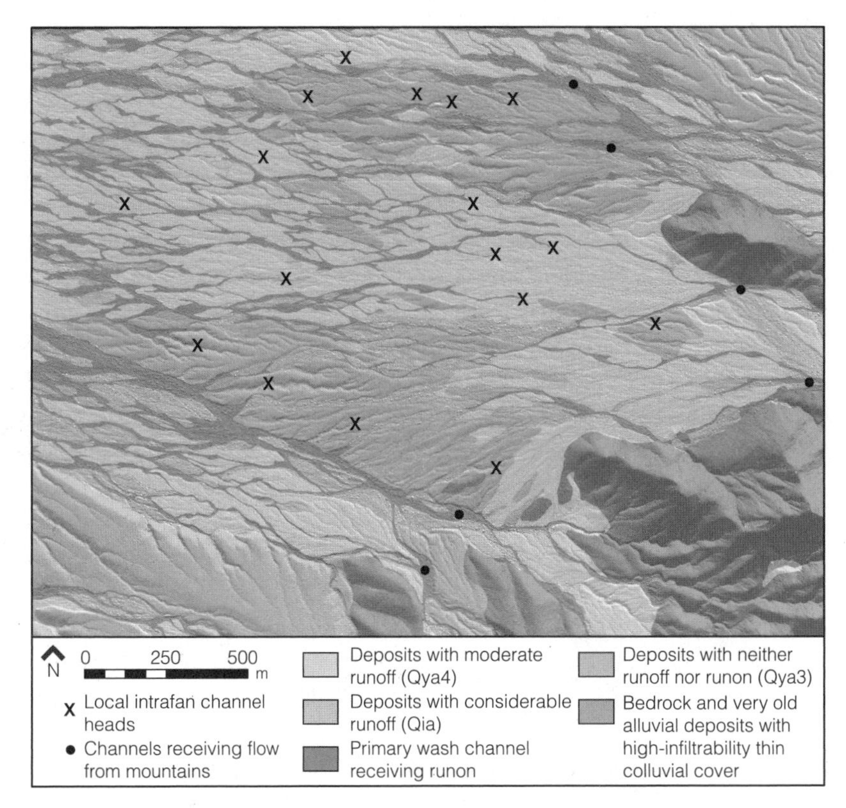

Fig. 11.10. Runoff Map derived from a detailed (1:10,000-scale) surficial geologic map of the Hayden piedmont on the west side of Providence Mountains, Mojave National Preserve. (D. M. Miller *unpublished data*). The age of surficial geologic units provides approximate infiltrability characteristics. A scenario rain event such as that in fig. 11.8 creates runoff at rapid and moderate rates, as indicated by the relationships in fig. 11.8. Channels receive local flow, indicated by an *x*, and also long-distance channel flow, indicated by solid circles, from areas higher on the piedmont. (See color plate following page 162.)

runoff under lower storm intensities than do young deposits. The partitioning by soils between infiltration into the soil column and runoff into channels is most common with summer storms. The "runon" water derived from runoff upstream is potentially available to plants with roots in and near channels, and we predict that summer-dormant plants such as *Ambrosia dumosa* (white bursage) respond less to the runon than do summer-active plants such as *Larrea.*

The observation that infiltration and runoff rates vary as a function of deposit age makes it possible to construct a runoff map with the surficial geologic map as a foundation (fig. 11.10; see also color plate). From this map, runoff generation maps for precipitation events can be created, and runon predictions can be compared with plant distributions and geomorphic features.

DISCUSSION

Surficial geology provides proxy information for soil moisture availability in the Mojave Desert. The physics of infiltration and evapotranspiration provide a means for quantifying transfer functions between geology and soil moisture under given climate conditions. Because geology is mappable and scalable, it provides a useful baseline dataset for understanding soil moisture and its ecological effects.

Many potential applications for the soil moisture properties derived from surficial geology maps remain to be explored. Runoff predictions for various climate scenarios can be compared with plant patterns, such as enhanced plant size along channel margins, to determine how plants make use of local runon and, perhaps, to make inferences about past climate trends. Infiltrability maps can be combined with climate forecasts to predict future changes in plant communities. Similarly, infiltrability maps can aid in determining what plant assemblages to use in active restoration projects. Study of the relations between soil properties and plant patterns at different scales could provide powerful predictors for plant communities.

The relations between infiltration rate, amount and location of soil moisture storage, and the canopy volume and photosynthetic activity of two shrubs (*Ambrosia* and *Larrea*) were examined by Hamerlynck et al. (2002). Results from this work indicate that *Larrea* utilizes both deep and shallow roots to maintain continuous photosynthetic activity, presumably by extracting shallow water during seasonal rainfall and deeper water during seasonal drought. *Ambrosia,* in contrast, is drought-dormant and utilizes shallow roots for rainy season activity. Canopy volume for these shrubs varies with soil development (deposit age): *Larrea* decreases in volume with deposit age (Bedford et al. *this volume*), whereas *Ambrosia* maintains a relatively constant canopy volume, but plant density increases with deposit age. These observations are explained by soil moisture models indicating that young deposits receive deep infiltration accompanied by shallow drying during summers, whereas soil horizons of old deposits retard infiltration and store soil moisture at shallow depths for longer time periods.

These results emphasize the idea that plant community structure in the Mojave Desert is more closely tied to soil development and water procurement mechanisms than to other environmental variables such as total precipitation and runoff. Our data for *Larrea* canopy volume, derived from surficial geology mapping and data collected over a broad part of the central Mojave Desert (Bedford et al. *this volume*), demonstrate that these relations hold regionally.

CONCLUSIONS

Desert ecosystems exhibit complex mosaics of plants that vary in cover and composition over distances of just a few meters. At the plot scale, relatively homogeneous areas can be studied to assess a variety of ecosystem components and processes, but extrapolating these measurements to areas beyond the plot is difficult because of the complexity of the landscape. Ecosystem properties are synchronized with surficial geology. Maps that combine process-driven information on deposit texture and pedogenesis can provide useful descriptions of soil moisture characteristics, and therefore can be used as indicators of other ecosystem properties and provide a means for scaling other ecosystem properties up from small areas of intensive study. This approach would allow ecological questions to be explored and a wide number of land management concerns to be addressed.

Monitoring studies, studies of biotic processes such as recovery from disturbance, and studies of the effects of climate change, will be improved by integrating surficial geology into the study design. In addition, understanding ecosystem responses to wildfire, invasions by exotic species, and diagnoses of past disturbance regimes may gain clarity by considering geologic substrates.

ACKNOWLEDGMENTS

Our understanding of the applications of surficial geology to desert ecology and geology have greatly benefited from guidance from and discussions with Jim Yount, Marith Reheis, Bob Webb, Chris Menges, Dennis McMahon, Phil Medica, Todd Esque, Lesley DeFalco, Kristin Berry, Jayne Belnap, Sue Phillips, and Rich Reynolds. We thank Kevin Ellett, Alan Flint, and Lorrie Flint for field infiltration measurements and for particle size analyses at the U.S. Geological Survey soil physics lab in Sacramento, California, and Leila Gass for database development and analysis. We also thank Kristin Berry and Chris Menges for careful reviews of a much earlier version of this paper, and Todd Caldwell and an anonymous reviewer for comments on this chapter.

REFERENCES

Bedford, D. R. 2003. *Surficial and bedrock geologic map database of the Kelso 7.5 minute quadrangle, San Bernardino County, California.* U.S. Geological Survey Open-File Report 03–501, scale 1:24,000. Online. Menlo Park, California. http://wrgis.wr.usgs.gov/open-file/of03–501.

Birkeland, P. W. 1990. *Soils and geomorphology.* Oxford University Press, New York, New York.

Blair, T. C., and J. G. McPherson. 1994. Alluvial fan processes and forms. Pages 354–402 *in* A. D. Abrahams and A. J. Parsons, editors. *Geomorphology of desert environments.* Chapman and Hall, London, England, UK.

Brooks, R. H., and A. J. Corey. 1964. Hydraulic properties of porous media. Hydrology Paper No. 3. Colorado State University, Fort Collins, Colorado.

Bull, W. B. 1991. *Geomorphic responses to climatic change.* Oxford University Press, New York, New York.

Childs, E. C., and N. Collis-George. 1948. Soil geometry and soil-water equilibria. *Discussions of the Faraday Society* **3**:78–85.

Ehleringer, J. T. 1985. Annuals and perennials of warm deserts. Pages 162–180 *in* B. F.

Chabot and H. A. Mooney, editors. *Physiological ecology of North American plant communities.* Chapman and Hall, New York, New York.

Forester, R. M., D. M. Miller, and V. A. Pedone. 2003. Ground water and ground-water discharge carbonate deposits in warm deserts. Pages 27–36 *in* R. E. Reynolds, editor. *Land of lost lakes: Proceedings of the 2003 Desert Symposium, April 18–21, 2003, Desert Studies Center, California State University.* Desert Studies Consortium, Fullerton, California.

Gardner, W. R. 1958. Some steady-state solutions of unsaturated moisture flow equations with application to evaporation from a water table. *Soil Science* **85**:228–232.

Gile, L. H., F. F. Peterson, and R. B. Grossman. 1966. Morphological and genetic sequences of carbonate accumulation in desert soils. *Soil Science* **101**:347–360.

Gillette, D. 1978. Tests with a portable wind tunnel for determining wind erosion threshold velocities. *Atmospheric Environment* **12**:2309–2313.

Green, W. H., and G. A. Ampt. 1911. Studies on soil physics: I. Flow of air and water through soils. *Journal of Agriculture Science* **4**:1–24.

Hamerlynck, E. P., J. R. McAuliffe, E. V. McDonald, and S. D. Smith. 2002. Ecological responses of two Mojave Desert shrubs to soil horizon development and soil water dynamics. *Ecology* **83**:768–779.

Harden, J. W., and E. M. Taylor. 1983. A quantitative comparison of soil development in four climatic regimes. *Quaternary Research* **28**:342–259.

Hillel, D. 1971. *Soil and water: physical principles and processes.* Academic Press, New York, New York.

Lancaster, N. 1994. Studies of late Quaternary eolian deposits of the Mojave Desert, California. Pages 172–175 *in* S. F. McGill and T. M. Ross, editors. *Geological investigations of an active margin.* Geological Society of America Cordilleran Section Fieldtrip Guidebook, Trip 8. San Bernardino County Museum Association, San Bernardino, California.

Lancaster, N., and V. P. Tchakerian. 2003. Late Quaternary eolian dynamics, Mojave Desert, California. Pages 231–249 *in* Y. Enzel, S. G. Wells, and N. Lancaster, editors. *Paleoenvironments and paleohydrology of the Mojave and southern Great Basin deserts.* Geological Society of America Special Paper No. 368. Geological Society of America, Boulder, Colorado.

Logie, M. 1982. Influence of roughness elements and soil moisture on the resistance of sand to wind erosion. Pages 161–173 *in* D. H. Yaalon, editor. *Aridic soils and geomorphic processes.* Proceedings of the International Conference of the International Society of Soil Science, March 29-April 4 1981, Jerusalem, Israel. Catena Supplement I. Catena Verlag, Reiskirchen, Germany.

Machette, M. N. 1985. Calcic soils of the southwestern United States. Pages 1–21 *in* D. L. Weide and M. L. Faber, editors. *Soils and Quaternary geology of the southwestern United States.* Special Paper No. 203. Geological Society of America, Boulder, Colorado.

McAuliffe J. R. 1994. Landscape evolution, soil formation, and ecological patterns and processes in Sonoran Desert bajadas. *Ecological Monographs* **64**:111–148.

McAuliffe, J. R., and E. V. McDonald. 1995. A piedmont landscape in the eastern Mojave Desert: examples of linkages between biotic and physical components. *San Bernardino County Museum Association Quarterly* **42**:53–63.

McDonald, E. V. 1994. The relative influences of climatic change, desert dust, and lithologic control on soil-geomorphic processes and soil hydrology of calcic soils formed on Quaternary alluvial-fan deposits in the Mojave Desert, California. Ph.D. dissertation. Earth and Planetary Sciences, University of New Mexico, Albuquerque, New Mexico.

McDonald, E. V. 2002. Numerical simulations of soil water balance in support of revegetation of damaged military lands in arid regions. *Arid Land Research and Management* **16**:277–290.

McDonald, E. V., L. D. McFadden, and S. G. Wells. 1995. The relative influence of climatic change, desert dust, and lithologic control on soil-geomorphic processes on alluvial fans, Mojave Desert, California: summary of results. *San Bernardino County Museum Association Quarterly* **42**:35–72.

McDonald, E. V., F. B. Pierson, G. N. Flerchinger, and L. D. McFadden. 1996. Application of a soil-water balance model to evaluate the influence of Holocene climate change on calcic soils, Mojave Desert, California, USA. *Geoderma* **9**:167–192.

McFadden, L. D., and P. L. K. Knuepfer. 1990. Soil geomorphology: the linkage of pedology and surficial processes. *Geomorphology* **3**:197–205.

McFadden, L. D., E. V. McDonald, S. G. Wells, K. Anderson, J. Quade, and S. L. Forman. 1998. The vesicular layer and carbonate collars of desert soils and pavements: formation, age and relation to climate change. *Geomorphology* **24**:101–145.

McFadden, L. D., J. B. Ritter, and S. G. Wells. 1987. Use of multiparameter relative-age methods for age estimation and correlation of alluvial fan surfaces on a desert piedmont, eastern Mojave Desert, California. *Quaternary Research* **32**:276–290.

Menges, C. M., E. M. Taylor, J. B. Workman, and A. S. Jayko. 2001. Regional surficial-deposit mapping in the Death Valley area of California and Nevada in support of ground-water modeling. Pages H151-H166 *in* M. N. Machette, M. L. Johnson, and J. L. Slate, editors. *Quaternary and late Pliocene geology of the Death Valley region: recent observations on tectonics, stratigraphy, and lake cycles* (Guidebook for the 2001 Pacific Friends of the Pleistocene Field Trip). U.S. Geological Survey Open-File Report 01-0051. Online. Reston, Virginia. http://pubs.usgs.gov/of/2001/ofr-01-0051/.

Miller, D. M., C. M. Menges, and M. McMackin. 2007. Geomorphology and tectonics at the intersection of Silurian and Death Valleys, Southern California: Fieldtrip road log. 7–49. *In* D. M. Miller and Z. C. Valin, editors. Geomorphology and Tectonics at the Intersection of Silurian and Death Valleys. U.S. Geological Survey Open-File Report 2007-1424.

Miller, E. E., and R. D. Miller. 1956. Physical theory for capillary flow phenomena. *Journal of Applied Physics* **27**:324–332.

Nimmo, J. R., J. A. Deason, J. A. Izbicki, and P. Martin. 2002. Evaluation of unsaturated-zone water fluxes in heterogeneous alluvium at a Mojave Basin site. *Water Resources Research* **38**:1215.

Noy-Meir, I. 1973. Desert ecosystems: environment and producers. *Annual Review of Ecology and Systematics* **4**:25–51.

Pavich, M. J., and O. A. Chadwick. 2004. Soils and the Quaternary climate system. Pages 311–330 *in* A. R. Gillespie, S. C. Porter, and B. F. Atwater, editors. *The Quaternary period in the United States: developments in Quaternary science.* Volume one. Elsevier, London, UK.

Peterson, F. F. 1981. Landforms of the Basin and Range province defined for soil survey. Technical Bulletin No. 28. University of Nevada, Nevada Agricultural Experiment Station, Reno, Nevada.

Purcell, W. R. 1949. Capillary pressures—their measurement using mercury and the calculation of permeability. *Transactions of the Society of Petroleum Engineers of the American Institute of Mining, Metallurgical, and Petroleum Engineers* **186**:39–48.

Quade, J., R. M. Forester, W. L. Pratt, and C. Carter. 1998. Black mats, spring-fed streams, and late-glacial-age recharge in the southern Great Basin. *Quaternary Research* **49**: 129–148.

Quade, J., M. D. Mifflin, W. L. Pratt, W. McCoy, and L. Burckle. 1995. Fossil spring deposits in the southern Great Basin and their implications for changes in water-table levels near Yucca Mountain, Nevada, during Quaternary time. *Geological Society of America Bulletin* **107**:213–230.

Reheis, M. C., J. W. Harden, L. D. McFadden, and R. R. Shroba., 1989. Development rates of late Quaternary soils, Silver Lake Playa, California. *Soil Science Society of America Journal* **53**:1127–1140.

Richards, L. A. 1931. Capillary conduction of liquids through porous mediums. *Physics* **1**:318–333.

Rodriguez-Iturbe, I. 2000. Ecohydrology: a hydrologic perspective of climate-soil-vegetation dynamics. *Water Resources Research* **36**:3–9.

Sabins, F. F. 1997. *Remote sensing, principles and interpretations.* Third edition. W. H. Freeman, New York, New York.

Schmidt K. M., and M. McMackin. 2006. Preliminary surficial geologic map of the Mesquite Lake 30' × 60' quadrangle, California and Nevada. U.S. Geological Survey Open-File Report 2006–1035. Online. Reston, Virginia. http://pubs.usgs.gov/of/2006/1035/.

Smith, S. D., C. S. Herr, K. L. Leary, and J. M. Piorowski. 1995. Soil-plant water relations in a Mojave Desert mixed shrub community: a comparison of three geomorphic surfaces. *Journal of Arid Environments* **29**:339–351.

Smith, S. D., and R. S. Nowak. 1990. Ecophysiology of plants in the intermountain lowlands. Pages 179–241 *in* C. B. Osmond, L. F. Pitelka, and G. M. Hidy, editors. *Plant biology of the basin and range.* Springer-Verlag, Berlin, Germany.

Thomas K. A., T. Keeler-Wolf, J. Franklin, and P. Stine. 2004. Mojave Desert ecosystem program: central Mojave vegetation mapping database. U.S. Geological Survey, Western Regional Science Center and Southwest Biological Science Center, Flagstaff, Arizona.

van Genucten, M. T. 1980. A closed-form equation for predicting the hydraulic conductivity of unsaturated soils. *Soil Science Society of America* Journal **44**:892–898.

Webb, R. H. 1982. Off-road motorcycle effects on a desert soil. *Environmental Conservation* **9**:197–208.

Webb, R. H., J. W. Steiger, and E. B. Newman. 1988. The response of vegetation to disturbance in Death Valley National Monument, California. U.S. Geological Survey Bulletin 1793. U.S. Government Printing Office, Washington D.C.

Whitford, W. G. 2002. *Ecology of desert systems.* Academic Press, San Diego, California.

Young, M. H., E. V. McDonald, T. C. Caldwell, S. G. Benner, and D. Meadows. 2004. Hydraulic properties of a desert soil chronosequence in the Mojave Desert, USA. *Vadose Zone Journal* **3**:956–963

Yount, J. C., E. R. Schermer, T. J. Felger, D. M. Miller, and K. A. Stephens. 1994. *Preliminary geologic map of Fort Irwin Basin, north-central Mojave Desert, California.* U.S. Geological Survey Open File Report 94–173, scale 1:24,000. Reston, Virginia.

APPENDIX 11.1

Summary of Common Surficial Geologic Deposits in the Mojave Desert

Alluvial fan and piedmont deposits are produced by deposition and transportation systems flanking mountains that route sediment eroded from the mountains toward valley axes. Slope generally decreases toward the valley (from proximal to distal fan) with an accompanying overall decrease in rock clast sizes and in degree of incision (channeling). The local relief of bar and swale morphology produced by multiple channels depositing and eroding sediment also decreases toward the distal fan. Proximal fans more commonly have deposits formed from debris flows and hyperconcentrated-flood deposits with very large clast sizes, channels, and levees. Distal fans, by contrast, chiefly have deposits formed by poorly channelized runoff and sheetflow, producing relatively thin, moderately size-sorted beds of fine gravel and sand. Two probable results of the changes in clast size and degree of incision from proximal to distal fan include decreases in bulk permeability and increases in vegetation cover because of more widely distributed water in shallower channels.

Piedmonts formed chiefly from a single lithologic source—grus-producing granite—form a special class of alluvial fan made up of relatively small and well-sorted particles. Because fine gravel and sand are present throughout grussy piedmonts, the piedmonts are broad, gently undulating, and exhibit little channel incision and less stabilization of surfaces, and therefore have less pedogenesis.

Valley-axis wash systems and integrated drainages collect surface discharge into through-flowing drainages that lie transverse to channels on fans. They differ from piedmont deposits in that they consist mainly of small clasts that have mixed sources from many mountains and piedmonts. The better size sorting and concentration of runoff for sustained periods contribute to an abundance of regularly and thinly bedded gravelly sand as the main deposit. Valley-axis washes commonly occur both as broad braided and incised narrow channels; these forms may alternate along a single wash. Overbank "floodplain" deposits are common along big wash systems; these mud layers contribute to generally lower permeability of wash deposits as compared to piedmont deposits. Another feature of valley-axis systems is complexes of fringing eolian sand dunes and sheets, because the big washes make sand available for wind transport and deposition in adjacent areas.

Playas and playa margins form distinctive geomorphic environments in basin interiors by virtue of their flatness and consequent combination of bedload and suspended-load deposition, which leads to very fine clast sizes. Playas commonly form the terminus for wash and piedmont surface discharge within a basin or local system of linked drainages, but in some places playas also form at nearly flat spots in through-going wash systems (such as Silurian Lake in eastern California). The fine sand, silt, and clay in playa deposits lead to low bulk permeability, shrink-swell behavior (commonly manifested as polygonal desiccation cracks), and (along with high alkalinity in some cases) a lack of vegetation. Near the margins of playas, coarser grain deposits may be interbedded with finer materials, which leads to greater permeability; these areas commonly have sparse vegetation.

Playas commonly are ringed by marginal zones of complex deposits formed from several processes: (1) inundation by periodic shallow lakes that form shoreline deposits, such as beach ridges and sand sheets; (2) sheetflow, which forms distal alluvial-fan wedges; and (3) eolian mud and sand deposited from the eroded playa bed, which creates dunes and sheets. These marginal deposits are both vertically stratified and laterally mingled and form a complex zone. As a result, the playa-margin environment is complex in terms of its sediment properties, landforms, and vegetation; it may be an environment extremely sensitive to changes in a variety of processes, such as surface runoff and rainfall, wind events, and changes in temperature.

Dunes and sheets of eolian sand form downwind of fluvially transported sand supplies that are not armored by processes such as chemical and biological soil crusting. Thick accumulations of eolian sand generally lie close to sources such as playas and large washes, where fluvial deposition repeatedly provides new material for eolian transport. Farther from these sand sources, and adjacent to less prolific sand sources, the eolian sand typically is mixed with alluvial fan gravel and sand to produce a shallow substrate with enriched sand. Large sand sources form generally east-oriented pathways of blowing sand (due to prevailing wind directions), such as the pathway from the Mojave River to the Kelso Dunes (Lancaster 1994; Lancaster and Tchakerian 2003).

Thick accumulations of eolian sand in the Mojave Desert are composed of unusually well-sorted sand with virtually no fines and gravel. The high sorting of eolian sand creates low compactability and very high bulk permeability and high, short-term water storage. However, water in eolian sand rapidly evaporates and, as a result, the deposits support sparse perennial vegetation. Annuals, such as grasses and several forbs, have adapted to take advantage of the brief periods of available water following rainfall events.

Wetland deposits form where the water table intersects the land surface, which is commonly in distal piedmont settings but also behind groundwater barriers such as bedrock barriers and faults. Deposits typically are fine-grained (sand to clay), light-colored, and carbonate-rich (Quade et al. 1995, 1998; Forester et al. 2003). Thin bedding, lateral persistence of bedding, and deposit texture are hallmarks. Ancient wetland deposits commonly form light-colored flat areas or dissected badlands with sparse vegetation.

APPENDIX 11.2

Summary of Piedmont Deposits and Map Units in the Mojave Desert

Young Deposits

Young deposits range from those actively inundated by sediment-carrying water (Qyal) to those with surfaces mostly stabilized and receiving infrequent inundations. The latter are generally divisible into deposits with surfaces inundated on a decadal to centennial scale (Qya2), deposits that are 3–6 kyr (thousands of years) old and rarely inundated (Qya3), and deposits that are 9–15 kyr (Qya4). The older two deposits typically display soil horizons such as weak, sandy Av, weak cambic (Bw) horizons, and calcic Stage I (calcic rinds under pebbles) or no calcic soil (see Gile et al. 1966 and Machette 1985 for more information on calcic soil stages); their surfaces typically are armored with unvarnished reg gravel (lag of gravel) or weakly varnished desert pavement, with pavement development increasing with age. Biological soil crusts composed of cyanobacteria, lichens, and mosses are common on all but the active deposits. Microrelief associated with depositional processes, such as bar and swale on alluvial fans, becomes progressively muted with time as swales gradually fill and the surface gravel layer homogenizes.

Intermediate-Age Deposits

Surfaces reach maximum stability and begin to degrade in the intermediate age range. Desert pavements, generally formed by tightly interlocking pebbles of nearly the same size that have uniform surface heights, dark varnish, and flat surfaces, are hallmarks of most intermediate-age alluvial fans. The surfaces generally are moderately to deeply incised with efficient water and sediment bypass in the incised channels. With progressive age, morphology changes from pavements with faint remnant bar and swale (Qia1), to flat pavements (Qia2), to crowned or rounded pavements (Qia3) between deep gullies. Soil horizonation is at a maximum stage of development: strongly developed Av horizons are almost entirely platy, hard silt and clay mixtures 5–15 cm thick; red Bt horizons are clay-rich and may be nearly a meter thick; and calcic horizons commonly reach Stage II and III, and less commonly are Stage IV. Intermediate-age deposits exhibit sparse perennial and annual vegetation and rare biological soil crusts. The greatly reduced infiltration also enhances surface runoff during precipitation events, which initiates channeling and degradation of the pavements.

Old Deposits

Eroded landforms and stripped soil horizons characterize old deposits (Qoa). Typical landforms are balenas (whalebacks) (Peterson 1981) that have narrow rounded ridge crests 5–15 m above neighboring channels. Remnants of A and B horizons are uncommon, but calcic horizons are 2–6 m thick with well-developed Stage IV and greater morphology, and constitute much of the exposed near-surface deposit. Degrading calcic horizons contribute carbonate chips and chunks to surface litter and weak pavements, creating a light color for many surfaces. Surfaces are relatively well vegetated despite decreased infiltration caused by the advanced stage of calcic horizon development near the surface of the deposit. Plants apparently take advantage of fracture permeability within the fragmental calcic horizon to capture sufficient moisture. Despite the deeply incised landforms, the commonly accordant crests of balenas generally define a relict alluvial fan or piedmont morphology.

TWELVE

Landscape-Scale Relationships Between Surficial Geology, Soil Texture, Topography, and Creosote Bush Size and Density in the Eastern Mojave Desert of California

DAVID R. BEDFORD, DAVID M. MILLER, KEVIN M. SCHMIDT, AND GEOFF A. PHELPS

Vegetation in arid and semiarid environments commonly exhibits mosaic behavior, whereby individual species, or clumps of individuals of like or dissimilar species, are separated by open ground. The mosaic nature of desert ecosystems leads to spatial heterogeneity in soil properties, runoff and erosion, and nutrients; these properties are key to dryland ecosystems (Tongway and Ludwig 1994; Schlesinger et al. 1996; Puigdefabregas et al. 1999; Ludwig et al. 2005). Because it is clear that the mosaic pattern is a manifestation of ecosystem function, it is therefore important to understand what controls the size and spacing of vegetation in arid ecosystems.

Arid ecosystems, and particularly shrublands in the Mojave Desert, are water limited (Sharifi et al. 1988), and it is a logical hypothesis that mosaic patterns may be determined by water dynamics. Two major controls on the water budget include precipitation inputs and the partitioning of precipitation to runoff or soil moisture. The partitioning of precipitation, as well as the location, magnitude, plant availability, and persistence of soil moisture, are largely a function of soil properties. Thus, the water budget will be a function of continuous changes in environmental gradients (e.g., climatic gradients of precipitation and evaporative potential), as well as potential discreet boundaries of physical conditions (e.g., different soil types). Furthermore, many soils exhibit intrinsic gradients of physical properties, which may differ depending on soil type.

Two ubiquitous features of the American Southwest's warm deserts have received much attention in the literature, and are the focus of this paper. The first is the evergreen shrub *Larrea tridentata* (creosote bush), extensive over a wide range of elevations in the Chihuahuan, Sonoran, and Mojave Deserts (Shreve 1964). *Larrea* can live for over 100 years, and recruitment and mortality are generally rare (Vasek 1980; Bowers et al. 1995; Cody 2000). Although *Larrea* can respond quickly to precipitation pulses as well as drought (Reynolds et al. 1999), its longevity suggests that *Larrea* characteristics are a function of the long-term hydrologic dynamics of the soils that they occupy. Therefore, we can use the characteristics of the soils on which *Larrea* exists in order to try to understand the long-term hydrologic characteristics and adaptations to water limitations.

The other ubiquitous feature in the Southwest is coalesced alluvial fans, also known as bajadas and piedmonts (Peterson 1981). Alluvial fans are formed at the base of upland regions, such as mountains or large hills, where sediments are deposited as sheet, channel, and debris flows as water flows exit the steeper upland regions and lose sediment transport capability (Bull 1977; Blair and McPherson 1994). Although the nature (e.g., lithology, clast size, amounts) of sediments derived from uplands, and the loci of deposition on the fan, change through time, most fans have remarkable similarities in character. Fan slope and area, trends in drainage area, and down-fan decreases in slope and particle size all tend to be related to one another (e.g., Blissenbach 1952; Harvey 1997). Alluvial fans also contain deposits of differing ages, due to migrations in depositional loci. Older deposits on alluvial fans are commonly stranded terraces, which provide stability to allow them to develop soil horizons through time with striking regional similarities in physical characteristics (Bull 1991; Menges et al. 2001).

Previous studies of vegetation patterns in the American Southwest have largely focused on the patterns of vegetation cover and species diversity along environmental gradients. More recently, research has focused on the characteristics of vegetation on different soil types. These two approaches, dubbed the "gradient" and "soil-mosaic" hypotheses, respectively, have largely existed separately, and have seldom been combined into a single view of the landscape at the regional scale.

Studies of vegetation along environmental gradients, conducted largely in the Sonoran Desert, have generally found that vegetation community and species density and cover change systematically along environmental gradients. At a broad scale on the order of entire desert regions, total cover—as well as cover, aggregation, and density of *Larrea*—increase with mean annual precipitation (Woodell et al. 1969; Solbrig et al. 1977). At the scale of large individual mountain-valley gradients, vegetation community (dominant types, function, etc.) tends to be a function of elevation and aspect (e.g., Whittaker and Niering 1965). Soil properties, and soil moisture amounts and stress, are commonly similar within vegetation communities, suggesting community organization around soil characteristics (e.g., Yang and Lowe 1956; Klikoff 1967; Halvorson and Patten 1974; Stein and Ludwig 1979; Wierenga et al. 1987; Parker 1991).

Most of the gradient-based studies have focused on a single alluvial fan, bajadas, or a small number of nearby alluvial fans, and on the gradient itself, and occasionally directly on soil texture changes that occur along the gradient. In general, total cover and species diversity (number of species, or diversity/richness indices) tend to decrease with distance from the mountain front along the alluvial fan (Shreve 1964; Solbrig et al. 1977; Phillips and MacMahon 1978; Key et al. 1984; Bowers and Lowe 1986; McAuliffe 1994; Parker 1995). Some authors have directly attributed the gradient effect to the fact that soil texture tends to become more fine-grained with increasing distance down a fan, and have generally found that sand and gravel content was a significant contributor to vegetation trends (Solbrig et al. 1977; Phillips and MacMahon 1978; Key et al. 1984). The effects of soil texture gradients on vegetation have been attributed to soil hydraulic properties, which affect infiltration and evapotranspiration rates (Klikoff 1967; Alizai and Hulbert 1970; Halvorson and Patten 1974; Clothier et al. 1977; Fernandez-Illescas et al. 2001).

The gradient-driven approach largely ignores the fact that along these gradients there are commonly different types of soils in varying proportions, with different ages and pedogenic effects, which greatly change soil hydraulic properties. Because of this, not all soils along a gradient respond to precipitation and evapotranspiration in the same way. Recently, researchers have begun investigating the control of soil type on vegetation characteristics.

The "soil-mosaic" approach recognizes that soils of different types are hydrologically distinct and vegetation is responding to differing soil and hydrologic characteristics. Differences are manifested in soil texture, structure (soil horizons), infiltration capacity, and rates of evapotranspiration. The lithology, age, and depositional environment of soils have been shown to control soil texture, hydrology, vegetation morphology, root structure and density, and community characteristics (Stein and Ludwig 1979; Wierenga et al. 1987; Parker 1991; McAuliffe 1994; McAuliffe and McDonald 1995; DeSoyza et al. 1997; Gile et al. 1998; Young et al. 2004; Miller et al. *this volume;* Stevenson et al. *this volume*). Studies have also shown that soils of different ages and deposit types, and thus different soil horizon characteristics, affect the sizes, distributions, and ecophysiological responses of plants (Cunningham and Burk 1973; Smith et al. 1995; Hamerlynck et al. 2000; Hamerlynck et al. 2002).

To date, we know of no papers examining the dual roles of soil type and environmental gradients on vegetation patterns in a regional context. We feel this approach is appropriate due to the regional similarities between soil-geomorphic surfaces on alluvial fans and the widespread nature of *Larrea.*

The factors that control vegetation are key to understanding ecosystem processes and to developing monitoring and management strategies. With an understanding of the controls on vegetation dynamics, field observations, combined with predictive maps of vegetation, can be used as baseline datasets. Such maps can also be used to better assess the net landscape-wide distributions of erosion, soil moisture, and nutrient and carbon pools. More importantly, this information is critical to developing the process-based ecological models required to anticipate the potential effects of disturbance and changes in external forcing such as changes in climate and nitrogen deposition. Predictive models of coupled ecological and physical systems become ever more important in light of the increased development and urbanization taking place in arid environments (e.g., Los Angeles, Las Vegas, Phoenix), as well as the potential increase in the pace of climate change.

The intention of this paper is to elucidate some of the drivers of *Larrea* size and density, such as surficial geology, soil texture, and topography. We hypothesized that soil type (as defined by the type of depositional process and the age of the surficial geologic deposit) and topographic variables will be important in describing the variability in size and spacing of *Larrea.* We further hypothesized that ecologically relevant aspects of soil variability will largely consist of soil texture and the derived hydrologic properties of the soil. Our hypotheses are based on the assumption that vegetation characteristics will be determined by the dynamics of soil water, which, although not measured here, can be inferred through soil hydraulic properties. We used relatively simple statistical methods to determine the contributions of topography, surficial geology, soil texture, and derived hydraulic properties in describing the variability in size and density of *Larrea* shrubs. Statistically significant relationships to properties that can be mapped can aid in the making of predictive maps and models of vegetation characteristics. These tools will increase our understanding of the processes involved and/or help to determine what type(s) of properties may be needed to further understand the spatial dynamics of vegetation.

METHODS

Study Area

The study was conducted in the Eastern Mojave Desert of California and Nevada, an area characterized by long north- to northwest-oriented mountain blocks and interior basins. Interior basins commonly consist of coalesced alluvial fans, and a central playa or through-going wash. The arid climate is dominated by winter precipitation. Vegetation on the piedmonts consists primarily of *Larrea* and the drought-deciduous *Ambrosia dumosa* (white bursage). Data were collected during surficial geologic field mapping for the Amboy (Bedford et al. 2006), Mesquite Lake (Schmidt and McMackin 2006), Ivanpah, Soda Mountains, and Owlshead Mountains 1:100,000 (30 × 60’) quadrangles (D. R. Bedford et al. *unpublished data*). Locations of sample sites are presented in figure 12.1.

Data Collection

Many types of data relating to surficial physical and biotic characteristics were recorded during the surficial geologic mapping exercises. Here we focus on a subset

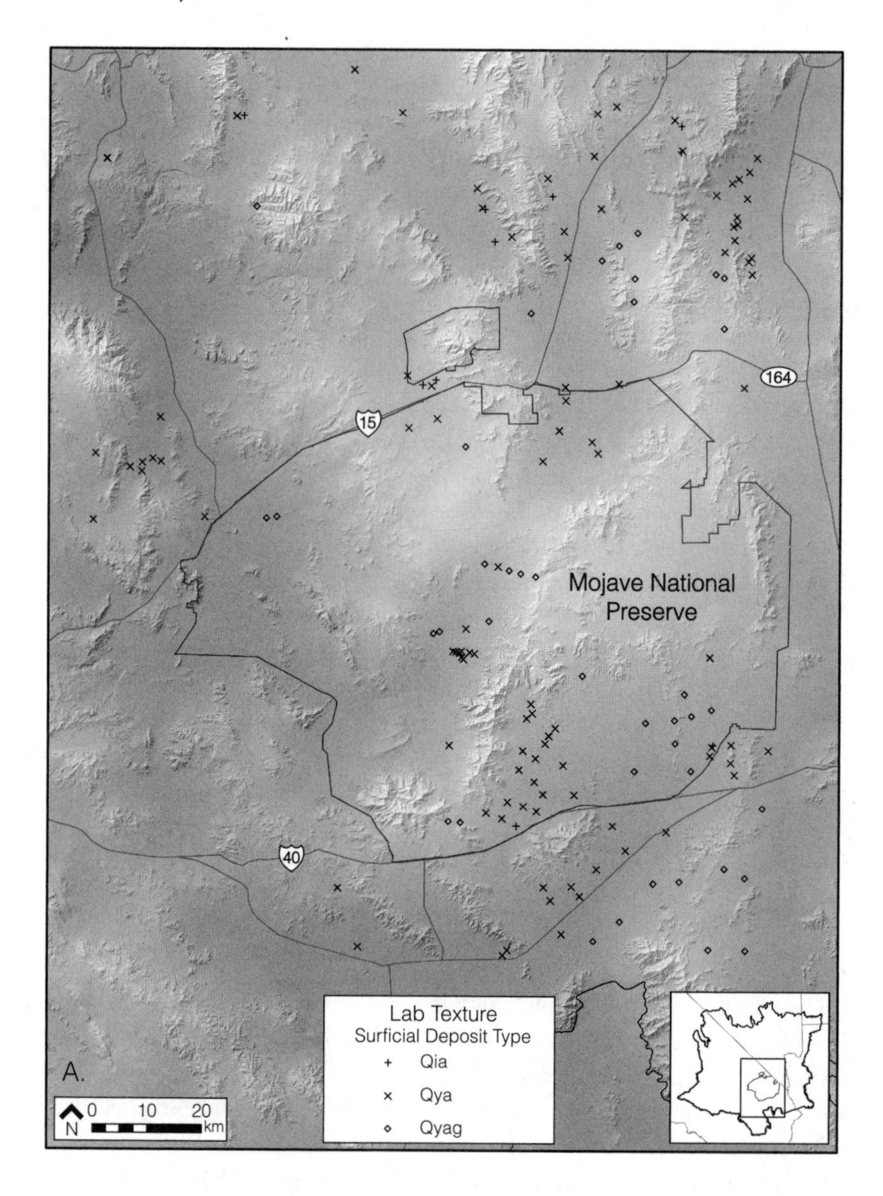

of data types, namely soil texture, *Larrea* size and density, and the type of surficial geologic deposit and its parent material. Surficial geologic deposits are classified according to the system described in the available map databases (Bedford et al. 2006; Schmidt and McMackin 2006; Bedford 2003), and are regionally correlated in Menges et al. (2001). Important ecological aspects of soils are discussed further in Miller et al. (*this volume*). Surficial geologic deposits describe depositional environment, geomorphic nature, and soil horizon characteristics. In this chapter, we focus

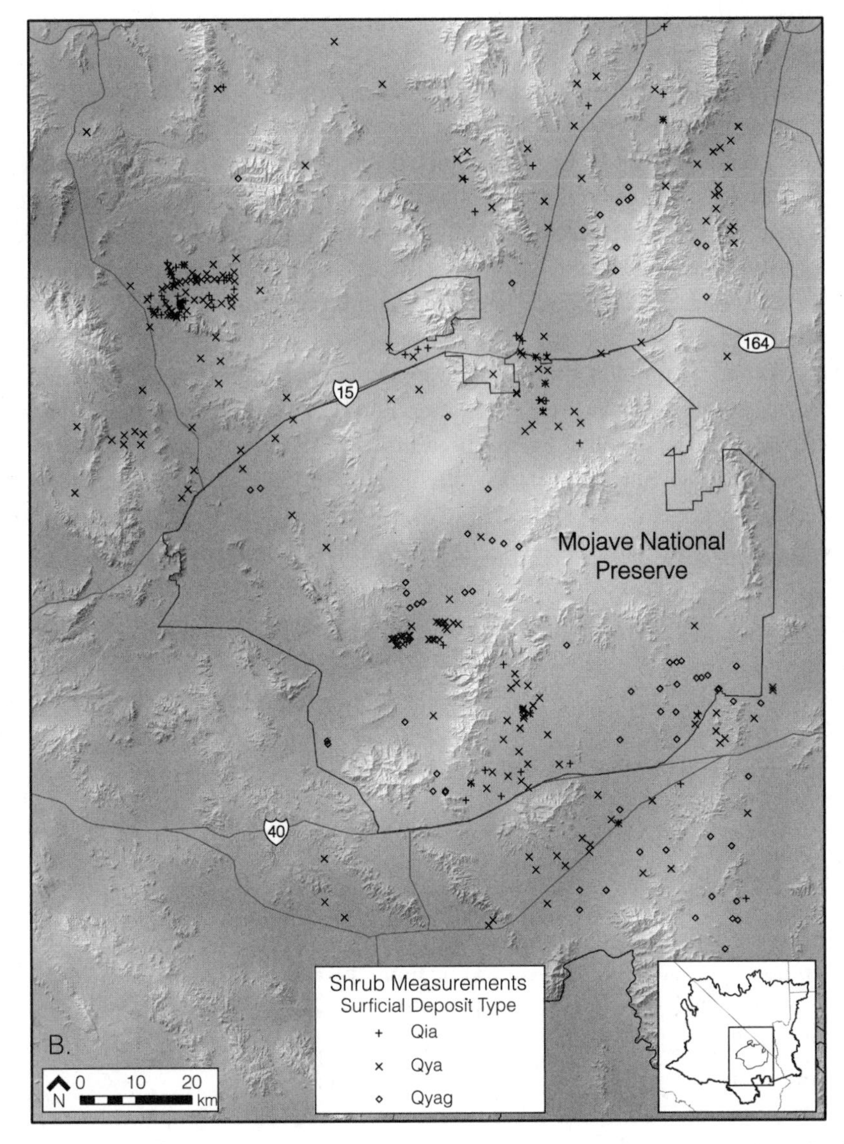

Fig. 12.1. Maps depicting the locations of measurement and sample sites for each surficial type examined in the eastern Mojave Desert. *A*, texture sampling locations. *B*, locations of shrub measurements and surficial-deposit descriptions.

on two prolific surficial geologic deposit types: Holocene and Pleistocene alluvial fan deposits. We use generalized designations of surficial geologic deposits to emphasize the broad importance of surficial geology in determining ecosystem function. However, more detailed surficial geology designations exist, and thus, more specific relationships may exist between surficial geology and *Larrea* patterns.

Within Holocene and Pleistocene age alluvial fan deposits, we focus on three surficial geologic deposits that are prolific in the study area. The first two types are younger Holocene deposits and are designated as units Qya and Qyag (fig. 12.2). Both generally lack strong soil horizons, and are characterized by weak to moderate silty vesicular (Av) horizons and weak Bw horizons with small accumulations of clay (Reheis et al. 1989; McDonald 1994; Bedford 2003). Further descriptions of soil characteristics are presented in Miller et al. (*this volume*). The primary difference between units Qya and Qyag is the parent material: the former is composed of mixed lithology clasts, and the latter is composed of moderately equal-sized grains of quartz and feldspar (grus), primarily derived from late Mesozoic and early Tertiary granitoids. Grus-weathering deposits are differentiated from other deposits, because they have a more homogenous grain-size distribution, and have differences in overall soil morphology and inset relationships.

On similar geomorphic surfaces, soils of unit Qya are much rockier, and tend to have slightly better developed soil horizons and more distinct inset relations than soils of unit Qyag (Bedford 2003). The third soil-geomorphic deposits that are analyzed are Late Pleistocene in age, and designated as Qia. These deposits consist of soils with a strong desert pavement, a moderate- to well-developed Av horizon, and moderate- to well-developed Bt horizon (Reheis et al. 1989; McDonald 1994; Bedford 2003) (fig. 12.2). A and B horizons generally include high amounts of silt and clay, respectively, which strongly alters soil hydraulic characteristics (Young et al. 2004). Formation of calcic horizons in this deposit also alters soil hydrology, as well as plant water use and rooting (Cunningham and Burk 1973; McDonald 1994; Gile et al. 1998). Unit Qia represents older deposits otherwise similar to the rocky Qya deposits. We hypothesized that perennial shrub characteristics differ among units Qya, Qyag, and Qia, and that the effects of soil texture and horizonation on the moisture budget will be correlated with *Larrea* characteristics.

Topographic attributes (elevation and aspect) are derived from U.S. Geological Survey 30 m digital elevation models (DEMs). Because aspect is not a linear variable, but rather a variable on a circular scale, we categorize aspect into the eight cardinal directions and treat it as a categorical variable.

At locations where we measured shrub size and density, soil texture (particle-size distribution) and derived hydraulic properties were determined in two ways. The first was laboratory analysis of paired geologic and biologic samples, and the second was the prediction of values from a GIS-based model. Thus, we essentially have two datasets: a small number ($n = 160$) of locations where we measured *Larrea* and quantified texture with laboratory analyses; and a larger number ($n = 393$, including the first 160) of locations where we measured *Larrea* and used a model to predict soil texture. Sample-derived data were collected from the interspaces between vegetation at three depth intervals (0–1 cm, 1–5 cm, and 5–10 cm) in all cases, and deeper in a limited number. The three depth intervals were selected as

Fig. 12.2. Photographs of typical examples of the three different surficial geologic deposit types and associated vegetation. *A*, surficial deposit type Qya; *B*, surficial deposit type Qyag and *C*, surficial deposit type Qia. The shovel handle in *B* and *C* is approximately 50 cm tall.

general pedogenic boundary breaks between the A, B, and potentially the C soil horizons. Samples were analyzed with dry-sieving and hydrometer techniques. Model-derived data was determined from a distributed model of soil texture based on the geomorphic parameters, which are described below. Although our lab and model data account for ten size fractions (three gravel fractions, five sand fractions, silt, and clay), we use only total gravel, and the sand, silt, and clay values for the fine (< 2 mm) fraction, because those (excluding gravel) are the size fractions used by most pedotransfer functions to calculate hydraulic properties. Software used to determine soil hydraulic properties requires that the proportions of sand, silt, and clay sum to 100%, and there were occasions when our laboratory or model methods gave values that did not sum to 100%. On these occasions, we adjusted our sand values to equal 100% minus the sum of silt and clay percentages. We acknowledge that gravel content has an effect on hydraulic properties but do not address the issue here.

Larrea size and spacing were determined at sites that have laterally persistent and fairly homogenous soil and plant characteristics. We measured the average width, height, spacing, and approximate error (~ standard deviation) of all *Larrea* shrubs within an area of 100 m^2, or larger if needed, to adequately describe widely spaced individuals. The goal was not to obtain exact measurements of individual characteristics or densities, but to rapidly describe the average conditions in many locations in the region.

Data Calculations

We used a GIS-based model to quantify the relative locations within an alluvial fan, as well as to spatially extrapolate soil texture. We calculated a fan position index from 30 m DEM values and a geologic map. The alluvial fan position index calculates the relative hydrologic location on an alluvial fan based on flow path distances from the mountain front (0 index value) to fan toe (1 index value) (fig. 12.3; see also color plate) (D. R. Bedford *unpublished data*).

The soil texture model used 30 m DEM-derived attributes of geomorphic sediment transport capability, which were calibrated using laboratory soil texture measurements. The model is based on the assumption that most alluvial fans are built by fluvial processes (thus excluding fans created by debris flows), and that the topography of the alluvial fans retains some characteristics of the fluvial environment in which the deposits were laid down. Thus, for deposits that are relatively young and have not been drastically altered by incision, the present topography is similar enough to the topography at the time of deposition that it can be used to infer sediment transport conditions. The model is built by spatially interacting sampled texture locations with DEM-derived values of sediment transport. Then a multiple regression model is developed for each of the sampled texture classes

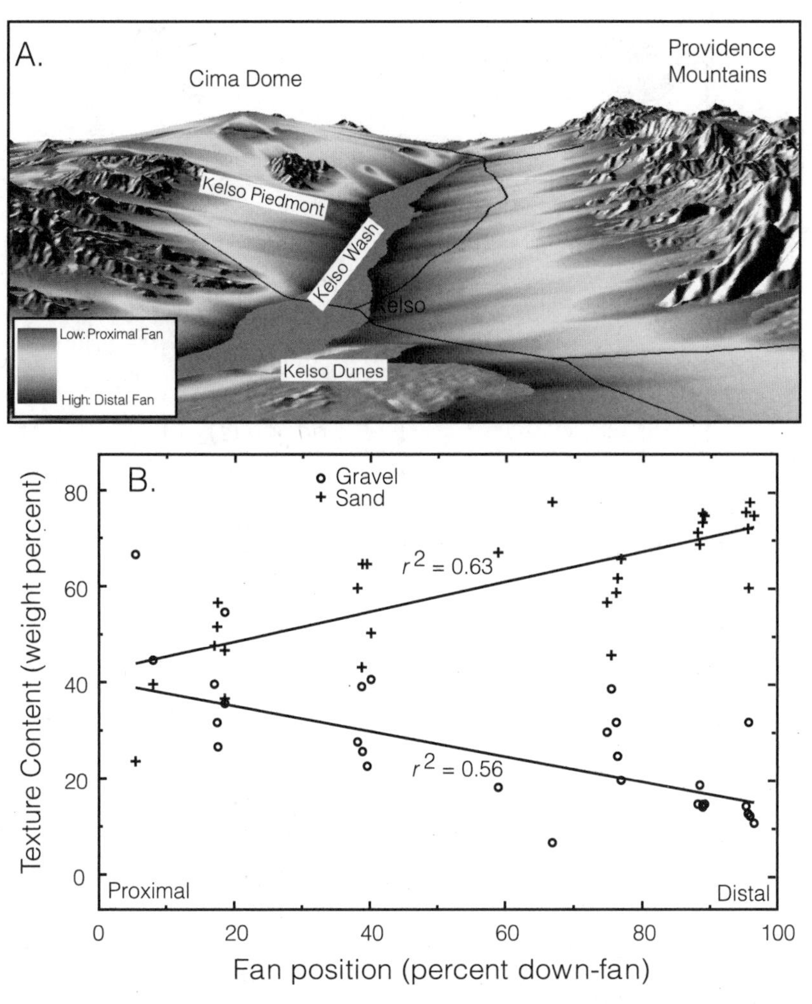

Fig. 12.3. A, a perspective view (looking towards the northeast) of the Kelso Valley, Mojave National Preserve, depicting the fan position index, which quantifies the relative location on an alluvial fan. *B,* relationships between alluvial fan position and selected soil particle sizes for samples on the Kelso piedmont are provided in this graph (*center-left* panel *A*). (See color plate following page 162.)

and for each depth interval for several ages of surficial geologic deposits. The regression models can then be applied back to the DEM-derived values, as well as a surficial geologic map, to achieve a spatially distributed soil texture model. We employ two traditional geomorphic attributes for describing sediment transport: drainage area and slope (Dietrich et al. 2003). We also use the fan position index, which accounts for the observation that soil texture tends to "fine" down

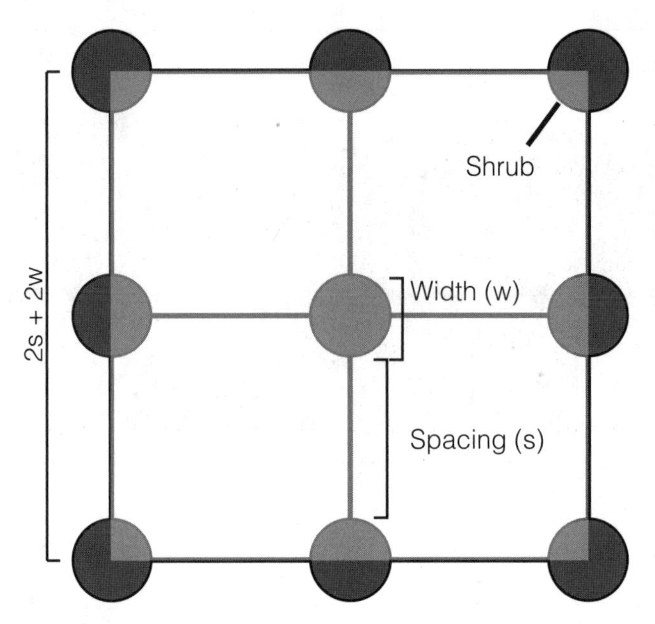

Fig. 12.4. Schematic diagram and equations for the method used to calculate *Larrea tridentata* (creosote bush) characteristics.

an alluvial fan (fig. 12.3), to directly test the gradient hypothesis. The soil texture model generally accounts for 50% of the observed, laboratory-determined, variability in particle-size distribution.

We determined size and density of *Larrea* by estimating the average width, height, and spacing of *Larrea* shrubs. We approximated canopy cover through the simple assumption of a semi-closest-packed set of individuals (fig. 12.4). We calculate percent cover, *Cover,* considering a set of nine individuals arranged in a cubic fashion:

$$Cover = 4\pi(W/2)^2/(2S+2W)^2 \quad (1)$$

where S is the average spacing (cm) between shrubs, and W is the average width (cm) of shrubs. Canopy volume is approximated based on the methods of Hamerlynck et al. (2002), where the volume of an individual is calculated as the volume of an inverted, truncated cone. For this calculation we assume that all *Larrea* shrubs have a basal diameter of 25 cm such that:

$$Volume = 1/3\pi\ H(W/2)^2 + 12.5^2 + (12.5 + W/2) \quad (2)$$

where H is the average height (cm) of shrubs, and W is as for (1). The choice of 25 cm for basal diameter is valid because it is within the range of average values for *Larrea,* and an initial investigation calculating volume with basal diameter values calculated from canopy height relationships in Hamerlynck et al. (2002) have a near-perfect linear correlation with volume calculated with 25 cm. Thus, basal diameter does not significantly vary enough to change the distribution of volume calculations. We calculate the volume fraction as the fraction percent of space occupied by canopy in the volume encompassing our calculations. In other words, we use our height, width, and spacing values to calculate a reference volume, then divide that number by the amount of canopy that occupies that space (fig. 12.4):

$$Volume\ fraction = (4 * Volume) / \left(H(2S + 2W)^2\right) \quad (3)$$

We use the value 4**Volume* because our cover calculations encompass the area covered by four complete *Larrea* (fig. 12.4). We calculate the volume fraction to approximate a local biomass measure, barring any changes in biomass density per unit canopy volume. Values for height, width, spacing, cover, volume, and volume fraction have approximately log-normal distributions and thus are log-base 10 transformed to approximate a normal distribution for statistical analysis.

We calculate soil hydraulic properties through the use of pedotransfer functions (PTFs). There are various forms of PTFs, ranging from those that consider soil texture class to those incorporating particle-size distributions and bulk density. A preliminary analysis of PTFs from two sources (Saxton et al. 1986; Schaap et al. 1998) using the methods described here suggested that those calculated by Rosetta v1.2 (see Schaap et al. 1998) were better able to describe our data. However, one limitation of most PTFs, including those used here, is their neglect of the effect of gravel content, which was as high as 94% by weight in our dataset. We calculated hydraulic properties for both the lab- and model-determined soil textures for each of the three depth horizons. By comparing the hydraulic properties, we were able to qualitatively test the ability of our model to represent soil properties relevant to *Larrea*. We calculated three general types of hydraulic properties: (1) traditional soil hydraulic properties; (2) properties describing rates and amounts of infiltration; and (3) properties describing water content under various negative water potentials (tensions). A complete listing of the hydraulic properties used is presented in table 12.1.

In addition to the basic hydraulic properties using PTFs, we calculated the amount of water that infiltrates before surface ponding (and presumably runoff) takes place during a rain event using a simplified version of the Green-Ampt equation:

$$Fp = \frac{\Psi_{mf} Ks(\Theta s - \Theta i)}{P - Ks} \quad (4)$$

where Fp is the amount of water (cm) that infiltrates before ponding, Ψ_{mf} is the water potential at the wetting front (kPa), Θs is the volumetric water content at

Table 12.1 Definitions and units of hydraulic properties analyzed for Mojave Desert geomorphic surfaces and deposits

Property type	Hydraulic property	Definition	Units
General	α	Moisture retention curve shape parameter	L^{-1}
	N	Moisture retention curve shape parameter	NA
Infiltration	Ks	Saturated Hydraulic Conductivity	$L\ T^{-1}$
	Ko	Matching point at saturation (~saturated conductivity)	$L\ T^{-1}$
	Fp	Infiltration amount before ponding	L
Water Content	Θr	Residual water content	$L^3\ L^{-3}$
	Θs	Saturated water content	$L^3\ L^{-3}$
	Θ10	Water content at –10 kPa	$L^3\ L^{-3}$
	Θ33	Water content at –33 kPa (~field capacity)	$L^3\ L^{-3}$
	Θ1500	Water content at –1500 kPa (~agricultural permanent wilting point)	$L^3\ L^{-3}$
	Θ6000	Water content at –6000 kPa	$L^3\ L^{-3}$
	Θ10000	Water content at –10000 kPa (~ *Larrea tridentata* permanent wilting point)	$L^3\ L^{-3}$
	Avail	Plant-available water at –10000 kPa	$L^3\ L^{-3}$

1 kPa = 0.01 bar

saturation, Θi is the initial volumetric water content before infiltration, P is the precipitation rate (cm s^{-1}), and Ks is the saturated hydraulic conductivity (cm s^{-1}, Mein and Larson 1973). We estimate Ψ_{mf} according to Campbell and Norman (1998):

$$\Psi_{mf} = \frac{2b+3}{b+3}\Psi e \quad (5)$$

where Ψe is the water potential at air entry (kPa), and b is the moisture retention curve shape factor. We assume that the N parameter calculated from Rosetta approximates b, and we use a PTF described by Saxton et al. (1986) for approximating Ψe from Θs. Because Fp is clearly a dynamic property that will vary with precipitation rate and initial conditions (and thus possibly environmental conditions), we calculate Fp using single values for P and Θi of 0.0001 cm s^{-1} (3.6 mm hr^{-1}) and 0.25, respectively. All other values are calculated from PTF values described in table 12.1.

We calculate water content at different water potentials (tension) using the van Genuchten (1980) method, for which we use the Rosetta calculated properties:

$$\Theta(\Psi) = \Theta r + \left(\frac{\Theta s - \Theta r}{\left[1+(\alpha\Psi)^N\right]^{1-1/N}} \right) \quad (6)$$

where $\Theta(\Psi)$ is the water content at the given tension, Θr is the residual water content, Θs is as above, α and N are curve-shape parameters of the moisture retention curve, and Ψ is the (positive) pressure potential of interest (van Genuchten 1980). We calculate water content at several tensions: -10 kPa, -33 kPa (the approximate field capacity), -1500 kPa (the approximate permanent wilting point for agricultural plants), -6000 kPa, and -10,000 kPa. The -6000 kPa and -10,000 kPa values represent water potentials close to the minimums observed in *Larrea* (e.g., Cunningham and Burk 1973; Halvorson and Patten 1974; Hamerlynck et al. 2000). We also describe the plant-available water at the traditional field capacity with the simple equation:

$$\Theta avail = \Theta(33\ kPa) - \Theta(10{,}000\ kPa) \quad (7)$$

We used traditional statistical techniques to explore relationships between *Larrea* size and density (response or dependent variables) and our hypothesized predictor variables of *Larrea* characteristics: surficial geologic unit, texture, hydraulic properties, and topography. We used ANOVA and multiple comparisons to detect significant categorical factors (i.e., surficial geologic unit and aspect). Scatterplots and Pearson and Spearman's rank correlation coefficients were used to identify linear and nonlinear relationships between *Larrea* and predictor variables. Significantly correlated variables were then input into stepwise regressions to develop predictive models of *Larrea* characteristics. If ANOVA and correlation/regression results suggested that subdividing the data by categorical variable would improve the correlation/regression, we developed regression models based on subdivided data. We determined the best potential predictor variables by ranking the absolute correlation coefficient values and their respective significance (*P* value). We present the top four best-correlated predictors of each of the *Larrea* size and density characteristics, and then input those predictors into stepwise regressions in order of decreasing correlation. We chose these simple statistical techniques as opposed to techniques such as regression of residuals or regression trees so as to maintain maximum control over the variables used. This potentially avoids building relationships that, while statistically important, are not physically obvious or based on predictors that cannot be spatially extrapolated.

RESULTS

One-way analysis of variance combined with Tukey's HSD multiple comparison method shows that many parameters of *Larrea* size and density are statistically different between surficial deposits (fig. 12.5). All values of log-transformed shrub measurements were significantly different, with the exception of width and spacing. However, the log-transformed calculated values that summarized our measurements were always different between surficial deposits. In summary, our data shows that *Larrea* cover, volume, and volume fraction are all highest on unit Qya, moderate on unit Qyag, and lowest on unit Qia, and that these differences are significant. Because of the apparent strong effect of surficial geologic deposit on vegetation size and density, we restricted further analysis to within each surficial geologic deposit type.

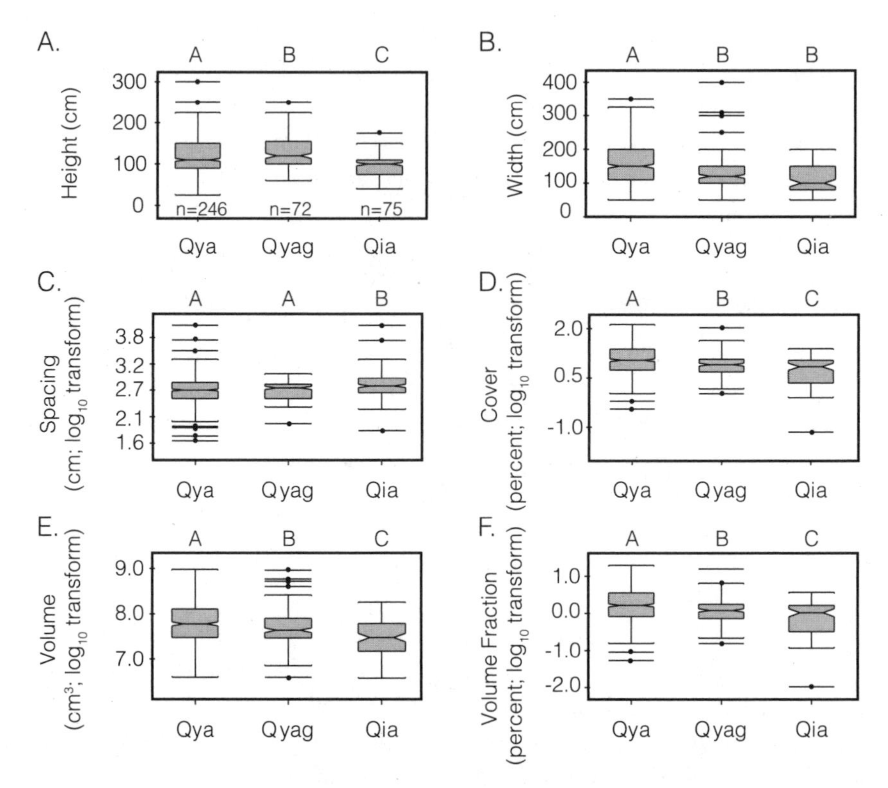

Fig. 12.5. Box plots of *Larrea tridentata* characteristics in comparison to surficial deposits revealing that the age of the deposit has a significant impact on most of these characteristics. The letters above the boxes indicate the results of multiple comparison tests: different letters denote units that are statistically different at the 95% confidence level. Note that plots *A* and *B*) are untransformed values, and plots *C–F* are comparisons between log-transformed values. Boxes are central quantiles (± 50%), notches are confidence bounds of the mean, and dots outside of box are outliers.

Aspect was not a significant factor at 95% or 90% confidence levels for any of the surficial geologic deposits, or for any *Larrea* characterization. This data may be biased by the northwest orientation of the mountain ranges (and thus the piedmonts) in the study area, which limits representation of some aspect classes. The northeast and southwest aspect classes are particularly underrepresented, and these classes would be expected to show the greatest extremes in factors such as evapotranspiration. Furthermore, piedmont slopes are generally less than 10°, so aspect may not have a large affect on precipitation (orographic effects) or evapotranspiration (solar exposure).

The four best correlated predictor variables for shrub size and density on each geologic deposit type are shown in table 12.2. We also included two general terrain indices, elevation, and the fan position index. The data did not display significant differences between Pearson correlation coefficients and Spearman rank correlation coefficients (not shown), suggesting that any relations between vegetation and our predictor variables are not significantly nonlinear, and that any poor correlations are due to relatively high spread in the data.

In general, laboratory-determined values for texture and derived hydraulic properties were well correlated with vegetation characteristics (table 12.2). It should be noted that locations with laboratory-determined texture data are much fewer than the total number of vegetation measurement locations. Units Qya and Qyag are well represented by lab samples (n = 109 and 42, respectively), while unit Qia only has ten lab measurements, and is therefore not likely to be accurately represented. In order to present statistics for a representative sample, table 12.2 also includes predictor variables determined only from the soil texture model (n = 75).

Larrea characteristics, regardless of surficial geologic deposit type, are most commonly negatively correlated with water contents at low potentials (high stress/tension), and positively correlated with elevation. The fan position index was only weakly correlated to *Larrea* characteristics, suggesting that the gradient hypothesis may not be entirely valid. Note that the volume fraction, the measure that includes the height, width, and spacing, is only weakly correlated to hydraulic properties, and moderately correlated with elevation.

Subdividing by surficial deposit type improved our correlation analysis, as suggested by our ANOVA results. For deposits Qya and Qyag, most correlations were statistically significant, and accounted for up to half of the observed variability. *Larrea* characteristics on deposit Qya were predominately correlated with elevation and water content at various tensions. For deposit Qyag, *Larrea* characteristics were correlated with soil texture, particularly sand and gravel at or near the surface, as well as with elevation.

Despite relatively high correlations between *Larrea* characteristics and predictor variables for surficial deposit Qia, many of the correlations were not significant. The significant correlations tended to be with soil texture, typically sand, gravel, and clay content. Elevation was significantly positively correlated with *Larrea* height, width, and cover on deposit Qia.

The fan position index was commonly not strongly nor significantly correlated with *Larrea* characteristics for any type of subdivision by surficial deposit. In nearly all cases, elevation was more strongly and significantly correlated with *Larrea* characteristics.

Table 12.3 presents results of stepwise linear regressions for *Larrea* characteristics. The regressions rarely resulted in all of the best-correlated predictors being used. Almost all of the regressions were significant, but only accounted for approximately one-third of the observed variability. In two cases for deposit Qia, the stepwise regressions did not provide any suitable regression model. As expected,

Table 12.2 Pearson correlation coefficients and P *values for soil property predictors of* Larrea tridentata *(creosote bush) size and density within surficial geologic deposits*

Unit	*Larrea* characteristic	Predictor	PCC	Predictor	PCC	Predictor	PCC	Predictor	PCC	Elevation Pearson	Fan position Pearson
All	Height	A_L Θ 1500[a]	−0.446	A_L Θ 6000[a]	−0.420	A_L Θ 33[a]	−0.404	A_L Θ 10000[a]	−0.396	0.382[a]	−0.133[a]
	Width	B_L Θ 1500[a]	−0.421	B_L Θ 33[a]	−0.409	B_L Θ 6000[a]	−0.399	B_L Θ 10000[a]	−0.381	0.226[a]	−0.096
	Spacing	A_LFp[a]	−0.217	A_L Θ _s[a]	−0.212	A_LT Θ _r[a]	0.203	A_L Θ 1500[a]	−0.194	0.251[a]	0.144[a]
	Cover	B_L Θ 10000[b]	−0.162	B_L Θ 6000[b]	−0.160	B_LClay[b]	−0.160	B_LFp[b]	−0.155	0.287[a]	0.003
	Volume	B_L Θ 1500[a]	−0.446	B_L Θ 33[a]	−0.434	B_L Θ 6000[a]	−0.428	A_L Θ 1500[a]	−0.427	0.241[a]	−0.124[a]
	Volume fraction	B_L Θ 10000[b]	−0.161	B_L Θ 6000[b]	−0.160	B_LClay[b]	−0.159	B_LFp[b]	−0.155	0.287[a]	0.003
Qya	Height	A_L Θ 1500[a]	−0.437	A_L Θ 33[a]	−0.429	A_LAvail33[a]	−0.412	A_LFp[a]	−0.397	0.349[a]	−0.183[a]
	Width	B_L Θ 1500[a]	−0.509	B_L Θ 33[a]	−0.490	B_L Θ 6000[a]	−0.484	A_L Θ 1500[a]	−0.473	0.205[a]	−0.169[a]
	Spacing	C_L Θ 1500[a]	−0.273	B_LSilt[a]	−0.254	A_LFp[a]	−0.245	C_L Θ 33[a]	−0.242	−.322[a]	0.099
	Cover	B_L Θ 10000[a]	−0.210	B_L Θ 6000[a]	−0.207	B_L Θ 1500[b]	−0.192	B_LSand[b]	0.186	0.358[a]	−0.039
	Volume	B_L Θ 1500[a]	−0.502	A_L Θ 1500[a]	−0.492	B_L Θ 33[a]	−0.488	B_L Θ 6000[a]	−0.475	0.220[a]	−0.207[a]
	Volume fraction	B_L Θ 10000[a]	−0.210	B_L Θ 6000[a]	−0.207	B_L Θ 1500[b]	−0.192	B_LSand[b]	0.186	0.358[a]	−0.039
Qyag	Height	C_MClay[a]	−0.369	B_LClay[b]	−0.329	C_M_[a]	−0.293	B_M Θ _r[a]	0.274	0.261[a]	0.083
	Width	A_LSand[a]	−0.478	A_LGravel[a]	0.473	B_LGravel[a]	0.439	C_MClay[a]	−0.407	0.461[a]	0.059
	Spacing	A_LSand[a]	0.330	B_L Θ 6000[b]	−0.327	A_LGravel[b]	−0.322	B_L Θ 1500[b]	−0.314	−0.191	−0.169
	Cover	A_LSand[a]	−0.523	A_LGravel[a]	0.516	B_LGravel[a]	0.467	B_LSand[a]	−0.401	0.447[a]	0.171
	Volume	A_LSand[a]	−0.447	A_LGravel[a]	0.439	C_MClay[a]	−0.421	B_LGravel[a]	0.407	0.427[a]	0.070
	Volume fraction	A_LSand[a]	−0.523	A_LGravel[a]	0.516	B_LGravel[a]	0.467	B_LSand[a]	−0.401	0.447[a]	0.171
Qia	Height	C_LSand[a]	0.760	C_LGravel[a]	−0.715	A_LN[a]	0.698	C_L Θ _r[a]	0.657	0.407[a]	−0.140
	Width	C_LSand[b]	0.591	A_LN[b]	0.588	C_LGravel	−0.527	C_MClay[a]	0.470	0.381[a]	0.119
	Spacing	A_LClay	−0.482	C_LSand	0.390	C_LGravel	−0.355	C_L Θ _r	0.339	−0.014	−0.296[a]
	Cover	B_MGravel[b]	0.331	C_L Θ 10000	−0.522	C_L Θ 6000	−0.506	C_LN	0.493	0.149	0.099
	Volume	C_LSand[a]	0.673	A_LN[a]	0.658	C_LGravel[b]	−0.617	B_LN[b]	0.562	0.386[a]	0.092
	Volume fraction	B_MGravel[b]	0.331	C_L Θ 10000	−0.522	C_L Θ 6000	−0.506	C_LN	0.493	0.149	0.098
Texture model only											
Qia	Height	C_M Θ 10[a]	−0.265	C_M α[b]	0.250	A_MClay[b]	−0.241	C_MClay	0.208	0.407[a]	−0.140
	Width	C_MClay[a]	0.470	C_M Θ 10[a]	−0.449	B_MGravel[a]	0.378	C_M α[b]	0.327	0.381[a]	0.119
	Spacing	A_MClay[a]	−0.389	A_MSilt[a]	−0.313	B_MGravel[a]	0.293	C_MGravel[a]	0.259	−0.014	−0.296[a]
	Cover	B_MGravel[b]	0.331	C_MGravel	0.271	B_MSand	−0.256	C_MSand	−0.250	0.149	0.099
	Volume	C_MClay[a]	0.494	C_M Θ 10[a]	−0.473	B_MGravel[a]	0.381	C_M α[a]	0.348	0.386[a]	0.092
	Volume fraction	B_MGravel[b]	0.331	C_MGravel	0.270	B_MSand	−0.256	C_MSand	−0.250	0.149	0.098

PCC = Pearson correlation coefficient; prefixes A, B, and C refer to soil depth intervals (A = 0–1 cm, B = 1–5 cm, C = 5–10 cm); prefixes L and M, following underscores, refer to lab-determined and texture model properties, respectively. [a] Significant at $P > 95\%$, [b] Significant at $P > 90\%$.

regressions generally performed better when the data was subset by surficial geologic deposit. In almost all cases, elevation was a significant contributor to *Larrea* characteristics, and generally only one or two other characteristics were significant, as chosen by the stepwise regressions. The common significant characteristics were texture and water content at various tensions.

Analysis of residuals (errors) via QQ plots suggests that residuals for all of the regressions (not shown) are approximately normally distributed. For surficial deposit Qya, residuals suggested homoscedasticity (i.e., evenly distributed variability), with the exception of height (and calculations considering height). This suggests that *Larrea* height shows heteroscedasticity, and therefore violates an assumption behind generalized linear models. This result is similar to that of Solbrig et al. (1977), who found that *Larrea* height distributions varied along a moisture gradient—in dry areas height distributions had a smaller peak (mean) with a narrow range, and in wet areas had a larger but less-defined peak with a wide range. Residuals for surficial deposits Qyag and Qia also appear to show heteroscedasticity, although this may be confounded by the relatively small sample set for these deposit types.

It should be noted that we have made no effort to address multicollinearity (i.e., nonindependence) of predictor variables, other than those detected by the stepwise regression algorithms. While commonly not an issue in making predictions, it can lead to problems when the goal is to understand the process behind a response variable. We have limited our interpretations of possible causes of variability to only one of the variables that are known to be perfectly or near perfectly correlated (e.g., sand and gravel content), and discuss correlation coefficient magnitudes and signs rather than multiple regression coefficients.

We focus interpretations on the performance of our predictors in describing the variability of *Larrea* cover and volume fractions because cover is commonly measured, and volume fraction approximates the total amount of vegetation on the landscape. These characteristics are also likely to be most relevant to descriptions of shrubland mosaics and ecosystem function. Furthermore, volume fraction includes height, width, and spacing measurements and may represent the most comprehensive summary of our measurements as well as *Larrea* characteristics.

DISCUSSION

We found significant differences in *Larrea* size and density among three extensive surficial geologic deposits in the study area, suggesting that surficial geology serves as a first-order determinant of *Larrea* patterns. We also found that within single categories of surficial geologic deposits, soil texture, elevation, and hydraulic properties significantly describe some of the *Larrea* size and density variability. We focus this discussion on the ecohydrologic implications of our findings and on ways in which the methodology could be improved.

Table 12.3 Stepwise regression results of the four best predictor variables (table 12.2) and elevation for Larrea tridentata *(creosote bush) characteristics in the Eastern Mojave Desert*

Unit	*Larrea* characteristic	r^2	*P*-value	Equation
All	Height	0.261	0.00	−14.31603(A_L Θ 1500) + 11.40958(A_L Θ 6000) + 3.8668 (A_L Θ 33) + 0.0001(Elevation) + 1.29282
	Width	0.247	0.00	−2.88188(B_L Θ 1500) + 0.00022(Elevation) + 2.16937
	Spacing	0.090	0.00	8.54541(A_L Θ _r) + −0.00024(Elevation) + 2.48185
	Cover	0.168	0.00	−0.09613(B_LFp) + 0.00071(Elevation) + 0.66347
	Volume	0.295	0.00	−4.5562(B_L Θ 1500) + −4.97093(A_L Θ 1500) + 0.00057(Elevation) +7.97592
	Volume fraction	0.168	0.00	−0.09604(B_LFp) + 0.00071(Elevation) −0.16562
Qya	Height	0.270	0.00	−28.94258(A_L Θ 1500) + 36.97483(A_L Θ 33) + −34.44458 (A_LAvail) + 0.272(A_LFp) + 0.00011(Elevation) + 0.73316
	Width	0.331	0.00	−2.60905(B_L Θ 1500) + −1.54711(A_L Θ 1500) + 0.00019(Elevation) + 2.29021
	Spacing	0.185	0.00	−0.15641(A_LFp) + −0.00038(Elevation) + 3.32342
	Cover	0.176	0.00	−5.58388(B_L Θ 6000) + 0.00068(Elevation) + 0.79166
	Volume	0.326	0.00	−5.58904(B_L Θ 1500) + −5.49815(A_L Θ 1500) + 0.00047(Elevation) + 8.19014
	Volume fraction	0.176	0.00	−5.57441(B_L Θ 6000) + 0.00068(Elevation) −0.03775
Qyag	Height	0.271	0.01	−6.06632(C_MClay) + −2.91937(B_LClay) + 2.64995
	Width	0.435	0.00	−7.40286(C_MClay) + 0.00034(Elevation) + 2.3587
	Spacing	0.122	0.05	0.47463(A_LSand) + 2.37472
	Cover	0.316	0.00	−1.01787(A_LSand) + 0.0005(Elevation) + 1.12192
	Volume	0.436	0.00	−20.99443(C_MClay) + 0.00078(Elevation) + 8.59079
	Volume fraction	0.316	0.00	−1.01772(A_LSand) + 0.0005(Elevation) + 0.29287
Qia	Height	0.672	0.04	−0.3337(C_LGravel) + 0.57743(A_LN) + 1.2046
	Width	0.350	0.09	0.58404(C_LSand) + 1.78399
	Spacing	0.497	0.02	−0.00069(Elevation) + 3.22944
	Cover	NA	NA	
	Volume	0.759	0.15	2.29644(C_LSand) + 5.76962(A_LN) + −5.55922(B_LN) + 0.00252(Elevation) + 3.97957
	Volume fraction	NA	NA	
Texture model only				
Qia	Height	0.227	0.00	−14.73007(C_M Θ 10) + 0.0002(Elevation) + 7.16866
	Width	0.284	0.00	−24.89862(C_M Θ 10) + 0.0002(Elevation) + 10.92752
	Spacing	0.151	0.00	−21.85357(A_MClay) + 3.36873
	Cover	0.110	0.05	4.14388(B_MGravel) −0.36687
	Volume	0.306	0.00	−70.0693(C_M Θ 10) + 0.00052(Elevation) + 32.51244
	Volume fraction	0.109	0.05	4.1387(B_MGravel) −1.19431

Our data best described *Larrea* characteristics for young alluvial deposits (i.e., deposits Qya and Qyag), and performed poorly in describing characteristics on older (Qia) deposits. The most likely explanation for this is that our characterization of soil properties only includes the top 15 cm of soil, while the majority of root systems occupy the top 40 cm of soil (Schwinning et al. *this volume*). For younger deposits, these properties are likely similar to those throughout the root systems, because the effects of soil development are relatively minor. For older surficial deposits (Qia), soil development has modified the upper and lower portions of the soil profile differentially, with accumulations of silt in Av horizons and clay and carbonate in B horizons, (McDonald et al. 1995). Therefore, our shallow characterization likely does not represent the deeper portions of the soil profile, where much of the plant roots occur (Gile et al. 1998; Stevenson et al. *this volume*). Also, our sampling of soil texture for the Qia deposits was limited to a small, and potentially nonrobust, statistical data set.

For surficial deposit type Qya, water content at low potentials and elevation described most *Larrea* characteristics. Interestingly, the data showed negative correlations between water content and *Larrea* size and density. Three possible explanations exist for this. The first is that relatively high water contents, especially at low water potential (ostensibly the normal condition) may limit the availability of oxygen, for which *Larrea* has a high requirement (Lunt et al. 1973). Second, Wilcox et al. (2004) showed that high water content is negatively correlated with fine root growth, suggesting that high water contents inhibit *Larrea*. This fits with the general description that *Larrea* inhabits well-drained (i.e., deep, gravelly) soils. Third, higher water content decreases water-use efficiency in *Larrea* (Smith et al. 1997), which may help to explain why our results suggest that increased water content decreases *Larrea* cover and volume fraction.

Larrea shrubs on Qyag surficial deposits appear to be sensitive to elevation and soil texture (particularly sand and gravel content) at the surface. Higher gravel contents are correlated with larger and denser *Larrea* (table 12.2) and are, again, indicative of well-drained soils with high infiltrability. Our parameterization of infiltration may not have been a good predictor, because gravel content was not considered in the pedotransfer functions used.

Using all of the available data to describe *Larrea* characteristics on Qia surficial deposits did not result in viable regression models for cover or volume fraction. Restricting the data to values determined with the soil texture model data, which permitted a larger sample size, gravel content was the best predictor of *Larrea* size characteristics.

Our data suggests that *Larrea* is larger and denser in areas with well-drained gravelly soils. This result fits in the larger view of desert ecology, in that water pulses (precipitation) must be concentrated and conserved, and that the magnitude of pulse, concentration, and conservation drives water and nutrient availability (e.g., Noy-Meir 1973; Austin et al. 2004; Schwinning and Sala 2004; Ludwig et al. 2005). Soils lacking strong horizons, and those with high gravel content, have high infiltration and conductivity rates (Saxton et al. 1986; Young et al. 2004), which allow water to infiltrate to deep moisture zones. Because evaporation largely occurs from shallow soil depths (Kurc and Small 2004), and because water stored at depth in gravelly soils is under less tension, water is more available for deep-rooted vegetation such as *Larrea.*

Elevation was commonly significantly positively correlated with *Larrea* cover and volume fraction. Inferring the ecohydrological effects of elevation on *Larrea* is difficult because elevation often serves as a proxy for many other variables, such as precipitation, temperature, radiation, freeze-thaw frequency, etc. In order to determine which of those processes are directly affecting *Larrea,* a similar study involving those datasets is advised.

Our data show that *Larrea* characteristics significantly vary between surficial geologic deposits, vary with elevation, and tend to reflect the gravel content of soils. This is likely a result of differing dynamics of soil-water contents and potentials in these different soil types. Our study suggests that *Larrea* size and density adhere to basic dryland ecohydrologic theory; gradients of elevation (more likely a proxy for precipitation, temperature, radiation, freeze-thaw) and soil type (particularly the degree of horizonation and gravel content) are the key drivers of this long-lived evergreen shrub.

However, our data only accounted for a relatively small portion of the variability. Volume fraction, the characteristic that included height, width and spacing measures, was poorly predicted by our data, and our predictions appear to be limited by our spacing measurement methods. This could be due to high amounts of intrinsic random variability in spacing (i.e., nonenvironmentally determined), or our data did not capture the important environmental variables that do describe variability in spacing. It is likely that there is a large amount of random variability in *Larrea* characteristics, and that the data violated traditional statistical assumptions (e.g., homoscedacity). Accepting that there may be large random variability in soil and *Larrea* characteristics, we would like to improve methods of predicting and understanding the nonrandom variability that may indicate the drivers of that variability. Improved methods could results in a clearer understanding of the drivers of *Larrea* variability, and the creation of predictive maps for understanding potential landscape change.

Two ways to enhance our understanding would be to improve our characterization of soil properties and to include other datasets. We can see three avenues for improving the datasets. The first would be to improve soil hydraulic property characterizations to include the gravel content. Our data shows that gravel is commonly correlated with *Larrea* characteristics, yet we know of no robust pedotransfer functions that include gravel content. The second way to advance the method would be to account for modifications of soil structure and texture that are commonly associated with shrubs (Jackson and Caldwell 1993; Seyfried and Wilcox 1995; Bhark and Small 2003). In other words, how do our samples, which were collected from the areas between shrubs, adequately relate to the soil directly below shrubs? Determining the relationships between soil characteristics in plant interspaces and under vegetation is clearly needed. A third way to improve our understanding of vegetation characteristics would be to consider the temporal characteristics of water amounts and potentials. The frequency and the magnitudes of water additions, and the lengths of water content and potentials, as determined from monitoring or spatio-temporal water balance models (e.g., Rodriguez-Iturbe et al. 1999; Hamerlynck et al. 2002; Rodriguez-Iturbe et al. 2006), may be a more appropriate dataset for analysis than the static soil-water properties used. Hence, we have installed instrumentation to record temporally continuous soil moisture and water potential data in deposits of varying age and pedogenic maturity, and will develop techniques to describe their spatial and temporal variability. Other datasets that are likely to contribute to variability, but were not analyzed, include the spatial variability in precipitation (both amount and intensity), and evapotranspiration.

The data presented here show that *Larrea* is apparently affected by the differences in soil properties and water dynamics that occur between different soil types associated with different surficial geologic deposits. If this is indeed true, then we might expect to see clear variations in the root systems and physiology of *Larrea* between these different soil types, since root dynamics are inherently tied to water acquisition. While recent advances have been made, we feel that more data is needed in order to understand the true mechanisms controlling *Larrea* size and density.

ACKNOWLEDGMENTS

We thank Eric McDonald, Chris Menges, Bob Webb, and Jim Yount for previous works in the region, as well as insightful discussions on the regional geology and vegetation. The U.S. Geological Survey soil physics lab in Sacramento, and particularly Kevin Ellett and Allan Flint, provided laboratory analysis and discussions of the soil texture data. Reviews by Scott Basset, one anonymous reviewer, and comments from the editors of this book provided suggestions that were very helpful. The Mojave National Preserve, a unit of the National Park Service, provided sampling permits and encouragement for our work. This work is funded by the National Cooperative Geologic Mapping Program and the Recoverability and Vulnerability of Desert Ecosystems Project of the Priority Ecosystems Studies of the U.S. Geological Survey.

REFERENCES

Alizai, H. A., and L. C. Hulbert. 1970. Effects of soil texture on evaporative loss and available water in semi-arid climates. *Soil Science* **110**:328–332.

Austin, A. T., L. Yahdjian, J. M. Stark, J. Belnap, A. Porporato, U. Norton, D. A. Ravetta, and S. M. Schaeffer. 2004. Water pulses and biogeochemical cycles in arid and semiarid ecosystems. *Oecologia* **141**:221–235.

Bedford, D. R. 2003. Surficial and bedrock geologic map database of the Kelso 7.5 minute quadrangle, San Bernardino County, California. U.S. Geological Survey Open-File Report No. 03-501. Denver, Colorado. Online. http://geopubs.wr.usgs.gov/open-file/of03-501/.

Bedford, D. R., D. M. Miller, and G. A. Phelps. 2006. Preliminary surficial geologic map database of the Amboy 30 × 60 minute quadrangle, California. U.S. Geological Survey Open-File Report No. 2006-1165. Denver, Colorado. Online. http://pubs.usgs.gov/of/2006/1165/.

Bhark, E. W., and E. E. Small. 2003. Association between plant canopies and the spatial patterns of infiltration in shrubland and grassland of the Chihuahuan Desert, New Mexico. *Ecosystems* **6**:185–196.

Blair, T. C., and J. G. McPherson. 1994. Alluvial fan processes and forms. Pages 354–402 *in* A. D. Abrahams and A. J. Parsons, editors. *Geomorphology of desert environments.* Chapman and Hall, London, England, UK.

Blissenbach, E. 1952. Relation of surface angle distribution to particle-size distribution on alluvial fans [Arizona]. *Journal of Sedimentary Petrology* **22**:25–28.

Bowers, J. E., R. H. Webb, and R. J. Rondeau. 1995. Longevity, recruitment and mortality of desert plants in Grand Canyon, Arizona, USA. *Journal of Vegetation Science* **6**:551–564.

Bowers, M. A., and C. H. Lowe. 1986. Plant-form gradients on Sonoran Desert bajadas. *Oikos* **46**:284–291.

Bull, W. B. 1977. The alluvial-fan environment. *Progress in Physical Geography* **1**:222–270.

Bull, W. B. 1991. *Geomorphic responses to climate change.* Oxford University Press, New York, New York.

Campbell, G. S., and J. M. Norman. 1998. *An introduction to environmental biophysics.* Springer, New York, New York.

Clothier, B. E., D. R. Scotter, and J. P. Kerr. 1977. Water-retention in soil underlain by a coarse-textured layer: theory and a field application. *Soil Science* **123**:392–399.

Cody, M. L. 2000. Slow-motion population dynamics in Mojave Desert perennial plants. *Journal of Vegetation Science* **11**:351–358.

Cunningham, G. L., and J. H. Burk. 1973. The effect of carbonate deposition layers (caliche) on water status of *Larrea divaricata. American Midland Naturalist* **90**:474–480.

DeSoyza, A. G., W. G. Whitford, E. MartinezMeza, and J. W. VanZee. 1997. Variation in creosote bush (*Larrea tridentata*) canopy morphology in relation to habitat, soil fertility and associated annual plant communities. *American Midland Naturalist* **137**:13–26.

Dietrich, W. E., D. G. Bellugi, L. S. Sklar, J. D. Stock, A. M. Heimsath, and J. J. Roering. 2003. Geomorphic transport laws for predicting landscape form and dynamics. Pages 103–132 *in* P. R. Wilcock and R. M. Iverson, editors. *Prediction in geomorphology. Geophysical Monograph Series.* Volume 135. American Geophysical Union, Washington, D.C.

Fernandez-Illescas, C. P., A. Porporato, F. Laio, and I. Rodriguez-Iturbe. 2001. The ecohydrological role of soil texture in a water-limited ecosystem. *Water Resources Research* **37**:2863–2872.

Gile, L. H., R. P. Gibbens, and J. M. Lenz. 1998. Soil-induced variability in root systems of creosote bush (*Larrea tridentata*) and tarbush (*Flourensia cernua*). *Journal of Arid Environments* **39**:57–78.

Halvorson, W. I., and D. T. Patten. 1974. Seasonal water potential changes in Sonoran Desert shrubs in relation to topography. *Ecology* **55**:173–177.

Hamerlynck, E. P., J. R. McAuliffe, E. V. McDonald, and S. D. Smith. 2002. Ecological responses of two Mojave Desert shrubs to soil horizon development and soil water dynamics. *Ecology* **83**:768–779.

Hamerlynck, E. P., J. R. McAuliffe, and S. D. Smith. 2000. Effects of surface and sub-surface soil horizons on the seasonal performance of *Larrea tridentata* (creosote bush). *Functional Ecology* **14**:596–606.

Harvey, A. M. 1997. The role of alluvial fans in arid zone fluvial systems. Pages 231–259 *in* D. S. G. Thomas, editor. *Arid zone geomorphology: process, form and change in drylands.* John Wiley and Sons, Chichester, England, UK.

Jackson, R. B., and M. M. Caldwell. 1993. Geostatistical patterns of soil heterogeneity around individual perennial plants. *Journal of Ecology* **81**:683–692.

Key, L. J., L. F. Delph, D. B. Thompson, and E. P. Vanhoogenstyn. 1984. Edaphic factors and the perennial plant community of a Sonoran Desert bajada. *Southwestern Naturalist* **29**:211–222.

Klikoff, L. G. 1967. Moisture stress in a vegetational continuum in the Sonoran Desert. *The American Midland Naturalist* **77**:128–137.

Kurc, S. A., and E. E. Small. 2004. Dynamics of evapotranspiration in semiarid grassland and shrubland ecosystems during the summer monsoon season, central New Mexico. *Water Resources Research* **40**:W09306.

Ludwig, J. A., B. P. Wilcox, D. D. Breshears, D. J. Tongway, and A. C. Imeson. 2005. Vegetation patches and runoff-erosion as interacting ecohydrological processes in semiarid landscapes. *Ecology* **86**:288–297.

Lunt, O. R., J. Letey, and S. B. Clark. 1973. Oxygen requirements for root growth in three species of desert shrubs. *Ecology* **54**:1356–1362.

McAuliffe, J. R. 1994. Landscape evolution, soil formation, and ecological patterns and processes in Sonoran Desert bajadas. *Ecological Monographs* **64**:111–148.

McAuliffe, J. R., and E. V. McDonald. 1995. A piedmont landscape in the eastern Mojave Desert: examples of linkages between biotic and physical components. *San Bernardino County Museum Association Quarterly* **42**:53–63.

McDonald, E. V. 1994. The relative influences of climatic change, desert dust, and lithologic control on soil-geomorphic processes and soil hydrology of calcic soils formed on Quaternary alluvial-fan deposits in the Mojave Desert, California. Ph.D. Dissertation. University of New Mexico, Albuquerque, New Mexico.

McDonald, E. V., L. D. McFadden, and S. G. Wells. 1995. The relative influences of climate change, desert dust, and lithologic control on soil-geomorphic processes on alluvial fans, Mojave Desert, California: summary of results. *San Bernardino County Museum Association Quarterly* **42**:35–42.

Mein, R. G., and C. L. Larson. 1973. Modeling infiltration during a steady rain. *Water Resources Research* **9**:384–394.

Menges, C. M., E. M. Taylor, J. B. Workman, and A. S. Jayko. 2001. Regional surficial-deposit mapping in the Death Valley area of California and Nevada in support of ground-water modeling. Pages 151–166 *in* M. N. Machette, M. Johnson, and J. L. Slate, editors. *Quaternary and late Pliocene geology of the Death Valley region: recent observations on tectonics, stratigraphy, and lake cycles.* U.S. Geological Survey Open-File Report No. 01–0051. Reston, Virginia.

Noy-Meir, I. 1973. Desert ecosystems: environment and producers. *Annual Review of Ecology and Systematics* **4**:25–52.

Parker, K. C. 1991. Topography, substrate, and vegetation patterns in the northern Sonoran Desert. *Journal of Biogeography* **18**:151–163.

Parker, K. C. 1995. Effects of complex geomorphic history on soil and vegetation patterns on arid alluvial fans. *Journal of Arid Environments* **30**:19–39.

Peterson, F. F. 1981. Landforms of the Basin and Range province defined for soil survey. Technical Bulletin No. 28. Nevada Agricultural Experiment Station, University of Nevada, Reno, Nevada.

Phillips, D. L., and J. A. MacMahon. 1978. Gradient analysis of a Sonoran Desert bajada. *Southwestern Naturalist* **23**:669–680.

Puigdefabregas, J., A. Sole, L. Gutierrez, G. del Barrio, and M. Boer. 1999. Scales and processes of water and sediment redistribution in drylands: results from the Rambla Honda field site in southeast Spain. *Earth-Science Reviews* **48**:39–70.

Reheis, M. C., J. W. Harden, L. D. McFadden, and R. R. Shroba. 1989. Development rates of late Quaternary soils, Silver Lake playa, California. *Soil Science Society of America Journal* **53**:1127–1140.

Reynolds, J. F., R. A. Virginia, P. R. Kemp, A. G. de Soyza, and D. C. Tremmel. 1999. Impact of drought on desert shrubs: effects of seasonality and degree of resource island development. *Ecological Monographs* **69**:69–106.

Rodriguez-Iturbe, I., V. Isham, D. R. Cox, S. Manfreda, and A. Porporato. 2006. Space-time modeling of soil moisture: stochastic rainfall forcing with heterogeneous vegetation. *Water Resources Research* **42**:W06D05.

Rodriguez-Iturbe, I., A. Porporato, L. Ridolfi, V. Isham, and D. R. Cox. 1999. Probabilistic modelling of water balance at a point: the role of climate, soil and vegetation. *Proceedings of the Royal Society A: Mathematical Physical and Engineering Sciences* **455**:3789–3805.

Saxton, K. E., W. J. Rawls, J. S. Romberger, and R. I. Papendick. 1986. Estimating generalized soil-water characteristics from texture. *Soil Science Society of America Journal* **50**:1031–1036.

Schaap, M., F. Leij, and T. van Genuchten M. 1998. Neural network analysis for hierarchical prediction of soil hydraulic properties. *Soil Science Society of America Journal* **62**:847–855.

Schlesinger, W. H., J. A. Raikes, A. E. Hartley, and A. F. Cross. 1996. On the spatial pattern of soil nutrients in desert ecosystems. *Ecology* **77**:1270–1270.

Schmidt, K. M., and M. R. McMackin. 2006. Preliminary surficial geologic map of the Mesquite Lake 30' × 60' quadrangle, California and Nevada. U.S. Geological Survey Open-File Report No. 2006–1035. Reston, Virginia.

Schwinning, S., and O. E. Sala. 2004. Hierarchy of responses to resource pulses in arid and semi-arid ecosystems. *Oecologia* **141**:211–220.

Seyfried, M. S., and B. P. Wilcox. 1995. Scale and the nature of spatial variability: field examples having implications for hydrologic modeling. *Water Resources Research* **31**: 173–184.

Sharifi, M. R., F. C. Meinzer, E. T. Nilsen, P. W. Rundel, R. A. Virginia, W. M. Jarrell, D. J. Herman, and P. C. Clark. 1988. Effect of manipulation of water and nitrogen supplies on the quantitative phenology of *Larrea tridentata* (creosote bush) in the Sonoran Desert of California. *American Journal of Botany* **75**:1163–1174.

Shreve, F. 1964. Vegetation of the Sonoran Desert. Pages 1–840 *in* F. Shreve and I. L. Wiggins, editors. *Vegetation and flora of the Sonoran Desert.* Stanford University Press, Stanford, California.

Smith, S. D., C. A. Herr, K. L. Leary, and J. M. Piorkowski. 1995. Soil-plant water relations in a Mojave Desert mixed shrub community: a comparison of three geomorphic surfaces. *Journal of Arid Environments* **29**:339–351.

Smith, S. D., R. K. Monson, and J. E. Anderson. 1997. *Physiological ecology of North American desert plants.* Springer-Verlag, Berlin, Germany.

Solbrig, O. T., M. A. Barbour, J. Cross, G. Goldstein, C. H. Lowe, J. Morello, and T. W. Yang. 1977. The strategies and community patterns of desert plants. Pages 67–106 *in* G. H. Orians and O. T. Solbrig, editors. *Convergent evolution in warm deserts: an examination of strategies and patterns in the deserts of Argentina and the United States.* Dowden, Hutchinson and Ross, Inc., Stroudsburg, Pennsylvania.

Stein, R. A., and J. A. Ludwig. 1979. Vegetation and soil patterns on a Chihuahuan Desert bajada. *American Midland Naturalist* **101**:28–37.

Tongway, D. J., and J. A. Ludwig. 1994. Small-scale resource heterogeneity in semi-arid landscapes. *Pacific Conservation Biology* **1**:201–208.

van Genuchten, M. T. 1980. A closed-form equation for predicting the hydraulic conductivity of unsaturated soils. *Soil Science Society of America Journal* **44**:892–898.

Vasek, F. C. 1980. Creosote bush: long-lived clones in the Mojave Desert. *American Journal of Botany* **67**:246–255.

Whittaker, R. H., and W. A. Niering. 1965. Vegetation of the Santa Catalina Mountains, Arizona: a gradient analysis of the south slope. *Ecology* **46**:429–452.

Wierenga, P. J., J. M. H. Hendrickx, M. H. Nash, J. Ludwig, and L. A. Daugherty. 1987. Variation of soil and vegetation with distance along a transect in the Chihuahuan Desert. *Journal of Arid Environments* **13**:53–63.

Wilcox, C. S., J. W. Ferguson, G. C. J. Fernandez, and R. S. Nowak. 2004. Fine root growth dynamics of four Mojave Desert shrubs as related to soil moisture and microsite. *Journal of Arid Environments* **56**:129–148.

Woodell, S. R. J., H. A. Mooney, and A. J. Hill. 1969. The behaviour of *Larrea divaricata* (creosote bush) in response to rainfall in California. *Journal of Ecology* **57**:37–44.

Yang, T. W., and C. H. Lowe. 1956. Correlation of major vegetational climaxes with soil characteristics in the Sonoran Desert. *Science* **123**:542.

Young, M. H., E. V. McDonald, T. G. Caldwell, S. G. Benner, and D. G. Meadows. 2004. Hydraulic properties of a desert soil chronosequence in the Mojave Desert, USA. *Vadose Zone Journal* **3**:956–963

THIRTEEN

Mojave Desert Root Systems

SUSANNE SCHWINNING AND MARK M. HOOTEN

The roots of desert plants in North America, including in the Mojave Desert, have been studied since the late 1800s (Cannon 1911). The initial focus of these studies was to establish a general understanding of the differences between root systems in arid settings and those in wetter environments (Rundel and Nobel 1991). Studies focused on agricultural or horticultural applications (Dayton 1931), and little attention was paid to the form and function of individual root systems. A new wave of Mojave Desert root studies began in the late 1960s, motivated by a need to understand the fate of radioactive isotopes released during nuclear weapons testing at the Nevada Test Site (NTS) (Wallace and Romney 1972; Wallace et al. 1974; Wallace et al. 1980). Though these studies ultimately failed to answer fundamental questions about the uptake and transport of radionuclides, they provided detailed descriptions and quantifications of individual root systems for common plant species.

When, in the 1980s, Yucca Mountain on NTS (fig. 18.4) was suggested as a repository for high-level nuclear waste, a new cycle of root research began, this time with a focus on the role of root systems in the hydrologic cycle of desert soils (e.g., Hessing et al. 1996). Of particular interest was (and still is) whether or how much water can percolate past the root zone and potentially corrode buried storage containers filled with radioactive waste products, which could contaminate groundwater systems. Thus, evaluating the interaction between desert plants and soils became a critical factor in assessing the safety of long-term storage of buried wastes in arid lands. Recent advances in hydrological research suggest that the sparse and inconspicuous vegetation of the Mojave, North America's driest desert, is responsible for regulating hydrologic fluxes on the scale of thousands of years (Walvoord et al. 2002; Scanlon et al. 2003).

Mojave Desert plants have also been ideal for testing some of the most innovative hypotheses in contemporary plant ecology regarding modes of communication between neighboring plants (Mahall and Callaway 1991). Results of recent root growth studies challenge our views of plants as largely passive site inhabitants. It has long been known that roots do not explore the soil randomly but have enhanced root growth in resource-rich soil microsites (Caldwell 1994). Roots of the same species appear to generate cooperative root system geometries that are more effective in nutrient and water extraction, while those of competing individuals appear to avoid one another. Some species are known to release chemical signals to which other roots respond by reducing root elongation rates, apparently to avoid overlap (Mahall and Callaway 1992). Nobody knows how pervasive such root signals are, but the implications of this phenomenon are potentially significant for entire plant populations and overall community structure. For example, Callaway and Aschehoug (2000) suggested that the success of some exotic plant invaders may, at least in part, be due to their chemical root signals, which strongly suppress the growth of neighbors in the invaded environment, but which elicit no response from the long-term neighbors in the invader's original environment.

There is no doubt that our understanding of root system organization is still in a formative state. A number of recent reviews have addressed the state of our knowledge on root ecology, notably a review by Schenk (2005) on the predictability of global patterns in root distribution, and one by Eissenstat (1997) on trade-offs in root form and function. Various commentaries and reviews point to deficiencies in our knowledge, especially regarding the measurement and function of fine roots (Pregitzer 2002; Waisel and Eshel 2002; Zobel 2003; Pierret et al. 2005). We will build upon these reviews to construct a coherent portrait of the root systems in the Mojave Desert's fluvial basins and to discuss management implications and questions for future research.

FORM VERSUS FUNCTION IN ROOT RESEARCH

Methods of Measuring Root Structure

A root system is a more heterogeneous collection of root types than meets the eye (Waisel and Eshel 2002). Even though individual roots may appear similar (e.g., in diameter), they can serve quite different functions. Structural roots (analogous to stems in a canopy) serve chiefly in providing structural support and in the transport of water and nutrients, while fine roots (analogous to leaves) serve primarily in resource uptake. There is possibly even greater specialization in roots adapted for nutrient and water uptake in an environment where these resources have unequal distribution. There are no sharp differences between transport and uptake roots; instead, there are varying degrees of efficacy for any single function. Structural and fine roots are genetically different, with structural roots extending rapidly and indeterminately, while fine roots grow for a limited time and usually remain short, though they can continue branching and are shed when the surrounding soil becomes resource depleted (Eissenstat and Yanai 1997; Pregitzer 2002).

Structural roots are easily excavated, and in some cases three-dimensional images of entire structural root systems have been graphically reconstructed from observations recorded during excavation (e.g., Gibbens and Lenz 2001). However, as much as 90% of fine roots (usually defined as < 0.2 mm in diameter) may be lost in the process of physically separating roots from soil (Pierret et al. 2005). More accurate methods for root quantification do exist [e.g., X-rays, computed axial tomography (CAT) scanning, nuclear magnetic resonance imaging (NMRI)] but are rarely used due to the high cost. All common methods for root excavation and quantification are destructive and provide only a snapshot of the root distribution at a single point in time. Thus, destructive methods of root quantification cannot describe the dynamics of root growth and cessation over time, which are essential for understanding long-term root function in an arid environment (Fernandez and Caldwell 1975).

While structural roots are durable over the long course of a plant's lifetime, fine roots appear to be specialized for rapid response to changing resource availability on the scale of days to months. For example, "rain roots" can appear near the soil surface within a day of rainfall events (Lauenroth et al. 1994; Nobel 1994), aiding in the uptake of near-surface soil water that might otherwise be rapidly lost by evaporation (Nobel 2002). If the soil stays wet, fine roots may continue to grow for weeks. Fine root growth may also follow the downward movement of a wetting front in the soil and cede as the front passes (Fernandez and Caldwell 1975). Typically, when the soil dries, rain roots and other fine structures (e.g., root hairs—filamentous outgrowths of the root epidermis that develop in the root elongation zone and greatly increase root surface area) are shed, leaving behind only roots that show a degree of resistance to drying due to the deposition of hydrophobic substances like suberin and lignin in the cell walls of the endodermis. However, even suberized roots may senesce when drought conditions become severe enough.

A dynamic system of fine roots allows plants to continually adjust to shifting soil water and nutrient sources, placing new roots where resources are most freely available. Because of this, species with fairly different coarse root structures can have similar patterns of fine root growth (Fernandez and Caldwell 1975) and resource uptake (Schwinning et al. 2005), particularly as measured in the vertical dimension of the soil (fig. 13.1). For example, Fernandez and Caldwell (1975) showed that the fine root growth of three dominant Great Basin shrubs shifted vertically downward over the course of a growing season, following the extractable

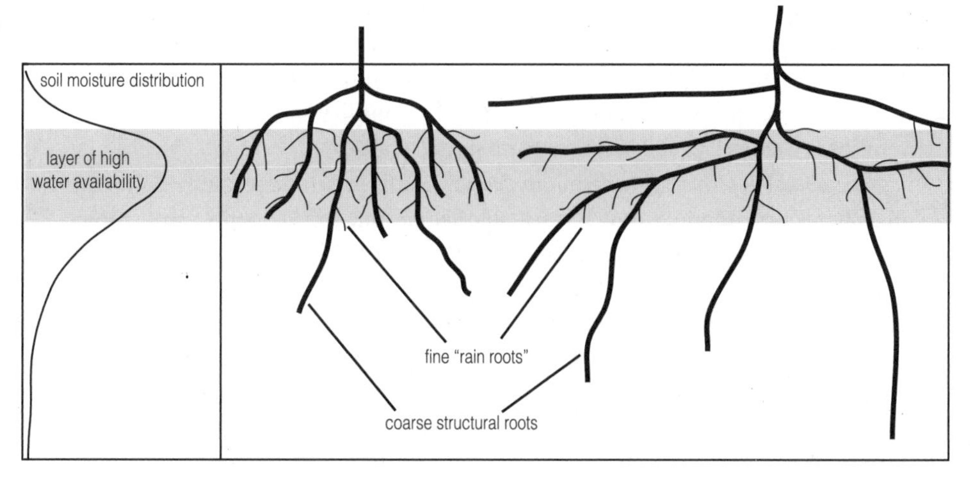

Fig. 13.1. Diagram of perennial plant species with different root structure and depth that can have similar vertical resource-uptake patterns, where both plants have fine root growth (indicated by the small, thin lines) on structural roots within the soil layer that has plant-extractable resources.

soil moisture as it receded from the surface. In this study, transparent root observation tubes called minirhizotrons were used to identify soil regions of enhanced fine root growth in situ. While this method allows nondestructive observation and quantification of root growth over time, one downside is that minirhizotrons provide extremely localized information on root dynamics and only a relative measure of fine root growth patterns (Wilcox et al. 2004).

Methods of Measuring Root Function In Situ

One of the least intrusive methods of measuring root function involves the rates of water transport; heat-pulse flowmeters can be attached to main stems above ground to measure the quantity and direction of water flow in plants (Cohen et al. 1981; Hultine et al. 2003). To quantify nutrient uptake, leaf area growth rates can be combined with measures of tissue concentrations of nitrogen (N) and phosphorus (P) to estimate the rate of nutrient uptake. Additional methods, using stable isotopes, are available for estimating the spatial origin of water and nutrients taken up by plants. White et al. (1985) pioneered a method that uses the natural differences in the stable isotope composition of water at different soil depths to identify the water sources used by plants. This method has since been widely applied in the Americas, though not, to our knowledge, in the Mojave Desert. In a recent refinement of this method, data on the vertical distribution of stable isotope ratios in soil water are combined with data on the distribution of soil-water potential (a measure of soil water's free energy which determines the direction of water transport and correlates with soil hydraulic conductivity) to estimate complete water extraction profiles for plants (Ogle et al. 2004). In another variant, water or nutrients with known, extreme (i.e., unnatural) stable isotope ratios are added to the soil to trace and quantify plant resource uptake from specific soil locations (Caldwell et al. 1985; Schwinning et al. 2002).

The overall lesson learned from in situ studies of root function is that patterns of resource uptake are extremely adaptable, largely due to the opportunistic growing patterns of the fine root system, though individual species may differ in the degree of fine root plasticity (Cui and Caldwell 1997). Patterns of resource uptake are not identical to the patterns of structural root distribution. However, the structural root distribution does constrain the space from where resource uptake is possible. For example, a more laterally extensive root system will be able to respond to water and nutrient pulses farther away from the main stem of the plant (Caldwell et al. 1996), and a deeper root system will be able to extract water and nutrients from deeper soil regions (Schwinning et al. 2005).

Linking Root Form and Function

The relationships among standing biomass, resource acquisition rates, and growth is central to any investigation of biological organization (Enquist 2002). The relationships between canopy photosynthesis, leaf area and placement, N content, and phenology are well understood (e.g., Aber et al. 1996). However, finding equivalent relationships among, for example, nutrient uptake rates and root area and placement is more difficult, not only because root form and function are more difficult to observe, but also because the soil is a more complex environment than the atmosphere, to which roots respond by undergoing more complex transformations than leaves typically undergo.

Young, actively growing roots have the highest capacity for water and nutrient uptake, aided by root hairs. These initially high water- and nutrient-uptake capacities can decline by an order of magnitude as roots grow older (Nobel et al. 1991), due to the loss of root hairs and the suberization of the outer cell layers of the root. However, not all root hairs are shorter-lived compared with the root. In shrubs of the Great Basin Desert, Fernandez and Caldwell (1975) observed that root hairs suberized along with the root epidermis, and persisted as long as the root itself. Suberization improves the longevity of roots, presumably by protecting the roots from adverse soil factors, but also reduces water uptake capacities. Suberization may not be uniform along a root and can be locally delayed under wet conditions in areas such as soil microsites that harbor water or in areas protected from evaporation, such as under rocks (North and Nobel 1998).

Roots can also adjust their nutrient uptake capacity in response to external ion concentrations (Ivans et al. 2003). In addition, if a plant's current demand for nutrients is low, or if its root system has access to higher nutrient concentrations elsewhere, uptake rates may remain low even if nutrient supply is enhanced (e.g., Duke and Caldwell 2000; Gebauer and Ehleringer 2000). By adjusting their nutrient uptake capacity in time and space according to nutrient supply and demand, plants presumably optimize resource acquisition.

Nutritive and nonnutritive chemical uptake by plants involves complex interactions between plant roots and soil under various physical and chemical conditions. Nutrients and water are absorbed by the root, which creates a boundary zone with reduced resource concentrations that draws resources through the soil toward the root via mass flow and diffusion (Jungk 2002). Nutrient mobility across this gradient is influenced by nutrient solubility (a function of water volume, oxygen content, and pH), the buffering capacity of soil (i.e., interactions with charged particulate surfaces), interactions between ions in solute, as well as soil texture and compaction. For example, cation uptake generally decreases with increasing pH (particularly above 8.0), while anion uptake increases, and vice versa (Neumann and Römheld 2002). In the pH range of 4.0–8.5 (common for soils of North America), highly soluble nutrients such as nitrate, calcium, magnesium, and sulfate are readily extracted and move with water by mass flow toward the roots (Jungk 2002; Neumann and Römheld 2002). Less soluble ions, of generally intermediate or low mobility in soil, such as iron (Fe^{2+} and Fe^{3+}), potassium (K^+), ammonium (NH_4^+), and inorganic phosphorus (P_i), are transported primarily by diffusion (Glass 2002; Jungk 2002; Neumann and Römheld 2002), a slow process compared with mass flow. Plant roots can influence the local nutrient uptake potential by means of exudates that alter the pH in the soil immediately surrounding the root, thus improving the availability of certain ionic species (Jungk 2002). Mass flow and diffusion of dissolved nutrients are also influenced by soil capillary potential and the tortuosity of the pore spaces, which are, in turn, affected by soil texture and the degree of compaction (Jungk 2002).

Cost-Benefit Considerations for Root Design

While the theory of water and nutrient transport is well developed, and mechanistically elaborate models of root function do exist (e.g., Sperry et al. 1998), these models are too cumbersome to be employed in the prediction of community and ecosystem processes. Ecosystem models often sidestep the mechanistic characterization of root function, relying instead on the assumption of a simple relationship between soil resource uptake, root/shoot ratio, and the vertical distribution of root biomass (reviewed in Weltzin et al. 2003). One methodology that has helped in the development of a general quantitative framework in ecophysiology is cost-benefit analysis. Applied to root function, this approach is promising for root research and the effort to better understand root organization.

The basis of all cost-benefit considerations is the adoption of a common "currency" used to compare the costs and benefits of plant structures. Ideally, this currency is fitness, but this is typically impractical, if not impossible, to measure. Perhaps carbon (C) is the second best choice as a common currency for plants, since many fitness components directly depend on C availability (e.g., biomass maintenance, growth, storage, and reproduction; Bloom et al. 1985). In this context, the "cost" of a root may be construed as the C requirement for building and maintaining the root, as well as for supporting root functions, such as nutrient transport, root exudation, and symbiotic associations. The "benefit" can be understood as the photosynthetic C "income" made possible by the existence of that root. The basic assumption is that natural selection would promote only those root growth patterns that optimize C capture efficiency. Eissenstat (1997) summarized the known root-related cost-benefit tradeoffs as follows:

1. When soil resources are more limiting, plants allocate a greater proportion of whole-plant C to root growth and maintenance. For example, plants growing in sites with higher water or nutrient limitations commonly have higher root/shoot ratios. This tradeoff applies not only to variation between sites, but also to phenotypic variation among plants within a site—ephemeral and drought-deciduous species should have smaller root/shoot ratios than evergreen species in the same site because the latter remain active during the more resource-limiting parts of the year. This generalization is widely applicable to desert plants (Smith et al. 1997).

2. Longer-lived (e.g., > 5 years) woody plants support a larger and more widespread system of structural roots, while shorter-lived (< 5 years) woody plants have smaller, compact root systems composed of thinner roots (Burgess 1995). The compact "herringbone" structure is better suited for the rapid and complete depletion of resources within the soil volume, while a spreading "dichotomous" structure leads to a more gradual depletion of soil resources via diffusion over longer distances (Fitter 2002).

3. Longer-lived roots have higher construction costs, lower maintenance costs, and reduced water and nutrient uptake efficiencies compared with shorter-lived roots. Many of these differences are due to variations in root thickness and in the degree of suberization and amount of tannin deposition in the cell walls. Thicker and more suberized roots are often more drought resistant, while tannins discourage herbivory. This allows roots to live longer, but also increases construction costs and reduces resource-capture efficiency. However, the lifetime C efficiency of longer-lived roots can still be high, because maintenance costs are low. In shorter-lived roots, construction costs may be relatively lower, but they require a major proportion of the lifetime resources of a shorter-lived root. Therefore, shorter-lived roots should be placed where resource levels are temporarily high, while longer-lived roots should be placed where the long-term prospect for resource uptake is high—for example, in competitor-free space in deeper soils that have more persistent water availability.

4. Roots that have high water-transport capacities relative to construction cost also have lower cavitation resistance (i.e., their ability to function under low suction pressures) and therefore a reduced range of operation with respect to soil-water potentials (Sperry and Hacke 2002). This trade-off should favor the construction of a denser, more costly xylem when most plant gas exchange takes place under relatively dry conditions (e.g., in drought-tolerant evergreen shrubs), and construction of a lighter wood with wide-diameter xylem elements and thinner cell walls when most gas exchange occurs under relatively wet conditions (e.g., in ephemerals or drought-deciduous shrubs). The tradeoff can also apply to roots of the same plant if roots are exposed to different soil moisture conditions. For example, in *Gutierrezia sarothrae* (broom snakeweed) the radial and axial hydraulic conductivities of deeper roots, which are exposed to more moderate soil-water potentials, are higher than those of the more suberized shallow roots, which are regularly exposed to much lower soil-water potentials (Wan et al. 1994).

5. Roots are shed when contributions to the plant's C income no longer justify maintenance costs. Fine roots are shed frequently, and longer-lived structural roots can also be shed during severe drought events when plants are in acute C deficit. This latter circumstance occurs when trade-offs associated with long-term C efficiency of longer-lived roots (which may still be high) are trumped by the immediate necessity for whole-plant survival. Short of shedding entire roots, plants can also shed just the outer layer (cortex) of the root, leaving the center (stele) intact. This reduces maintenance costs without losing all uptake and transport capacity, or the capacity to deploy new laterals when conditions improve (Jupp and Newman 1987).

6. Mycorrhizal associations increase nutrient uptake capacities of roots but also increase root maintenance costs. This symbiosis is more valuable for plants with coarse root systems in nutrient-poor environments. Mycorrhizal associations are usually reduced in fine roots and with high soil nutrient levels (Titus et al. 2002).

As we review the known rooting patterns of Mojave Desert plants, we will return to these cost-benefit principles and evaluate how well they describe Mojave Desert plants.

PATTERNS OF ROOT DISTRIBUTION

General Rooting Characteristics in Desert Biomes

In ecosystem models, the maximal rooting depth of plant communities is one of the more uncertain parameters, but has potentially significant effects on soil development, soil moisture fluxes (including moisture availability to plants), nutrient cycling, and community composition (e.g., Kleidon and Heimann 1999). Several recent studies have therefore focused on understanding prevailing vertical rooting patterns found within major bionomic associations on a global scale

(Canadell et al. 1996; Jackson et al. 1996; Schenk and Jackson 2002). In these global comparisons, desert biomes do not have the deepest root systems (Schenk and Jackson 2002), but they do have deeper root distributions than most other biomes. In addition, perennial plants in deserts have an average of 50% of their root biomass in the top 30 cm, as opposed to perennial plants in temperate grasslands and boreal forests, which have an average of 83% of their root biomass in the top 30 cm, and perennial plants in tundra, which have 93% (Jackson et al. 1996). This pattern might at first appear contrary to the typical view of deserts as places where water and nutrients are concentrated near the soil surface because of low rainfall and high evaporation. However, high soil surface temperatures in deserts tend to be lethal to roots, and generally low soil resource concentrations may drive desert plants to explore a greater portion of the soil volume, including marginal resource reservoirs at greater depth. Therefore, increasing water limitation in the environment tends to increase the average maximum rooting depth of shrubs relative to the size of their canopies (Schenk and Jackson 2002).

Soil resource levels are also quite variable horizontally in water-limited ecosystems, due to variable surface characteristics, biogeochemical conditions, and soil surface exposure (Breshears and Barnes 1999). Schenk and Jackson (2002) noted generally reduced lateral root growth by plants in arid environments compared with those in more mesic environments. However, lateral root spread and root response to horizontal resource heterogeneity (both temporally and spatially) vary dramatically among species. The lateral root spread of some woody perennials in the Mojave Desert is impressive, with lateral root lengths of up to 9 m, and commonly between 1 and 3 m (Hooten and Myles 2006).

Plant Communities in the Mojave Desert and their Overall Rooting Patterns

Compared with other North American deserts, the Mojave Desert has a relatively low diversity of perennial plants and summer annuals, and a fairly high diversity of winter annuals (Beatley 1969; Rundel and Gibson 1996). Among the perennial plants, woody shrubs, cacti, yuccas, and perennial grasses are dominant, while perennial forbs are relatively sparse. Nonsucculent trees are nearly all phreatophytic (with roots that obtain water from a permanent ground supply) and restricted to areas of shallow groundwater along desert riparian areas and near springs. Trees such as *Juniperus* ssp., *Pinus edulis,* and *Pinus monophylla* (juniper and pinyon) may be found at higher elevations (typically above 1,700 m) in the Mojave Desert, yet these species are typically considered landmark species for the Great Basin Desert bionomic province.

In an extensive survey of the northern Mojave Desert flora on the NTS, Ostler et al. (2000) listed 28 species of shrubs and trees useful for describing shrubland alliances and identified only three shrubland alliances in the northern Mojave Desert biome. Alliances were identified based on cluster analyses that required

75% similarity of species composition between landform units that were internally consistent with respect to soils, slope, geology, hydrology, vegetation, and resident animal species, and showed notable differences in plant species composition compared with neighboring units. One plant species may clearly dominate any given site, although one or more additional species may frequently occur. The strongest indicator of the Mojave Desert bionomic province was found to be the creosote bush–white bursage shrubland alliance, so named due to the predominance of the two species.

About one-third of the above-ground standing biomass of Mojave Desert plants is composed of these two species—*Larrea* and *Ambrosia* (Turner and Randall 1989). Of the total area of the Mojave Desert, 70% is inhabited by this alliance (Lathrop and Rowlands 1983), which typically occurs on gentle to moderately sloping bajadas, 800–1600 meters elevation, with soils composed of coarse, sandy loams. Within this alliance, *Larrea* and *Ambrosia* comprise, on average, 12.7% and 43.1%, respectively, of the shrub abundance, with considerable local deviation around the mean. Other perennial shrub species in the alliance with abundances over 5% are *Ephedra nevadensis* (Mormon tea, 7%), *Krameria erecta* (littleleaf ratany, 6.7%), and *Lycium pallidum* (pale desert-thorn, 5.2%).

Other Mojave Desert alliances on the NTS include the *Lycium shockleyi-Lycium pallidum* and *Atriplex confertifolia-Ambrosia* shrublands, each comprising about 1–2% of Mojave Desert plant communities. On average, the *Lycium shockleyi-Lycium pallidum* alliance is composed of 32.0% *Lycium shockleyi* (Shockley's thornbush), 29.2% *L. pallidum,* and 14.7% *Ambrosia,* and occurs exclusively in lowlands on clayey loams and loamy sand. The *Atriplex-Ambrosia* alliance (29.6% *Atriplex,* 20.3% *Ambrosia,* 10.1% *Ephedra*) occurs over the same wide elevation range as the *Larrea-Ambrosia* association, but predominantly on loamy soils (Ostler et al. 2000).

These distribution patterns illustrate the relatively low species diversity of the vast majority of the Mojave Desert and suggest that much of the rooting patterns of Mojave Desert communities can be studied by concentrating on relatively few key species. (For a more thorough review of the shrubland alliances of the NTS, see Ostler et al. 2000 and Hooten and Myles 2006.)

The most extensive observational analysis on root distributions of Mojave Desert shrubs was done in the late 1960s and early 1970s in Rock Valley on the NTS by Wallace et al.(1980), who excavated the top 50 cm of soil under 48 individual shrubs composed of 9 species. In their study, the top 50 cm were chosen because the greatest mass of coarse roots were located at those depths at the study site, which reportedly had an extensive hardpan (caliche) layer close to the soil surface. Other studies also reported the greatest abundance of structural and permanent roots in the top 1 m of soil (Wallace and Romney 1972; Gibbens and Lenz 2001). However, at another site in Rock Valley, without a restrictive caliche layer, Wallace and Romney (1972) found an abundance of roots well below 1 m.

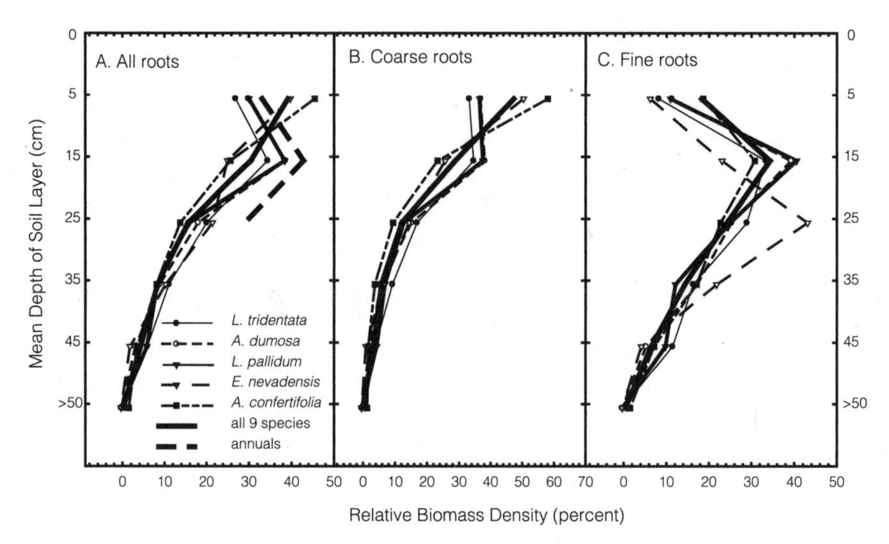

Fig. 13.2. Graphs comparing relative vertical root allocation of five different species [*Larrea tridentata* (creosote bush), *Ambrosia dumosa* (white bursage), *Lycium pallidum* (wolfberry), *Ephedra nevadensis* (Mormon tea), and *Atriplex confertifolia* (shadscale), all species combined, and a sampling of annuals. Note that *Ephedra* is the only species with maximum fine root density between 20 and 30 cm (data from Wallace et al. 1980).

The overall root distribution of the creosote bush–white bursage alliance is dominated by the coarse root fraction (i.e., roots > 2 mm diameter), which comprises roughly 64%–78% of total root biomass (Wallace et al. 1980) (fig. 13.2). Between 25% and 50% of all coarse root biomass is in the top 10–20 cm of the soil and coarse root density drops off approximately exponentially with an average decline of 5% per centimeter. Fine root biomass (i.e., roots < 2 mm diameter) peaks in the 10–20 cm soil depth interval for most species and drops off below this in an approximately linear fashion in relation to depth. Wallace et al. (1980) speculated, and Nobel (2002) confirmed, that the upper 10 cm of soil may frequently be too hot and dry to support water or nutrient uptake by fine roots, while the 10–20 cm layer has more moderate temperatures and is still shallow enough to be wetted relatively frequently by larger precipitation events.

Wallace and Romney (1972) found that while the relative vertical allocation of shallow root biomass was quite similar among shrub species, communities differed both in overall root density and in the horizontal placement of roots with respect to canopy and interspaces. In a *Larrea-Ambrosia* community, the top 30 cm of soil had four times more root biomass between than underneath shrub canopies (fig. 13.3). At another site, dominated by *L. pallidum,* root biomass below the shrub canopies was three times greater than in the interspaces, while an *Ambrosia*-dominated site had more evenly distributed, horizontally placed roots

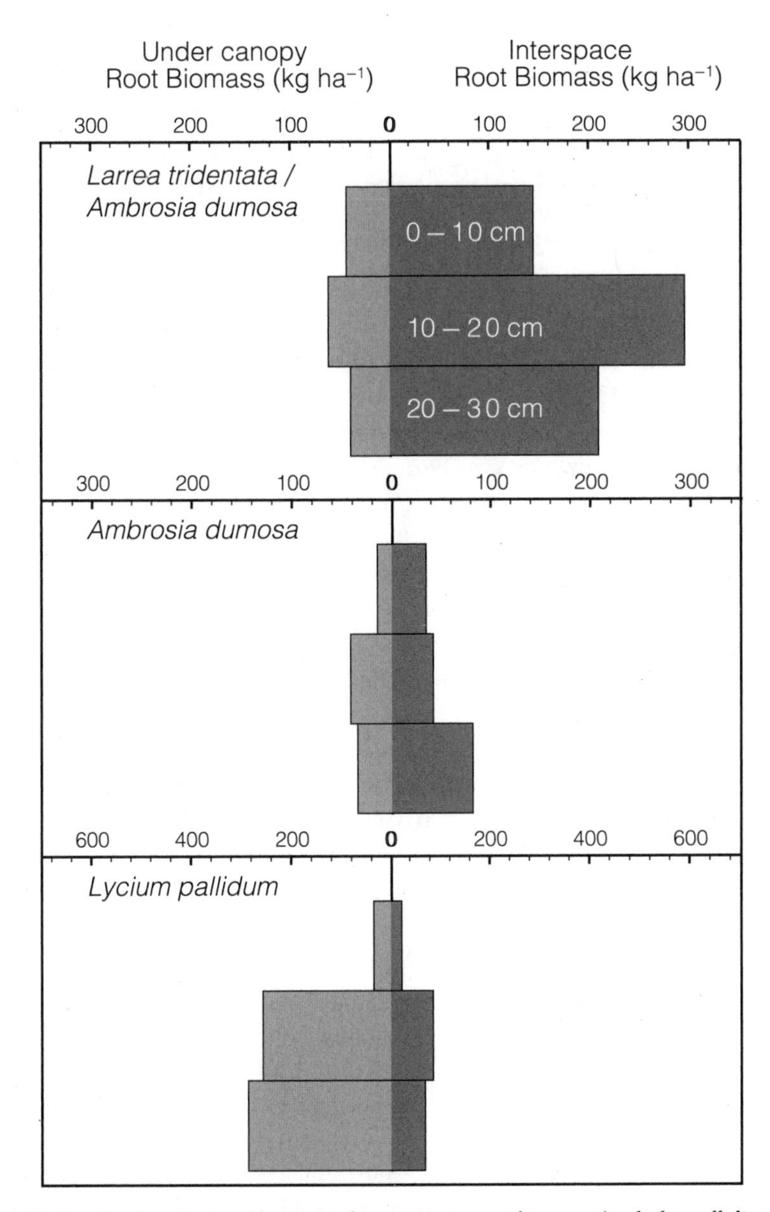

Fig. 13.3. Graph of vertical and horizontal variation in root biomass (including all diameter classes) depicted for three sites dominated by different species: *Larrea tridentata-Ambrosia dumosa, Ambrosia dumosa,* and *Lycium pallidum.* Only the site dominated by *Lycium* had greater biomass under the plant canopy (data from Wallace and Romney 1972).

than at either of the two other sites. These differences suggest that shrub species may differ more in their horizontal than in their vertical root distributions, the latter being greatly constrained by steep resource gradients. Thus, to better understand the root structure and function of Mojave Desert plants, we have to consider the full three-dimensional configuration of the root systems.

Rooting Patterns of *Larrea tridentata*

The most common shrub of the Mojave, and arguably the most studied, is *Larrea tridentata,* a longer-lived sclerophyllous evergreen. It is thought that individuals of this species can live for centuries and, in some sites, perhaps more than a thousand years (McAuliffe 1988). The root distribution of this species has been studied in the Mojave (Wallace and Romney 1972; Stevenson et al. *this volume*), the Chihuahuan (Chew and Chew 1965; Brisson and Reynolds 1994; Gibbens and Lenz 2001) and the Sonoran Deserts (Yeaton et al. 1977). All excavations were conducted on basin and bajada soils, which are typically coarse gravelly loams, alluvial in nature, and with variable prominence of caliche layers at various depths. Parent materials are typically limestone sediments.

Excavations portray the root system of *Larrea* in a generally consistent manner, with an extensive system of lateral roots, typically at 15–40 cm depth, depending on the location of the caliche layer (lateral roots stay above this layer). Stevenson et al. (*this volume*), working in dry sites of the Central Mojave Desert, found that root systems on young Holocene soils were deeper but had relatively less lateral spread than root systems developed in older Pleistocene soils.

Lateral roots grow outward at very low angles and even upward toward the soil surface, typically extending 3 m, and occasionally up to 4.5 m, from the central trunk (Gibbens and Lenz 2001). Several vertical sinker roots originate from laterals and penetrate deeply through breaks in caliche layers and highly compacted argillic horizons (Gibbens and Lenz 2001) (fig. 13.4). This was observed by Stevenson et al. (2006) in the Mojave Desert, though he deemed them rare.

At a relatively flat, sandy loam site in the Jornada Experimental Range, Gibbens and Lenz (2001) found taproots as deep as 5 m, though most roots ended at about 3 m (fig. 13.4). At a site where the slope of the terrain was 2%—slightly steeper than at the Gibbens and Lenz (2001) study site—Gile et al. (1998) found that taproots penetrated to no more than 2 m and thought that this was the result of increased runoff and reduced infiltration. Wallace and Romney (1972) sketched a *Larrea* root system from the NTS with essentially the same characteristics as those described by Gile et al. (1998), with sinker roots that reached into a gravelly layer 1.68 m deep (fig. 13.5), while Hooten and Myles (2006) observed *Larrea* roots below 3 m on the NTS. Thus, it appears that *Larrea* has the potential to grow roots to 5 m or more, but that individual site characteristics determine the actual depth, which is most likely constrained by the depth of water infiltration.

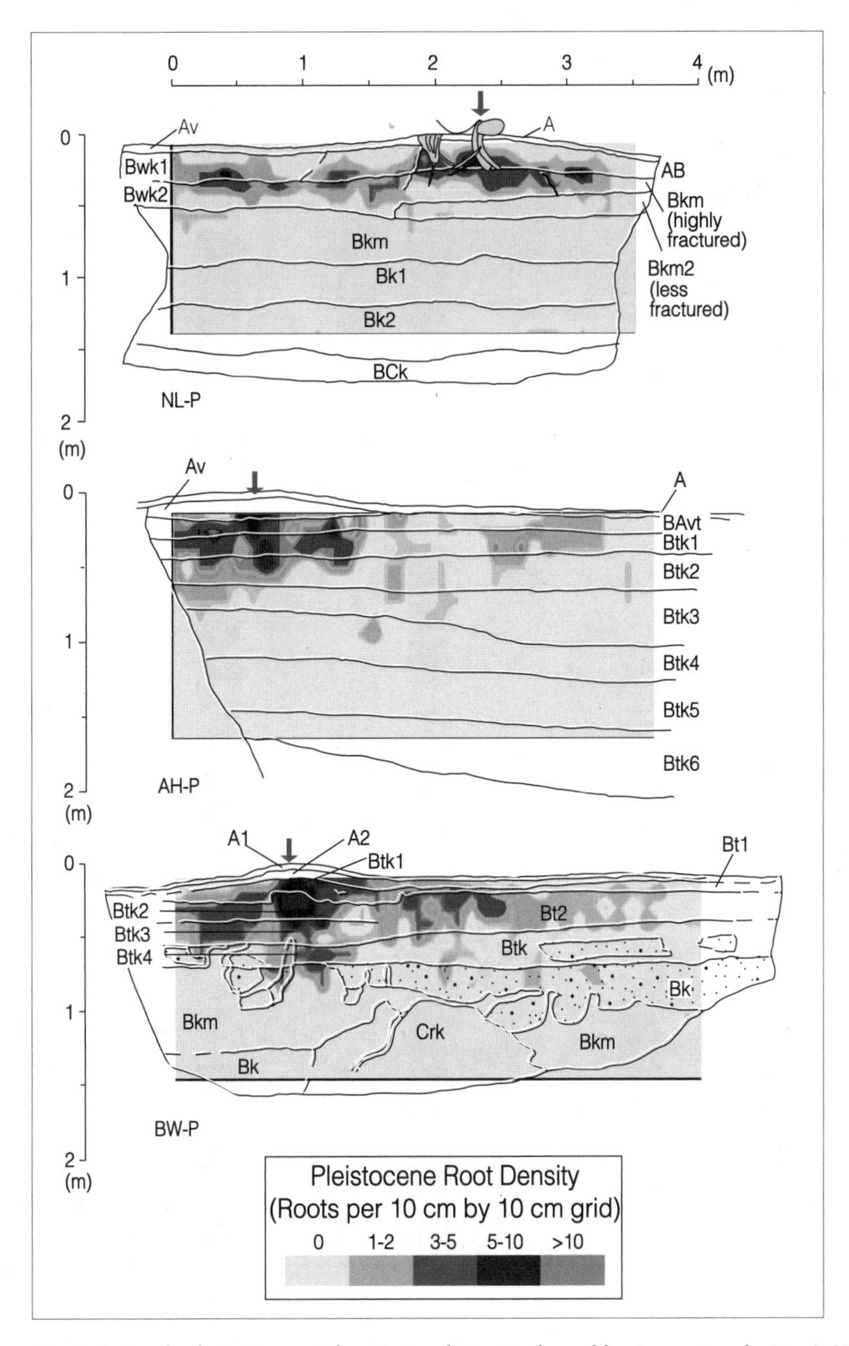

Fig. 14.5. Graphs depicting root density in relation to the soil horizons at each site. *A,* Holocene surface; and *B,* Pleistocene surface. Rooting depth was significantly greater in Holocene soils, whereas lateral root growth (% of roots observed in interspaces) was greater in Pleistocene soils. Horizons restrictive to root penetration (primarily cemented petrocalcic [Bkm] horizons) are highlighted in orange. The location of the root crown at each site is indicated by a red arrow. (See color plate following page 162.)

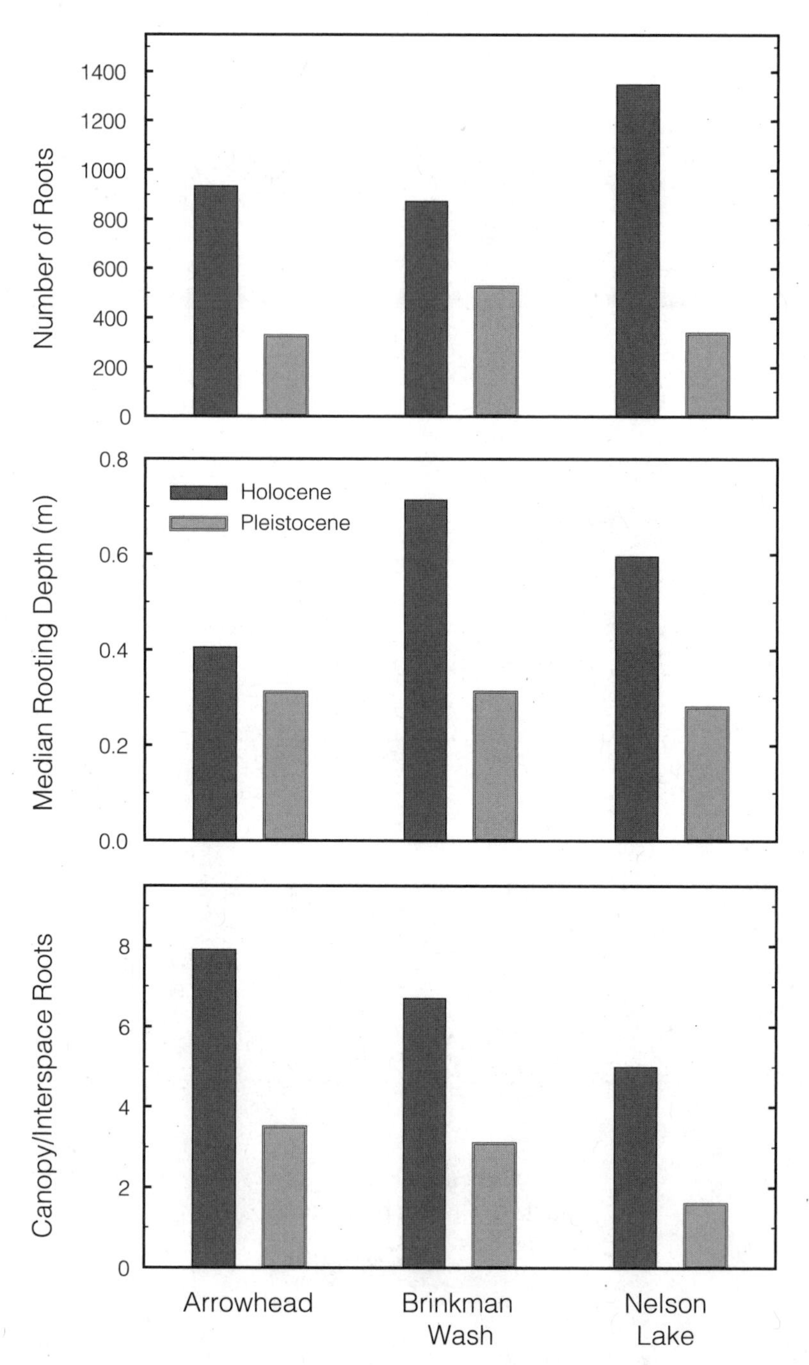

Fig. 14.6. Graphs depicting the total number of roots, median rooting depth, and under-canopy to interspace root ratio observed in Holocene versus Pleistocene soils at each site. When grouped by soil age, all parameters showed significant differences between Holocene and Pleistocene soils ($n = 3$, $P < 0.05$).

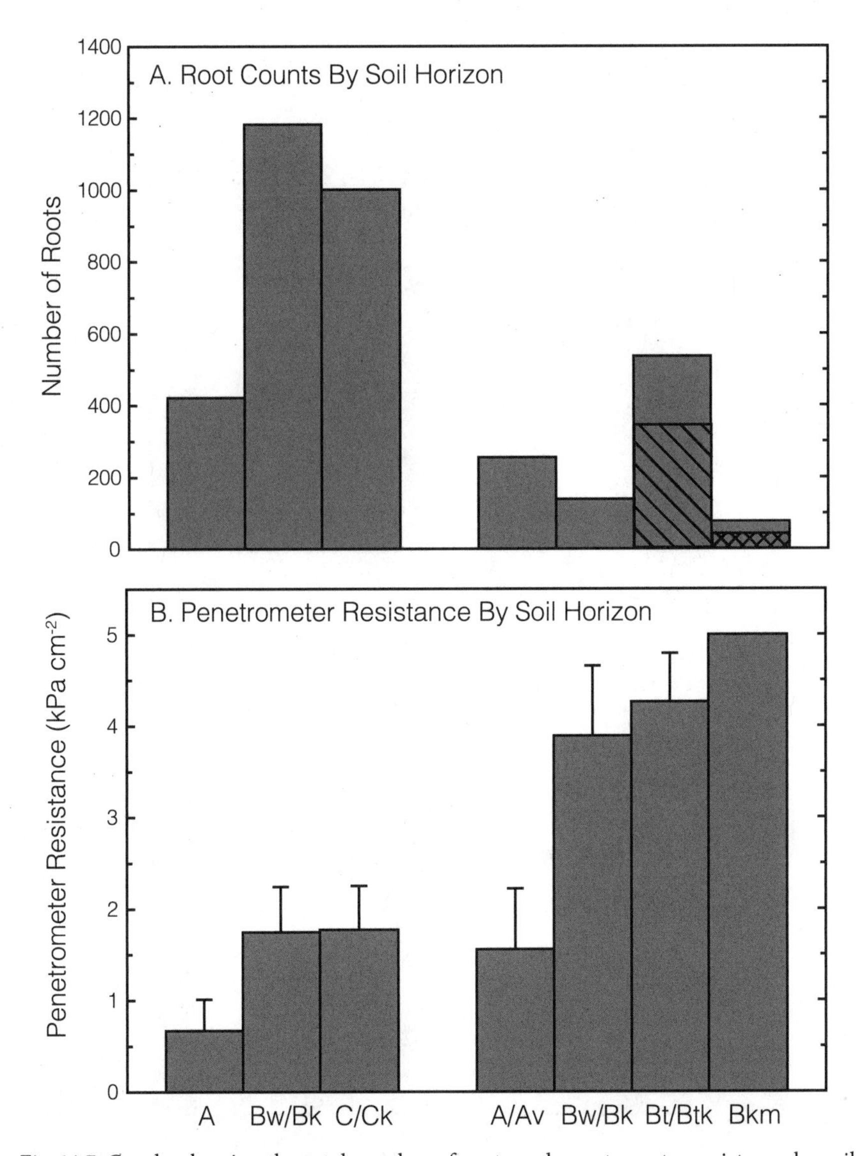

Fig. 14.7. Graphs showing the total number of roots and penetrometer resistance by soil horizon type (horizons arranged by depth from the surface). Cross hatches represent roots that occurred in fractured Bkm (petrocalcic) horizons. The majority of roots in Pleistocene Bt/Btk horizons were found in the upper 30 cm of these horizons (indicated by hatches). Error bars represent the standard deviation of penetrometer readings ($n = 10$). The maximum recordable penetrometer reading was 5.0.

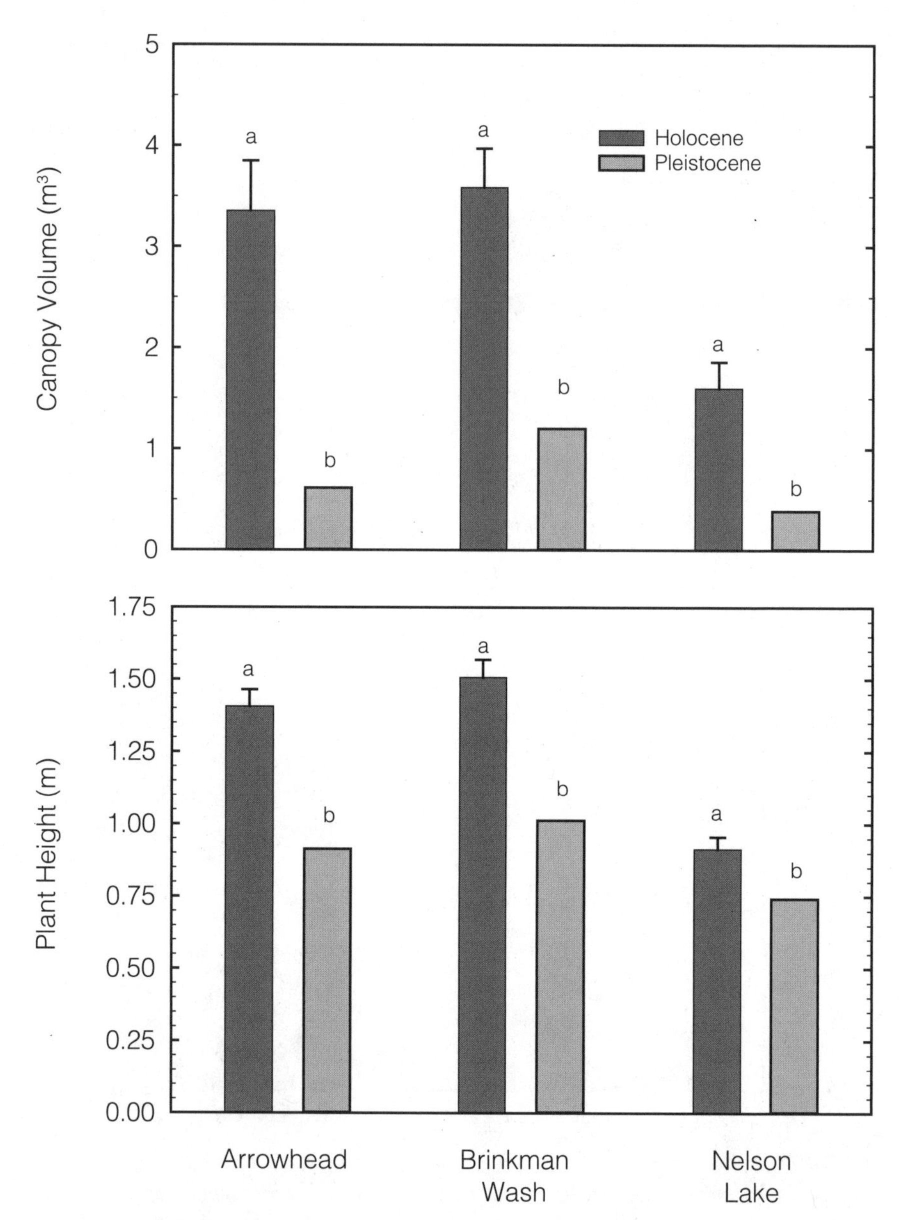

Fig. 14.8. Graphs showing the mean *Larrea tridentata* canopy volume and plant height for Holocene versus Pleistocene soils at each site. Error bars represent the standard error of the mean. The presence of different letters at a site indicates a significant difference between Holocene and Pleistocene soils within that site ($P < 0.05$).

two-thirds of all roots observed in Bkm horizons. Roots did occasionally follow breaks (fissures, crotovina, etc.) into, or even through, the petrocalcic horizon, and some roots were found in less dense soil horizons below the cemented layers. Of the overall root count, fine roots made up 92%, medium roots made up 6%, and coarse roots constituted 2%. Coarse roots (> 5 mm in diameter) were generally restricted to depths less than 30 cm and to areas below the plant canopy (data not shown). In Holocene soils, total root densities were also generally greatest below the plant canopy.

Mean *Larrea* canopy volume was significantly greater on each of the three Holocene soils compared with their respective Pleistocene soils (approximately five times greater at the AH and NL sites and three times greater at the BW site), and varied from a low of 0.39 m^3 at the NL-P site to a high of 3.6 m^3 at the BW-H site (fig. 14.8). Plant height followed a similar pattern: It was 1.25 times greater at the NL Holocene site, and 1.5 times greater at the AH and BW Holocene sites, compared with their respective Pleistocene sites. Plant density was remarkably similar at all sites (approximately 300 plants ha^{-1}), except the AH-P site, where the density was approximately 500 plants ha^{-1}.

DISCUSSION

Soil-Root Dynamics

Where a petrocalcic horizon was present (at the BW-P and NL-P sites), *Larrea* rooting depth was restricted, with roots primarily growing laterally above the cemented horizon (Bkm, stage IV). Penetrometer results confirmed the high resistance of the petrocalcic horizon to root penetration. Our results also support those of Gile et al. (1998), who showed that the rooting depth of *Larrea* in Chihuahuan Desert sites was restricted by the development of a petrocalcic horizon.

The AH Pleistocene site did not have a petrocalcic horizon, but had a similar median rooting depth to the two Pleistocene sites that did. The AH Pleistocene soil comprised mostly argillic horizons (the whole profile below 14–24 cm depth), and nearly two-thirds of the roots found in these argillic horizons were located in their upper 30 cm. Penetrometer resistance in these horizons (Bt and Btk) was high but variable. These results indicate that the intrinsic structure (subangular, blocky) of these clay-rich argillic horizons allowed some root penetration but only to a limited depth (unlike the petrocalcic horizons, which acted as an abrupt, continuous barrier).

Canopy hydraulic conductivity also had the lowest value at the AH Pleistocene site, suggesting that water flux may be as important in root restriction as strong physical barriers such as petrocalcic horizons. McDonald et al. (1996) and Young et al. (2004) demonstrated that soil moisture in older, well-developed soils is primarily restricted to the surface soil horizons, and that desert pavement restricts water movement into the soil. The formation of calcic horizons occurs due to the evaporation of moisture at the wetting front (the leading edge of soil water moving downward through soil), which results in precipitation of the salts within the soil water solution. The depth at which this occurs depends upon the amount of water entering the soil, the hydraulic conductivity of the soil (a function of soil parent material, particle-size distribution, and structure), and the amount of soil water lost to evaporation and transpiration.

Hamerlynck et al. (2000, 2002) demonstrated that *Larrea* performance (as indicated by plant-water potentials and photosynthetic rate) was inferior in older soils with argillic (Bt) horizons compared with younger soils without argillic horizons. Soil conductivity data from our study indicate that the hydraulic conductivity of Pleistocene soils was lower than that of Holocene soils, and likely prevents soil water from penetrating very far into the soil profile during limited-duration precipitation events.

Among the Holocene soils, which offered little physical resistance to downward root growth, the two sites that had the greatest hydraulic conductivity (BW-H and NL-H) also had the deepest median rooting depth. The AH-H site had the shallowest rooting depths and also the lowest conductivity of any Holocene soil. The AH site also had the smallest difference in average rooting depth between Holocene and Pleistocene soils (~ 30% deeper in the Holocene soil compared with the Pleistocene soil, as opposed to ~ 100% deeper at the other two sites), as well as the smallest difference in undercanopy hydraulic conductivity between the Holocene and Pleistocene sites.

Over landscape scales, Bedford et al. (*this volume*) concluded that variations in surface soil texture correlate with variations in *Larrea* properties on recent geological deposits but not on older ones. Our results support these findings: In young (Holocene) soils, soil properties are relatively constant with depth so that surface texture is a valid indicator of hydraulic conductivity, rooting depth, and plant size. On older (Pleistocene) deposits however, soil properties vary dramatically with depth, and the subsurface characteristics often control hydrology and root morphology; therefore, there is a weaker relationship between surface soil texture and various plant characteristics.

As reported by Caldwell et al. (2006), interspace hydraulic conductivity measurements were similar to or slightly higher than the undercanopy measurements, with the exception of the site with the strongest desert pavement development—NL-P site—where the undercanopy conductivity was greater. It is unclear whether differences between interspace and undercanopy soil hydrology directly affect root distribution (particularly the lateral spread of roots), posing an interesting question for further research. In contrast to this study, the majority of studies have concluded that water flux is greater under canopies (Lyford and Qashu 1969; Dunkerley 2000; Shafer et al. 2007). This discrepancy may be due to the use of different methodologies: Unlike other methods, the infiltrometer used here strictly measures near-saturated soil conductivity ($h < 0$), and not conductivity under ponded, saturated ($h \geq 0$) conditions. Also, the infiltrometer method does not account for stem flow and root channel macropore flow, which have been shown to be important in plant mounds when saturated conditions develop due to high-intensity precipitation (Devitt and Smith 2002). The slight decrease in conductivity between the interspace and the undercanopy area that we observed may be due to the loss of soil structure under the plant canopy, which could have resulted in a loss of preferential paths of water flow through the soil. At the NL-P site, where the pattern was reversed, the well-formed desert pavement acted as a barrier to water infiltration in the interspace, while the minimal soil structure associated with the undercanopy soil increased the water conductivity.

During the course of the study, we observed virtually no overlap of *Larrea* root systems, and found few roots in the interspaces even in Pleistocene soils, where the lateral spread of roots was greater. In contrast, personal observations in a less arid Mojave Desert site (the Nevada Test Site in western Nevada) revealed significant root densities in the interspaces. Intraspecific competition between *Larrea* plants is thought to be strong, as *Larrea* roots inhibit growth of other *Larrea* roots, as well as those of other species. Mahall and Callaway (1992) and Brisson and Reynolds (1994) reported that *Larrea* root systems in the Chihuahuan Desert had minimal overlap with each other, while in contrast, Gibbens and Lenz (2001) reported that *Larrea* roots were intermingled both with each other and with the roots of other species at some sites in the Chihuahuan Desert.

The prevailing opinion seems to be that roots grow toward nutrient patches (moisture as well as mineral nutrients) in the soil (Eissenstat and Caldwell 1988; Jackson and Caldwell 1989). Several studies involving *Larrea,* however, have revealed no relationship, or an inverse relationship, between soil moisture and root growth (Obrist and Arnone 2003; Wilcox et al. 2004). These findings may be due to the reported affinity of *Larrea* roots to well-oxygenated conditions (Lunt et al. 1973). The interactions between soil moisture and root morphology, and the competitive interactions of *Larrea,* appear complex, and further comparative studies of these interactions are needed.

Plant-Soil Interactions

Our data indicate that *Larrea* in Holocene soils were not only deeper rooted, but were taller and had a larger canopy than plants in well-developed Pleistocene soils. As there was no evidence of disturbance, or indication that the populations were of different age classes, we assume that the differences in size and rooting depth were not due to differences in plant age. The advantage of greater rooting depth has recently been debated; McCulley et al. (2004) present evidence that in arid and semiarid ecosystems, deeper roots may be more important for the uptake of mineral nutrients than for the uptake of moisture. They mention that there is little evidence of significant available soil water deep in the profiles of arid soils. From a soils perspective, this makes sense because calcic horizons form at the average depth (over centuries to millennia) of the wetting front in these arid soil profiles. During intense precipitation events water may be able to flow through cracks and fissures in the petrocalcic horizon, particularly when these horizons are discontinuous or fractured; but this probably occurs less in the Mojave Desert than in other arid systems such as the Chihuahuan Desert, where many of the root morphology studies were carried out. This is because the petrocalcic horizon (where present) is generally thicker and more continuous in the Mojave Desert.

While it seems obvious that larger plants would have more roots and deeper rooting systems, we suggest that it is the constraints (both physical and hydrological) put on the root systems by the soil environment that are largely controlling plant performance (and thus size) and not the other way around. Admittedly, this is a difficult assertion to prove with our limited data set (and with other unaccounted-for variables, such as differential salt content in the soils), but other studies, particularly the work of Hamerlynck et al. (2000, 2002), would appear to corroborate our conclusions.

The process of soil morphological development occurs over long time spans (many thousands of years) and vegetation is one of the factors influencing the process. The degree to which individual plants modify the soil is an intriguing question given their short life span compared with the age of the soil. The formation of plants mounds—features associated with shrubs in arid environments—is an example of the extent to which individual plants can modify their surrounding environment. These "islands of fertility" are thought to form from the accumulation of eolian material around the plant in the same manner that snowdrifts form from the disturbance of wind around an object. McAuliffe and McDonald (2006) noted that rodent burrowing around the plant may be another important factor in plant mound development, and speculated that the disturbance to soil microtopography caused by plant mound formation may be preserved for thousands of years.

Based on a study conducted across a bajada in the Jornada Basin of New Mexico, Reynolds et al. (1999) suggested that the size of the plant mound was related to the size of the plant. Plant mound size was larger for the Holocene in comparison to the Pleistocene surfaces at the NL site (data not shown) and, although it was not measured at the other two sites, we observed that mound size in general was larger on the Holocene soils in comparison to the Pleistocene soils. While we agree that there is a definite correlation between plant size and plant mound size, we would assert that the interaction of soil morphology and hydrology with root development is also a controlling factor in plant mound development. There is an obvious feedback mechanism in which the larger the plant, the larger the effects on the soil. However, we suggest that a plant's ability to modify the soil is greater in younger, less-developed soils than in older, well-developed soils.

SUMMARY AND RELEVANCE

Schenk et al. (2003) noted that in arid systems, where base levels of many resources are low and competitive interactions are high, subtle differences in edaphic characteristics may cause large differences in the spatial distribution of immediately adjacent plants. We suggest that soil morphological development is one such spatially variable edaphic characteristic that is often overlooked. Bedford et al. (*this volume*) and Miller et al. (*this volume*), however, demonstrate that advances in geographic information systems and spatial modeling can extend the application of complex soil information from a local scale to a landscape scale.

The constraints placed on root systems (both physical and hydrological) by soil horizon development can result in substantial differences in root morphology between plants on younger, less-developed soils and those on older, better-developed soils (as shown in fig. 14.5); and these differences affect plant performance. The differences we observed in the rooting patterns of *Larrea* (particularly the greater lateral spread of roots in Pleistocene soils) are also likely to affect the complex competitive interactions both between individual *Larrea* plants and between *Larrea* and other species.

On a practical level, our results suggest that soil morphology and hydrology should be considered in arid land management planning. The effects of land use intensification or disturbance are likely to be very different on older, well-developed soils (where soil properties often change dramatically with depth) in comparison to younger soils with little profile development. In restoration and revegetation activities, the common practice of planting shrubs in boreholes provides an example of an activity for which the degree of soil morphological development may affect restoration success. Borehole planting is likely to be more effective in younger, less-developed soils (where roots and moisture tend to penetrate relatively deeply into the soil), than in older, well-developed soils, where the borehole is likely to intersect strongly developed subsurface horizons, placing the majority of roots in a portion of the soil (beneath petrocalcic or argillic horizons) where soil moisture is scarce. Results presented here suggest that alternative strategies, such as planting so that roots are placed above the petrocalcic (or argillic) horizon, may prove more successful and increase plant survival rates.

In conclusion, certain soil characteristics have long been linked to plant species distributions and boundaries. However, the importance of soil morphological development (and its affect on soil hydrologic properties) is often overlooked. Our data suggest that soil morphological features and soil hydrologic properties, both a function of soil age, play a key role in determining *Larrea* root distribution and canopy size—both directly, by obstructing deep root growth, and indirectly, by modifying soil water distribution.

ACKNOWLEDGMENTS

We wish to thank Ruth Sparks, Matt Hamilton, and the Fort Irwin Integrated Training Area Management crew for logistical and field support. Brian Boyd and Don Geisinger of the Desert Research Institute provided lab and/or field assistance. Sophie Baker and Lynn Fenstermaker from the Desert Research Institute assisted greatly with the preparation of the manuscript. The Army Research Office (Contract No. DAAD19-03-1-0159) and the Center for Arid Land Management at the Desert Research Institute contributed financial support for the project. We also wish to thank the individuals who reviewed this chapter and provided comments to improve the chapter, as well as the book editors.

REFERENCES

Anderson, K., S. Wells, and R. Graham. 2002. Pedogenesis of vesicular horizons, Cima Volcanic Field, Mojave Desert, California. *Soil Science Society of America Journal* **66**: 878–887.

Ankeny, M. D., M. Ahmed, T. C. Kaspur, and R. Horton. 1991. Simple field method for determining unsaturated hydraulic conductivity. *Soil Science Society of America Journal* **55**:467–470.

Ben-Dor, E., and A. Banin. 1989. Determination of organic matter content in arid-zone soils using a simple "loss on ignition" method. *Communications in Soil Science and Plant Analysis* **20**:1675–1695.

Blake, G. R., and K. H. Hartage. 1986. Bulk density. Pages 363–375 *in* A. Klute, editor. *Methods of soil analysis, part 1: physical and mineralogical methods.* Second edition. American Society of Agronomy, Madison, Wisconsin.

Bohm, W. 1979. Profile wall methods. Pages 49–76 *in* W. Bohm, editor. *Methods of studying root systems.* Springer, New York, New York.

Brisson, J., and J. F. Reynolds. 1994. The effect of neighbors on root distribution in a creosote (*Larrea tridentata*) population. *Ecology* **75**:1693–1702.

Caldwell, T. G., E. V. McDonald, and M. H. Young. 2006. Soil disturbance and hydrologic response at the National Training Center, Ft. Irwin, California. *Journal of Arid Environments* **68**:456–472.

Casper, B. B., and R. B. Jackson. 1997. Plant competition underground. *Annual Review of Ecology and Systematics* **28**:545–570.

Cunningham G. L., and J. H. Burk. 1973. The effect of carbonate deposition layers ("caliche") on the water status of *Larrea divaricata. American Midland Naturalist* **90**:474–480.

Devitt, D. A., and S. D. Smith. 2002. Root channel macropores enhance downward movement of water in a Mojave Desert ecosystem. *Journal of Arid Environments* **50**:99–108.

Dreimanas, A., 1962. Quantitative gasometric determination of calcite and dolomite by using Chittick apparatus. *Journal of Sedimentary Petrology* **32**:520–529.

Dunkerley, D. 2000. Hydrologic effects of dryland shrubs: defining the spatial extent of modified soil water uptake rates at an Australian desert site. *Journal of Arid Environments* **45**:159–172.

Eissenstat, D. M., and M. M. Caldwell. 1988. Seasonal timing of root growth in favorable microsites. *Ecology* **69**:870–873.

Gee, G. W., and J. W. Bauder. 1986. Particle-size analysis. Pages 383–412 *in* A. Klute, editor. *Methods of soil analysis, part 1: physical and mineralogical methods.* Second edition. American Society of Agronomy, Madison, Wisconsin.

Gibbens, R. P., and J. M. Lenz. 2001. Root systems of some Chihuahuan Desert plants. *Journal of Arid Environments* **49**:221–263.

Gile, L. H., R. P. Gibbens, and J. M. Lenz. 1998. Soil-induced variability in root systems of creosote bush (*Larrea tridentata*) and tarbush (*Flourensia cernua*). *Journal of Arid Environments* **39**:57–78.

Hamerlynck, E. P., J. R. McAuliffe, E. V. McDonald, and S. D. Smith. 2002. Ecological responses of two Mojave Desert shrubs to soil horizon development and soil water dynamics. *Ecology* **83**:768–779.

Hamerlynck, E. P., J. R. McAuliffe, and S. D. Smith. 2000. Effects of surface and subsurface soil horizons on the seasonal performance of *Larrea tridentata* (creosote bush). *Functional Ecology* **14**:596–606.

Jackson, R. B., and M. M. Caldwell. 1989. The timing and degree of root proliferation in fertile-soil microsites for three cold-desert perennials. *Oecologia* **81**:149–153.

Lunt, O. R., J. Letey, and S. B. Clark. 1973. Oxygen requirements for root growth in three species of desert shrubs. *Ecology* **54**:1356–1362.

Lyford, F. P, and H. K. Qashu. 1969. Infiltration rates as affected by desert vegetation. *Water Resources Research* **5**:1373–1376.

Machette, M. N. 1985. Calcic soils of the southwestern United States. Pages 1–21 *in* D. L. Weide, editor. *Soils and Quaternary geology of the southwestern United States.* Special Paper No. 203. Geological Society of America, Boulder, Colorado.

Mahall, B. E., and R. M. Callaway. 1992. Root communication mechanisms and intracommunity distribution of two Mojave Desert shrubs. *Ecology* **73**:2145–2151.

McAuliffe, J. R. 1988. Markovian dynamics of simple and complex desert plant communities. *American Naturalist* **131**:459–490.

McAuliffe, J. R., and E. V. McDonald. 1995. A piedmont landscape in the eastern Mojave Desert: examples of linkages between biotic and physical components. *San Bernardino County Museum Association Quarterly* **42**:55–63.

McAuliffe, J. R., and E. V. McDonald. 2006. Holocene environmental change and vegetation contraction in the Sonoran Desert. *Quaternary Research* **65**:204–215.

McCulley, R. L., E. G. Jobbagy, W. T. Pockman, and R. B. Jackson. 2004. Nutrient uptake as a contributing explanation for deep rooting in arid and semi-arid ecosystems. *Oecologia* **141**:620–628.

McDonald, E. V., F. B. Pierson, G. N. Flerchinger, and L. D. McFadden. 1996. Application of a soil-water balance model to evaluate the influence of Holocene climate change on calcic soils, Mojave Desert, California, USA. *Geoderma* **74**:167–192.

McFadden, L. D., E. V. McDonald, S. G. Wells, K. Anderson, J. Quade, and S. L. Forman. 1998. The vesicular layer and carbonate collars of desert soils and pavements: formation, age and relation to climate change. *Geomorphology* **24**:101–145.

McFadden, L. D., S. G. Wells, W. J. Brown, and Y. Enzel, 1992. Soil genesis on beach ridges of pluvial Lake Mojave: implications for Holocene lacustrine and eolian events in the Mojave Desert, southern California. *Catena* **19**:77–97.

McFadden, L. D., S. G. Wells, and M. J. Jercinovich, 1987. Influences of eolian and pedogenic processes on the origin and evolution of desert pavements. *Geology* **15**:504–508.

Obrist, D., and J. A. Arnone. 2003. Increasing CO_2 accelerates root growth and enhances water acquisition during early stages of development in *Larrea tridentata. New Phytologist* **159**:175–184.

Perroux, K. M., and I. White. 1988. Designs for disk permeameters. *Soil Science Society of America Journal* **52**:1205–1215.

Reynolds, J. F., R. A. Virginia, P. R. Kemp, A. G. de Soyza, and D. C. Tremmel. 1999. Impact of drought on desert shrubs: effects of seasonality and degree of resource island development. *Ecological Monographs* **69**:69–106.

Rhoades, J. D. 1996. Salinity: electrical conductivity and total dissolved solids. Pages 417–435 *in* D. L. Sparks, editor. *Methods of soil analysis, part 3: chemical methods.* American Society of Agronomy, Madison, Wisconsin.

Schenk, H. J., C. Holzapfel, J. G. Hamilton, and B. E. Mahall. 2003. Spatial ecology of a small desert shrub on adjacent geological substrates. *Journal of Ecology* **91**:383–395.

Schlesinger W. H., J. A. Raikes, A. F. Hartley, and A. E. Cross. 1996. On the spatial pattern of soil nutrients in desert ecosystems. *Ecology* **77**:364–374.

Shafer, D. S., M. H. Young, S. F. Zitzer, T. G. Caldwell, and E. V. McDonald. 2007. Impacts of interrelated biotic and abiotic processes during the past 125,000 years of landscape evolution in the northern Mojave Desert, Nevada. *Journal of Arid Environments.*

Smith, S. D., R. K. Monson, and J. E. Anderson. 1997. *Physiological ecology of North American desert plants.* Springer-Verlag, Berlin, Germany.

Soil Survey Staff. 1998. *Soil survey manual.* Eighth edition. Soil Conservation Service. U.S. Government Printing Office, Washington, D.C.

Wells, S. G., L. D. McFadden, and J. C. Dohrenwend, 1987. Influence of late Quaternary climatic changes on geomorphic and pedogenic processes on a desert piedmont, eastern Mojave Desert, California. *Quaternary Research* **27**:130–146.

Wells, S. G., L. D. McFadden, J. Poths, and C. T. Olinger. 1995. Cosmogenic He-3 surface-exposure dating of stone pavements: implications for landscape evolution in deserts. *Geology* **23**:613–616.

Western Regional Climatic Center. 2005. Database. Online. Division of Atmospheric Sciences, Desert Research Institute, Reno, Nevada. http://www.wrcc.dri.edu.

Wilcox, C. S., J. W. Ferguson, G. C. J. Fernandez, and R. S. Nowak. 2004. Fine root growth dynamics of four Mojave Desert shrubs as related to soil moisture and microsite. *Journal of Arid Environments* **56**:129–148.

Young, M. H., E. V. McDonald, T. G. Caldwell, S. G. Benner, and D. G. Meadows. 2004. Hydraulic properties of a desert soil chronosequence in the Mojave Desert, USA. *Vadose Zone Journal* **3**:956–963.

APPENDIX 14.1

Soil Physical and Chemical Properties of Sites at Fort Irwin/National Training Center

	Horizon	Depth cm	Penetrometer Resistance[1] kg cm^{-2} (sd)	Bulk Density g cm^{-3}	Gravel[2] %	Sand %	Silt %	Clay %	Soluble Salts mg kg^{-1}	$CaCO_3$ %	LOI[3] %
				Arrowhead (AH) Site							
Holocene Soil											
Interspace											
	Av	2	0.4 (0.3)	1.51	60	68	24	8	228	0.4	2.1
	Bwk	14	0.7 (0.5)	1.68	45	70	24	6	194	0.5	1.2
	BCk	30	0.7 (0.6)	1.61	51	77	17	6	NA	0.7	1.0
	Ck1	43	0.8 (0.7)	1.60	50	75	19	7	137	0.6	1.1
	Ck2	56	1.8 (0.4)	1.85	60	80	14	6	121	0.5	0.9
	Ck3	81	2.2 (0.3)	1.84	56	85	10	6	82	0.5	0.7
	Btk1b	109	2.6 (0.6)	1.75	62	76	16	9	97	0.6	1.0
	Btk2b	149	3.4 (0.4)	NA	65	79	12	9	82	0.5	0.8
Canopy											
	A1	3	0.2 (0.1)	1.23	56	74	21	5	563	0.5	3.4
	A2	17	0.5 (0.4)	1.45	49	70	22	8	451	0.6	2.8
	ACk	39	0.8 (0.6)	1.62	52	69	21	10	351	0.4	2.0
	Ck1	63	1.6 (0.6)	1.50	62	76	16	8	182	0.6	1.4
	Ck2	90	2.2 (0.3)	NA	67	79	13	8	140	0.7	1.1
	Ck3	110	2.0 (0.4)	1.94	50	89	6	5	124	0.9	0.6
	Ck4	154	3.0 (0.4)	NA	77	76	15	9	59	0.5	0.9
Pleistocene Soil											
Interspace											
	Avk	6	2.5 (1.2)	1.77	29	38	47	14	225	1.9	2.0
	BAvk	14	2.5 (1.1)	1.71	16	38	43	19	307	1.5	1.5
	Btk1	24	3.8 (0.6)	1.99	64	37	34	30	171	0.4	2.4
	Btk2	58	4.5 (0.4)	2.02	84	44	26	30	2690	0.1	2.4
	Btk3	86	4.5 (0.4)	1.72	68	64	14	22	2810	0.3	1.7
	Btk4	107	4.8 (0.4)	2.10	62	87	6	8	187	0.2	0.9
	Btk5	137	5.0 (0.0)	NA	69	85	8	7	666	0.2	0.8
	Btk6	180	5.0 (0.0)	NA	65	85	7	8	653	0.2	0.7

Arrowhead (AH) Site (*continued*)											
	Horizon	Depth cm	Penetrometer Resistance[1] kg cm^{-2} (sd)	Bulk Density g cm^{-3}	Gravel[2] %	Sand %	Silt %	Clay %	Soluble Salts mg kg^{-1}	$CaCO_3$ %	LOI[3] %
Canopy											
	A	4	0.8 (0.8)	1.52	42	56	33	11	346	0.4	2.6
	Av	16	2.4 (0.9)	1.65	34	46	39	15	396	0.9	1.9
	BAvt	23	3.2 (1.3)	1.28	38	37	43	21	342	1.6	2.6
	Btk1	35	3.7 (0.6)	1.91	50	44	24	32	200	0.5	2.3
	Btk2	51	3.9 (0.3)	1.78	83	48	21	31	73	0.2	2.8
	Btk3	69	4.3 (0.5)	1.76	64	73	10	17	112	0.3	1.4
	Btk4	99	4.7 (0.4)	1.76	69	71	16	13	79	0.2	1.1
	Btk5	131	5.0 (0.0)	1.93	71	84	8	8	48	0.2	1.0
	Btk6	146	5.0 (0.0)	NA	74	74	14	12	51	0.1	1.2

[1](SD) = Standard deviation (n = 10)
[2]Weight percent > 2 mm
[3]LOI = loss-on-ignition
NA = Not analyzed

Brinkman Wash (BW) Site											
	Horizon	Depth cm	Penetrometer Resistance[1] kg cm^{-2} (sd)	Bulk Density g cm^{-3}	Gravel[2] %	Sand %	Silt %	Clay %	Soluble Salts mg kg^{-1}	$CaCO_3$ %	LOI[3] %
Holocene Soil											
Interspace											
	A1	2	0.9 (0.4)	1.51	28	84	13	3	405	0.6	1.2
	A2	8	2.1 (0.6)	1.68	23	81	15	4	233	0.8	0.8
	Bw1	28	1.6 (0.9)	1.71	22	82	13	5	264	0.4	0.6
	Bw2	69	1.2 (0.7)	1.62	33	87	9	4	168	0.5	0.5
	Ck	98	1.0 (0.6)	1.68	31	90	7	3	196	0.8	0.4
	C	139	0.9 (0.3)	NA	25	96	2	2	167	0.9	0.2
Canopy											
	A1	2	0.1 (0.1)	1.56	21	83	14	3	995	0.5	2.3
	A2	10	0.3 (0.2)	1.33	17	80	16	5	717	0.7	1.8
	BA	20	1.3 (0.4)	1.62	21	82	14	5	593	0.8	1.4
	Bw	40	1.4 (0.6)	1.57	20	81	14	5	525	0.7	1.0
	Bwk	87	1.5 (0.4)	1.57	26	87	10	4	329	0.7	0.6
	BCk	102	1.1 (0.4)	1.64	27	80	18	2	688	2.0	0.7
	Ck	132	1.6 (0.3)	1.67	27	91	7	2	869	1.2	0.5
	C	167	0.8 (0.3)	NA	30	95	4	1	746	0.9	0.3

	Horizon	Depth cm	Penetrometer Resistance[1] kg cm^{-2} (sd)	Bulk Density g cm^{-3}	Gravel[2] %	Sand %	Silt %	Clay %	Soluble Salts mg kg^{-1}	$CaCO_3$ %	LOI[3] %
Pleistocene Soil											
Interspace											
	A1	3	1.1 (0.7)	2.00	27	78	11	11	438	0.4	1.8
	A2	8	1.7 (0.5)	1.68	25	76	10	14	367	0.5	1.2
	Bt1	17	3.4 (1.0)	1.77	18	68	8	25	421	0.5	1.6
	Bt2	39	4.7 (0.5)	1.91	34	69	4	27	312	0.4	1.3
	Btk	64	4.5 (0.5)	1.63	41	82	5	13	270	0.9	0.8
	Bk	96	4.5 (0.7)	1.59	37	89	4	7	281	1.5	0.5
	Bkm	126	5.0 (0.0)	1.55	51	84	9	8	340	3.8	1.0
	Crk	154	5.0 (0.0)	1.74	76	86	9	5	177	NA	0.9
Canopy											
	A1	2.5	1.1 (0.7)	1.24	21	80	10	11	1030	0.6	2.7
	A2	12	1.7 (0.5)	1.50	10	76	11	12	757	0.7	2.1
	Btk1	21	3.4 (1.0)	1.74	9	69	11	20	995	1.3	2.0
	Btk2	37	4.2 (0.8)	1.80	19	63	11	26	1140	1.2	2.1
	Btk3	54	4.5 (0.4)	1.88	26	75	3	22	1140	1.1	1.2
	Btk4	70	4.0 (0.4)	1.74	26	86	3	11	959	1.0	0.6
	Bk/Bkm	101	5.0 (0.0)	1.68	16	84	10	6	1450	11.9	0.9
	Bkm	122	5.0 (0.0)	1.53	28	80	13	7	4120	19.6	2.1
	Bk	161	4.8 (0.3)	1.78	42	76	18	6	3080	16.2	1.7

[1](SD) = Standard deviation (n=10)
[2]Weight percent > 2 mm
[3]LOI = loss-on-ignition
NA = Not analyzed

Nelson Lake (NL) Site

	Horizon	Depth cm	Penetrometer Resistance[1] kg cm^{-2} (sd)	Bulk Density g cm^{-3}	Gravel[2] %	Sand %	Silt %	Clay %	Soluble Salts mg kg^{-1}	$CaCO_3$ %	LOI[3] %
Holocene Soil											
Interspace											
	C1		2.8 (0.6)	1.61	32	76	17	7	279	1.0	1.2
	C2	18	1.8 (0.8)	1.67	30	79	14	8	304	0.9	0.7
	A	35	2.0 (0.7)	1.58	5	71	15	14	364	2.0	0.8
	Bwk	63	1.8 (0.8)	1.53	7	72	19	9	338	1.7	0.7
	BCk	87	1.5 (0.4)	1.62	21	68	23	9	467	1.6	0.8
	CBk	107	0.8 (0.4)	1.68	36	74	17	9	516	0.5	0.7

	Nelson Lake (NL) Site (*continued*)										
	Horizon	Depth cm	Penetrometer Resistance[1] kg cm^{-2} (sd)	Bulk Density g cm^{-3}	Gravel[2] %	Sand %	Silt %	Clay %	Soluble Salts mg kg^{-1}	$CaCO_3$ %	LOI[3] %
Canopy											
	C1	5	0.8 (0.4)	1.38	0	92	7	2	287	0.5	0.8
	C2	20	0.9 (0.5)	1.43	0	74	16	11	486	0.6	0.7
	C3	48	1.3 (0.6)	1.40	3	78	16	7	577	0.6	1.2
	C4	55	1.4 (0.7)	1.46	5	77	16	7	561	0.7	1.0
	AC	67	2.1 (0.4)	1.52	4	76	16	8	555	0.8	1.0
	Bwk1	87	2.2 (0.6)	1.60	13	73	16	12	702	1.7	0.7
	Bwk2	107	1.6 (0.5)	NA	12	74	18	7	1450	1.7	0.8
	BCk	130	3.3 (0.5)	NA	13	70	22	8	817	1.1	0.7
	CBk	142	4.1 (0.6)	1.78	35	73	20	8	1110	0.5	0.6
Pleistocene Soil											
Interspace											
	Av	5	2.0 (0.5)	1.66	11	58	29	13	377	3.8	0.9
	Bwk1	20	4.0 (0.6)	1.60	17	57	19	25	549	5.6	1.0
	Bwk2	34	3.4 (0.6)	1.46	37	63	21	15	571	16.7	1.2
	Bkm	67	5.0 (0.0)	1.78	33	79	15	7	413	17.7	0.9
	Bk1	98	4.6 (0.5)	1.94	36	85	11	4	362	11.2	0.9
	Bk2	128	4.4 (0.5)	1.72	28	86	9	5	669	15.0	0.8
	BCk	158	4.2 (0.6)	NA	47	92	4	4	1090	3.8	0.7
Canopy											
	A	4	1.2 (0.5)	1.45	12	66	21	13	574	3.9	0.7
	AB(Bwk)	19	2.2 (0.6)	1.42	4	61	21	18	716	7.1	1.8
	Bkm1	37	5.0 (0.0)	1.56	48	75	15	10	755	22.1	1.0
	Bkm2	47	5.0 (0.0)	1.29	34	76	18	6	818	28.7	1.0
	Bkm/Bk	72	5.0 (0.0)	1.50	29	78	18	4	608	26.9	0.7
	Bk1	103	4.1 (0.7)	1.66	28	79	17	5	5900	23.5	0.7
	Bk2	131	3.9 (0.6)	NA	36	82	13	5	5360	15.1	0.7
	BCk (Bk3)	196	3.5 (0.5)	1.40	33	91	6	3	3260	5.0	0.7

[1](SD) = Standard deviation (n=10)
[2]Weight percent > 2 mm
[3]LOI = loss-on-ignition
NA = Not analyzed

PART IV

Recovery, Restoration, and Ecosystem Monitoring

DAVID M. MILLER, DEBRA L. HUGHSON, AND ROBERT H. WEBB

Abandoned desert landscapes bear highly visible scars and commonly have elevated dust production, increased water erosion, and diminished environmental quality for plants and habitat for animals. Faced with questions about how to rehabilitate abandoned landscapes, managers must choose among active restoration, natural recovery, or some combination of the two. Active-restoration techniques have evolved considerably in the last ten years (Weigand and Rodgers *this volume*), focusing on re-creating habitat for plants and animals, and diminishing visual impact. Natural recovery, once thought to require enormous amounts of time in desert ecosystems, has recently been shown to occur in time periods of decades to centuries, although complete restoration of species composition may still require millennia in certain types of plant assemblages (Webb et al., Natural Recovery, *this volume*). Hybrid combinations, where highly visible parts of disturbances are actively restored to stop future impacts and the remaining areas are left to recover naturally, are an evolving area of landscape restoration.

In this section, chapters explore the three approaches with an emphasis on the scientific viability and economic cost of each, with a secondary emphasis on various disturbance scenarios that might require specific treatments. Severe disturbance, such as roads, townsites, and military camps, in the past have recovered to varying degrees by several natural processes (Webb et al., Natural Recovery, *this volume;* Hereford *this volume*). The disturbances affected soil processes such as diminished water-holding capacity and nutrient cycling and destroyed vegetation, creating visual scars on the landscape. Using a variety of metrics for soil and biotic systems, Webb et.al. (Natural Recovery *this volume*) show that natural recovery is not only measurable in less than a century, but that depending on the recovery goal (e.g., ecosystem function, plant species composition, plant cover, soil function), significant recovery can occur in a few decades. Hereford (*this volume*) demonstrates similar visual recovery in a few decades by a mechanism of shallow burial by alluvial materials, which apparently promotes plant recruitment by providing restored soil function. These studies indicate that in some cases, natural recovery is a viable restoration method if time is not of the essence.

In contrast, Weigand and Rodgers (*this volume*) summarize several decades of investigations into active restoration, outlining methods and the significant costs. They conclude that for many disturbances a combination of active restoration at points where visual or functional recovery is needed quickly and natural restoration for the remainder of the disturbance provides a cost-effective management tool. They also describe a likely near-future in which human pressures are likely to be far more intensive than those of today.

These studies of natural and active recovery are empirical approaches that involve considerable uncertainty. A complimentary approach is to develop mechanistic understanding of disturbance and recovery processes to better predict the most effective recovery methods. These studies are initial forays that may lead to mechanistic models that provide improved understandings of root distributions (Schwinning and Hooten *this volume;* Stevenson et. al. *this volume*), soil characteristics (Miller et. al. *this volume;* Bedford et al. *this volume*), and climate predictions (Redmond *this volume;* Smith et al. *this volume*), can lead to precisely targeted recovery methods. For example, Hereford et al. (2006) hypothesize that natural recovery is a function of climatic variability, and that drought may play a central role in determining the course of recovery and, by analogy, active restoration.

Long-term studies are particularly important to the evaluation of the potential for recovery and restoration in the Mojave Desert, where variation can be extreme and impacts are long-term (Webb et al., Long-Term Data, *this volume*). Very few long-term studies have occurred in the Mojave Desert, mostly because they cease or become dormant when funding or the investigator's career ends. One exception is Rock Valley on the Nevada Test Site (NTS), where sustained research on somewhat atypical Mojave Desert ecosystems occurred from the late 1950s through the mid-1980s (Rundel and Gibson 1996) with some ongoing work.

The work of Dr. Janice C. Beatley, a UCLA researcher, well illustrates the legacy of long-term data on the NTS and within the Mojave Desert. In 1962, Beatley established permanent plots for the purpose of studying the effects of ionizing radiation from above-ground nuclear testing on various aspects of desert ecosystems. Her network of 68 permanent ecological plots ranged from 1000 to 2400 m in elevation, encompassing the transition region between the Mojave and Great

Basin deserts as well as the elevational extent of the Mojave Desert alone (Beatley 1980; Webb et al. 2003). The focus of her research changed after the above-ground test ban treaty of 1963, after which she characterized the perennial plant associations of the Northern Mojave and transition Great Basin Desert and established the basis for assessing long-term changes in the ecosystem (Beatley 1975, 1980). Her research lay dormant for 25 years before other researchers revisited her plots (Webb et al., Long-Term Data, *this volume*).

For some long-term studies, protection of the site is imperative. The NTS is a highly secure facility because of its stature as the nation's nuclear testing facility. Military bases, such as Fort Irwin/National Training Center (Stevenson et al. *this volume*), also have secure boundaries, but long-term research on undisturbed desert habitats takes a distant priority to the primary mission of military training. Field stations and reserves, such as the Sweeney Granite Mountain Field Station, the Zzyzx facility, and Boyd Deep Canyon, provide at least minimal protection, but they only represent a fraction of the habitats and organisms in this desert region. Wilderness areas provide pristine lands for study, and are valued by scientists, but these sites are too restrictive for some kinds of studies and remain susceptible to impacts such as grazing and vandalism. The tradeoff between site protection and representation of large areas is one that needs to be explored for future research areas committed to long-term studies.

A difficult concept that is necessary to understand for recovery decisions is the interrelation between temporal-spatial scales of resource management practices and temporal-spatial scales of natural processes. Landscape-scale processes commonly are observed on small plots over a few years to a few decades at most, but the processes are driven by climate changes regionally to globally over decadal to millennial time scales and by biotic processes that are decadal to centennial in scale. Can plot-level data acquired within short timeframes be extrapolated to reflect what is occurring over longer time periods on a landscape scale? Can land management plans based on plot-scale observations adequately assess the various likelihoods of alternative futures? Our ability to accurately reflect landscape processes by scaling up using plot data will largely determine the success of resource management plans and practices. Conversely, separating the effects of long-term versus short-term processes and plot-level experimental effects must address analogous downscaling problems. Understanding the relations between scales, both spatial and temporal, and ecosystem processes, and accounting for scale-dependence in our restoration and monitoring efforts, are one of the most significant challenges to managing for sustainability.

REFERENCES

Beatley, J. C. 1975. Climates and vegetation pattern across the Mojave/Great Basin Desert transition of southern Nevada. *American Midland Naturalist* **93**:53–70.

Beatley, J. C. 1980. Fluctuations and stability in climax shrub and woodland vegetation of the Mojave, Great Basin and Transition Deserts of southern Nevada. *Israel Journal of Botany* **28**:149–168.

Hereford, R., R. H. Webb, and C. Longpré. 2006. Precipitation history and ecosystem response to multidecadal precipitation variability in the Mojave Desert and vicinity, 1893–2001. *Journal of Arid Environments* **67**:13–34.

Rundel, P. W., and A. C. Gibson. 1996. *Ecological communities and processes in a Mojave Desert ecosystem: Rock Valley, Nevada.* Cambridge University Press, New York, New York.

Webb, R. H., M. B. Murov, T. C. Esque, D. E. Boyer, L. A. DeFalco, D. F. Haines, D. Oldershaw, S. J. Scoles, K. A. Thomas, J. B. Blainey, and P. A. Medica. 2003. *Perennial vegetation data from permanent plots on the Nevada Test Site, Nye County, Nevada.* U.S. Geological Survey Open-File Report No. 03–336. Online. Tucson, Arizona. http://www.werc.usgs.gov/lasvegas/ofr-03-336.html.

FIFTEEN

Natural Recovery from Severe Disturbance in the Mojave Desert

ROBERT H. WEBB, JAYNE BELNAP, AND KATHRYN A. THOMAS

Severe disturbances, where all or most of the above-ground vegetation is removed, are increasing throughout the Mojave Desert region (fig. 15.1) (Lovich and Bainbridge 1999; Lei *this volume*). When faced with the rehabilitation of these severely disturbed sites, land managers must choose between active restoration, which is generally expensive and often unsuccessful (Rodgers and Weigand *this volume*), or natural restoration, which is inexpensive but often very slow. The initial post-disturbance conditions of a severely disturbed site can have a large effect on the rate and course of natural recovery. The climatic conditions following abandonment (Hereford et al. 2006; Redmond *this volume*) and the encroachment of non-native species (Brooks et al. 2004; Brooks *this volume*) also affect natural recovery processes.

Most severely disturbed sites in the Mojave Desert are abandoned without rehabilitation (Wallace et al. 1980), and these sites are ideal for the study of natural recovery processes, particularly if postabandonment disturbances are minimal. In this paper, we consider natural recovery processes in two major types of severely disturbed sites: (1) those where vegetation removal is accompanied by soil compaction, and (2) those where vegetation removal is accomplished with little, if any, soil compaction. Abandoned mining towns (Webb and Wilshire 1980; Webb et al. 1988 Knapp 1992a) and areas used for World War II training exercises (Prose 1985; Prose and Metzger 1985; Steiger and Webb 2000; Prose and Wilshire 2000; Belnap and Warren 2002; Hereford *this volume*) provide ideal areas for studying the natural recovery of soil and vegetation at sites where soils have been compacted (fig. 15.2). Severe disturbances that involve vegetation removal with little, if any, soil compaction include fire (Brooks 1999; Brooks and Matchett 2003; Brooks and Minnich 2006), dryland agriculture (Carpenter et al. 1986), and above-ground nuclear detonations (Shields et al. 1963; Wallace and Romney 1972).

EFFECTS OF SEVERE DISTURBANCE

When first abandoned, locations where military camps, towns, pipeline corridors, heavy grazing, off-road vehicle areas, and roads once existed are generally left

Fig. 15.1. Aerial photograph of a view west along a natural gas pipeline in the Eastern Mojave Desert showing several types of severe disturbances. These disturbances occurred about 1990 and have several geometries ranging from abandoned access roads (compacted, no subsurface disturbance) to pipeline trenches (compacted with subsurface disturbance). An active road crosses the view in the distance (photograph courtesy of Michael Collier).

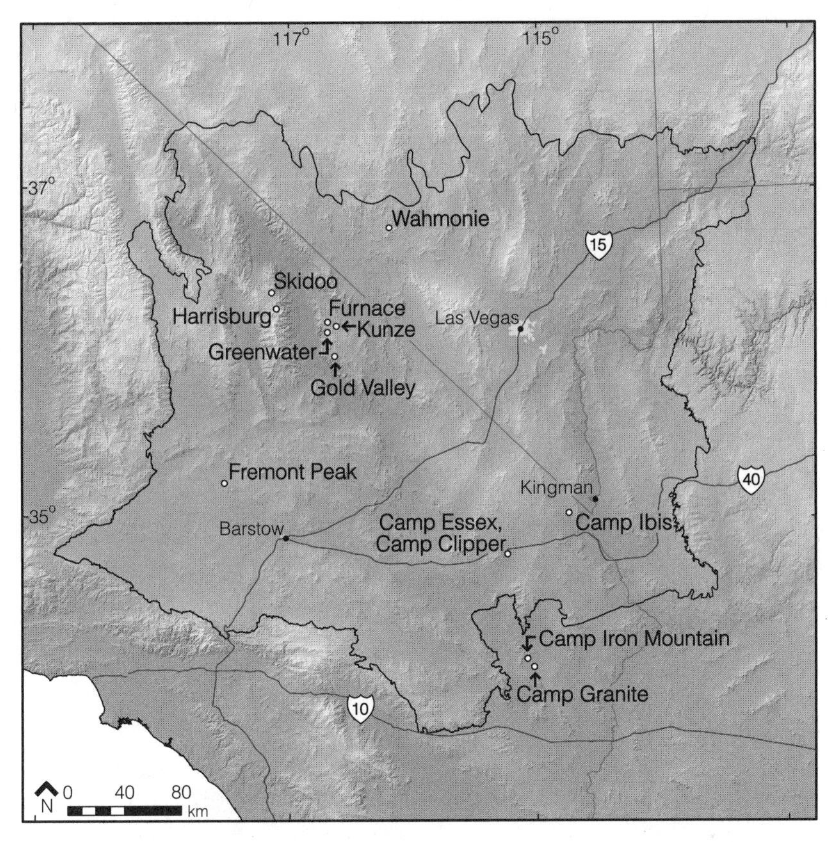

Fig. 15.2. Map of the Mojave Desert ecoregion showing the locations of sites examined for disturbance recovery in the Mojave Desert.

with little or no vegetation and severely compacted soil. For instance, off-road vehicle use causes significant compaction and surface disruption with as few as 1–10 vehicle passes (Davidson and Fox 1974; Vollmer et al. 1976; Wilshire and Nakata 1976; Wilshire et al. 1978; Eckert et al. 1979; Webb 1983; Matchett et al. 2004; Lei *this volume*). Other recreational land uses, including hiking and bicycling, also cause significant soil compaction (Lei 2004 *this volume*). Domestic livestock causes soil compaction as well, especially when concentrated near stock tanks or wells (Scholl 1989; Brooks et al. 2006).

High levels of soil compaction or disruption result in reduced seed bank and soil function, including a loss of biological soil crusts (Belnap and Eldridge 2003); a loss of soil structure; reduced water infiltration; and a reduction in carbon (C) and nutrient inputs, availability, and cycling rates. Given the lack of components that would normally stabilize the soil surface (e.g., rocks, crusts, vegetation), these sites often contribute considerable fugitive dust to the environment (Campbell

Table 15.1 Summary of recovery rates for several metrics of ecosystem recovery on low-slope xerophytic sites in the Mojave Desert

Metric of ecosystem recovery	Minimum recovery time (years)	Maximum recovery time (years)	Research needs
Visual appearance	20	> 1,000	Visual recovery should be evaluated over a range of geomorphic surfaces and vegetation assemblages.
Biological soil crusts, cyanobacteria	20	50	The interdependence of biological soil crust recovery and surficial particle size, soil nutrients, perennial plants, and climate should be evaluated.
lichens/mosses	100	> 1,000	
Surface soil compaction (0–60 mm)	70	140	Current data emphasizes sandy loams and loamy sands at intermediate elevations. The curvilinear trajectory may be a function of climatic fluctuations (wetting-drying cycles). Interrelation with plant recovery needs to be quantified.
Subsurface soil compaction (0.3–0.5 m)	Unknown	> 1,000	Little is known about recovery rates for subsurface compaction or whether they are important to overall recovery.
Annual vegetation	Unknown	Unknown	The nonnative vegetation affect on recovery of native annual vegetation needs to be better quantified.
Total cover of perennial vegetation	20	80	A better understanding of recovery trajectories for different disturbances on different geomorphic surfaces is needed.
Density of perennial vegetation	Unknown	> 1,000	The processes leading to reduction in average plant size are unknown and effects recovery rates.
Cover and species composition of perennial vegetation	80	> 1,000	This, the most important metric of recovery, allows comparison of natural recovery rates with active restoration. More information on key abiotic and biotic factors that control recovery is needed.
Biomass and productivity of perennial vegetation	Unknown	Unknown	Little is known about changes in biomass and productivity in the recovery process, or what key abiotic and biotic factors control recovery.
Rodents	Unknown	Unknown	The linkage between rodent populations and species composition changes in annual and perennial vegetation, or key abiotic or biotic factors, is needed.
Reptiles	Unknown	Unknown	Population changes likely are linked to species composition changes in annual and perennial vegetation or key abiotic or biotic factors.

1972; Gillette and Adams 1983). Soil loss via water erosion can be 10–20 times greater on slopes at these sites compared with similar undisturbed slopes (Iverson 1980; Iverson et al. 1981; Hinckley et al. 1983).

Because of lowered soil nutrient availability, vascular plants are likely to provide less nutrition to animals, and lowered vegetation cover provides reduced animal habitat.

With all types of disturbance, the shape and size of the disturbed area greatly affects recovery processes. Military training areas are often large, broad polygons (e.g., Fort Irwin) (McCarthy 1996), whereas utility corridors are often long and narrow. Disturbance geometry is important because narrow strips (fig. 15.1), may recover more quickly than broad areas, owing to a closer proximity to seed sources and dispersers.

Of the severe disturbances that do not significantly compact soils, fire is the most common in the Mojave Desert. Fire is increasingly common in all southwestern deserts (Callison et al. 1985; Brown and Minnich 1986; Minnich 1995), particularly in areas where recent invasions of nonnative annuals provide large amounts of easily ignited fuel (Brooks 1999). The spatial variability of fire intensity (Brooks and Matchett 2003) is the most important determinant as to the amount of disturbance at a site, as even the hottest fires will leave a landscape consisting of burned and unburned patches. Postfire changes to burned patches include loss of vegetation and biological soil crusts, loss or reduction of seedbanks, hydrophobic soils, decreased organic material, and a variable soil nutrient response which can include a reduction, increase, or no change in many soil nutrients [e.g., nitrogen (N), potassium (P)] [Lei 1999, 2001] The unburned patches will retain most of their original character, although heat from the adjacent burned patches may alter soil characteristics to some degree. As with other types of disturbance, the geometry of burned areas is an important factor. Because fires generally affect large tracts of land, the number, distribution, size, and shape of burned patches will determine proximity to seed and inoculation sources, as well as affect existing seed dispersal mechanisms.

METRICS AND MODELS OF RECOVERY

Recovery from severe disturbance is evaluated using quantitative techniques (e.g., total cover of perennial plants) and qualitative techniques (e.g., visual recovery of disturbances). Quantitative assessments of ecosystem recovery rely on comparing an ecosystem attribute in a recovering area with the same attribute under either predisturbance or adjacent ambient conditions and in some cases under conditions at the time of abandonment (table 15.1). For example, Webb et al. (1986) indexed soil compaction on active roads (to simulate conditions at the time of abandonment) and compaction in undisturbed areas to create a linear recovery model. Using a model to compare the recovering condition to a control condition generally requires the assumption that the recovery process is linear or curvilinear (usually logarithmic or a power function). However, many studies have shown this assumption, especially when applied to biotic resources, generally predicts recovery rates that are too long.

Quantitative assessment of recovery is complicated by at least three observed responses to severe disturbance. First, in some cases, an ecosystem attribute (e.g., annual plant biomass) in a recovering area will greatly exceed its status in ambient conditions. This "overshooting" of recovery greatly complicates the calculation of recovery time and usually involves a change in species composition (K. A. Thomas and R. H. Webb *unpublished data*). Second, some attributes such as rodent populations depend on other ecosystem attributes for recovery (such as annual and perennial vegetation); separation of those interrelated attributes is difficult, if not impossible. Finally, most metrics of the Mojave Desert ecosystem fluctuate with rainfall and extreme climatic events, particularly drought (Hereford et al. 2006). Whereas some metrics of recovery, such as the status of biological soil crusts, soil compaction, and perennial vegetation, appear to be relatively independent of each other, they all are likely dependent on climate.

VISUAL RECOVERY

Visual recovery of disturbed landscapes may be the highest priority for land managers. Visual recovery is linked to recovery of biological soil crusts and perennial vegetation, and is also evident in the substrate and topography. If an area has sufficient visual recovery, it attracts less additional disturbance. For example, one of the most effective ways to rehabilitate a road may be to obscure its intersection with an active thoroughfare, thus discouraging future use. Visual recovery is expected to occur more quickly on younger, more active geomorphic surfaces than on older geomorphic surfaces because eolian and fluvial processes cause greater erosion and deposition on younger surfaces, which obscures evidence of the initial disturbance (Hereford *this volume*). Reestablishment of stream channel networks following severe disturbances (Nichols and Bierman 2001) is one of the primary enabling mechanisms for visual recovery in the Mojave Desert.

Recovery of disturbances such as at the Skidoo townsite, which was built on a young geomorphic surface (Webb et al. 1988), indicate that the process of visual recovery can be relatively swift in the higher elevations of the Mojave Desert (fig. 15.3). This is in contrast to lower elevations where desert pavements are present, because disturbances such as roads and tracks are visually apparent too long to estimate recovery (Belnap and Warren 2002). Skidoo appears visually recovered in less than a century, in part because the undisturbed vegetation in this valley consists mostly of shorter- and intermediate-lived plant species. However, a closer inspection of the townsite reveals that building excavations, berms, and other surface disturbances remain readily apparent.

Linear disturbances such as roads may visually recover more quickly than nonlinear disturbances such as townsites. In 1979, Webb (1982, 1983) installed a series of motorcycle tracks near Fremont Peak in the Western Mojave Desert (fig. 15.4). In 1999, the 200-pass motorcycle track could not be located without a photograph (fig. 15.4C), indicating visual recovery in less than 20 years. In a landscape consisting of a patchwork of geomorphic surfaces and vegetation assemblages, visual recovery is highly variable (fig. 15.5). One important aspect of visual recovery of narrow disturbances is the edge effect that results in decreased competition along the edge of a disturbance and allows increased plant growth (Johnson et al. 1975). This expanded plant size may help to visually obscure the disturbance.

Overleaf:

Fig. 15.3. Photographs showing recovery of perennial vegetation at the Skidoo townsite in the Panamint Mountains of Death Valley National Park. In 1906, two prospectors found a rich ledge of gold on the north end of the Panamint Mountains. The ensuing strike created one of Death Valley's most prosperous mining claims and a town named Skidoo, which had a maximum population of 500 and lasted ten years. *A,* photograph taken in the winter of 1906 (note the snow in the right foreground), shows a few buildings and the streets and avenues carved through the low-stature vegetation, which had been dominated by *Grayia spinosa* (spiny hopsage), *Lycium andersonii* (wolfberry), and *Artemisia spinosa* (budsage). An arrow shows the location of Montgomery Street on the east side of town. (Yeager and Woodward, courtesy of Death Valley National Park, number 2230). Skidoo was abandoned in 1916. *B,* by the time this photograph was taken, on August 17, 1960, the streets and avenues were still readily visible after forty-four years of recovery, particularly Montgomery Street (arrow). (R. H. Wauer, courtesy of Death Valley National Park, number 2382). On May 7, 1984, the amount of recovery was dependent on the amount of disturbance (Webb et al. 1988). *C,* the east end of Montgomery Street (*right side, in the distance*) had recovered more than the downtown area (*left midground*), which had less vegetation recovery and significant soil compaction. The species in the downtown area include *Chrysothamnus viscidiflorus* and *C. nauseosus* (rabbitbrush), both short-lived shrubs normally associated with washes, and *Artemisia tridentata* (big sagebrush), a shrub that is more common at higher elevations. *Grayia* and *Lycium* were common in the townsite, although not as numerous as in the ambient undisturbed vegetation. (R. M. Turner). *D,* on May 11, 1999, eighty-three years after abandonment, the streets and avenues of Skidoo townsite were no longer visible from this camera station. The former Montgomery Street (*left side*) has recovered its total perennial cover and species composition. In the downtown area, differences persist between vegetation in the townsite and the surrounding undisturbed vegetation, but the relative amount of *Grayia* and *Lycium* had increased, indicating that recovery was progressing. Compaction remains measurable in the downtown area. (D. Oldershaw, Stake 1081).

A.

B.

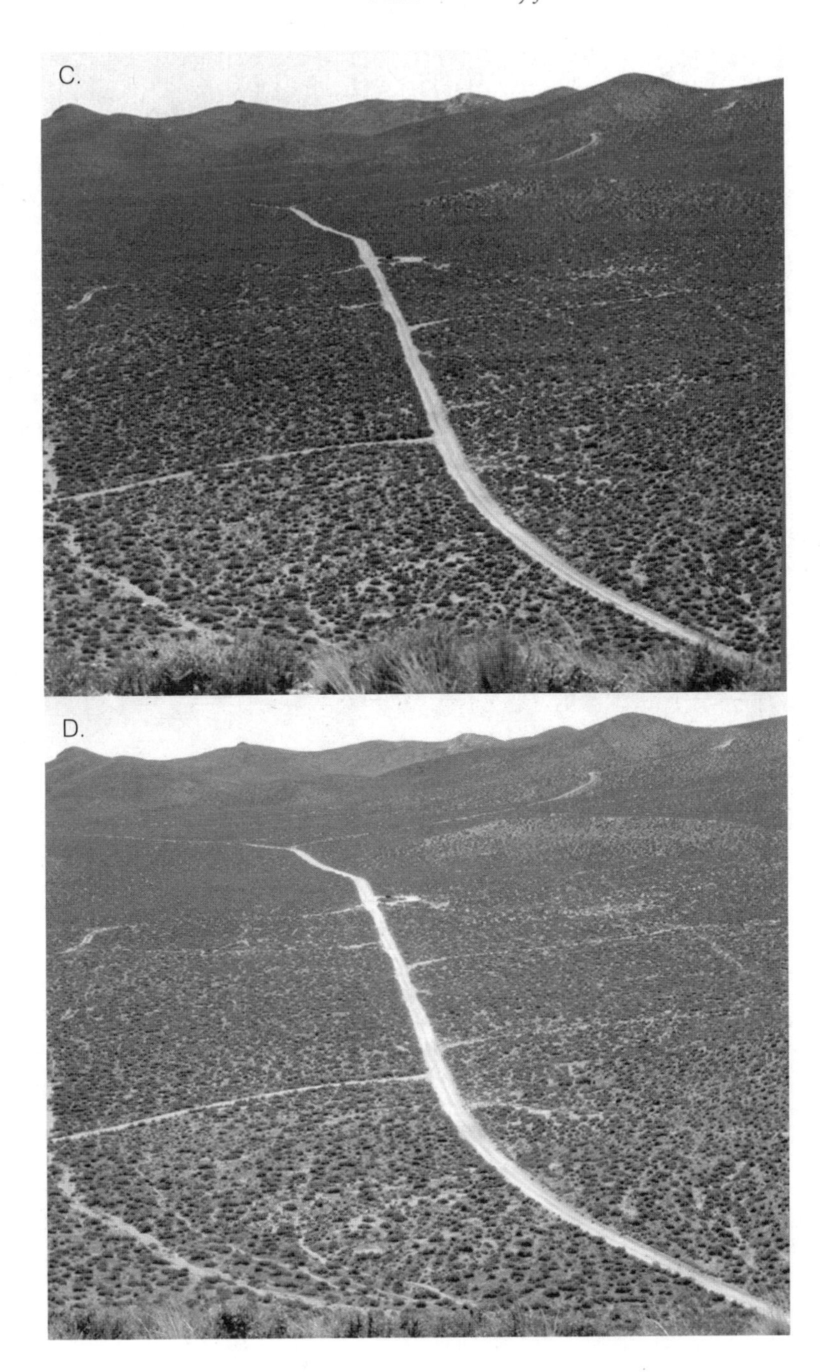
C.
D.

A.
B.
C.

SOILS

Desert Pavement

Desert pavements consist of a layer of rocks loosely fitted together and overlying a fine-grained soil horizon (Cooke et al. 1993). The underlying horizon, called the Av horizon, forms by accumulation of eolian material on poorly sorted sediments (Wells et al. 1985, 1995; Reheis et al. 1995; McFadden et al. 1998; Miller et al. *this volume*), and the subsequent effects of pedogenic modification cause the larger rocks to move to the surface. Wetting and drying of the soil surface help keep the rocks suspended above the finer Av horizon (Cooke 1970).

The Av horizon exerts strong control on pavement formation and stability (Wells et al. 1995). Desert pavements are easily disrupted by shear forces, whether caused by wheels, hooves, or feet. Shear forces displace pavement-forming rocks, which changes the size of the particles that may be left to reform a new pavement. Disturbance exposes the underlying Av horizons (Wilshire 1983; Gilewitch 2004), which are readily eroded by wind and water.

Desert pavements typically require thousands of years to develop, although many may be less than 11,000 years old (Quade 2001). Because recovery times are longer than "centuries," natural recovery of desert pavements is not a viable

Facing page:
Fig. 15.4. Photographs of desert vegetation west of Fremont Peak in the Western Mojave Desert record the results of a disturbance experiment. To document the effect of motorcycle traffic on desert soils, a controlled study was initiated on April 1, 1979, near Fremont Peak in the Mojave Desert. *A*, taken prior to disturbance show that the plants in the foreground are *Atriplex polycarpa* (cattle spinach), and scattered *Larrea tridentata* (creosote bush) appear in the middle distance, partially obscuring the parked truck. (R. H. Webb). *B*, taken on April 1, 1979, several hours after a motorcycle and rider made 200 passes down a fixed track. The camera station is about 0.3 m from the original location. Soil compaction in the 200-pass track was nearly the same as that measured in nearby active roads, indicating a high level of disturbance. Although the course was set up to destroy as few shrubs as possible, several were destroyed, including a small *Atriplex* in the foreground, the stump of which is apparent on the edge of the track. (R. H. Webb). *C*, on April 14, 1999, the track was virtually impossible to detect visually from the surrounding undisturbed desert after twenty years. A slight color change betrays the track location and is caused by a difference in the amount of biological soil crusts. In fact, we could not have found the track without the old photograph and a lot of searching. However, soil measurements indicate that the compaction levels have only reduced by half, and the soil is still highly compacted at 25 cm depth. At lower-elevation sites with few perennial species, recovery rates can be relatively fast. Remarkably, the stump of the small *Atriplex* is still present in the foreground, indicating that decomposition rates are extremely slow at this site. (D. Oldershaw, Stake 3396).

Fig. 15.5. Photograph of the Skidoo pipeline across the northern Harrisburg Flats in the Panamint Mountains, California, 2004. The pipeline corridor, abandoned no later than 1942, not visible in the foreground, is apparent in the middle distance (*lower arrow*), then disappears across a young geomorphic surface with active sheetwash (*upper arrow*). This type of visual recovery is common where the pipeline corridor crosses young geomorphic surfaces.

restoration option (Belnap and Warren 2002). Changes do occur following disturbances, and a limited rock cover can become reestablished (Webb and Wilshire 1980; Belnap and Warren 2002). In at least some cases, the difference in the size of rocks covering the soil surface creates a visual contrast between disturbed sites and desert pavements, contributing to a lack of visual recovery.

Biological Soil Crusts

Biological soil crusts are important because they stabilize desert surfaces, thereby reducing wind and water erosion. They also contribute N and C to soils and soil food webs, which increase nutrient availability to vascular plants (Belnap and Lange 2003). Disturbance can profoundly affect the cover, species composition, and physiological functioning of biological soil crusts. Undisturbed crusts have a high biomass of late-successional species of cyanobacteria, lichens, and mosses. Most disturbances, regardless of whether they are a result of trampling, off-road vehicles, annual grass invasion, or fire, generally result in a greatly simplified crust community, dominated by a low biomass of early-successional cyanobacteria (Belnap and Eldridge 2003). The loss of the late-successional species and the lowered biomass lessen the crust contribution to soil stability and fertility. This, in turn, results in loss of soils and less nutritious plants for wildlife. In addition, the loss of these dark colored, late-successional species increases the albedo of the soil, which reduces soil temperatures.

Sequence of Colonization

After disturbance severe enough to remove all crust organisms, large filamentous cyanobacteria (*Microcoleus vaginatus, Microcoleus chthonoplastes*) are generally the first photosynthetic organisms to appear at a disturbed site (fig. 15.6), likely because of their large size and mobility, which enables them to anchor large soil particles and minimizes the potential for burial (Belnap and Gardner 1993). Once soils are somewhat stabilized, the smaller and less mobile cyanobacteria (e.g., *Nostoc, Scytonema, Calothrix,* and *Schizothrix*) and green algae (e.g., *Chlorococcum* and *Chlorella*) appear. Because lichens and mosses require very stable soil surfaces for growth, they follow the smaller cyanobacteria.

Throughout the deserts of the western United States, N-fixing cyanolichens (*Collema tenax* and *Collema coccophorum*) are generally the first lichen species to appear, likely because their photobiont, *Nostoc,* commonly occurs in desert soils, and they are able to readily reproduce by spores. *Collema* is followed by the chlorolichen *Placidium* (whose phycobiont is fairly rare in Mojave soils) and the moss *Syntrichia* (Belnap and Warren 2002; Belnap and Eldridge 2003). In lower-elevation areas of the Mojave Desert, soil moisture availability limits the number of lichen and moss species that can colonize an area, and thus there may be no or only a few mid- and late-successional species present even in undisturbed areas. In higher elevation areas, soil moisture availability is sufficient to support a relatively high diversity of late-successional lichens and mosses (e.g., *Fulgensia, Psora, Toninia*). It should be kept in mind, however, that the conditions that allowed the establishment of a particular species in the past may no longer be present. Thus, it

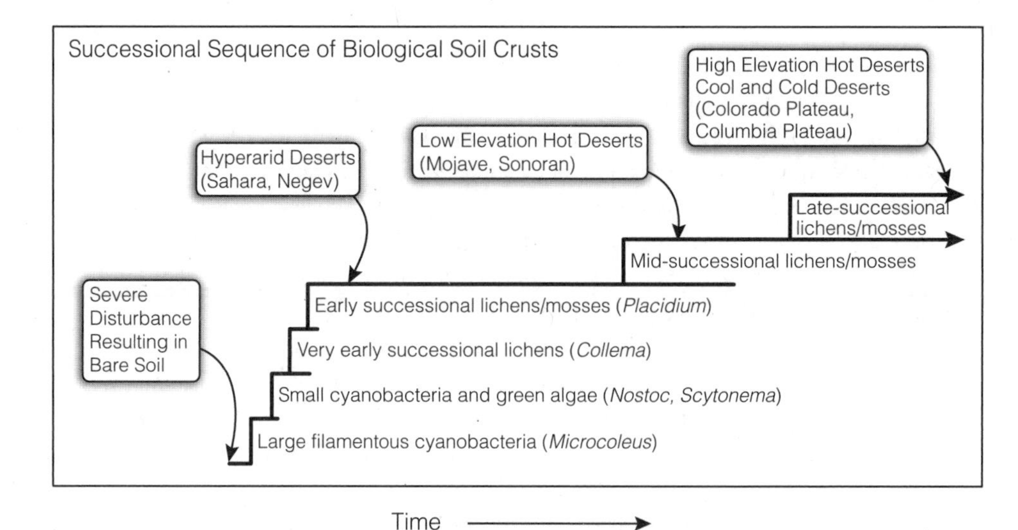

Fig. 15.6. Schematic diagram showing the colonization sequence and estimated recovery times for species found within biological soil crusts in the western United States. Arrows indicate the maximal development of crusts for a given climate zone. The length of line indicates relative time for recovery of each success ional group.

may not be possible to regain the species richness of a biological crust in a specific location once the species are extirpated.

Factors Affecting Recovery Rates

Climate Regimes and Soils: Crustal organisms are metabolically active only when wet, and recovery is faster in areas where soil moisture is highest for the longest period of time. Soil moisture availability is determined by precipitation, the air temperature when precipitation occurs, and soil texture. Areas with lower precipitation or higher air temperatures during storms—which causes higher evaporation rates—have less time with sufficient soil moisture for crust activity, and thus recover more slowly than areas with higher total precipitation or cool/cold season precipitation.

Recovery of crust components is faster in fine- than in coarse-textured soils. Fine-textured soils have inherently greater moisture-holding capacity, and a greater ability to form physical crusts, which contributes to surface stability and facilitates biological crust recovery. High salt content and other factors that increase physical crusting will also increase recovery rates. In addition, biological crusts on soils high in P and manganese, which promote soil crust organisms, should recover faster than on nutrient-poor soils (Bowker et al. 2005).

Frequency, Severity, Size, and Type of Disturbance: Recovery rates of biological crusts are highly dependent on the frequency, severity, size, and type of disturbance. Because fewer organisms are likely to survive frequent and/or severe disturbances, recovery will take longer than when disturbances are infrequent or moderate. Because crust organisms need light to photosynthesize, disturbances that bury crusts are more destructive than those that only crush the organisms. Areas with active erosion also recover more slowly. For example, vehicle tracks generally have slow recovery rates because they can channel water and wind (Wilshire 1983; Belnap 1996). In addition, the timing of the disturbance is important: when dry, crust organisms are brittle and the connections they make between soil particles are easily crushed; when wet, crust organisms are more flexible. If it rains soon after the disturbance, the crust organisms are more likely to reestablish themselves in place rather than being blown away.

Repeated disturbance will generally keep crusts at an early-successional stage (e.g., cyanobacteria-dominated) by preventing colonization of lichens or mosses, thus slowing recovery rates (fig. 15.7) (Belnap 1996). Recovery is generally faster when disturbance crushes the crusts in place, preserving inoculum, rather than removing soil and all crustal material. In contrast, because recolonization of disturbed areas occurs mostly by material washing or blowing in from adjacent, less disturbed areas, the rate of lichen recovery is much slower at sites with a large disturbed area relative to perimeter area (Eldridge and Ferris 2000).

Condition of Adjoining Substrate: In general, recovery of biological crusts is slower if soils adjacent to disturbed areas are destabilized; sediment transported out of disturbed areas buries crusts on adjacent soils and/or sandblasts nearby surfaces (Belnap 1995; McKenna-Neumann et al. 1996; Leys and Eldridge 1998). In addition, the presence of well-developed crusts will provide a source of inoculum for the disturbed area.

Vascular Plant Structure: Crusts generally recover more quickly under shrub canopies than in adjacent plant interspaces, probably due to greater soil moisture and fertility. This was demonstrated at a lower elevation site near Needles, California, where recovery of lichens 50 years after complete removal was much faster under shrubs than on the adjacent interspace (36% vs. 4% recovery, respectively) (Belnap and Warren 2002). However, it should be kept in mind that lichen and moss cover is not always the most developed under shrub canopies, as many of these organisms can prefer to colonize the shrub interspaces.

The recovery of biological soil crusts is also likely an interactive process with the recovery of soil structure and vascular plants. Eldridge and Greene (1994) developed a conceptual model in which crust development is strongly linked to the restoration of vital soil processes. As primary colonizers of recovering sites, biological soil crusts enhance microsite soil structure and fertility and provide safe sites for seeds. In turn, established vascular plants help stabilize soils, provide shade, and reduce wind speeds at the soil surface, providing conditions

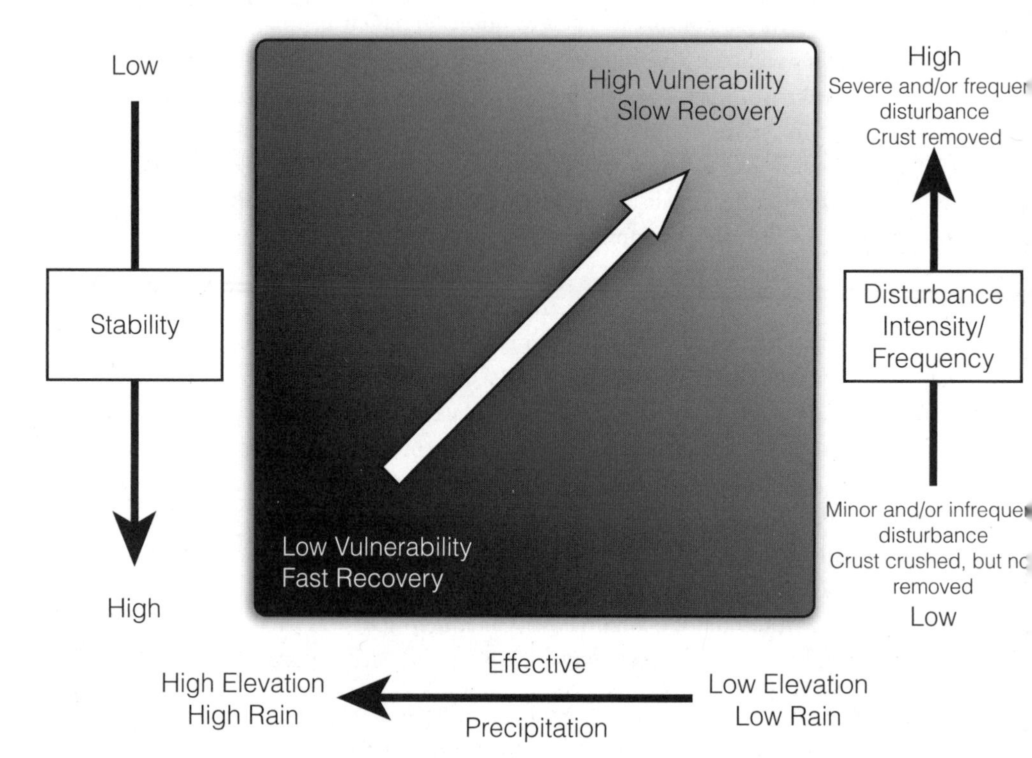

Fig. 15.7. Schematic diagram depicting vulnerability and recoverability of biological soil crusts that depend on gradients of site stability, effective precipitation, and disturbance regimes. Crusts at sites with the greatest stability, greatest effective precipitation, and lowest disturbance frequency or intensity will be less impacted (*light shading*) than sites with lower stability, less effective precipitation, and higher disturbance frequency or intensity (*dark shading*). Similarly, recovery time is faster in areas of low vulnerability and slower where vulnerability is higher.

conducive to further diversification of the biological soil crusts (Wood et al. 1982; Danin et al. 1989).

Nitrogen and Carbon Fixation: Recovery of N and C fixation after disturbance is dependent on the species composition, biomass, and physical structure of the crust. The recovery of soil N inputs is dependent on recovery rates of N-fixing cyanobacteria (e.g., *Nostoc, Scytonema*) and cyanolichens (e.g., *Collema, Heppia*). In addition, N fixation in crusts can require anaerobic microzones in the soils, which may be dependent on the buildup of cyanobacterial biomass (Belnap 1996). Recovery of C fixation is dependent on the biomass and species composition of the biological crusts, and C input to soils will increase as crust biomass increases.

In addition, lichens and mosses fix more C per unit of surface area than cyanobacteria; therefore, as their cover increases, so does C input to the soils.

Surface Albedo: Restoration of surface albedo and soil temperature depends on having the predisturbance cover and biomass of cyanobacteria, lichens, and mosses present. Whereas cyanobacteria form a darkish matrix in which other components are embedded, mosses and lichens are much darker (Belnap 1993). Consequently, recovery of surface albedo will be limited by the factors that control recovery of lichens and mosses.

Estimation of Recovery Rates

We documented recovery rates for biological crust organisms at 53 sites in the Mojave Desert where the elapsed time since disturbance ranged from 19 to 82 years, and the disturbance types included tank tracks, townsites, and pipeline construction. We used a linear model of recovery and compared recovering sites to nearby undisturbed sites. Most of the recovering sites exhibited complete recovery of cyanobacteria, but lichens had recovered at only 38 sites. At three sites, the earliest-successional lichen *Collema* showed complete recovery, while estimates of recovery time at the other 12 sites ranged from about 200 years to no observable recovery. The early-successional *Placidium* was found at 7 sites; of these, 1 site had recovered fully, 1 site was estimated to require a millennium for recovery, and 5 sites had even less recovery. No recovery of the late-successional lichen *Heppia* was seen at any sites. The late-successional lichen species *Aspicilia cinerea* showed no recovery at lower elevation sites, but recovery time of *Aspicilis repens* was estimated at 160 years for a higher elevation site. At that same higher elevation site, the late-successional species *Psora, Toninia,* and *Fulgensia* had estimated recovery rates in the range of 170–500 years.

Soil Compaction

Previous studies have shown that recovery of compacted surface soils requires 70–140 years in the Mojave and Great Basin deserts (Webb and Wilshire 1980; Webb et al. 1986; Knapp 1992b; Webb 2002; Webb and Thomas 2003). One study concluded that soil heterogeneity obscured any overall patterns of compaction recovery (Bolling and Walker 2000), indicating the importance for researchers to control for geomorphic surface and particle-size distribution. Amelioration of soil compaction is a complex process, and several factors affect recovery rates. Reviews of the relevant studies on the processes and rates of compaction recovery (Webb et al. 1986; Knapp 1992b; Webb 2002) suggest that clay-mineral expansion during wetting and drying cycles (Yaalon and Kalmar 1972) is the most important process that loosens compacted soils, followed by freeze-thaw cycles and bioturbation. Freeze-thaw cycles are likely important only at higher elevations in the Mojave

Desert. Bioturbation, which includes loosening by plant roots and burrowing animals, may be more important than physical processes in loosening compacted subsurface soils.

The trajectory of compaction recovery is an important management consideration (Webb and Wilshire 1980). Although a linear relation can be used to estimate full recovery time, this trajectory does not represent the rapid changes in soil physical properties that occur immediately following abandonment. The linear trajectory may also be less useful because some researchers have found that low levels of soil compaction may enhance plant growth (Lathrop and Rowlands 1983), and thus the ecosystem effects of compaction may not last as long as the compaction itself.

Although compaction recovery near the soil surface is fairly well understood, knowledge of recovery at depth is very limited. Webb (1982) notes that the maximum soil compaction occurs at a depth of 0.0–0.1 m in desert soils, but compaction typically extends to 0.50 m or deeper. Prose (1985) and Prose and Metzger (1985) concluded that subsurface soil compaction persisted considerably longer than compaction at the surface. Such persistence would be expected because freeze-thaw loosening would not occur deeper than about 0.05 m, and wetting and drying cycles are less frequent in subsurface horizons. Penetrometer data (Webb and Wilshire 1980; Webb et al. 1988) indicate that subsurface compaction might persist longer, but variability in penetration resistance at depth precludes definite conclusions. As shown in figure 15.8, penetrometer data from the Wahmonie townsite suggests that although recovery appears to have occurred above 0.20 m, residual compaction may remain to a depth of about 0.25 m.

Decomposition, Nutrient Cycling, and Soil Fauna

Many studies have shown that soil levels of C and plant-available N and P are naturally low in desert regions (Whitford 2002), and N and P may limit plant productivity more than water in some areas of the Mojave Desert (Penning de Vries and Djiteye 1982). Soil disturbances produced by off-road vehicles, livestock grazing, and mining can affect soil C and other nutrients via alteration of soil structure, removal of vascular and nonvascular vegetation, and disruption of soil food webs. Compressional disturbances crush soil aggregates, which is where most soil biota reside and where most soil nutrient transformations occur (Coleman et al. 2004). Loss of pore space reduces habitable space for soil biota and can directly limit species richness and abundance, as well as the mobility of soil biota and thus their ability to utilize resources (Belnap 1995). Compaction reduces the amount of water and air that enters the soil, which reduces the biotic activity and nutrient transformations that are dependent on water or oxygen (Webb and Wilshire 1980). In contrast, due to reduced oxygen concentrations, anaerobic nutrient

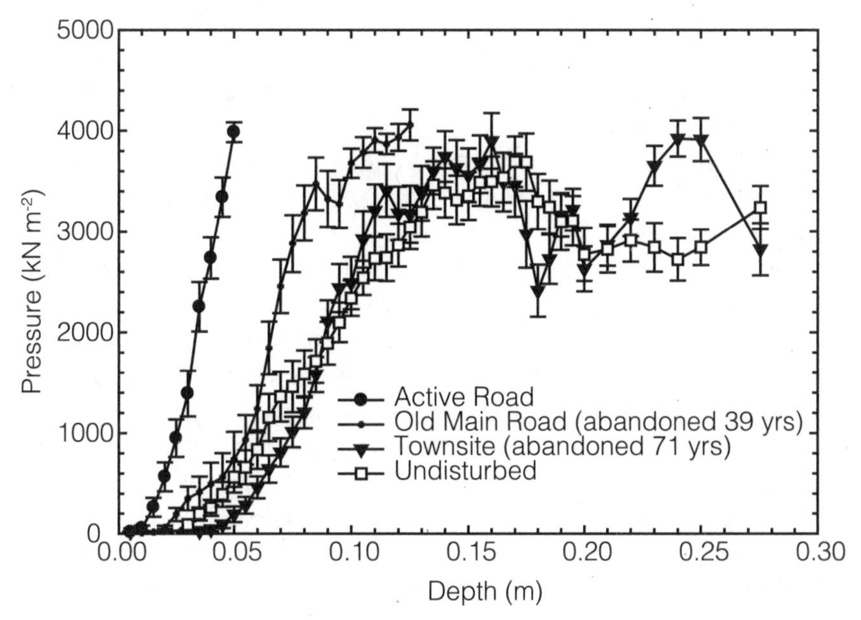

Fig. 15.8. Graph showing penetrometer curves for compaction treatments within the Wahmonie townsite, located on the Nevada Test Site, were measured in 1999. The curves show that it took seventy years for soil compaction to return to levels comparable with undisturbed soils. The error bars represent standard error of estimate for thirty replicates, and the instrument is not capable of measuring strengths > 4000 kN m^{-3}. The soil-moisture content at the time of the measurements was 6.4% and represented moderately wet conditions.

transformations (e.g., nitrification) may be enhanced by compaction, although the need for soil moisture may offset any nutrient increase.

Compressional disturbances can also kill soil microbes outright. Because desert soil biota are C-limited, the reduction of biological soil crusts and plant biomass that is concomitant with severe disturbance (both through compaction and direct removal) reduces C inputs to the soil, thereby reducing the abundance of soil biota. The change in soil temperature via surface albedo changes also alters nutrient transformation rates (Belnap 1995). In addition, changes in plant species composition have the potential to alter decomposition and nutrient cycling rates due to decreased soil temperature, lowered input and quality of litter and plant root exudates, altered soil food web structure, and increased erosion, all of which result in a loss of soil C and other nutrients (Belnap 1995; Wardle 2002; Adl 2003; Neff et al. 2005).

Recovery of decomposition and nutrient cycling rates after disturbance requires recovery of compacted soils, vascular and nonvascular vegetation, and soil biota. There have been almost no studies on recovery of soil biota, decomposition, or nutrient cycling after disturbance in the Mojave Desert, although a few studies have addressed recovery of N cycles using isotopes and N mineralization potentials (the rate at which N is made bio-available). These studies show that N cycles are disrupted by both mechanical disturbances and fire, and that recovery is measured in multiples of decades (J. Belnap *unpublished data;* T. C. Esque *personal communication*).

Although studies have shown that severe disturbance can disrupt decomposition and nutrient cycling because of changes in soils and vegetation (Belnap 1995; Whitford 2002), there is no information on how soil biota or processes recover relative to recovery of the ecosystem components on which their recovery depends. For example, severe compaction reduces the abundance and richness of soil biota. However, no data show how these factors are affected by the different levels of compaction that occur as soils recover. Similarly, as compaction is alleviated, soil temperatures change, infiltration rates increase, and vascular and nonvascular vegetation recover. As with compaction recovery, there are no studies addressing how vegetative cover relates to soil temperature as the recovery proceeds. Although the plant composition, cover, and density have a large influence on decomposition and nutrient cycling rates, the amount of vegetation recovery required for the recovery of these ecosystem processes is unknown. As it is unlikely that all ecosystem attributes need to fully recover in order for decomposition and nutrient cycling rates to recover, it is impossible to estimate recovery times without additional data.

ANNUAL VEGETATION

Little is known about the effect of severe disturbances on native annuals in the Mojave Desert, in part because nonnative annuals have become pervasive and the interannual variability of ephemeral plant growth is high (Lathrop and Rowlands 1983; Brooks and Matchett 2003). Although increasing levels of soil compaction are known to decrease annual plant biomass (Adams et al. 1982), some species appear to benefit from low levels of soil compaction (Lathrop and Rowlands 1983). In particular, soil compaction may decrease the cover but increase the density of native annual plants (Adams et al. 1982). This may occur because of a trade-off between decreased soil macropores, which facilitates moisture retention, and increased soil strength, which hinders root elongation. As a result, more plants may germinate, but each plant may be smaller.

Recovery of desert annuals consists of decreasing population density and biomass in response to increasing perennial vegetation. In areas recovering from severe disturbance, annuals—whether native or nonnative—are initially more numerous than in nearby undisturbed areas (Hunter 1995a). Hunter (1995a) showed that from 1989 to 1994, while perennial vegetation was still far from full recovery in disturbed areas of the Nevada Test Site, the total biomass of annuals was approximately the same in the disturbed and ambient undisturbed areas. However, species diversity is typically greater in disturbed than in ambient undisturbed areas. Because annual plant populations strongly fluctuate in response to winter rainfall (Beatley 1967), and because of the pervasive effects of nonnative species, phytometrics of annuals are poor indicators of recovery.

PERENNIAL VEGETATION

Natural revegetation of abandoned sites in the Mojave Desert is relatively slow (Vasek 1980; Webb and Wilshire 1980; Webb et al. 1988; Lovich and Bainbridge 1999), but significant recovery can occur in as little as a few decades (Webb et al. 1988). Regardless of whether the disturbance involves soil compaction or not, the species that initially become established after abandonment typically are shorter-lived ruderal species (Webb and Wilshire 1980; Lathrop 1983; Prose et al. 1987; Webb et al. 1988; Angerer et al. 1994; Lei 1999). As recovery proceeds, these species are eventually replaced by longer-lived species, although the precise trajectory of species composition in a recovering site remains unknown. It is not clear that the concept of succession applies to the recovery of desert vegetation, as it implies biotic control of species composition changes (Lathrop and Rowlands 1983; Webb et al. 1988; Knapp 1992a), and many of the changes in desert perennial plant recovery are abiotically driven.

Whether or not soil compaction was severe at the time of abandonment likely affects the reestablishment of perennial plants (Webb et al. 1988), and the variability in observed responses is high. Also, the presence of large numbers of nonnative annual species, particularly in burned areas, may result in a different assemblage of perennial vegetation, which reduces the rate of recovery to prefire conditions (Brooks et al. 2004).

The amount of soil disruption is also extremely important to the course of vascular plant recovery. If the islands of fertility that often exist beneath shrubs (particularly *Larrea tridentata*) are intact after disturbance, the recovery of total cover, irrespective of species composition, may be accelerated (Wallace et al. 1980). In addition, if desert pavements are removed and the Av horizon is disrupted, moisture penetration may be higher than on undisturbed pavements (Gilewitch 2004), which partially explains why some scraped areas, notably the Tonopah and Tidewater Railroad corridors across the eastern Silurian Valley, have higher density and cover of perennial vegetation than the adjacent pavemented surface—after less than a century of recovery (R. H. Webb *unpublished data*).

Previous Studies of Vegetation Recovery

Natural recovery of perennial vegetation has been measured in numerous different types of severe disturbances in the Mojave Desert. Studies of noncompacting disturbances include agricultural fields that were cleared and abandoned (Webb et al. 1983; Carpenter et al. 1986), burned areas (Minnich 1995; Lei 1999, 2001), and areas affected by above-ground nuclear detonation (Wallace and Romney 1972; Hunter 1995b). Studies of areas with significant soil compaction and/or disruption include military maneuver areas (Lathrop 1983; Prose and Metzger 1985; Prose et al. 1987; Steiger and Webb 2000; Prose and Wilshire 2000), ghost towns (Webb and Wilshire 1980; Webb et al. 1988; Webb and Thomas 2003), construction sites (Angerer et al. 1994), utility corridors (Vasek et al. 1975a, 1975b; Lathrop and Archbold 1980a, 1980b), levees built to divert overland flow from roadways (Schlesinger and Jones 1984), and off-road vehicle areas (Vollmer et al. 1976; Wilshire et al. 1978). This aggregation of studies allows a comparison of recovery rates for perennial vegetation as a function of the disturbance type.

Total Perennial Cover

Total perennial cover is perhaps the most important metric used to describe a recovering desert ecosystem and allows direct comparison between natural recovery and active restoration. Several studies have estimated natural recovery rates for total perennial cover. For instance, Angerer et al. (1994) estimated that total perennial cover at Yucca Mountain (Nevada Test Site) would, on average, recover in 20 years. Webb and Thomas (2003) found that, on average, total vegetation cover requires about 80 years for recovery in 16 Mojave Desert ghost towns.

Figure 15.9 shows recovery curves for total perennial cover of sites with a variety of disturbance types following abandonment. The curves are based on data from 50 measurements along pipeline or other underground utility corridors (Webb et al. 1988; K. A. Thomas *unpublished data*), 34 measurements in cleared and abandoned areas (Webb et al. 1988), 26 measurements in abandoned townsites (Webb and Thomas 2003), 17 measurements in burned areas (R. A. Minnich *unpublished data*), and 9 measurements in abandoned roads. Using a curve-fitting algorithm, only the recovery of roads follows a linear trajectory; recovery of other types of disturbances follows a curvilinear relation, indicating that total perennial cover is 60%–80% recovered in about 30 years (fig. 15.9).

Three examples illustrate exceptions to this general pattern of recovery. Hunter (1995b) reported that several areas on the NTS disturbed by above-ground nuclear testing were "conspicuously bare" more than 30 years after disturbance. These sites, at the northern limits of the Mojave Desert, have some chenopods [*Grayia spinosa* (spiny hopsage), *Atriplex canescens* (fourwing saltbush), *Atriplex confertifolia* (shadscale), and *Ceratoides lanata,* (winterfat)] and perennial grasses [*Achnatherum speciosum* (desert needlegrass), and *A. hymenoides,* (Indian ricegrass)], which are strongly affected by drought. Along the Mojave pipeline, completed 10 years before measurements of recovery, some plots dominated by the chenopod *Atriplex polycarpa* had far greater cover than in the undisturbed plots

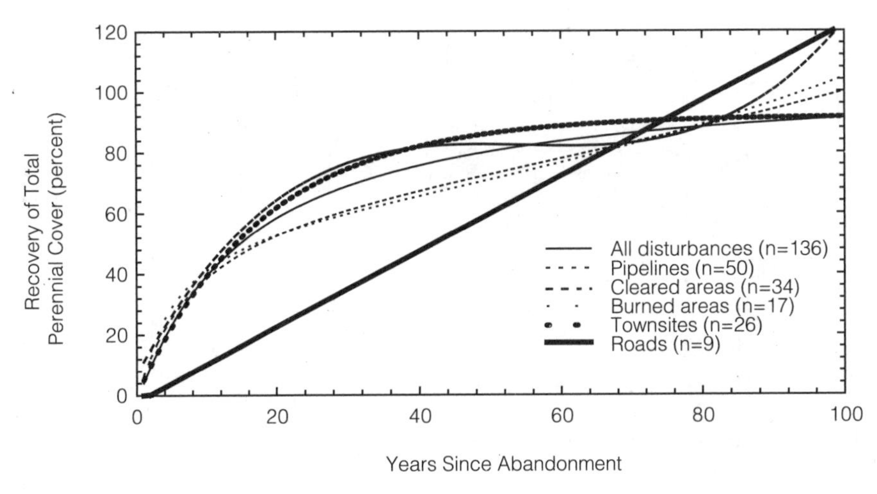

Fig. 15.9. Graph depicting the best-fit recovery models of total perennial cover following abandonment from a variety of disturbance types in the Mojave Desert.

(K. A. Thomas *unpublished data*). Finally, in some areas along the Tonopah and Tidewater Railroad corridor, disturbance of desert pavement resulted in an increased cover of *Larrea* and *Ambrosia dumosa* (white bursage) after 94 years of recovery, relative to the adjacent reference areas, presumably because of increased water infiltration into the disturbed soils (R. H. Webb *unpublished data*). These examples show that the course of recovery is strongly affected by species composition at a given site, as well as disturbance type, soil properties and climate.

Vegetation Density

For most plant assemblages, vegetation density is a difficult attribute to measure because many Mojave shrub species are clonal, complicating the process of defining an individual plant. Also, any shrub that is capable of sprouting from the root crown tends to create extremely large individuals, a phenomenon that is not described by vegetation density. *Larrea* in particular resprouts from damaged root crowns, which can more or less fully recover within five years of being crushed by a vehicle (Gibson et al. 2003), and tends to form large individuals in abandoned townsites (Webb et al. 1988).

At the Wahmonie townsite, which was abandoned in 1928 (Webb and Wilshire 1980) and measured for vegetation density after 51 years in 1979, the total perennial plant density was approximately 75% of the density in nearby undisturbed plots (fig. 15.10). In 1979, *A. speciosum* had the highest density in both the townsite and the undisturbed area. However, as a result of the severe drought of 1989–1991, the density of this species had decreased by more than half when re-measured in 1999. Compared with 1979, the total density in the undisturbed area decreased in both

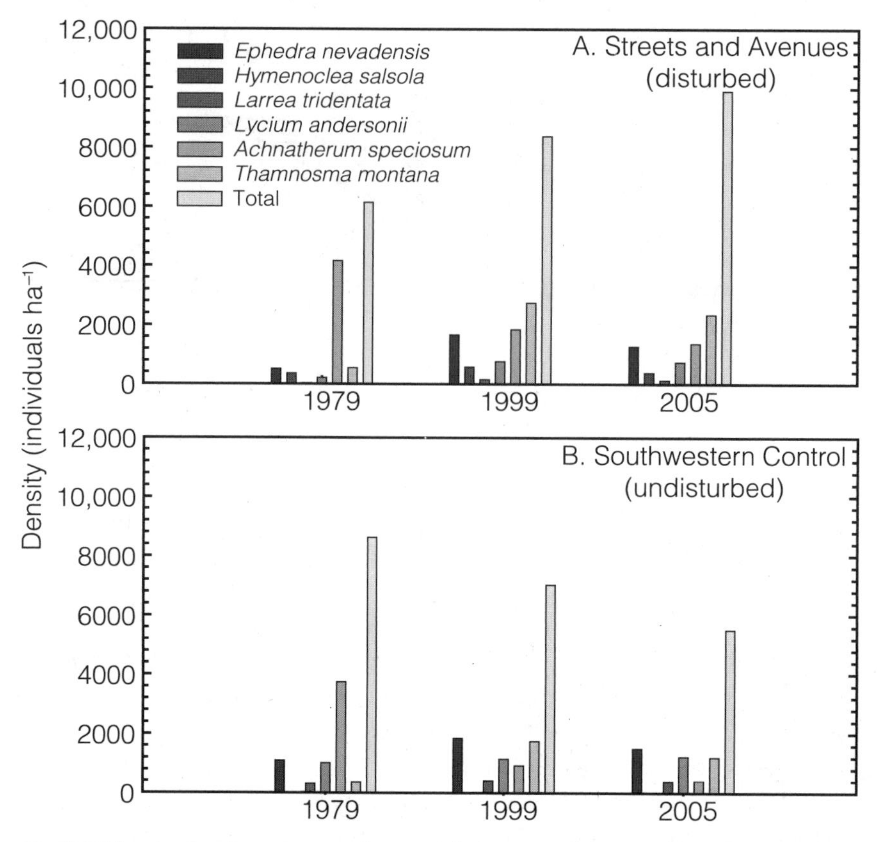

Fig. 15.10. Graphs showing recovery of perennial vegetation density at the Wahmonie townsite on the Nevada Test Site. By comparing the bars in *A,* streets and avenues (disturbed sites), with *B,* southwestern control (undisturbed sites), the differences in species representation between these sites can be determined.

1999 and 2005, largely because of the decline in the perennial grass *Achnatherum.* In the recovering townsite, total plant density increased from 1979 to 1999 and 2005 and surpassed the density of perennial vegetation in the undisturbed assemblage because the increase in shrubs offset the decline in *Achnatherum* (fig. 15.10). The overall density of woody perennial plants (all species combined) in the disturbed and undisturbed areas was essentially equal.

However, the density of individual woody perennial species at the Wahmonie townsite had not recovered completely by 2005. Shorter-lived perennials, such as *Hymenoclea salsola* (burrobush) and *Thamnosma montana* (turpentine broom), had higher densities in the townsite, whereas longer-lived perennials, particularly *Lycium andersonii* (Anderson thornbush), *Larrea,* and *Ephedra nevadensis* (Mormon tea), were less numerous (fig. 15.10). Of 15 sites measured in 7 Mojave Desert ghost towns established in the first 3 decades of the twentieth century, only one—Harrisburg (Webb et al. 1988)—appeared to be completely recovered in terms of the perennial species density.

Species Composition by Cover

Models of secondary succession based on perennial cover have been used empirically to predict full recovery times of "less than a century" (Lathrop and Archbold 1980a, 1980b), "many centuries" (Vasek et al. 1975a), "several millennia" (Vasek 1980), and "thousands of years" (Webb and Wilshire 1980; Webb et al. 1988) in the Mojave Desert. However, considering the long lifespans of some species in the Mojave Desert (Bowers et al. 1995), the recovery process necessarily requires extended periods to allow longer-lived species to displace shorter-lived ones.

Returning to the Wahmonie example, the cover of individual species has changed and cover of some species fluctuated both up and down between 1963 and 2005 (fig. 15.11). The cover of *Larrea,* the dominant perennial in the undisturbed area, increased from 1963–2005 in both the recovering townsite and the adjacent undisturbed vegetation. The cover of *Ephedra,* one of the dominants, increased by over 100% in the undisturbed area and fluctuated by 50% from 1963 to 2005, due to drought pruning and mortality. Cover of *Grayia, Lycium,* and *Achnatherum* all significantly decreased in both the disturbed and undisturbed sites between 1975 and 1999, presumably because of the drought from 1989 through 1991. Whereas the shorter-lived *Thamnosma* increased by about the same amount in the disturbed and the undisturbed areas from 1963 to 2005, the cover of *Hymenoclea,* a species largely absent from the undisturbed area, peaked in the 1970s and declined through 2005 in the disturbed site (fig. 15.11).

This example demonstrates both the validity and shortcomings of a succession-based model of desert vegetation recovery. Most models of secondary succession are based on an initial establishment of shorter-lived ruderal species (Webb et al. 1988), as was seen at the Wahmonie townsite with the initial establishment of *Hymenoclea* and *Achnatherum.* Over time, however, the composition of this community shifted towards an assemblage dominated by *Larrea, Ephedra,* and other longer-lived species. This change was presumably driven by a high drought-induced mortality of the shorter-lived species, whereas the longer-lived perennials experienced a lower drought-induced mortality. This differential mortality occurred in both the recovering townsite as well as the undisturbed reference area, which created a moving target for assessing recovery. Current succession models are predicated on an ultimate stable state, and this data show that undisturbed assemblages of perennial vegetation in the Mojave Desert are not necessarily stable.

Climatic variability plays a large role in quantitative assessment of recovery of Mojave Desert perennial vegetation (Hereford et al. 2006; Webb et al. *this volume*). Whereas longer-lived species, such as *Larrea* and *Coleogyne ramosissima*

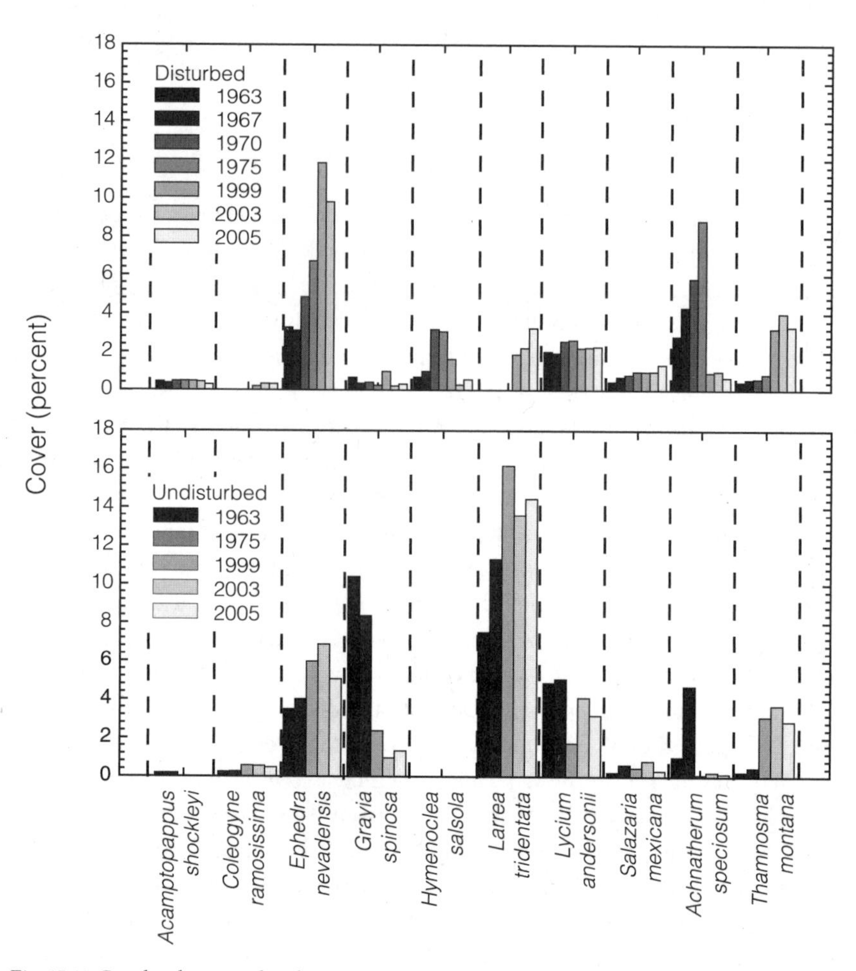

Fig. 15.11. Graphs showing the change in percent cover between disturbed and undisturbed sites, plotted for different perennial species at the Wahmonie townsite on the Nevada Test Site.

(blackbrush), have not experienced large mortality episodes in response to historical droughts, shorter-lived species, especially *Achnatherum* and *Grayia,* have fluctuated considerably. Recovery goals for perennial vegetation should therefore be based on a comparison with nearby undisturbed vegetation, rather than with the exact conditions before the beginning of disturbance (Webb et al. 1988).

Biomass and Productivity

Little data are available for changes in assemblage biomass and productivity that occur during recovery from severe disturbance. Hunter (1995b) noted large fluctuations in plant volume on undisturbed plots and in nearby areas affected by above-ground nuclear detonations. Over an 8-year period, biomass decreased significantly in one undisturbed area, owing to the effects of the severe 1989–1991 drought, confounding any interpretation of recovery in nearby disturbed plots. In contrast, another paired disturbed-undisturbed plot (T4) (Hunter 1995b) had a 270% increase in plant volume within the blast area, and a 28% decrease in plant volume within the undisturbed area. Because these study areas are dominated by species that are strongly affected by drought, these results may not reflect what might occur in areas dominated by other species.

VERTEBRATES

Recovery of vertebrate populations following severe disturbances in the Mojave Desert has received little study. Here, we only report on the recovery of reptiles and rodents. Understanding the recovery of these animals is difficult, as their populations fluctuate strongly with climate and annual plant production (Beatley 1969, 1976; Turner et al. 1970, 1982). Vertebrate populations are also dependent on perennial vegetation for habitat, and fluctuations in perennial plants may also induce fluctuations in animal populations. As a result, recovery of rodents and reptiles may be inextricably linked to recovery of perennial vegetation, particularly total cover and biomass (Saethre 1995).

Long-term fluctuations in rodent populations have been studied primarily on the NTS (Saethre 1995; Rundel and Gibson 1996). Persistent drought from 1989 to 1991 caused large fluctuations in some rodent species, but no directional trend was detected. In disturbed areas, the total number of animals present was about the same as in adjacent undisturbed areas, but species diversity was lower. Merriam's kangaroo rat (*Dipodomys merriami*) was the most common species in disturbed areas, whereas the undisturbed areas had a larger, more evenly mixed variety of species. Saethre (1995) linked recovery of rodents to recovery of perennial vegetation habitat, which provides cover for some rodent species reluctant to use areas without shelter from predators.

Most studies of Mojave Desert reptile populations have occurred on the NTS using the short-lived and ubiquitous side-blotched lizard (*Uta stansburiana*) (Woodward 1995). Populations of this species, measured annually from 1988 through 1994, were lowest during 1990, in the middle of the severe drought (Woodward 1995). The large temporal and spatial variability in side-blotched lizard populations make generalizations difficult, but population models indicate complex linkages with climate and plant productivity variables (Turner et al. 1982). In plots disturbed about 40 years before measurement, the side-blotched lizard was the only species present, compared with undisturbed sites with more diverse lizard populations. Woodward (1995) concluded that recovery of lizard populations in severely disturbed areas takes longer than 20 years. Because lizard populations are linked to habitat, recovery of lizards and other reptiles are likely dependent on recovery of perennial vegetation.

DISCUSSION AND CONCLUSIONS

Recovery of severely disturbed lands in the Mojave Desert is a complicated process involving a number of ecosystem attributes. Because these diverse attributes, which range from annual plants to perennial vegetation and vertebrate populations, fluctuate considerably in response to climatic variability, measurements of recovery must be made with reference to nearby undisturbed conditions; under changing environmental conditions, restoration to the predisturbance state is largely irrelevant, if not impossible. While many metrics of recovery are interdependent and difficult to evaluate in isolation, measurement of some ecosystem attributes—notably soil compaction, biological soil crusts, and perennial vegetation—may provide solid benchmarks for quantitative estimation of natural recovery rates.

Other than soil compaction and biological soil crusts, recovery of soil properties has not been well studied. Major questions remain as to the importance of soil flora and fauna to the recovery process, and whether decomposition-rate recovery is important. Because biological soil crusts help stabilize surfaces, fast recovery of crusts is important to minimize soil erosion. Likewise, amelioration of soil compaction helps reduce soil erosion by increasing infiltration. The influence of soil compaction on the recovery of annual and perennial vegetation remains an open question, as it appears that many species may become established in highly compacted soil, while other species may be harmed from the lowered infiltration and water-holding capacity.

Assessing the recovery of total perennial cover is complicated because the species composition of recovering sites is often considerably different from the nearby undisturbed vegetation (Webb et al. 1988). Recovery of total perennial cover is also affected by drought frequency (Hereford et al. 2006). Frequent droughts kill shorter-lived species, such as *Hymenoclea*, *Achnatherum*, and *Atriplex*, which are common in disturbed sites, whereas only extreme drought kills longer-lived species such as *Larrea*. Differences in the types of existing vegetation assemblages affect the rate of recovery because of the life-history strategies of the constituent species.

Recovery of severely disturbed areas, in terms of changes in soil compaction and total cover of perennial vegetation, generally requires about a century in the Mojave Desert. This recovery time varies with patch geometry; broad patches recover more slowly than narrow linear disturbances. Species composition of the ambient vegetation also affects the generalized recovery rate, as close proximity of seed sources may increase the recovery rate for some shorter-lived species (e.g.,

H. salsola). Because of the high variability in total perennial cover, the influence of residual soil compaction on vegetation recovery remains largely unknown, although the trajectory of recovery in roads appears to be linear and the trajectory of recovery in other types of disturbed areas appears to be curvilinear. The full recovery time required for any disturbed desert area is dependent upon the type of soil, the age of the geomorphic surface, the size and shape of the disturbance, the magnitude of soil compaction, the reestablishment rate of vegetation, the species composition of the ambient undisturbed vegetation, the amount and temporal distribution of rainfall, the severity of the winter climate, and the amount of postabandonment disturbance. Plant assemblages with few perennial species and low total cover may recover more quickly than more complex assemblages (Webb et al. 1988).

ACKNOWLEDGMENTS

Mimi Murov and Diane Boyer helped with the field work and digital photography. The authors thank Denis Kearns of the Bureau of Land Management, Debra Hughson of the National Park Service, and David Miller of the U.S. Geological Survey for carefully reviewing the manuscript. This work was funded as part of the Recoverability and Vulnerability of Desert Ecosystems Project of the Priority Ecosystem Studies of the U.S. Geological Survey.

REFERENCES

Adams, J. A., L. H. Stolzy, A. S. Endo, P. G. Rowlands, and H. B. Johnson. 1982. Desert soil compaction reduces annual plant cover. *California Agriculture* **36**:6–7.

Adl, S. M. 2003. *The ecology of soil decomposition.* CABI Publishing, Wallingford, UK.

Angerer, J. P., W. K. Ostler, W. D. Gabbert, and B. W. Schultz. 1994. *Secondary succession on disturbed sites at Yucca Mountain, Nevada.* Report No. EGG 11265–1118, UC-702. EG&G Energy Measurements Las Vegas, Nevada.

Beatley, J. C. 1967. Survival of winter annuals in the northern Mojave Desert. *Ecology* **48**:745–750.

Beatley, J. C. 1969. Dependence of desert rodents on winter annuals and precipitation. *Ecology* **50**:721–724.

Beatley, J. C. 1976. Rainfall and fluctuating plant populations in relation to distributions and numbers of desert rodents in southern Nevada. *Oecologia* **24**:21–42.

Belnap, J. 1993. Recovery rates of cryptobiotic crusts: inoculant use and assessment methods. *Great Basin Naturalist* **53**:89–95.

Belnap, J. 1995. Surface disturbances: their role in accelerating desertification. *Environmental Monitoring and Assessment* **37**:39–57.

Belnap, J. 1996. Soil surface disturbances in cold deserts: effects on nitrogenase activity in cyanobacterial-lichen soil crusts. *Biology and Fertility of Soils* **23**:362–367.

Belnap, J., and D. Eldridge. 2003. Disturbance and recovery of biological soil crusts. Pages 363–383 *in* J. Belnap and O. L. Lange, editors. *Biological soil crusts: structure, function, and management.* Ecological Studies 150. Second edition. Springer-Verlag, Berlin, Germany.

Belnap, J., and J. S. Gardner. 1993. Soil microstructure in soils of the Colorado Plateau: the role of the cyanobacterium *Microcoleus vaginatus. Great Basin Naturalist* **53**:40–47.

Belnap, J., and O. L. Lange. 2003. *Biological soil crusts: structure, function, and management.* Ecological Studies 150. Second edition. Springer-Verlag, Berlin, Germany.

Belnap, J., and S. D. Warren. 2002. Patton's tracks in the Mojave Desert, USA: an ecological legacy. *Arid Land Research and Management* **16**:245–258.

Bolling, J. D., and L. R. Walker. 2000. Plant and soil recovery along a series of abandoned desert roads. *Journal of Arid Environments* **46**:1–24.

Bowers, J. E., R. H. Webb, and R. J. Rondeau. 1995. Longevity, recruitment, and mortality of desert plants in Grand Canyon, Arizona, USA. *Journal of Vegetation Science* **6**:551–564.

Bowker, M. A., J. Belnap, D. W. Davidson, and S. L. Phillips. 2005. Evidence for micronutrient limitation of biological soil crusts: potential to impact aridlands restoration. *Ecological Applications* **15**:1941–1951.

Brooks, M. L. 1999. Alien annual grasses and fire in the Mojave Desert. *Madroño* **46**:13–19.

Brooks, M. L., C. M. D'Antonio, D. M. Richardson, J. B. Grace, J. E. Keeley, J. M. DiTomaso, R. J. Hobbs, M. Pellant, and D. Pyke. 2004. Effects of invasive alien plants on fire regimes. *BioScience* **54**:677–688.

Brooks, M. L., and J. R. Matchett. 2003. Plant community patterns in unburned and burned blackbrush (*Coleogyne ramosissima* Torr.) shrublands in the Mojave Desert. *Western North American Naturalist* **63**:283–298.

Brooks, M. L, J. R. Matchett, and K. H. Berry. 2006. Effects of livestock watering sites on plant communities in the Mojave Desert, USA. *Journal of Arid Environments* **67**: 125–147.

Brooks, M. L., and R. A. Minnich. 2006. Southeastern deserts bioregion. Pages 391–414 *in* N. G. Sugihara, J. W. van Wagtendonk, K. E. Shaffer, J. Fites-Kaufman, and A. E. Thode, editors. *Fire in California's Ecosystems.* University of California Press, Berkeley, California.

Brown, D. E., and R. A. Minnich. 1986. Fire and creosote bush scrub of the western Sonoran Desert, California. *American Midland Naturalist* **116**:411–422.

Bury, R. B., R. A. Luckenbach, and S. D. Busack. 1977. *Effects of off-road vehicles on vertebrates in the California Desert.* Wildlife Research Report No. 8. U.S. Fish and Wildlife Service, Washington, D.C.

Callison, J., J. D. Brotherson, and J. E. Bowns. 1985. The effects of fire on the blackbrush (*Coleogyne ramosissima*) community of southwestern Utah. *Journal of Range Management* **38**:535–538.

Campbell, C. E. 1972. Some environmental effects of rural subdividing in an arid area: a case study in Arizona. *Journal of Geography* **71**:147–154.

Carpenter, D. E., M. G. Barbour, and C. J. Bahre. 1986. Old field succession in Mojave Desert scrub. *Madroño* **33**:111–122.

Coleman, D. C., D. A. Crossley, and P. F. Hendrix. 2004. *Fundamentals of soil ecology.* Second edition. Elsevier Academic Press, Burlington, Massachusetts.

Cooke, R. U. 1970. Stone pavements in deserts. *American Association of Geographers Annals* **60**:560–577.

Cooke, R. U., A. Warren, and A. Goudie. 1993. *Desert geomorphology.* University College London Press Limited, London, UK.

Danin, A, Y. Bar-Or, I. Dor, and T. Yisraeli. 1989. The role of cyanobacteria in stabilization of sand dunes in southern Israel. *Ecologia Mediterranea* **XV**:55–64.

Davidson, E., and M. Fox. 1974. Effects of off-road motorcycle activity on Mojave Desert vegetation and soil. *Madroño* **22**:381–412.

Eckert Jr., R. E., M. K. Wood, W. H. Blackburn, and F. F. Peterson. 1979. Impacts of off-road vehicles on infiltration and sediment production of two desert soils. *Journal of Range Management* **32**:394–397.

Eldridge, D. J., and J. M. Ferris. 2000. Recovery of populations of the soil lichen *Psora crenata* after disturbance in arid South Australia. *Rangeland Journal* **21**:194–198.

Eldridge, D. J., and R. S. B. Greene. 1994. Microbiotic soil crusts: a review of their roles in soil and ecological processes in the rangelands of Australia. *Australian Journal of Soil Research* **32**:389–415.

Gibson, A. C., M. R. Sharifi, and P. W. Rundel. 2003. Resprout characteristics of creosote bush (*Larrea tridentata*) when subjected to repeated vehicle damage. *Journal of Arid Environments* **57**:411–429.

Gilewitch, D. A. 2004. The effect of military operations on desert pavement. Pages 243–258 *in* D. R. Caldwell, J. Ehlen, and R. S. Harmon, editors. *Studies in military geography and geology.* Kluwer Academic Publishers, Dordrecht, the Netherlands.

Gillette, D. A., and J. Adams. 1983. Accelerated wind erosion and prediction of rates. Pages 97–109 *in* R. H. Webb and H. G. Wilshire, editors. *Environmental effects of off-road vehicles.* Springer-Verlag, New York, New York.

Hereford, R., R. H. Webb, and C. Longpré. 2006. Precipitation history and ecosystem response to multidecadal precipitation variability in the Mojave Desert region, 1893–2001. *Journal of Arid Environments* **67**:13–34

Hinckley, B. S., R. M. Iverson, and B. Hallet. 1983. Accelerated water erosion in ORV-use areas. Pages 81–96 *in* R. H. Webb and H. G. Wilshire, editors. *Environmental effects of off-road vehicles.* Springer-Verlag, New York, New York.

Hunter, R. B. 1995a. Status of ephemeral plants on the Nevada Test Site, 1994. Pages 183–244 *in* R. B. Hunter, compiler. *Status of the flora and fauna on the Nevada Test Site, 1994.* Report No. DOE/NV/11432–195. U.S. Department of Energy, Nevada Operations Office, Las Vegas, Nevada.

Hunter, R. B. 1995b. Status of perennial plants on the Nevada Test Site, 1994. Pages 245–348 *in* R. B. Hunter, compiler. *Status of the flora and fauna on the Nevada Test Site, 1994.* Report No. DOE/NV/11432–195. U.S. Department of Energy, Nevada Operations Office, Las Vegas, Nevada.

Iverson, R. M. 1980. Processes of accelerated pluvial erosion on desert hill slopes modified by vehicular traffic. *Earth Surface Processes* **5**:369–388.

Iverson, R. M., B. S. Hinckley, R. H. Webb, and B. Hallet. 1981. Physical effects of vehicular disturbances on arid landscapes. *Science* **212**:915–917.

Johnson, H. B., F. C. Vasek, and Y. Yonkers. 1975. Productivity, diversity, and stability relationships in Mojave Desert roadside vegetation. *Bulletin of Torrey Botany Club* **102**: 106–115.

Knapp, P. A. 1992a. Secondary plant succession and vegetation recovery in two western Great Basin Desert ghost towns. *Biological Conservation* **60**:81–89.

Knapp, P. A. 1992b. Soil loosening processes following the abandonment of two arid western Nevada townsites. *Great Basin Naturalist* **52**:149–154.

Lathrop, E. W. 1983. Recovery of perennial vegetation in military maneuver areas. Pages 265–277 *in* R. H. Webb and H. G. Wilshire, editors. *Environmental effects of off-road vehicles.* Springer-Verlag, New York, New York.

Lathrop, E. W., and E. F. Archbold. 1980a. Plant response to Los Angeles aqueduct construction in the Mojave Desert. *Environmental Management* **4**:137–148.

Lathrop, E. W., and E. F. Archbold. 1980b. Plant response to utility right of way construction in the Mojave Desert. *Environmental Management* **4**:215–226.

Lathrop, E. W., and P. G. Rowlands. 1983. Plant ecology in deserts: an overview. Pages 113–152 *in* R. H. Webb and H. G. Wilshire, editors. *Environmental effects of off-road vehicles.* Springer-Verlag, New York, New York.

Lei, S. A. 1999. Postfire woody vegetation recovery and soil properties in blackbrush (*Coleogyne ramosissima* Torr.) shrubland ecotones. *Journal of the Arizona-Nevada Academy of Sciences* **32**:105–115.

Lei, S. A. 2001. Postfire seed bank and soil conditions in a blackbrush (*Coleogyne ramosissima* Torr.) shrubland. *Bulletin of the Southern California Academy of Sciences* **100**: 100–108.

Lei, S. A. 2004. Soil compaction from human trampling, biking, and off-road motor vehicle activity in a blackbrush (*Coleogyne ramosissima*) shrubland. *Western North American Naturalist* **64**:125–130.

Leys, J. F., and D. J. Eldridge. 1998. Influence of cryptogamic crust disturbance to wind erosion on sand and loam rangeland soils. *Earth Surface Processes and Landforms* **23**: 963–974.

Lovich, J. E., and D. Bainbridge. 1999. Anthropogenic degradation of the southern California desert ecosystem and prospects for natural recovery and restoration. *Environmental Management* **24**:309–326.

Matchett, J. R., L. Gass, M. L. Brooks, A. M. Mathie, R. D. Vitales, M. W. Campagna, D. M. Miller, and J. F. Weigand. 2004. *Spatial and temporal patterns of off-highway vehicle use at the Dove Springs OHV Open Area, California.* U.S. Geological Survey, Sacramento, California.

McCarthy, L. E. 1996. Impact of military maneuvers on Mojave Desert surfaces: a multiscale analysis. Ph.D. dissertation, University of Arizona, Tucson, Arizona.

McFadden, L. D., E. V. McDonald, S. G. Wells, K. Anderson, J. Quade, and S. L. Forman. 1998. The vesicular layer and carbonate collars of desert soils and pavements: formation, age and relation to climate change. *Geomorphology* **24**:101–145.

McKenna-Neuman, C., C. D. Maxwell, and J. W. Boulton. 1996. Wind transport of sand surfaces crusted with photoautotrophic microorganisms. *Catena* **27**:229–247.

Minnich, R. A. 1995. Wildland fire and early postfire succession in Joshua Tree woodland and blackbrush scrub of the Mojave Desert of California. *San Bernardino County Museum Association Quarterly* **43**:99–105.

Neff, J. C., R. Reynolds, J. Belnap, and P. Lamothe. 2005. Multi-decadal impacts of grazing on soil physical and biogeochemical properties in southeast Utah. *Ecological Applications* **15**:87–95.

Nichols, K. K., and P. R. Bierman. 2001. Fifty-four years of ephemeral channel response to two years of intense World War II military activity, Camp Iron Mountain, Mojave Desert, California. *Reviews in Engineering Geology* **XIV**:123–136.

Penning de Vries, F., and M. A. Djiteye, editors. 1982. *La productivite des pasturages Saheliens.* Centre for Agricultural Publishing and Documentation, Backhuys Publishers, Wageningen, the Netherlands.

Prose, D. V. 1985. Persisting effects of armored military maneuvers on some soils of the Mojave Desert. *Environmental Geology and Water Science* **7**:163–170.

Prose, D. V., and S. K. Metzger. 1985. *Recovery of soils and vegetation in World War II military base camps, Mojave Desert.* U.S. Geological Survey Open-file Report No. 85–234. Menlo Park, California.

Prose, D. V., S. K. Metzger, and H. G. Wilshire. 1987. Effects of substrate disturbance on secondary plant succession; Mojave Desert, California. *Journal of Applied Ecology* **24**:305–313.

Prose, D. V., and H. G. Wilshire. 2000. *The lasting effects of tank maneuvers on desert shrubs and intershrub flora.* U.S. Geological Survey Open-File Report No. OF 00–512. Menlo Park, California.

Quade, J. 2001. Desert pavements and associated rock varnish in the Mojave Desert: how old can they be? *Geology* **29**:855–858.

Reheis, M. C., J. C. Goodmacher, J. W. Harden, L. D. McFadden, T. K. Rockwell, R. R. Shroba, J. M. Sowers, and E. M. Taylor. 1995. Quaternary soils and dust deposition in southern Nevada and California. *Geological Society of America Bulletin* **107**:1003–1022.

Rundel, P. W., and A. C. Gibson. 1996. *Ecological communities and processes in a Mojave Desert ecosystem: Rock Valley, Nevada.* Cambridge University Press, New York, New York.

Saethre, M. B. 1995. Small mammal populations on the Nevada Test Site, 1994. Pages 75–148 *in* R. B. Hunter, compiler. *Status of the flora and fauna on the Nevada Test Site, 1994.* Report No. DOE/NV/11432–195. U.S. Department of Energy, Nevada Operations Office, Las Vegas, Nevada.

Schlesinger, W. H., and C. S. Jones. 1984. The comparative importance of overland runoff and mean annual rainfall to shrub communities of the Mojave Desert. *Botany Gazette* **145**:116–124.

Scholl, D. G. 1989. Soil compaction from cattle trampling on a semiarid watershed in northwest New Mexico. *New Mexico Journal of Science* **29**:105–112.

Shields, L. M., P. V. Wells, and W. H. Rickard. 1963. Vegetational recovery on atomic target areas in Nevada. *Ecology* **44**:697–705.

Steiger, J. W., and R. H. Webb. 2000. *Recovery of perennial vegetation in military target sites in the eastern Mojave Desert, Arizona.* U.S. Geological Survey Open-File Report No. OF 00–355. Tucson, Arizona.

Turner, F. B., G. A. Hoddenbach, P. A. Medica, and J. R. Lannom. 1970. The demography of the lizard, *Uta stansburiana* Baird and Girard, in southern Nevada. *Journal of Animal Ecology* **39**:505–519.

Turner, F. B., P. A. Medica, K. W. Bridges, and R. I. Jennrich. 1982. A population model of the lizard *Uta stansburiana* in southern Nevada. *Ecological Monographs* **52**:243–259.

Vasek, F. C. 1980. Early successional stages in Mojave Desert scrub vegetation. *Israel Journal of Botany* **28**:133–142.

Vasek, F. C., H. B. Johnson, and G. D. Brum. 1975b. Effects of power transmission lines on vegetation of the Mojave Desert. *Madroño* **23**:114–130.

Vasek, F. C., H. B. Johnson, and D. H. Eslinger. 1975a. Effects of pipeline construction on creosote bush scrub vegetation of the Mojave Desert. *Madroño* **23**:1–13.

Vollmer, A. T., B. G. Maza, P. A. Medica, F. B. Turner, and S. A. Bamberg. 1976. The impact of off-road vehicles on a desert ecosystem. *Environmental Management* **1**:115–129.

Wallace, A., and E. M. Romney. 1972. *Radioecology and ecophysiology of desert plants at the Nevada Test Site.* Report No. TID-25954. National Technical Information Service, Springfield, Virginia.

Wallace, A., E. M. Romney, and R. B. Hunter. 1980. The challenge of a desert: revegetation of disturbed desert lands. *Great Basin Naturalist Memoirs* **4**:216–225.

Wardle, D. A. 2002. *Communities and ecosystems: linking the aboveground and belowground components.* Monographs in Population Biology 34, Princeton University Press, Princeton, New Jersey.

Webb, R. H. 1982. Off-road motorcycle effects on a desert soil. *Environmental Conservation* **9**:197–208.

Webb, R. H. 1983. Compaction of desert soils by off-road vehicles. Pages 51–79 *in* R. H. Webb and H. G. Wilshire, editors. *Environmental effects of off-road vehicles.* Springer-Verlag, New York, New York.

Webb, R. H. 2002. Recovery of severely compacted soils in the Mojave Desert, California, USA. *Arid Lands Research and Management* **16**:291–305.

Webb, R. H., J. W. Steiger, and E. B. Newman. 1988. The effects of disturbance on desert vegetation in Death Valley National Monument, California. U.S. Geological Survey Bulletin No. 1793. U.S. Government Printing Office, Washington, D.C.

Webb, R. H., J. W. Steiger, and H. G. Wilshire. 1986. Recovery of compacted soils in Mojave Desert ghost towns. *Soil Science Society of America Journal* **50**:1341–1344.

Webb, R. H., and K. A. Thomas. 2003. Recoverability of severely disturbed soils and vegetation in the Mojave Desert, California, USA. Pages 283–290 *in* A. S. Alsharhan, W. W. Wood, A. S. Goudie, A. Fowler, and E. M. Abdellatif, editors. *Desertification in the third millennium.* Swets and Zeitlinger (Balkema) Publishers, Dordrecht, the Netherlands.

Webb, R. H., and H. G. Wilshire. 1980. Recovery of soils and vegetation in a Mojave Desert ghost town, Nevada, USA. *Journal of Arid Environments* **3**:291–303.

Webb, R. H., H. G. Wilshire, and M. A. Henry. 1983. Natural recovery of soils and vegetation following human disturbance. Pages 279–302 *in* R. H. Webb and H. G. Wilshire, editors. *Environmental effects of off-road vehicles.* Springer-Verlag, New York, New York.

Wells, S. G., J. C. Dohrenwend, L. D. McFadden, B. D. Turrin, and K. D. Mahrer. 1985. Late Cenozoic landscape evolution on lava flow surfaces of the Cima volcanic field, Mojave Desert, California. *Geological Society of America Bulletin* **96**:1418–1529.

Wells, S. G., L. D. McFadden, J. Poths, and C. T. Olinger. 1995. Cosmogenic ^{3}He surface-exposure dating of stone pavements: implications for landscape evolution in deserts. *Geology* **23**:613–617.

Whitford, W. G. 2002. *Ecology of desert systems.* Academic Press, London, UK.

Wilshire, H. G. 1983. The impact of vehicles on desert soil stabilizers. Pages 31–50 *in* R. H.

Webb and H. G. Wilshire, editors. *Environmental effects of off-road vehicles.* Springer-Verlag, New York, New York.

Wilshire, H. G., and J. K. Nakata. 1976. Off-road vehicle effects on California's Mojave Desert. *California Geology* **29**:123–132.

Wilshire, H. G., S. Shipley, and J. K. Nakata. 1978. Impacts of off-road vehicles on vegetation. Pages 131–139 *in* K. Sabol, editor. *Transactions of the forty-third North American Wildlife and Natural Resources Conference, Phoenix Arizona.* Wildlife Management Institute, Washington, D.C.

Wood, M. K., W. H. Blackburn, R. E. Eckert Jr., and F. F. Peterson. 1982. Influence of crusting soil surfaces on emergence and establishment of crested wheatgrass, squirreltail, Thurber needlegrass, and fourwing saltbush. *Journal of Range Management* **35**:282–287.

Woodward, B. 1995. Status of reptiles on the Nevada Test Site, 1994, and summary of work 1987–1994. Pages 1–73 *in* R. B. *Hunter, compiler. Status of the flora and fauna on the Nevada Test Site, 1994.* Report No. DOE/NV/11432–195. U.S. Department of Energy, Nevada Operations Office, Las Vegas, Nevada.

Yaalon, D. H., and D. Kalmar. 1972. Vertical movement in an undisturbed soil: continuous measurement of swelling and shrinkage with a sensitive apparatus. *Geoderma* **8**: 231–240.

SIXTEEN

Active Restoration for the Mojave Desert

JAMES WEIGAND AND JANE RODGERS

Disturbances to soil and vegetation can create lasting and highly visible legacies on the desert landscape (fig. 16.1). Effects of disturbances, such as the installation of energy pipelines or transmission corridors, military activity, agriculture, mining, and off-road vehicle travel and other recreational pursuits, often remain long after the activity has ceased. Tracks of old roads remain clearly visible and attract curious travelers. Abandoned mines become recreational destinations, visits to which sometimes end with bodily injuries and fatalities. Land managers, faced with land management mandates and public pressure, frequently find that natural recovery processes are too slow (Webb et al., Natural Recovery, *this volume*) or are prevented by repeated disturbance. In such cases, they may choose to take physical action to restore natural processes and native habitat.

Funding availability, political priorities, and scientific knowledge are frequently the most important considerations in deciding how and when to implement active restoration efforts. Maintaining or re-creating the character of congressionally designated wilderness areas, recovering abandoned mining areas, erasing signs of unauthorized vehicle trails, and protecting critical species habitat require that land managers spend scarce federal government funds. Political concerns often manifest as conflicting interests that can delay or even prevent restoration projects.

In addition, while the scientific knowledgebase is increasing, managers still have no comprehensive source of information on the methods, techniques, and results of past restoration projects in the Mojave Desert. There are, however, several well-documented, long-term restoration projects that, when analyzed, can provide invaluable information for planning future efforts. In recent years, land managers and scientists have been working together to document and monitor the methods and results of restoration projects, and have taken an innovative experimental approach that will facilitate the success of future projects.

However, to ensure that active restoration is cost-effective and meets the goals of land managers, several issues need addressing. The many challenges to successful restoration, including the unique characteristics of the urban-desert interface, the interconnectivity of riparian areas, and the increasing occurrence of wildfires, must be considered. Increasing public awareness needs to become a priority to facilitate political support, funding, and volunteer efforts. Projects need to be implemented on a bioregional scale and utilize an interdisciplinary approach. Managers and scientists with restoration experience must be retained by agencies for long-term maintenance and monitoring of restoration sites as well as to pass knowledge on to others. In addition, practitioners must continue to explore innovative and adaptive techniques for lowering the cost and enhancing the success of restoration efforts.

PROJECT PLANNING AND SITE ANALYSIS

Active restoration can be implemented several ways, depending on the timeframe, the objectives and scale of the project, and the amount of available funding, tools, and access. Projects in the Mojave Desert have diverse goals and objectives. In some cases, work focuses on restoring predisturbance soil conditions. To date, most desert restoration projects have focused on accelerating revegetation, largely of desert shrub and grass species.

The first step to creating a successful project is careful planning and analysis. When choosing between active and natural restoration, the first question to ask is whether the site is likely to recover on its own within the desired timeframe. Using a recovery and restoration timetable, such as the one produced by Lovich (1999), can be useful (appendix 16.1). Once it is determined that active restoration is preferable, many more issues must be taken into account. In addition to financial and political concerns, planners must consider local public values, possible threats to life or property (e.g., wildfire), the presence of endangered and threatened species, the potential affects on resource availability, and air and water quality. Managers must also ensure that subsequent disturbance can be prevented. It is also preferable to locate a paired, undisturbed site to refer to in determining restoration goals; however, these are often difficult to find nearby (Bainbridge 2000).

During the planning phase careful analyses must be used to collect information for the project plan. With any analysis, several factors must be considered. The physical characteristics of the disturbed site and the surrounding area, including soil type, sediment transport, elevation, exposure, slope, climate, and natural and anthropogenic disturbance histories should be carefully documented. Analyses should also include interactions among desert ecosystem processes and

Fig. 16.1. Aerial photograph of the Second Los Angeles Aqueduct Retrospective Restoration Study Site in Kern County, California, August 2001. After forty years, the corridor, which crosses nearly level terrain, remains starkly apparent from 7300 m altitude. The obvious buffer zone around the utility road, indicated by the solid line, demarcates the zone of restored vegetation, and beyond that is undisturbed vegetation. The sinuous road in the lower right of the photograph, indicated by the dash line, is the route of the first Los Angeles aqueduct, constructed between 1908 and 1913.

their effects on urban areas, wildlands, and the urban-wildland interface. With any type of analysis, it is essential to collaborate with land managers and landowners to create an effective restoration plan.

Landscape Analysis

With landscape-based analysis, the area of concern is defined by public priorities and ecosystem interactions, as well as by site conditions. Landscape analysis considers how restoration projects will affect and be affected by the surrounding landscape and view-shed. This type of analysis can clarify how changes in disturbance regimes have resulted in current landscape conditions, and what techniques and options are available for active restoration. Another important consideration is the aesthetic and financial priorities of the local community. These landscape values are human-imposed, and concern the picturesque quality of a natural landscape, as well as concerns for the value of real estate and the viability of agriculture. These are often major issues where natural desert landscapes lie within walking and driving distance of cities such as Palm Springs (Riverside County) and Ridgecrest (Kern County), where officials have collaborated with the Forest Service and the Bureau of Land Management (BLM) to manage the vistas for these cities.

Watershed Analysis

A second type of analysis is a watershed-based approach in which the area of concern is defined by geomorphology and hydrology. Watershed analyses, which involve biological and physical assessment of a drainage area, are an increasingly common practice in California and provide key guidance documentation for preparing restoration prescriptions (Shilling et al. 2005). The purpose of watershed analysis is to facilitate the coordination of stakeholders and to provide information on the socioeconomic processes within an area, as well as the biological and physical characteristics. Downstream effects and interactions across property lines can make or break restoration efforts. Without the collaborative process of watershed assessment and the resulting shared objectives, even the best-conceived restoration prescriptions can fail.

Cost-Benefit Analysis

A third approach, which should be used in planning every restoration project, is cost-benefit analysis. The lack of resources and public funds means that public land managers are accountable for the cost effectiveness of a restoration project. A cost-benefit analysis of possible prescriptions is a necessary step and facilitates an explicit evaluation of the goals. This type of analysis provides a comparison of estimated costs and benefits side by side on a timeline. Increasingly, labor and travel costs are limiting factors. Training and supervisory costs are also important when working with unskilled (but often enthusiastic) volunteers.

A cost-benefit analysis can be the deciding factor in determining a specific restoration prescription when two or more prescriptions are nearly equal in their environmental benefits. The U.S. Geological Survey (USGS) is currently developing a restoration cost-benefit analysis tool through the Recoverability and Vulnerability of Desert Ecosystems Project (see the Desert Science Database at http://www.dmg.gov/projects.php). This tool will assist managers in evaluating natural versus active recovery, and provide a cost-benefit model for restoration planning.

ACTIVE RESTORATION METHODS

Plants

Salvaging Mature Plants: While desert habitat is invariably lost in the process of urbanization, the land disturbance affords restoration project managers with an

Fig. 16.2. Photographs documenting an abandoned road that was closed at Joshua Tree National Park. *A,* was taken before treatment and shows a devegetated, barren roadway. *B,* was taken immediately after treatment with rock and mulch. The road has been transformed and is well camouflaged from future disturbance (photographs courtesy of the National Park Service).

easily and inexpensively acquired supply of mature vegetation for restoration projects. In road construction, for example, the Nevada Department of Transportation promotes, as a standard practice, the donation of native plant material along with the native topsoil (Nevada Department of Transportation 2005).

Salvaging plant material can significantly reduce revegetation costs and has the added bonus of immediate revegetation on a site. The cost of plant salvage depends on plant size, location, and time of year. Depending upon vertical height and crown size, large *Yucca brevifolia* (Joshua tree) and *Yucca schidigera* (Mojave yucca) can cost as much as $425 per tree to transplant. Large specimens can be temporarily maintained in earthen berms or wooden tree boxes (fig. 16.2), which requires supplemental irrigation. Salvage efforts may be most appropriate in high-use areas that require immediate landscaping, such as visitor centers, campgrounds, road closures, and sites necessary for wildlife habitat.

While most perennial species can be salvaged, some are more cost effective than others. Shallow-rooted species, such as yuccas, cacti, and perennial bunchgrasses are easy to harvest (Schrenk 2002). As part of large-scale road realignment, Joshua Tree National Park (JTNP) in Riverside County, California, salvaged plants extensively between 1999 and 2004. The survival rates were generally high (table 16.1). However, small unprotected plants were subject to increased grazing pressure from rodents and insects—the survival rate for *Pleuraphis rigida* (galleta grass) 18 months after transplanting was only 37% due to herbivory. *Yucca brevifolia* and *Y. schidigera* transplanted during the project had survival rates of 83% and 95%, respectively, after three years (Joshua Tree National Park *unpublished data*). *Yucca brevifolia* produced new roots after four months in wooden boxes. The mortality of the salvaged plants was due to overwatering, soil erosion, and wind. Experimental salvage of *Juniperus californica* (California juniper) revealed that individuals boxed with root-balls intact and wrapped in burlap had 100% survival, while those boxed with bare roots had complete mortality (Cox and Rodgers 2003; Joshua Tree National Park *unpublished data*).

Grass salvage at JTNP has had mixed results—*Pleuraphis, Achnatherum speciosum* (desert needlegrass), and *Achnatherum hymenoides* (ricegrass) salvaged and stored in a nursery had a 50% survival rate when outplanted. Mortality may have resulted from root rot in heavy compacted soils with poor drainage.

Seeding

Since prehistory, people have collected seeds in the Mojave Desert. Managers can take advantage of bountiful seed years; high, well-timed precipitation often results in bumper crops of viable seed from *Larrea tridentata* (creosote bush), *Coleogyne ramosissima* (blackbrush), and other dominant perennial species. One advantage of using seeds is that agencies can amass native seed stocks for producing container plants or for direct application at restoration sites.

Table 16.1 Survivorship by species for Joshua Tree National Park Project No. 173 (Year 3)

Species	Total planted	Percent survival 3 years of post-planting
Acacia greggii	148	75.0%
Atriplex canescens	5	0.0%
Chilopsis linearis	4	100.0%
Coleogyne ramosissima	94	28.7%
Echinocereus engelmannii	7	57.1%
Echinocereus spp.	3	66.0%
Echinocereus triglochidiatus	75	77.3%
Encelia farinosa	33	81.8%
Ephedra californica	4	100.0%
Ephedra nevadensis	60	56.7%
Eriogonum fasciculatum	23	60.9%
Grayia spinosa	98	51.0%
Hymenoclea salsola	5	100.0%
Juniperus californica	1	0.0%
Larrea tridentata	1	0.0%
Lycium andersonii	11	9.1%
Lycium cooperiu	10	10.0%
Opuntia basilaris	23	82.6%
Opuntia echinocarpa	332	48.8%
Opuntia ramosissima	202	63.9%
Opuntia stanlyi	27	70.4%
Prunus fasciculata	34	79.4%
Salazaria mexicana	9	77.8%
Tetradymia spinosa	3	0.0%
Yucca brevifolia	782	54.1%
Yucca schidigera	478	83.5%
Total survival	2,472	61.8%

Direct seeding has had mixed results in desert environments, and natural precipitation is often the deciding factor between success and failure. Even with supplemental water, seeding trials of *Ambrosia dumosa* (white bursage), *Encelia* ssp. (brittlebush), and *Brickellia incana* (wooly brickellbush) at the Fort Irwin National Training Center (NTC) showed no signs of germination (Mason 2001). Projects carried out by the Soil Ecology and Restoration Group (SERG) at San Diego University had similar results, with low germination attributed to seeds lost to wind and herbivory, and unpredictably low levels of precipitation. Field trials using seeds pre-treated by rinsing and soaking in water and thiourea greatly improved germination rates of *Larrea* and *Ambrosia* at the NTC (Ostler et al. 2002). However, many treatments, particularly optimal germination temperatures, are species-specific and thus make successful seeding more difficult and costly.

Tailoring the microenvironment of the planted seeds may also affect germination rates. Anderson and Ostler (2002) experimented with two types of mulches (straw and gravel) and found that the germination densities in the mulches did not differ after a wet year. In the third year, however, straw mulch provided denser shrub survivorship than gravel mulch, and both had significantly higher survivorship than in unmulched treatment sites. In a second set of trails at the NTC in 1999, plastic mulching yielded the highest immediate germination response.

Trials often have poor results because seeding is timed to adhere to funding schedules rather than to coincide with precipitation. The ideal revegetation scenario would be to stockpile native seed and plant propagules during dry years until years of reasonably predictable high precipitation, such as during El Niño events (Bainbridge 2003). On the other hand, some seed does not store well for long periods, and seed storage guides should be consulted prior to undertaking such a program.

Container Planting

Many land managers use container plants for revegetation projects, and several nurseries have been established to provide container plants to agencies such as the National Park Service (NPS), Department of Defense, California State Parks, and the BLM. While container planting can provide immediate, genetically appropriate native vegetation to a disturbed site, these plants often require physical protection and irrigation, which can be problematic in remote sites without access to water.

Large-scale container planting projects have occurred mostly on military and NPS lands. The NTC has used thousands of container plants over the past ten years with mixed results (Soil Ecology and Restoration Group 2005). At JTNP, container plant projects include revegetation of abandoned mine lands, borrow pits, road edges, and closed roads. Over 1,500 tall pots of 23 species were planted between 1988 and 1992 as part of the Cottonwood Road Federal Highways project. Assessment of 1,000 of the plants in 1995 showed a survival rate for each species ranging from 70%–100% (Joshua Tree National Park *unpublished data*). Survival rates appear to be the result of supplemental watering during drought years (1988–1990) and high precipitation during and following outplanting years (1991 = 205 mm, 1992 = 346 mm, 1993 = 309 mm). In Pinto Basin at JTNP, *Larrea* survival rates were 80% one year postplanting ($n = 133$); rainfall between planting and monitoring was 81 mm. Many nurseries maintain unpublished files on propagation of desert species from transplantings, cuttings, and seedings (Joshua Tree National Park *unpublished data*).

Use of nursery stock is not always appropriate at backcountry sites. Closed roads and hill climbs in the Rock House area of JTNP were revegetated in October 1999 and February 2001, and of the 261 shrubs planted, only 25 were alive in May 2003. The rough roads to this isolated site created chronic vehicle problems, and supplemental watering was only possible 1–3 times per year (Joshua Tree National Park *unpublished data*).

Fences and Other Physical Barriers

Most restoration sites in sparsely vegetated desert environments require physical barriers to facilitate restoration success. Barriers can be exclusionary and prevent continued disturbance at a site from grazing or vehicle trespass (Brooks

1995, 1999). Temporary fencing can be developed to exclude specific species, such as rabbits, rodents, lizards, and deer, which might eat or damage vegetation. Grantz and Vaughn (2001) recommend barriers to modify sediment transport, such as wind fences, silt fences, and furrows. In Little Morongo Canyon at the southern boundary of the Mojave Desert, earthen berms were installed to reduce surface runoff during high rainfall events. These berms proved ineffective, however, during intense rainfall in 2004–2005 and may have exacerbated flooding and erosion.

Barriers can be constructed from a variety of materials. In addition to earthen berms, materials can include weed-free straw bales, rock, gabions, and vegetation. Rock walls and dams can slow the rate of water flow and erosion on steep slopes, but may also have detrimental effects because of the reduced overland flow to surfaces downslope. The BLM Ridgecrest Field Office frequently uses check dams on closed roads and slopes to reduce erosion. Gabions, wicker or wire mesh baskets filled with mud balls or rocks, can facilitate revegetation while blending in with the surrounding terrain. Straw bales, in addition to preventing erosion, can restrict motorized vehicles from unauthorized roads. Vegetation is very effective as a barrier for reducing dust in wind-prone areas. Most information about live vegetation as a barrier comes from work done in the Antelope Valley, Los Angeles County, where Grantz and Vaughn (2001) found that *Atriplex canescens* (fourwing saltbush) appears to be the best choice for rapid live fencing (Grantz 1998; Grantz et al. 1998; Grantz and Vaughn 2001). *Proposopis glandulosa* (mesquite), *Larrea*, and *Isomeris arborea* (bladderpod) windbreaks planted and watered monthly at the NTC and Marine Corps Air Ground Combat Center in Twentynine Palms, California, between 1997 and 1998 had high survival rates eight months after planting. No data was available on the long-term effectiveness of these projects.

Any time that barriers and fencing are installed as a permanent feature, efforts should be made to reduce its visual impact.

Ground surface treatments

Soil Manipulation: While soil can be used to create barriers, it can also be manipulated in other ways to facilitate the success of restoration projects. In the Mojave Desert, soil has been re-contoured to concentrate the flow of water to planted areas, to reduce erosion and sediment transport by flattening and terracing hill slopes, and to catch moisture and seeds in surface depressions called pits. In addition, restoration projects have used redistributed topsoil and decompaction methods to facilitate the natural regeneration of shrubs, grasses, and forbs.

Soil Salvage: Many disturbed sites have lost important soil surface structure, nutrient content, and, in some cases, topography. All of these factors can exacerbate erosion. Because organic material is very limited in arid climates, soil salvage can be very useful in accelerating recovery. Topsoil can be salvaged from construction sites, in tandem with vegetation, which is very cost effective. If project goals include the conservation of topsoil, this can be done by stripping the top 100–150 mm of soil. Soil salvage conserves soil organic matter and surface bulk density. When applying salvaged topsoil, the natural contour should be restored to reduce future soil erosion and lessen the contrast with adjacent areas.

Decompaction: Severely disturbed desert soils may be highly compacted, which reduces water infiltration and nutrient cycling (Webb et al. *this volume*). Ripping (deep plowing) and harrowing (surface plowing), are methods used in the Mojave Desert to alleviate compaction. Decompaction on slopes has had varying success in desert restoration projects (Montalvo et al. 2002; Scoles and DeFalco 2003). Although ripping may improve soil bulk density at the surface, it can also break up a caliche layer, unnaturally desiccate the soil, and make shrub regeneration nearly impossible. Decompaction can also lead to increased erosion. Soils with high silt content, for example, are easily eroded by winds. Hydromulch, or matting, can help stabilize the soil. As with any soil-disturbing activity, invasive plant species may become established; monitoring and implementation of an integrated weed management program may be necessary.

Mulching: Between 2002 and 2005, the BLM completed revisions to the California Desert Conservation Area Plan (Bureau of Land Management 1999). The amendments determined which off-highway vehicle (OHV) trails will remain open as part of a permanent route network—thousands of miles of OHV routes are slated for closure. The cost of container stock and work crews to revegetate these routes is beyond the institutional capacity and financial means of the BLM. Because of this, managers must optimize natural processes and target restoration funds efficiently across large areas. Mulching is a common technique and currently the primary method used to facilitate natural revegetation and halt travel on closed routes. Mulching roughens soil surfaces and accelerates the regeneration of native shrub and bunchgrass species. Mulching can improve water infiltration and retention, reduce wind velocity at the soil surface, reduce soil temperature, lessen runoff, protect the soil from raindrop splash, and reduce chemical crusting of the soil (Munshower 1994).

Mulching can also minimize the visual differences between the closed trail and the adjacent native vegetation, and thus discourage vehicle riders from veering off of clearly marked, designated OHV routes (fig. 16.3). Beyond the line of sight from the designated OHV route, restoration proceeds at natural rates without additional restoration efforts.

Horizontal Mulching: Horizontal mulching makes use of vegetation to cover surfaces of OHV trail beds so that formerly bare surfaces no longer attract the notice of passersby on vehicles. Shrub and tree branches can be laid along slope

Fig. 16.3. Photographs documenting restoration along a trail in the Jawbone-Butterbredt Area of Critical Environmental Concern in Kern County, California. *A,* was taken in the autumn of 2005 before vertical mulching practices. *B,* was taken in the autumn of 2005 after vertical mulching practices (photographs courtesy of Ronald Gartland, BLM California Desert District archives).

contours and over disturbed soil. This type of mulching has been shown to increase soil fauna, increase the rates of biological processes, and reduce moisture loss due to evaporation, in addition to the advantages listed above (Ludwig and Tongway 1996; Tongway and Ludwig 1996).

Vertical Mulching: Vertical mulching consists of straw, sticks, or brush, tied in bunches and buried upright in the soil. Materials that have been used include broom corn, straw, reeds, and brush from existing vegetation. Material selection depends on the severity of disturbance at the site, other restoration methods being utilized, the project's goals in terms of aesthetics, and the availability and cost of materials (Bainbridge 1994). Vertical mulching slows runoff and increases infiltration by allowing water to seep into the soil around stems of the buried material. Fairbourn (1975) found that vertical mulch can increase soil moisture more than 20%. Vertical mulching also provides shade and cover for seedlings and below-ground organic matter as it decomposes (Bainbridge 1994).

Bureau of Land Management staff have buried bunches of rice straw halfway into road beds to imitate the appearance of desert bunch grass. Because of the improvisational and creative nature of vertical mulching, this method needs to be documented more fully and communicated more widely. Ecologists with the BLM and USGS are presently studying the long-term effectiveness of vertical mulching at both deterring unauthorized OHV travel and promoting natural revegetation. Joshua Tree National Park has summarized their vertical mulch and closed road techniques in several unpublished protocol handbooks.

Rock Mulching: Rocks can also be used to obscure disturbed areas and restore ecological function. Rock mulch promotes water retention from dew and rainfall, and creates a rough surface with pockets that capture seeds and can facilitate germination. Like other mulches, rocks reduce soil moisture evaporation and wind and water erosion (Walker and Powell 2001). Rocks that resemble the native material can be used to camouflage disturbed sites by restoring the texture and color of the ground surface.

Early restoration projects at the BLM Barstow Field Office and JTNP used variations on these mulching techniques, and involved experimenting with rock mulches ranging in size from coarse gravel to large boulders. Restoration crews have improved their techniques by using brooms to smooth the surfaces between the trail beds and the undisturbed surfaces and by partially burying larger rocks. At Lake Mead National Recreation Area (NRA), rock stains and soil colorings, such as Permeon™, have also been used to obscure tire tracks on desert pavement.

Mulching to Sequester Excessive Nitrogen. A major concern to ecologists in the Mojave Desert is the increasing nitrogen deposition from urban sources (Allen et al. *this volume*). Results from studies in other parts of California show that mulching with a combination of sucrose and straw/wood chips provides carbon and increases the carbon to nitrogen (C:N) ratio (Cione et al. 2002; Corbin and

D'Antonio 2004). One useful effort would be to see whether a shift in the C:N ratio can be used to give native annuals and grasses a competitive advantage as a means to suppress nonnative plants, especially *Bromus* ssp. (brome grasses) and *Erodium cicutarium* (redstem stork's bill) (Brooks 1998).

Continuing support for diverse mulching techniques on BLM lands in the California portion of the Mojave Desert comes from the California Department of Parks and Recreation Off-Highway Motor Vehicle Recreation Division (OHMVRD). California state law has promulgated that at least $7 million of OHMVRD funds be applied to restoration projects, a considerable portion of which is dedicated to the closing and restoration of redundant vehicle routes in the Mojave Desert.

Irrigation

Restoration publications generally highlight restoration work conducted in temperate ecosystems, where annual precipitation is fairly predictable and outplanting and irrigation of plants is easier than in arid ecosystems. Because water can determine the outcome of a restoration project, understanding hydrology is essential for successful restoration projects in the Mojave Desert.

Many restoration projects in the Mojave Desert occur in remote areas where access is limited and materials and equipment sources are distant. Nearby water sources often do not exist, or are restricted to use by wildlife. In these cases, five options are available: (1) use a restoration prescription that does not require water other than natural rainfall, (2) include physical structures, such as berms and rainwater catchments, to increase infiltration and the amount of precipitation that reaches the site, (3) construct an irrigation system to convey water and supply water to the site, (4) use water-retaining gels, and (5) a combination of methods.

Irrigation Systems: Irrigation systems are usually part of a restoration prescription when container-grown plants are used. Irrigation is often the most expensive component of a restoration project. Travel to remote sites incurs fuel, labor, and vehicle maintenance costs. Experience at JTNP shows that transplants may require irrigation for as long as two years.

The Soil Ecology and Restoration Group has experimented with irrigation techniques such as surface watering using berms, pitting, and treeshades, as well as deep pipe drip systems and porous capsule drip systems. They have also studied the effectiveness of systems adapted from other desert cultures, such as wick irrigation, used in India (Bainbridge and Virginia 1991), and clay pot irrigation, used for centuries in China and Pakistan (Bainbridge 2001).

When using an irrigation system, the amount of water and the scheduling of water deliveries need to be efficient and timely. Because labor and truck delivery costs can be prohibitively high, temporary irrigation systems are sometimes necessary.

Water-Absorbing Soil Amendments: Soil amendments can also be used to increase the amount of water available to plants. One type of water-absorbing amendment is a polyacrylamide-based polymer that slowly releases water to the plant at the root zone. These polymers, also referred to as root-watering crystals, planting gels, and water-retention granules, are long-chained organic compounds that molecularly bind water molecules. The water is released as soil bacteria decompose the molecular chains (Fischer 2004). These gels may hold up to as much as 1,000 times their weight in water (Staton 1998).

The water supplied by the gels reduces the physiological shock to transplanted stock, promotes the growth of seedling root systems, and aids plants in withstanding desert droughts. Both fine- and medium-textured gel granules are available. Bare-root seedlings are usually dipped in a mixture of fine granules and water before planting. Previously saturated, medium-textured gels are then mixed with desert soils poured into the planting hole. Subsequent watering or irrigating will replenish the water held by the gels.

Polyacrylamide, however, can potentially contaminate groundwater as it degrades to an acrylamide monomer, which is toxic to humans, animals, and plants (Jha 2004). An alternative nontoxic gel is DriWater®, which is 98% water and 2% cellulose and alum—food-grade ingredients. Generally, when used as a soil amendment, a one-quart gel pack is placed in a vertical tube near the plant root (Fischer 2004). According to the manufacturer, this product lasts up to 90 days, as opposed to the longer-lasting polymers mentioned above. Gel packs, however, are susceptible to animal consumption and removal.

DriWater® has been used at the Lake Mead NRA and the Mojave National Preserve. Joshua Tree National Park conducted a small study ($n = 27$, sample size was limited due to budget constraints) of DriWater® effectiveness in 2004. Gel pack tubes were filled at the time of planting and 3, 6, and 9 months postplanting. After 2 years, 20 of 27 nursery shrubs planted were still alive. Soil erosion caused 3 of the 7 deaths. In a nearby planting, also in 2004, 61 of 75 shrubs transplanted with gel pack watering tubes survived 2 years postplanting, and 35 of 35 shrubs died when just receiving surface water. Plants without gel packs received supplemental water once every six weeks in the summer of 2004, and only once in the summer of 2005. Gel pack plants received supplemental water once in the summer of 2005 (Joshua Tree National Park *unpublished data*).

THE HISTORY OF MOJAVE DESERT RESTORATION

Beginning in the mid-1800s the human population in the Mojave Desert began to grow as explorers, immigrants, and pioneers moved west (Hughson *this volume*). For decades, the residents of the newly created agricultural and mining communities held no sense of long-term stewardship. When key resources became depleted, people simply moved on, leaving abandoned towns and mining sites—areas which now provide invaluable information about natural recovery processes

(Wells 1961; Webb and Wilshire 1980; Webb and Newman 1982; Webb et al. 1988). A general perception of the Mojave Desert as desolate and worthless continued well into twentieth-century and accounts, in part for the creation of military training bases and hazardous waste dumps (Reith and Thomson 1993).

In the past 25 years the human population has exploded (Lovich and Bainbridge 1999; Hughson *this volume*). Only recently has the value of preserving the natural state of the Mojave Desert become a priority. Active restoration is a relatively recent concept in the Mojave Desert; threads of knowledge have only emerged in the last 50 years, and no long-term, finely tuned plan for sustained use or active restoration has been formulated.

Los Angeles Aqueduct Restoration, 1968–1972

The first carefully documented active restoration project in the Mojave Desert followed construction by the Los Angeles Department of Power and Water of a second aqueduct barrel to transport water from Owens Lake and Oiyee Reservoir in southern Inyo County, California, across BLM land to Los Angeles. During construction, bulldozers bladed a 40–45 m right-of-way on both sides of the aqueduct access road, which left areas scraped of topsoil, occasional hills of piled topsoil, and strips of soil studded with cement rubble at the edges of the bladed areas. Aerial photographs 35 years after construction still reveal a stark contrast in soil surface color between the undisturbed and disturbed areas (fig. 16.1).

To mitigate the damage, several revegetation experiments were conducted using seeds, transplanting, soil ripping, and irrigation. The species selected for revegetation included *Larrea* and *Ambrosia,* which dominated the area, and other native shrubs, including *Ericameria nauseosa* (rubber rabbitbrush), *Hymenoclea salsola* (cheesebush), *Lepidospartum squamatum* (California scalebroom), *Grayia spinosa* (spiny hopsage), *Krascheninnikovia lanata* (winterfat), *Atriplex polycarpa* (cattle saltbush) and *A. canescens.* The success of the experiments varied with species, and was generally low (Lovich and Bainbridge 1999). The overall conclusion of these experiments, according to Kay and Graves (1983), is that revegetation efforts should focus on the visually dominant species—*Larrea,* in this case.

A series of 24 Mojave Revegetation Notes were published between 1970 and 1972, and again after return visits in 1979 and 1988, which describes and evaluates the restoration methods and species selection. Sites with topsoil removed fared poorly or failed altogether in the first decade, a time of severe drought in California. However, the second decade prompted rapid natural revegetation of shrubs on scraped sites (Kay 1988). The soil ripping created linear strips that appeared to favor nonnative annual *Schismus* grasses. Shrubs on hills of piled topsoil, however, showed notable long-term success. Complex vegetation islands consisted of unusual combinations of pioneer and late-seral species from the original species mix.

Revegetation at Joshua Tree National Park

Restoration efforts at JTNP first began in the 1970s, when park rangers began removing nonnative *Tamarix ramosissima* (salt cedar) from remote springs and oases. In 1986, JTNP staff founded the Center for Arid Lands Restoration (CALR), a plant nursery and information hub for restoration technology. Initially, the CALR served a large-scale realignment project along 7 miles of road from Interstate 10 to the JTNP Cottonwood Visitor Center. Funded by a small grant from the Federal Highways Program, the CALR became the first facility in the Mojave Desert region to grow native Mojave Desert species for restoration projects. Since its establishment, the CALR has supplied hard-to-get container stock to the San Bernardino National Forest, the BLM's California Desert District field offices, California State Parks, Marine Corps Air Ground Combat Center, Fort Irwin National Training Center (NTC), local nonprofit organizations, as well as to Death Valley National Park and the Mojave National Preserve. Databases maintained by JTNP on propagation methods and requirements for 90 perennial species are available through CALR to restoration practitioners (Joshua Tree National Park *unpublished data*).

In the late 1990s, park managers tried direct seeding of abandoned borrow pits with poor results. Insufficient seed application rates, combined with a lack of reliable moisture, rodent predation, and continued slope erosion were the likely culprits. As often happens with limited funds and high staff turnover, the outcomes of these trials were not studied further, and this could still be a viable method at select sites.

Between 1989 and 1999, most JTNP revegetation and mine reclamation projects used container stock of various sizes and species. Careful monitoring and documentation of germination and outplanting has made it possible for JTNP to share successful methods for seed collection, propagation, planting, and site maintenance. In addition, invention of the "tall pot" has improved long-term shrub survival in the harsh conditions of the Mojave Desert. Designed to focus growth on roots rather than shoots, the original tall pot was made of Schedule 35 PVC pipe, 152 mm in diameter, cut to 760 mm in height, and enclosing 13.6 L by volume. The tall pot and its successors (in dimensions appropriate to specific shrub species) have enabled the nursery to produce desert shrubs with extensive, healthy root systems. Emphasis on the development of deep roots has consistently proved much more effective for plant survival. Species such as *Hymenoclea* and *Ambrosia* can become very leggy and fleshy in the nursery, and they are highly subject to herbivory by lizards, ants, and rodents; all outplants are protected with chicken wire for up to 3 years postplanting to prevent herbivory.

Past projects in the lower, drier areas of the park relied heavily on the use of *Larrea* because of its ability to survive with minimal watering and extremely limited natural precipitation. Other early seral species successfully used in harsh conditions included *Hymenoclea, Ambrosia,* and *Isomeris arborea.*

In the late 1990s, JTNP staff inventoried and assessed closed roads and abandoned mine lands parkwide, particularly along park boundaries and points of access. Spatial and tabular databases allowed managers to prioritize restoration practices and initiate them as funds became available. After several years of applying various methods, vertical mulching, pitting, and recontouring proved to be the most cost-effective. Ultimately, 167 miles of road were closed to vehicle traffic, and 14 sites were rehabilitated 200 feet or less from the intersection with an open road (Joshua Tree National Park *unpublished data*). Refinement of methods continued as JTNP began extensive road realignment and parking lot expansion within Joshua tree woodland communities. Concerns over nonnative species invasion, off-road vehicle activity along road edges, and the loss of numerous large *Yucca brevifolia* and *Y. schidigera* created new restoration opportunities. Operations shifted from the exclusive use of container stock to salvaging live plants wherever possible. Plants unsuitable for live salvage were harvested for use as mulch.

In addition to these projects, JTNP has collected over 15 years of information on germination and survivorship of planted species. Germination information is also housed at the CALR; a good deal of this information has been added to the latest revision of the U.S. Department of Agriculture's Woody Plant Seed Manual at http://www.nsl.fs.fed.us/wpsm/.

The Soil Ecology and Restoration Group at San Diego State University

The Soil Ecology and Restoration Group (SERG) at San Diego State University began conducting research in the Mojave Desert in 1991, and has been at the forefront of restoration at challenging Mojave Desert sites, in particular by undertaking novel applications in restoring OHV hill climb areas at Red Rock Canyon State Park in Kern County, and in revegetating upland sites heavily impacted from military maneuvers at the NTC. In addition to innovative irrigation systems adapted from ancient desert cultures, SERG has also developed techniques for efficiently channeling and capturing rainfall to supply those systems (Edwards et al. 2000). In addition, they emphasize the importance of summarizing past land use and disturbance as part of developing successful restoration prescriptions.

Unlike other restoration groups, SERG has emphasized improvements to soil fertility through management of microbial ecology and soil-plant processes. The group combines information about specific sites with academic research and produces valuable knowledge on below-ground ecological processes. Graduate students have provided new information on all aspects of soil, with special emphasis in two areas: (1) the effects of nodulation and soil productivity on *Prosopis glandulosa* var. *torreyana* (honey mesquite) (Darby 1993; Kay 1989; Thomas 1995) and

Larrea (Sorensen 1993); and (2) the effects of metals, such as selenium, on nitrogen fixation in *Astragalus crotalariae* (Salton milkvetch) (Kramer 1989).

The SERG faculty, staff, and students have also created a useful online extension service that introduces and illustrates restoration technologies. A series of SERG Restoration Bulletins outlining desert restoration techniques is available online at http://www.sci.sdsu.edu/SERG/techniques.html. By carefully documenting restoration projects and making the technologies and results available to the public, SERG scientists provide a valuable and rare source of information to both interested individuals and public land managers.

UNIQUE CHALLENGES IN ACTIVE RESTORATION IN THE MOJAVE DESERT

The Urban-Desert Interface

The human imprint on the Mojave Desert has increased dramatically in the past 25 years and shows no signs of abating soon (Lovich and Bainbridge 1999). The human population in the cities of San Bernardino County, California, quadrupled between 1980 and 2005 (State of California 2006), from 81,749 to 326,020. Similarly, over the same period, Clark County, Nevada, grew in population from 463,087 to 1,796,380 (Hardcastle *undated*).

The urban-desert interface, where the wild landscape and the anthropogenic landscape meet, is particularly vulnerable to repeated disturbance. Public lands nearest urban areas are subject to the greatest anthropogenic impacts and have the greatest need for restoration. Due to continuing disturbance, restoration at a given site may have to be repeated. Maintaining the pristine appearance of the landscape within scenic viewsheds is important because it contributes to the value of real estate and to a sense of community identity and land stewardship.

An issue of increasing concern, which is affected by restoration at the urban-desert interface, is human health and well-being. Vegetation cover and soil stability are important for reducing dust emissions and conserving air quality worldwide; disturbance of soil and vegetation cover generates airborne dust. Numerous mineral, fungal, and pollen allergens have adverse effects around the world. Illnesses associated with desert dust include silicosis (Hirsch et al. 1974), bronchial asthma (Ezeamuzie et al. 2000), and allergic rhinitis (Dowaisan et al. 2000; Behbehani et al. 2004).

No published research studies exist on the effects of desert dust on the health of Mojave Desert communities; most documentation of the affects to human health from desert dust comes from China, Israel, and Kuwait. Studies in Kuwait (Al-Dowaisan et al. 2004) of desert fungal spores and plant pollen point to desert plants in the Chenopodiaceae family and mesquites (genus *Prosopis*) as triggers for widespread asthma in Kuwaiti communities. Chenopodiaceae species are frequently a major, if not dominant, component of shrub colonization of disturbed sites at the urban-desert interface in the Mojave Desert. Recent research from Spain (Colas et al. 2005; Garde et al. 2005) and Kuwait (Al-Dowaisan et al. 2004) also underscores the allergenic sensitivity of residents in semiarid communities to the pollen of *Salsola* ssp. (Russian thistles or tumbleweeds), invasive exotic species in the Chenopodiaceae family, also commonly found on disturbed sites in the Mojave Desert.

The role of desert restoration at the urban-desert interface has not received attention from public land managers, although they are aware of the impacts at the interface as precursors to expanded impacts from illegal vehicle travel and noxious weed invasions. Joshua Tree National Park, which has the poorest air quality of all national parks in the United States, also has housing development adjacent to some of its boundaries. Designing robust desert vegetation communities along the west and south margins of JTNP to effectively ward off nonnative plant invasions and, at the same time, improve air quality, is now a priority.

Riparian Restoration

Freshwater streams and accompanying riparian forests and woodlands are rare in the Mojave Desert. Their structural complexity provides habitat for multiple vertebrate species, including the federally listed Mohave tui chub (*Gila bicolor mohavensis*), arroyo toad (*Bufo microscaphus californicus*), Southwestern Willow Flycatcher (*Empidonax traillii* ssp. *extimus*), and Least Bell's Vireo (*Vireo bellii pusillus*). The Colorado, Virgin, Mojave, and Amargosa rivers are the major watersheds, having at least partially perennial streamflow. Restoration and maintenance of the native biological diversity along these rivers is critical.

Efficient restoration and habitat management in these desert riverine ecosystems is a challenge for federal land management agencies. The Mojave River watershed, for example, stretches from the San Bernardino National Forest to the Mojave National Preserve. Multiple federal, state, county, and city agencies, as well as interest groups with a stake in Mojave River restoration, must collaborate for meaningful watershed management. There is the possibility that restoration projects will be implemented piecemeal, and thus be subject to greater likelihood of long-term failure.

The complexity of perennial stream restoration can be seen in the history of the Mojave River segment in the BLM Afton Canyon Area of Critical Environmental Concern. Since 1992, the BLM Barstow Field Office has targeted exotic *Tamarix* ssp. populations and restoration of native plant communities comprised of *Populus fremontii* (Fremont cottonwood), *Salix* ssp. (willows), and *Prosopis glandulosa* (honey mesquite) and *Prosopis pubescens* (screwbean mesquite). Innovative and complex methods utilized herbicides, pole plantings, shrub seeding, and prescribed fire (Egan 1999; West 1996) to reduce *Tamarix* cover and capacity to recolonize after initial removal on 121 hectares.

Multiple partners, such as American Forests and Quail Unlimited, contributed funding to support restoration work in Afton Canyon. The conversion of vegetation back to native species was successful until January and February 2005. On January 11, 2005, the U.S. Army Corps of Engineers, Los Angeles District, authorized an all-time maximum release of 470 $m^3 s^{-1}$ of water from the Mojave River Dam on the north side of the San Bernardino Mountains. Downstream, the Silverwood Reservoir also released water. The resulting inundation of the Afton Canyon riparian woodlands negated, in the course of a few days, the progress of more than 10 years of restoration. Almost immediately, the Mojave Desert Resource Conservation District was applying for funding to promote reestablishment of native riparian vegetation.

Because infrequent flash floods direct the disturbance regimes in riparian areas, restoration efforts are sometimes erased. Maintaining streams and riparian vegetation in native conditions requires repeated and rapid management responses if the natural functioning of desert rivers is a goal. Short-term restoration accomplishments between catastrophic floods must be the basis for determining the success of restoration in Mojave Desert riparian areas.

Wildland Fire

Other research addresses the connection between nonnative plants and the accumulation of fine fire fuels (Brooks and Pyke 2001; Dudley *this volume;* Brooks *this volume*). The increasing human presence in the Mojave Desert merits a closer analysis of the fire frequency on federal public lands since 1980, particularly at the urban-desert interface and in higher elevation woodlands within National Forests—in particular the Inyo, San Bernardino, Sequoia, and Toiyabe National Forests. Many wildland fires originate at higher elevations and spread downslope to BLM or national park lands with pinyon-juniper or Joshua tree woodlands.

The structural complexity of higher elevation Mojave Desert woodlands make them important habitats for species such as Gray Vireo (*Vireo vicinior*), which is designated as sensitive by the BLM, and can be found on the northern slopes of the San Bernardino Mountains in California. Fires in 1994 and 1999, originating from the San Bernardino National Forest, spread onto BLM lands to the north. California juniper woodlands burned during the fires and, to date, no juniper regeneration has been noted in the Juniper Flat Area of Critical Environmental Concern or elsewhere in the burned areas. Focused restoration of desert woodlands and the desert/chaparral interface will be key to reconstructing nesting habitat for Gray Vireos and other neotropical migrant species. Collaboration across agency management boundaries in the San Bernardino Mountains and elsewhere in the Mojave Desert is a key part of conservation biology and restoration response to wildland fire.

Wildlife and Rare Plant Habitats

Specific active restoration measures and prescriptions do not exist for wildlife habitat for most species of concern in Mojave Desert ecosystems. Most conservation measures proposed thus far are general and passive in nature. Fencing critical habitat and reliance on natural rates of soil and vegetation recuperation are currently the principal focus for wildlife habitat and rare plant management. Public expectations that wildlife biologists will improve the habitats of threatened and endangered species appears to exceed current agency capabilities. Greater attention to the processes, composition, and structure of Mojave Desert ecosystems as factors in population biology is warranted. Now is the time to synthesize information from research studies on wildlife species such as the desert tortoise (*Gopherus agassizii*) and to develop practices to enhance the supply of annual plant species known to provide high-quality forage (e.g., Jennings 2002; Martin and van Devender 2002; Oftedal et al. 2002). Management of air quality, in particular of dust from OHV trails, may also be critical to maintaining viable populations of rare plants such as *Astragalus jaegerianus* (Lane Mountain milkvetch). Wildlife biologists and botanists have been slow to specify which features of soil, vegetation, and topography (all of which are focal elements in restoration projects), should be enhanced to recreate Mojave Desert habitats. Hypotheses from wildlife biologists and rare plant specialists need to be translated into adaptive management experiments. Recently, USGS wildlife biologists have begun analyzing the distribution of spring annual plant species to guide efforts to augment seed bank supplies of species especially favored by desert tortoises.

A few examples do exist where restoration ecologists have put into practice specific restoration actions to benefit wildlife. At JTNP, restoration ecologists relocated dead and downed yucca species in road construction projects to adjacent undisturbed areas to maintain habitat for the desert night lizard (*Xantusia vigilis*) (Cox and Rodgers 2003). These microhabitat specialists live exclusively among the stems of *Yucca* and *Agave* species (Bezy 1989). Vertical mulch distribution and structure were applied in desert tortoise habitat, which also benefited more common species such as the desert spiny lizard (*Sceloperus magister*) and kangaroo rats (*Dipodomys* ssp.). Shallow pits to collect seeds and water were created to promote higher propagation of the annual forbs favored by desert tortoise. However, there is no data on the effectiveness of this technique (Joshua Tree National Park *unpublished data*).

Plant ecologists and wildlife biologists are now discussing ways that restoration of vegetation benefits plants that are critical for population rebound. Only now are managers and scientists discussing specific restoration practices for improving desert tortoise forage and sources of seeds for rare mammals such as the Mohave ground squirrel (*Spermophilus mohavensis*). Unexpectedly, restoration measures to reduce the proportion of nonnative and less nutritious forb and grass species in desert tortoise critical habitat may require the application of sugar. This treatment in steppe and chaparral ecosystems has been shown to increase the C:N ratio, and thus renders nitrogen deposition from air pollutants unavailable to invasive plant species that otherwise thrive in nitrogen-enriched sites. This shift in the C:N ratio then makes sites more suitable for the native legume species that provide better nutrition for desert tortoises.

Wilderness Areas

The 1994 California Desert Protection Act designated nearly 1.42 million hectares of BLM land in the California desert as wilderness, as well as designating national park status for Death Valley National Park, Joshua Tree National Park, and the Mojave National Preserve. Restoration in federally designated wilderness is often constrained by regulations that exclude the use of motorized equipment. However, many agencies are now recognizing that restoration requires flexibility when it comes to working in wilderness areas. The problem of restoration in wilderness is particularly acute where wilderness areas are close to Mojave Desert cities such as Ridgecrest and Yucca Valley. Over the decade since passage of the Act, the California Off-Highway Motor Vehicle Recreation Commission has funded restoration of OHV trails inside BLM wilderness areas that adjoin other BLM lands with designated OHV routes. Some of the most challenging restoration sites remain unaddressed, however, because the technology for restoring disturbed, steep slopes has not been worked out, given the constraints on the tools permitted in wilderness areas. A restoration analysis of abandoned mine lands in JTNP produced a list of tools that included, at a minimum, small generators, power augers, welding equipment, and helicopters (Joshua Tree National Park *unpublished data*).

MANAGING FOR SUCCESSFUL ACTIVE RESTORATION

Management of restoration programs is constantly growing and improving. The following sections outline some directions for management that could amplify the benefits of ecosystem restoration efforts and encourage creativity and efficiency in acquiring knowledge and applications for desert restoration.

Increase Public Awareness and Environmental Education

Most people who now live in the Mojave Desert have not lived there very long—they do not always have a strong connection with the land. Involving people in restoration projects enhances a sense of place in several ways. Volunteers become informed supporters of restoration projects and knowledgeable restoration practitioners. In urban-desert interface areas, at sites most in need of intensive restoration, volunteers can provide site preparation and irrigation as revegetation projects become established. Without public involvement, public land managers will never have enough labor and financial support to make restoration successful. Community stewardship on the part of volunteers is essential. Ecological restoration can serve as a tool for informing people about Mojave Desert ecosystems and for creating a modern, indigenous knowledge base.

Making restoration methods and technology a core part of curriculums in high schools and community colleges could also inform Mojave Desert residents, and young people in particular, about the function and management of ecosystems. Victor Valley Community College in Victorville, California, for example, offers students a flexible curriculum in natural resource management with an emphasis on Mojave Desert ecosystems. Courses include nursery management, plant propagation, soil rehabilitation, mine reclamation, and restoration project design. Such educational programs build expertise in local communities. Students such as these are also ideal candidates for public agencies, which in the past have had high turnover rates.

Create a Bioregional Plan

A comprehensive, bioregional restoration and conservation plan is needed to create common goals, consolidate resources, and to offset adverse impacts due to development. Questions to be addressed include: Where can people conserve the natural landscape and its resources, restore damaged lands, and maintain the complexity, diversity, and resilience of the Mojave Desert? Will cost-benefit analysis be the sole determinant? What are the explicit and implicit assumptions of economic decisions?

The bioregional plan could make restoration more efficient because it would transcend boundaries and present restoration issues at multiple spatial scales. Efforts on the part of public land managers can reinforce one another and maximize overall benefits to watersheds and landscapes. The inclusion of state parks would make planning more comprehensive and gain greater participation. Private conservation organizations can add expertise, political leverage, and resources to facilitate community awareness and involvement.

Conserve Information and Experience

The public expects land management agencies to be proficient in both the basic science and specific applications of restoration. Given that desert restoration has a short history in the Mojave Desert, managers and restoration practitioners have much to learn. Conserving institutional memory about ecosystem restoration for land management agencies is critical to accumulating and transmitting knowledge. Documenting restoration projects fully, returning to monitor the outcomes, and widely communicating the results are critical to keeping agencies aware and learning. Rapid turnover of agency staff makes conservation of institutional memory haphazard and makes agencies vulnerable to making and repeating mistakes, thus inviting public criticism.

The following example illustrates another reason to conserve information and personnel. Kay (1988) showed that the failure of initial restoration actions at some sites along the second Los Angeles aqueduct resulted from a severe drought in the mid-1970s. However, by 2005, Berry et al. (2005) found that, in time, sites had regenerated, with canopy area comparable to that of nearby undisturbed vegetation. The initial goal of restoring shrub canopy cover was met, but at a rate not in concert with restoration expectations. While canopy cover may be the same, the composition of the plant species remains different. If the monitoring, first established by Kay and continued until 1988, had been extended to 2005, we might have a better understanding of the relationship between natural revegetation and climate. Ongoing monitoring of restoration in the medium- and long-term (> 5 years) is important to understanding restoration planning at multiple time scales. Measures of success or failure may indicate quite different results at different times.

Solutions to the problem of conserving institutional memory are not difficult to implement. First, professional and interdisciplinary training of restoration ecologists is essential. Rather than treating restoration in the Mojave Desert as an orphan activity, agencies can learn from forest restoration on less arid public lands. Structured thinking and the incorporation of experimental treatments into the design of restoration projects promote learning. While many projects have been undertaken on public lands, few practitioners have published the results of their efforts. It is crucial that individuals share their successes and failures if large-scale, active restoration is to be successful.

Practice Adaptive Management

Adaptive management experiments are critically important to advancing practical knowledge of desert restoration methods. Adaptive management is defined as a systematic approach to continually improving restoration policies and practices by learning from the outcomes of past and current projects. Prescriptions are designed so that, even if they fail, they will provide useful information for the future. The advantages of adaptive management are: (1) it addresses gaps in available scientific and technical information; (2) it provides examples of the success or failure of tested practices; and (3) it provides a basis for statistically validating and objectively evaluating restoration results. Incorporating tests of hypotheses and technologies in restoration spurs innovation, efficiency, and rapid learning by agencies and the concerned public.

Tend Soil Fungi and Soil Bacteria to Facilitate Restoration

Attention to soil physical properties and chemistry in restoration appears to increase revegetation success. However, much of the Mojave Desert, and in particular the lands managed by the BLM, has not been surveyed for soil characteristics. The lack of available information about local soils and the lack of land managers with expertise in soil science, microbiology, and hydrology presents a continuous challenge. An area currently neglected in restoration projects is management of subsurface biotic resources to enhance soil fertility, soil-water relations, and soil formation.

For example, arbuscular-mycorrhizal (AM) fungi stabilize desert soils and enhance plant acclimatization and growth by reducing drought stress and facilitating nutrient acquisition (Carrillo García et al. 1999; Requeña et al. 2001). Arbuscular-mycorrhizal propagules develop around plant roots and promote "islands of fertility" that speed the pace of colonization and increase the survival of desert plants (Azcón Aguilar et al. 2003). Work by U.S. Department of Agriculture mycologists in Corvallis, Oregon, and soil scientists at the Universidad Autónoma de Baja California Sur in La Paz have contributed to an understanding of these dynamics in the Sonoran Desert. Bacteria and fungi both facilitate fertility and adaptive colonization of desert plants, particularly cacti species, in rocky areas (Carrillo García et al. 2001; Puente et al. 2004a, 2004b). Titus et al. (2002) have started to analyze AM fungi and their role in the establishment of native shrubs in the Mojave Desert. The potential for more efficient restoration in the most difficult sites in the Mojave Desert may improve with mastery of desert microbial ecology.

DISCUSSION AND CONCLUSIONS

While active desert restoration is a large financial and management commitment, there are some cases where accelerating the restoration process is worthwhile. For instance, investing in obscuring a road junction halts off-highway vehicle disturbance and allows natural recovery to take place on the remaining stretch of abandoned road. Disguising access points leading to abandoned mines or former encampments improves safety and visitor protection on public lands. Restoration "landscaping," in conjunction with construction activities, can control traffic in fragile areas and may reduce the potential for invasion by nonnative plant species.

The most successful projects utilize new and innovative materials and techniques, and often new or adapted inventions. There is a wide body of information on past restoration efforts in the Mojave Desert. However, much of it consists of unpublished materials found at agency offices. It is important that this information be made available to ecologists and land managers. A "Mojave Desert Restoration Bibliography" would be of great use to practitioners in the field. While many management prescriptions may be new or cutting-edge in the Mojave, restoration has occurred far longer in other arid regions of the globe, such as China,

Tibet, and Mongolia. Gathering information from around the world would be an invaluable addition to the existing literature.

Federal public land managers are now the principal trustees for conservation of Mojave Desert natural resources. Increasingly, federal land managers must meet regional expectations of excellent air quality, attractive landscapes, and access to recreation opportunities, as well as national expectations for conservation of biological diversity and intact ecosystems. Meeting the public's often contradictory expectations will require more intensive land stewardship on the part of federal land managers. Restoration of sites, watersheds, and landscapes will increasingly command management attention to offset the impacts of humans on federal public lands.

ACKNOWLEDGMENTS

Contributions to this chapter were made by many. Thanks to Michelle Cox for providing valuable data from her work at Joshua Tree National Park. Thanks also to our excellent reviewers: David Miller, Jill Heaton, and Leslie DeFalco. And greatest thanks to those innovators in the field who try and try again to heal our richly diverse, wildly beautiful, upland desert ecosystems.

REFERENCES

Al-Dowaisan, A., N. Fakim, M. R. Khan, N. Arifhodzic, R. Panicker, A. Hanoon, and I. Khan. 2004. *Salsola* pollen as a predominant cause of respiratory allergies in Kuwait. *Annals of Allergy, Asthma and Immunology* **92**:262–267.

Anderson, D. C., and W. K. Ostler. 2002. Revegetation of degraded lands at U.S. Department of Energy and U.S. Department of Defense installations: strategies and successes. *Arid Land Research and Management* **16**:197–212.

Azcón Aguilar, C., J. Palenzuela, A. Roldan, S. Bautista, R Vallejo, and J. M. Barea. 2003. Analysis of the mycorrhizal potential in the rhizosphere of representative plant species from desertification-threatened Mediterranean shrublands. *Applied Soil Ecology* **22**:29–37.

Bainbridge, D. A. 1994. Restoration in the California Desert vertical mulch for site protection and revegetation. Online. Soil Ecology and Restoration Group, San Diego State University, San Diego, California. http://www.sciences.sdsu.edu/SERG/techniques/erosion.html.

Bainbridge, D. A. 1999. Desert restoration—steps toward success. Online. Alliant International University, United States International College of Business, San Diego, California. http://www.ecocomposite.org/restoration/desert.htm. Accessed April 15, 2006.

Bainbridge, D. A. 2000. Understanding disturbance. Online. Alliant International University, United States International College of Business, San Diego, California. http://www.ecocomposite.org/restoration/disturbance.htm. Accessed May 14, 2006.

Bainbridge, D. A. 2001. Buried clay pot irrigation: a little known but very efficient traditional method of irrigation. *Agricultural Water Management* **48**:79–88.

Bainbridge, D. A. 2003. New hope for desert lands. Online. *Arid Southwest Lands Habitat Restoration Conference, March 5, 2003, Palm Springs, California.* Desert Managers Group, Barstow, California. http://www.westernecology.com/pdfs/wed-01-bainbridgeLI.pdf.

Bainbridge, D. A., R. MacAller, M. Fidelibus, R. Franson, A. C. Williams, and L. Lippitt. 1995. A beginner's guide to desert restoration. Online. National Park Service, Denver Service Center, Denver, Colorado. http://www.westernecology.com/pdfs/bainbridgebook.pdf.

Bainbridge, D. A., and R. A. Virginia. 1991 Irrigation for remote sites: SERG restoration bulletin No. 6. Online. Soil Ecology and Restoration Group, San Diego State University, San Diego, California. http://www.sciences.sdsu.edu/SERG/techniques/Irrigation.pdf.

Bauder, E. T., and A. Larigauderie. 1991. *Rehabilitation success and potential of Mojave and Colorado Desert sites: final report.* Contract No. 4–55–0074. California Department of Parks and Recreation, Off-Highway Motor Vehicle Recreation Division, Sacramento, California.

Behbehani, N., N. Arifhodzic, M. al-Mousawi, S. Marafie, L. Ashkanani, M. Moussa, and A. Al-Dowaisan. 2004. The seasonal variation in allergic rhinitis and its correlation with outdoor allergens in Kuwait. *International Archives of Allergy and Immunology* **133**: 164–167.

Berry, K. J., J. F. Weigand, B. L. Kay, and D. Kearns. 2005. Restoration in the Mojave Desert, California, USA: a retrospective look—large-scale recovery and facilitated restoration along the Los Angeles aqueducts. Poster presented at Ecological Restoration—A Global Challenge. Seventeenth Conference of the Society for Ecological Restoration International, September 12–18, 2005, Zaragoza, Spain. Society for Ecological Restoration International, Tucson, Arizona.

Bezy, R. L. 1989. Night lizards: the evolution of habitat specialists. *Terra* **28**:29–34.

Brooks, M. L. 1995. Benefits of protective fencing to plant and rodent communities of the western Mojave Desert, California. *Environmental Management* **19**:65–74.

Brooks, M. L. 1998. Competition between alien annual grasses and native annual plants in the Mojave Desert. *American Midland Naturalist* **144**:92–108.

Brooks, M. L. 1999. Effects of protective fencing on birds, lizards, and black-tailed hares in the western Mojave Desert. *Environmental Management* **23**:387–400.

Brooks, M. L., and D. A. Pyke. 2001. Invasive plants and fire in the deserts of North America. Pages 1–14 *in* K. E. M. Galley, and T. P. Wilson, editors. *Proceedings of the invasive species workshop: the role of fire in the control and spread of invasive species, Fire Conference 2000, the First National Congress on Fire Ecology, Prevention, and Management, November 27-December 1, 2000.* Miscellaneous Publication No. 11. Tall Timbers Research Station, Tallahassee, Florida.

Bureau of Land Management. 1999. The California Desert: conservation area plan 1980 as amended. Online. U.S. Department of the Interior, Bureau of Land Management, California Desert District, Riverside, California. http://www.blm.gov/ca/pdfs/cdd_pdfs/CA_Desert_.pdf.

Carrillo-García, Á, Y. Bashan, and G. J. Bethlenfalvay. 2001. Resource-island soils and the survival of the giant cactus, cardon, of Baja California Sur. *Plant and Soil* **218**:207–214.

Carrillo García, Á., J.-L. León de la Luz, Y. Bashan, and G. J. Bethlenfalvay. 1999. Nurse plants, mycorrhizae, and plant establishment in a disturbed area of the Sonoran Desert. *Restoration Ecology* 7:321–335.

Cione, N. K., P. E. Padgett, and E. B. Allen. 2002. Restoration of a native shrubland impacted by exotic grasses, frequent fire, and nitrogen deposition in southern California. *Restoration Ecology* **10**:282–288.

Colás, C., S. Monzón, M. Venturini, A. Lezaun, M. Laclaustra, S. Lara, and E. Fernández Caldas. 2005. Correlation between Chenopidiacea/Amaranthacea pollen counts and allergic symptoms in *Salsola kali* monosensitized patients. *Journal of Investigatory Allergology and Clinical Immunology* **15**:254–258.

Corbin, J. D., and C. M. D'Antonio. 2004. Can carbon addition increase competitiveness of native grasses? A case study from California. *Restoration Ecology* **12**:36–43.

Cox, M. C., and J. E. Rodgers. 2003. Arid land restoration methods at Joshua Tree National Park. Page 59 *in Assembling the pieces: restoration, design, and landscape ecology.* Fifteenth annual conference of the Society for Ecological Restoration International, November 19–22, 2003, Austin, Texas. Society for Ecological Restoration International, Tucson, Arizona.

Darby, M. M. 1993. Distribution of rhizobia and VA mycorrhizal fungi in two desert soils and their effects on the growth of *Prosopis glandulosa* var. *torreyana* (mesquite) at varying nutrient levels. MS thesis. San Diego State University, San Diego, California.

Dowaisan, A., S. Al-Ali, M. Khan, Z. Hijazi, M. S. Thomson, and C. I. Ezeamuzie. 2000. Sensitization to aeroallergens among patients with allergic rhinitis in a desert environment. *Annals of Allergy, Asthma and Immunology* **84**:433–438.

Edwards, F. S., D. A. Bainbridge, T. A. Zink, and M. F. Allen. 2000. Rainfall catchments for restoration in the Mojave Desert. *Ecological Restoration* **18**:100–103.

Egan, T. B. 1999. Afton Canyon riparian restoration project: fourth year status report. Online. U.S. Department of the Interior, Bureau of Land Management, Barstow, California. http://www.cwss.org/1999/130–144.pdf.

Ezeamuzie, C. I., M. S. Thomson, S. Al-Ali, A. Dowaisan, M. Khan, and Z. Hijazi. 2000. Asthma in the desert: spectrum of the sensitizing aeroallergens. *Allergy* **55**:157–162.

Fairbourn, M. L. 1975. Field evaluation of microwatersheds and vertical mulch systems. Pages 233–243 *in* G. W. Frasier, editor. *Water Harvesting Symposium, March 26–29, 1974, Phoenix Arizona.* Report No. USDA-ARS W-22. U.S. Department of Agriculture, Agricultural Research Service, Berkeley, California.

Fischer, R. A. 2004. Using soil amendments to improve riparian plant survival in arid and semi-arid landscapes. ERDC Technical Note No. EMRRP-SR-44. (Online). Environmental Laboratory, U.S. Army Engineer Research and Development Center, Vicksburg, Mississippi. http://el.erdc.usace.army.mil/elpubs/pdf/sr44.pdf.

Garde, J., A. Ferrer, V. Jover, J. A. Pagan, C. Andreu, A. Abellan, R. Felix, J. M. Milan, M. Pajaron, A. J. Huertas, J. R. Lavin, and F. de la Torre. 2005. Tolerance of a *Salsola kali* extract standardized in biological units administered by subcutaneous route. *Allergología et Immunopathología* **33**:100–104.

Gartland, R., and J. F. Weigand. 2003. *Management plan for restoring soils and vegetation on illegal OHV routes in the West Rand Mountains Area of Critical Environmental Concern.* U.S. Department of the Interior, Bureau of Land Management, Ridgecrest Field Office, Kern County, California.

Grantz, D. A. 1998. Revegetation techniques and fugitive dust in the western Mojave Desert. Contract No. 94–337. Air Resources Board, Research Division, California Environmental Protection Agency, Sacramento, California.

Grantz, D. A., and D. L. Vaughn. 2001. Factors in plant survival for revegetation in the Antelope Valley for particulate matter control. Air Resources Board, Research Division, California Environmental Protection Agency, Sacramento, California.

Grantz, D. A., D. L. Vaughn, R. J. Farber, B. Kim, L. Ashbaugh, T. Van Curen, and R. Campbell. 1998. Wind barriers suppress fugitive dust and soil-derived airborne particles in arid regions. *Journal of Environmental Quality* **27**:946–952.

Hardcastle, J. *Undated.* Nevada county population estimates July 1, 1986 to July 1, 2005. Prepared for the Nevada Department of Taxation. Reno, Nev.: The Nevada State Demographer's Office. 20 pp.

Hirsch, M., J. Bar-Ziv, E. Lehmann, and G. M. Goldberg. 1974. Simple siliceous pneumoconiosis of Beduoin females in the Negev Desert. *Clinical Radiology* **25**:507–510.

Jennings, W. B. 2002. Diet selection by the desert tortoise in relation to the flowering phenology of ephemeral plants. *Chelonian Conservation and Biology* **4**:353–358.

Jha, M. N. 2004. An evaluation of degradation, movement, and fate of polyacrylamide (PAM) in the environment. Online. Poster presented at the Environmental Protection Agency Science Forum 2004. Healthy Communities and Ecosystems, June 1–3, 2004, Washington, D.C. U.S. Environmental Protection Agency, Washington D.C. http://www.epa.gov/ord/scienceforum/2004/oeiposter/Jha_Mitra.pdf.

Kay, B. L. 1988. Artificial and natural revegetation of the second Los Angeles aqueduct. *Mojave Revegetation Notes* **24**:1–32.

Kay, I. S. 1989. Early nodulation of mesquite (*Prosopis glandulosa* var. *torreyana*) by indigenous Sonoran Desert rhizobia: the role of contrasting bacterial life history strategies. MS thesis. San Diego State University, San Diego, California.

Kay, B. L., and Graves, W. L. 1983. Pages 315–324 *in* R. H. Webb and H. G. Wilshire, editors. *Environmental effects of off-road vehicles: impacts and management in arid regions.* Springer-Verlag, New York, New York.

Kramer, N. E. 1989. The effect of selenium on nitrogen fixation in the Se-accumulating desert legume *Astragalus crotalariae.* MS thesis. San Diego State University, San Diego, California.

Lovich, J. 1999. Human-induced changes in the Mojave and Colorado Desert ecosystems: recovery and restoration potential. Pages 529–531 *in* M. J. Mac, P. A. Opler, C. E. Puckett-Haecker, and P. D. Doran, editors. *The status and trends of the nation's biological resources.* Volume two. U.S. Geological Survey, Reston, Virginia.

Lovich, J., and D. Bainbridge. 1999. Anthropogenic degradation of the southern California desert ecosystem and prospects for natural recovery and restoration. *Environmental Management* **24**:309–326.

Ludwig, J. A., and D. J. Tongway. 1996. Rehabilitation of semiarid landscapes in Australia. II. Restoring vegetation patches. *Restoration Ecology* **4**:398–406.

Marble, J. R. 1985. *Techniques of revegetation and reclamation of land damaged by off-road vehicles in the Lake Mead National Recreation Area.* Report No. 027/03. University of Nevada, Department of Biological Sciences, Cooperative National Park Resources Study Unit, Las Vegas, Nevada.

Martin, B. E., and T. R. van Devender. 2002. Seasonal diet changes of *Gopherus agassizii* (desert tortoise) in desert grassland of southern Arizona and its behavioral implications. *Herpetological Natural History* **9**:31–42.

Mason, J. 2001. Restoration of damaged lands at the national training center. Online. *In* Restoration and Reclamation Review. Student Online Journal. Volume seven. Department of Horticultural Science, University of Minnesota, St. Paul, Minnesota. http://horticulture.coafes.umn.edu/vd/h5015/01papers/mason.htm. Accessed May 14, 2006.

Montalvo, A. M., McMillan, P. A., and E. B. Allen. 2002. The relative importance of seeding method, soil ripping, and soil variables on seeding success. *Restoration Ecology* **10**: 52–67.

Munshower, F. F. 1994. *Practical handbook of disturbed land revegetation.* Lewis Publishers, Boca Raton, Florida.

Nevada Department of Transportation. 2005. I-15 landscape and aesthetics corridor plan: I-15 from Primm to Mesquite. Online. Nevada Department of Transportation, Carson City, Nevada. http://www.ndothighways.org/documents/i-15_corridor_plan_part1.pdf.

Oftedal, O. T., S. Hillard, and D. J. Morafka. 2002. Selective spring foraging by juvenile desert tortoises (*Gopherus agassizii*) in the Mojave Desert: evidence of an adaptive nutritional strategy. *Chelonian Conservation and Biology* **4**:341–352.

Ostler, W. K., D. C. Anderson, D. B. Hall, and D. J. Hansen. 2002. *New technologies to reclaim arid lands user's manual.* Report No. DOE/NV/11718–731. Online. U.S. Department of Energy, National Nuclear Security Administration, Nevada Operations Office, Las Vegas, Nevada. http://www.osti.gov/energycitations/servlets/purl/804924-KSqIyz/webviewable/804924.pdf.

Puente, M. E., Y. Bashan, C. Y. Li, and V. K. Lebsky. 2004a. Microbial populations and activities in the rhizoplane of rock-weathering desert plants. I. Root colonization and weathering of igneous rocks. *Plant Biology* **6**:629–642

Puente, M. E., C. Y. Li, and Y. Bashan. 2004b. Microbial populations and activities in the rhizoplane of rock-weathering desert plants. II. Growth promotion of cactus seedlings. *Plant Biology* **6**:643–650.

Racin, J. A. 1988. *Revegetating desert cut slopes with containerized native shrubs: final report.* Report No. FHWA/CA/TL-88/09. Online. California Department of Transportation, Sacramento, California. http://www.dot.ca.gov/newtech/researchreports/1981–1988/88–09.pdf.

Reith, C. C., and B. M. Thomson. 1993. *Deserts as dumps? The disposal of hazardous materials in arid ecosystems.* University of New Mexico Press, Albuquerque, New Mexico.

Requeña, N., E. Perez Solis, C. Azcón Aguilar, P. Jeffries, and J. M. Barea. 2001. Management of indigenous plant-microbe symbioses aids restoration of desertified ecosystems. *Applied and Environmental Microbiologoy* **67**:495–498.

Schrenk, A. V. 2002. Successful native plant salvaging. Oral presentation at the 87th Annual Meeting of the Ecological Society of America and the 14th Annual International Conference of the Society for Ecological Restoration, August 4–9, 2002, Tucson, Arizona. Ecological Society of America, Washington, D.C.

Scoles, S., and L. DeFalco. 2003. *The revegetation of disturbed areas associated with roads at Lake Mead National Recreation Area, Nevada.* U.S. Geological Survey, Western Ecological Research Center, Las Vegas Field Station, Henderson, Nevada.

Shilling, F., S. Sommarstrom, R. Kattelmann, B. Washburn, J. Florsheim, and R. Henly. 2005. *The California watershed assessment manual.* Volume one. Online. California Resources Agency, Sacramento, California. http://cwam.ucdavis.edu/Manual_chapters .htm.

Soil Ecology and Restoration Group. 2005. National Training Center, Fort Irwin, California. Online. Soil Ecology and Restoration Group, San Diego State University, San Diego, California. http://www.sciences.sdsu.edu/SERG/restorationproj/mojave%20desert/ ft_Irwin.html.

Sorensen, N. S. 1993. Arid land revegetation: effects of tree shelters and irrigation on establishment of three shrubs with emphasis on the physiological ecology of *Larrea divaricata* (= *L. tridentata*). MS thesis. San Diego State University, San Diego, California.

State of California. *Undated.* Report 84 E-4: population estimates for California counties and cities, January 1, 1976 through January 1, 1980 Online. State of California, Department of Finance, Sacramento, California. http://www.dof.ca.gov/HTML/ DEMOGRAP/ReportsPapers/Estimates/E4/E4-70-80/E4CALL.HTM#tab76to80.

State of California. 2006. E-1: population estimates for cities, counties and the state with annual perçent change: January 1, 2005 and 2006. Online. State of California, Department of Finance, Demographic Research Unit. Sacramento, California. http://www.dof .ca.gov/HTML/DEMOGRAP/ReportsPapers/Estimates/E1/E-1text.asp.

Staton, J. C. 1998. Heat and mass transfer characteristics of desiccant polymers. MS thesis. Virginia Polytechnic Institute and State University, Department of Mechanical Engineering, Blacksburg, Virginia.

Thomas, P. M. 1995. Molecular analysis of bacteria that symbiotically fix nitrogen with the desert tree legume, *Prosopis glandulosa* (mesquite). Ph.D dissertation. University of California, San Diego, and San Diego State University, San Diego, California.

Titus, J. H., P. J. Titus, R. S. Nowak, and S. D. Smith. 2002. Arbuscular mycorrhizae of Mojave Desert plants. *Western North American Naturalist* **62**:327–334.

Tongway, D. J., and J. A. Ludwig. 1996. Rehabilitation of semiarid landscapes in Australia. I. Restoring productive soil patches. *Restoration Ecology* **4**:388–397.

Walker, L. R., and E. A. Powell. 2001. Soil water retention on gold mine surfaces in the Mojave Desert. *Restoration Ecology* **9**:95–103.

Webb, R. H., and E. B. Newman. 1982. Recovery of soil and vegetation in ghost towns in the Mojave Desert, southwestern United States. *Environmental Conservation* **9**:245–248.

Webb, R. H., J. W. Steiger, and E. B. Newman. 1988. The response of vegetation to disturbance in Death Valley National Monument, California. U.S. Geological Survey Bulletin No. 1793. U.S. Government Printing Office, Washington, D.C.

Webb, R. H., and H. G. Wilshire. 1980. Recovery of soils and vegetation in a Mojave Desert ghost town, Nevada, USA. *Journal of Arid Environments* **3**:291–303.

Wells, P. V. 1961. Succession in desert vegetation on streets of a Nevada ghost town. *Science* **134**:670–671.

West, B. R. 1996. Prescribed burning and wildfire (fire as a tool in saltcedar management). *In* S. Stenquist, editor. *Proceedings, Saltcedar Management and Riparian Restoration Workshop, September 17–18, 1996, Las Vegas, Nevada.* U.S. Department of Interior, Fish and Wildlife Service, Portland, Oregon.

APPENDIX 16.1

Additional Resources for Restoration of Disturbed Sites in the Mojave Desert

A recovery and restoration timetable, produced by the U.S. Geological Survey, is a useful tool when choosing between active restoration and natural recovery (http://biology.usgs.gov/s+t/SNT/noframe/gb151.htm).

Bauder and Larigauderie (1991) provided an early overview of the technologies and results of restoration efforts in the Mojave Desert.

The Desert Lands Restoration Task Force (DLRTF), under the Desert Manager's Group (DMG), developed a guide for restoration practices titled "A Beginner's Guide to Desert Restoration" (Bainbridge et. al. 1995).

Similar guides and decision-making tools have been developed at Joshua Tree National Park (Joshua Tree National Park *unpublished data*), the BLM Ridgecrest Field Office (Gartland and Weigand 2003), and the Soil Ecology and Restoration Group (Bainbridge 1999, http://www.serg.sdsu.edu/SERG/index.html).

Racin (1988) published a report on revegetation of slopes along US Highway 395 in Inyo and Mono Counties, California, with native plant seedlings.

SEVENTEEN

Alluviation and Visual Removal of Landscape Disturbances, Camp Ibis–Piute Wash, Eastern Mojave Desert

RICHARD HEREFORD

Alluvial activity, driven by runoff-producing precipitation, redistributes sediment across the landscape where it accumulates on low-lying depositional sites, such as stream channels, alluvial fans, and at the base of hillslopes. Deposition of alluvium, or of eolian sand for that matter, has the potential to cover and remove visual evidence of trails, tracks, and other low-relief anthropogenic landscape disturbances. Another type of visual removal of comparatively limited spatial extent results from the erosion of landscape disturbances. The effectiveness of these processes of landscape restoration is poorly understood, particularly in arid landscapes, where the frequency and spatial extent of sediment movement are unknown. This study addresses the role of climate, alluviation, and erosion in historical landscape change and their effects on partial environmental recovery from the disturbances of World War II–era military operations.

The study area is 23 km northwest of Needles, California, in Piute Valley at 540 m elevation (fig. 17.1; see also color plate). Camp Ibis extends for 5.4 km along U.S. Highway 95, and Piute Wash is an ephemeral, south-flowing tributary of the Colorado River that passes about 1.5 km east of the former camp. Evidence of fluvial activity in Piute Valley is abundant, despite receiving only 150 mm of precipitation annually. Many—essentially countless—washes, gullies, and rills dissect the weakly consolidated Pleistocene sedimentary deposits of Piute Valley, and fluvial activity is active and ongoing, as demonstrated by a variety of recent archeological material found in flood debris (cans, bottles, and other objects discarded along Highway 95) as well as discontinuous World War II–era roads and tank tracks (fig. 17.2). Used intensively for only two years in the early 1940s, the Camp

Ibis-Piute Wash area is an excellent place to examine the rate and spatial extent of alluvial processes in a desert landscape.

Earlier studies of World War II–era military camps, discussed in another section of this chapter, dealt with the physical and biological properties of soil and with vegetation dynamics in small areas of several ha, such as individual tank tracks or short channels within small catchments. For the most part, the objectives were to identify World War II–era disturbances of soils and vegetation, to measure their effects, and to estimate recovery rates. The present study takes a different approach. A relatively large area (~ 50 km^2) was mapped geologically to portray sites of recently active alluvial deposition. The mapping reveals that across relatively large areas, tracks and other anthropogenic disturbances undergo visual removal through deposition of alluvium. The physiographic conditions in the Camp Ibis-Piute Wash area are similar to other large alluvial valleys. Observations by the author show that visual removal of disturbances is typical of several other alluvial valleys throughout the Central Mojave Desert.

MAPPING METHODS

The deposits of Piute Wash, its tributaries, and valley-margin sheetwash were mapped using Global Positioning System (GPS) instrumentation. Using Geographic Information System (GIS) software, these data were compiled on black and white digital orthophotographic-quarter quadrangles (DOQQs) produced from photographs taken May 26, 1994. The DOQQs cover one-quarter of a standard U.S. Geological Survey 7.5' topographic quadrangle, giving field-verified resolution at a scale of about 1:4000. After field examination, the DOQQs and GIS software were also used to map the Camp Ibis road system, although the entire system was not physically traversed. Archival aerial photography of Camp Ibis taken in April 1943 (table 17.1) was used to map cleared and graded occupational sites, such as tent sites and parking areas.

BACKGROUND

Camp Ibis was constructed in the winter of 1942–1943 and was abandoned in March 1944; it was one of 14 facilities in eastern California and western Arizona comprising the Desert Training Area (DTA). Established by Major General

Table 17.1 Aerial photography used in analysis of the Camp Ibis/Piute Wash area

Date	Photograph no.	Source	Approximate scale
Apr 1943	6–51 strip 4	R. H. Webb (2001, personal communication)	1:37,000
Jul 25, 1943	US16MS	*Ibid.*, Birschoff (2001, p. 67)	Low-angle oblique
Apr 25, 1953	GS YP 2–17	US Geological Survey	1:50,000
Nov 15, 1979	GS VETR 1–27	*Ibid.*	1:39,000
May 26, 1994	NAPP 881–262	*Ibid.*	1:40,000

Fig. 17.1. Landsat image of Piute Valley (outlined in white) in southeastern Clark County, Nevada, and eastern San Bernardino County, California. The solid white line is the drainage divide of Piute Wash upstream of the Camp Ibis area. The black rectangle indicates the study area. (See color plate following page 162.)

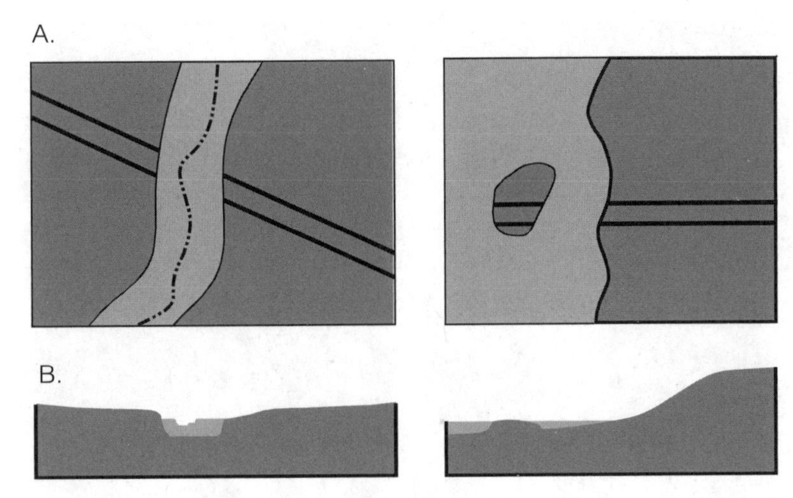

Fig. 17.2. Schematic diagrams showing in map view (*A*) and cross-section (*B*) the burial and visual removal of trails and tracks at stream crossings (*left*) and in low-lying areas at the base of hillslopes (*right*) at Camp Ibis in the Eastern Mojave Desert. These relations are widely observed in the Mojave Desert and involve foot trails and wagon, automobile, and tank tracks.

George S. Patton Jr., the facilities were used to train troops for battle in the deserts of North Africa. Camp Ibis was designed to house a full division, up to 15,000 soldiers. Three armored divisions trained there separately from March to November 1943 and, while occupied, the camp was effectively a small tent city replete with hundreds of vehicles, including tanks and heavy earth-moving equipment (fig. 17.3). By late 1943, the campaign in North Africa was ending and desert combat training was no longer necessary. Troops were withdrawn from the DTA in early 1944, it was declared surplus on March 16, 1944, and ownership of the land was returned to the U.S. Department of Interior, Bureau of Land Management. The facilities were quickly closed and by May 1, 1944, all material and equipment had been removed (Bischoff 2000). Since then, Camp Ibis has been largely undisturbed except for light recreational use and construction of a natural gas pipeline. The heaviest use is near the historical marker on U.S. Highway 95 (fig. 17.4). Elsewhere, most camp roads are rarely traveled, except for three roads used for occasional access to Piute Wash and the east side of Piute Valley.

Previous Studies

A substantial and diverse literature, reviewed by Lovich and Bainbridge (1999), treats the effects of human activity on the Mojave Desert ecosystem. Two such activities that severely affect the ecosystem are use of off-road vehicles and military training operations; both were widespread in the Mojave Desert during and

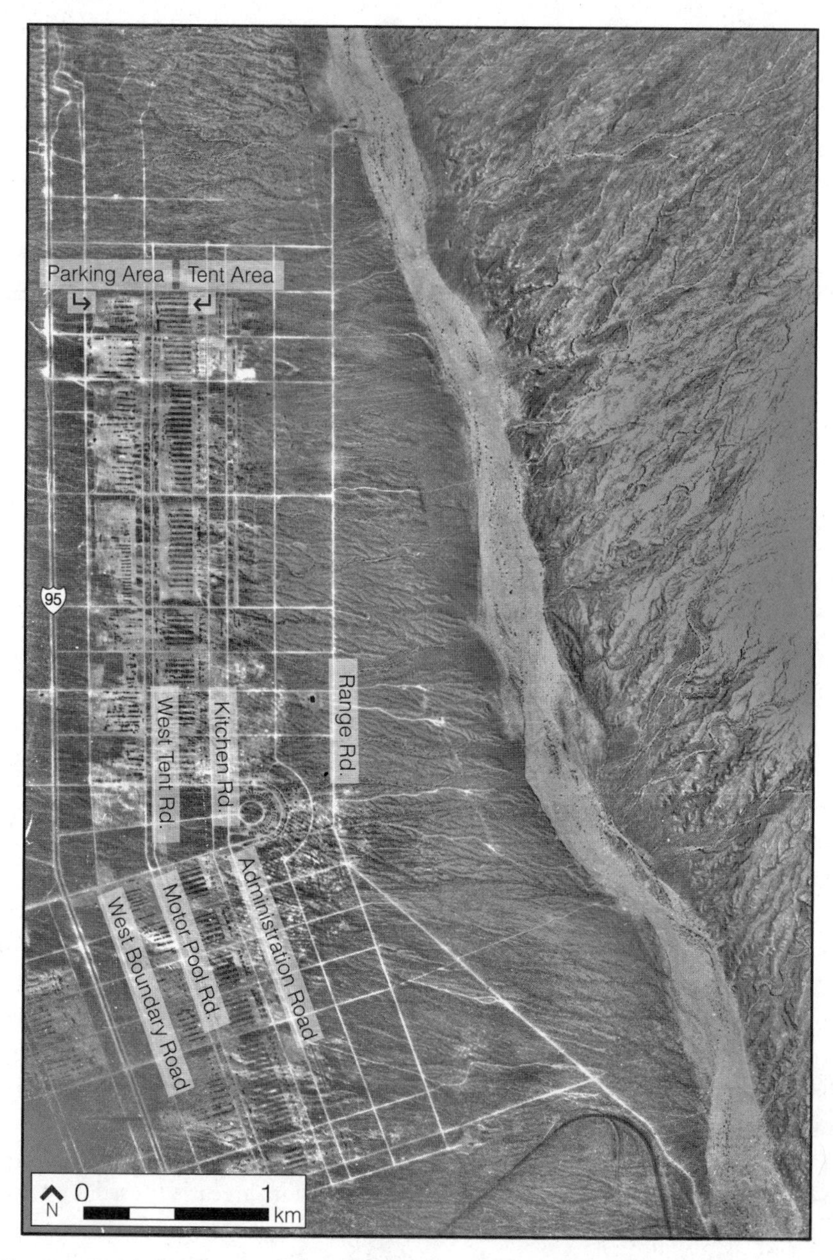

Fig. 17.3. Aerial photograph of Camp Ibis and Piute Wash (east of the camp), taken while the camp was occupied, April 1943. Note the many tents located on the eastern side of the West Tent Road and the numerous vehicles in the parking area west of the Motor Pool Road. Labeled localities are from Prose and Metzger (1985) and Bischoff (2000). The photograph was digitally processed to remove defects in the original image.

after World War II (Prose 1986). Generally, the environmental effects of these activities are soil compaction and vegetation destruction. Published rates of recovery, as measured by loss of soil compaction and by reestablishment of vegetation, range widely from several decades to several millennia (Lathrop 1983; Prose and Metzger 1985; Lovich and Bainbridge 1999; Belnap and Warren 2002; Webb 2002). Much of this variability is related to the age of the disturbed geomorphic surface, such that older disturbed surfaces recover slowly compared with younger surfaces (Steiger and Webb 2000). Old geomorphic surfaces occupy elevated topographic positions in an alluvial valley; they are not sites of widespread alluviation. Thus, on older, high-level surfaces, visual removal of disturbances by fluvial activity is spatially restricted to relatively small channels.

At Camp Ibis and other nearby World War II–era military camps, the specific effects of off-road vehicle use and vegetation removal have been described in several studies (Iwasiuk 1979; Lathrop 1983; Prose 1985; Prose and Metzger 1985; Prose et al. 1987; Prose and Wilshire 2000). These studies, which were summarized by Prose and Wilshire (2000), considered the environmental effects of tank maneuvers and construction and use of campsites, roads, and armored vehicle parking areas. At Camp Ibis, even after 40 years of recovery (as of 1983), heavily compacted soils and reduced vegetation cover and diversity were persistent effects of military operations. Generally, the density of vegetation recovered to 30%, 58%, and 80% of predisturbance levels in the road grid, roads in the tent areas, and tent sites, respectively (fig. 17.3) (Prose and Metzger 1985). Reestablishment of vegetation resulted in a partial loss of long-lived species, particularly *Larrea tridentata* (creosote bush), and an increase in short-lived species such as *Encelia frutescens* (green brittlebush).

Increased soil compaction from vehicular activity was identified as a major factor affecting the composition of the regenerated perennial plant cover, its density, and species composition (Prose et al. 1987). Webb (2002), however, pointed out that alluvial deposition on disturbed surfaces could eliminate compacted soil as a problem for plant establishment. Although Prose (1985) reported that soils remain compacted after burial, this subsurface compaction evidently did not preclude the reestablishment of vegetation at Camp Ibis.

The effects of road and walkway construction on the drainage patterns of short, ephemeral streams were studied 100 km southwest of Camp Ibis at Camp Iron Mountain by Nichols and Bierman (2001), who found that channels had not returned to their pre-impact shape and size after 44 years of recovery. However, the relatively minor recovery noted by them appears to be related to erosion. Thirty-five km south-southeast of Camp Ibis, in Chemehuevi Valley, Belnap and Warren (2002) examined the recovery of biological soil crusts in short segments of tank tracks on desert pavement. They found that the cyanobacterial component had recovered to 45%–65% of predisturbance levels, although full recovery of the most sensitive species may require several thousand years.

Geomorphic Setting

Piute Wash flows south through Piute Valley from its headwaters in the southern McCullough and Highland mountain ranges northwest of Searchlight, Nevada. In the central and southern portion of the drainage basin, Piute Wash drains the Piute Range and the Newberry and Piute mountains (fig. 17.1). Drainage-basin area upstream of the map area is 1400 km^2, and the slope of the channel through the mapped area is 10 m km^{-1}. West of Piute Wash, Piute Valley slopes gently to the east on the Homer Mountain piedmont at 23 m km^{-1}. East of Piute Wash, the valley slopes southwest at 27 m km^{-1} on the Dead Mountains piedmont (fig. 17.1). The channel of Piute Wash is incised 2–12 m into late Tertiary (?) to Pleistocene valley-fill deposits underlying the piedmonts of Homer Mountain and the Dead Mountains. On undisturbed surfaces, *Larrea* is the dominant species in the plant community.

The post-1943 deposits mapped here occur in active channels, on alluvial fans, and at the base of hillslopes. Elsewhere in the desert region, active channel deposits similar to those at Camp Ibis were mapped as late Holocene unit Q4 by Steiger and Webb (2000) and Lashlee et al. (2002). Fresh alluvium, lacking soil development, extends downslope from the base of the low hillslopes adjoining Piute Wash into active channels that extend to the main wash. This indicates that most of the post-1943 sediment in tributary channels is derived from erosion of the valley-fill deposits underlying the Homer Mountain and Dead Mountains piedmonts (fig. 17.1). Near the surface, these older deposits are unconsolidated to weakly consolidated granule to cobble gravelly sand that, according to Prose and Metzger (1985), consists of 60% granitic clasts from the Newberry Mountains and 40% volcanic clasts derived from the Piute Range. Undisturbed piedmont surfaces typically do not have well-developed pavements or Av horizons; this lack of development suggests that the surfaces are relatively young. In steep hillslopes, thin, resistant, and discontinuous layers of pedogenic carbonate are sparingly present a few meters below the surface.

Results

Visual Removal by Erosion at Camp Ibis: During construction of Camp Ibis, more than 100 km of roads were graded and 3.5 km^2 of undisturbed desert was smoothed (fig. 17.3). The 1943 aerial photographs of Camp Ibis (table 17.1) show that vegetation in the smoothed areas was completely removed except for in swales and unused areas in the road grid. Moreover, the east-flowing washes that cross the camp at a high angle appear to have been partly to completely filled in during grading of tent sites and parking areas. In the 1953 aerial photography, shrubby vegetation was not present in the cleared areas, whereas in the 1979 photography, scattered shrubs were present. Many of the washes have reestablished channels across the cleared areas, as suggested by field observations and the 1994 aerial photography.

Presently, the graded roads and cleared areas are still visible, but because the roads are washed out, they are faint and discontinuous (fig. 17.4; see also color plate), and vegetation has reestablished itself in places in the road grid to densities approaching undisturbed desert (Prose 1985; Prose and Metzger 1985). In the 1953 aerial photography, the rectangular road grid was continuous and appeared fresh, and the cleared areas were well defined. In the 1979—and particularly the 1994—photography, the grid was fragmented with numerous discontinuities; today the roads are largely impassable. Only 107 km of the original road system remains, which is estimated to be 80% of its original length. This type of visual removal is related to erosion rather than to the burial of the landscape disturbance.

Alluviation and Visual Removal in Piute Wash

During occupation of Camp Ibis, 11 gunnery ranges on the east side of Piute Valley were used for artillery training (Bischoff 2000). Hundreds, if not thousands, of military vehicles crossed Piute Wash, presumably from the Range Road (fig. 17.3), to reach the gunnery ranges, but field observations during this study show that tracks are not present in Piute Wash. As discussed below, the tracks were evidently buried by alluviation in Piute Wash, as shown in figure 17.2.

Piute Wash is a braided stream with 2 to 5 active channels floored by sandy pebble to cobble-size gravel. The active braided channels are separated by braid bars (elevated sandy deposits between channels) 20–150 cm above the channel floor. Pebble to small-boulder gravel is typical of the upstream end of the bars, whereas the downstream two-thirds is silty sand. The mapped area of the braided channel and bars is 2.5 km^2, which is essentially the area of the braid bars, as the channels are too small to map separately. The average thickness of the alluvium exposed in cutbanks is at least one meter; thus, the volume of sediment deposited in the post-1943 channel is at least 2.5 million m^3. This alluvium covers the floor of the channel that was active when Camp Ibis was in use.

The braid bars are vegetated with *Larrea*, *Acacia greggii* (catclaw), *Psorothamnus spinous* (smoke tree), and rare *Salix exigua* (sand bar willow). The root collars of many of these plants are beneath the surface of the braid bars, indicating they were partly buried by sediment after germination. *Chrysothamnus paniculatus* (rabbitbrush) is abundant on low bars in the active channel and on the margin of braid bars. The presence of these plants readily identifies the post-1943 channel deposits. Of these plants, only *Larrea* is present outside the channel.

The changes in Piute Wash since 1943 are well illustrated in a series of aerial

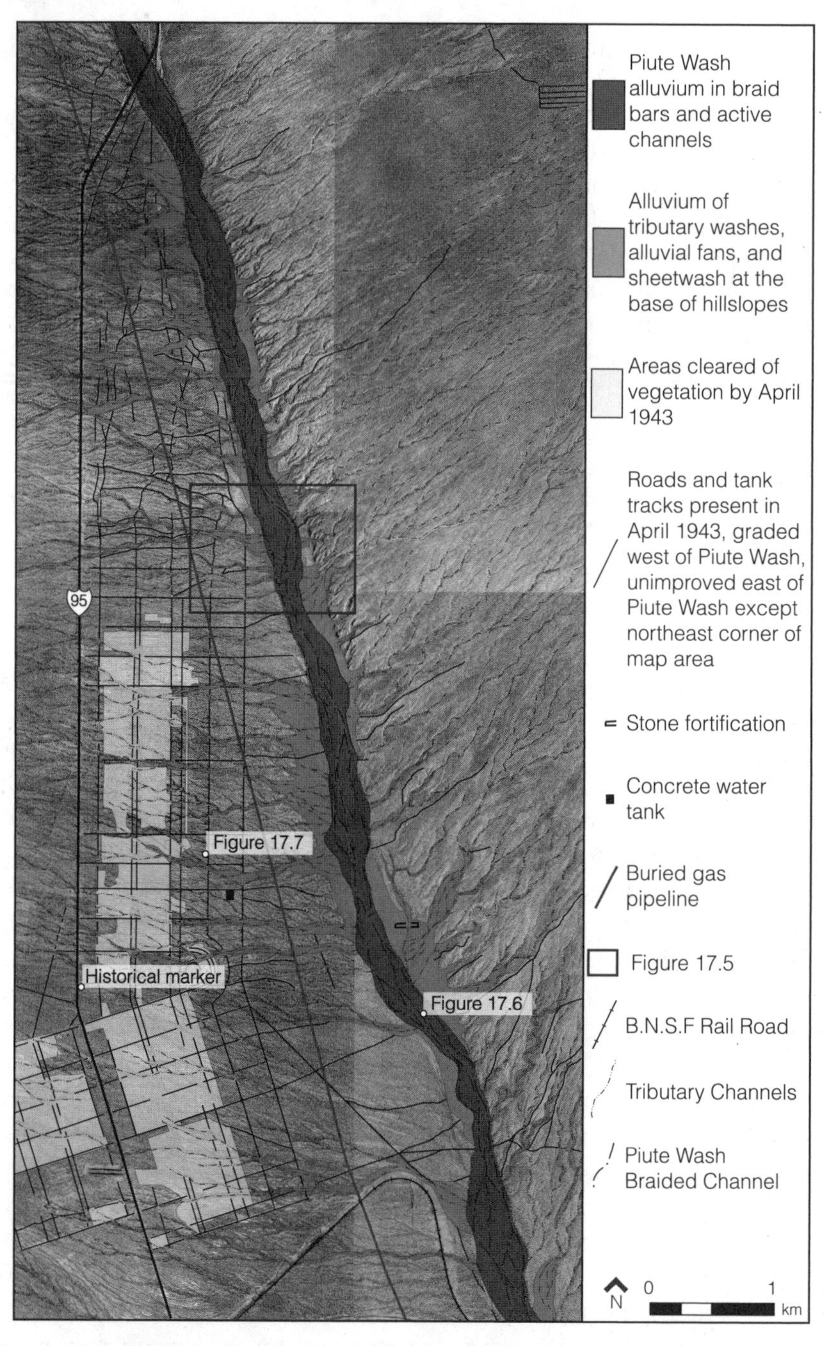

Fig. 17.4. Aerial photograph of the Camp Ibis–Piute Wash area in the Eastern Mojave Desert showing the post–1943 alluvial deposits along Piute Wash and its tributaries, the remnants of the Camp Ibis road system, and smoothed and cleared areas. The thin, solid black lines indicate roads that are faint and discontinuous. The black rectangle indicates the area shown in fig. 17.5. The cutbank exposure is shown in fig. 17.6. Sediment from the excavated pond is depicted in the cross-section presented in fig. 17.7 (photograph is courtesy of the Nevada Historical Society, California Photographs, photograph number CA 576). (See color plate following page 162.)

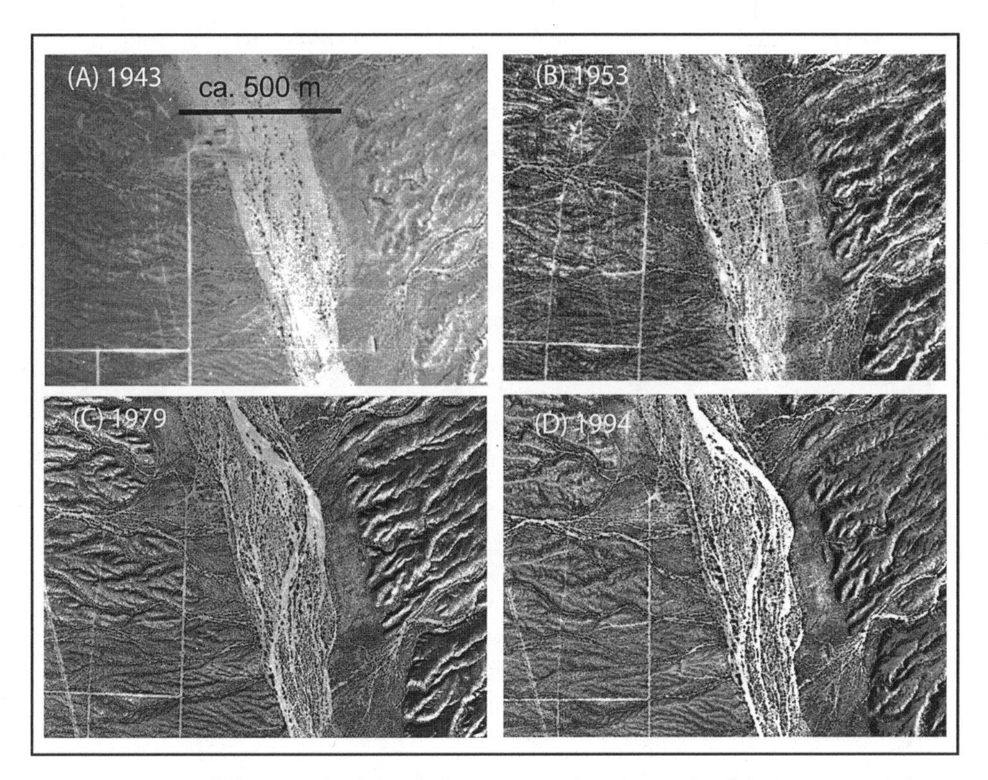

Fig. 17.5. Aerial photographs showing changes occurring in Piute Wash between 1943 and 1994. The area covered in the photographs is outlined with a red box in fig. 17.4.

photographs (table 17.1). A portion of each photograph portraying a reach of the wash upstream of the main camp (rectangle in fig. 17.4) is shown in figure 17.5. In 1943, Piute Wash had high albedo, it had only a few poorly formed braid bars with sparse brushy vegetation, and it lacked well-defined braided channels (figs. 17.3 and 17.5a). These conditions suggest the wash recently underwent a series of large floods. In 1953, the channel had lower albedo, increased vegetation, and no well-defined braided channels (fig. 17.5b). Also, tracks in a grid-like pattern, similar to those on the east side of the wash, were present in the channel; these were probably made after April 1943 and before closure of Camp Ibis in 1944. By 1979, the channel had shifted to the east, the tracks were no longer present in the wash, several braid bars and channels had developed, and brushy vegetation had increased substantially (fig. 17.5c). The tracks on the east side of the wash near the base of the hillslope (fig. 17.5b) were no longer present, as they were covered by sheetwash deposits (fig. 17.4). Fifteen years later, in 1994, the braided channels had high albedo and the channel had shifted farther to the east, while the density of brushy vegetation did not change appreciably (fig. 17.5d).

Briefly, interpretation of the sequential aerial photography suggests that most of the vegetation spread into the channel after 1943, promoting deposition and formation of braid bars, as well as visual removal of tracks and other vehicular disturbances. Moreover, except as discussed above, the channel was mostly unaffected by large, destructive floods from after 1943 until at least 1979. The high albedo of the braided channels in the 1994 photographs suggests the channel had recently been altered by flooding not long before 1994. Thus, between 1979 and 1994, one or more large floods altered the channel, reactivating earlier braided channels and eroding the east bank of the wash.

Alluvial Deposits and Runoff History of Piute Wash

The alluvium underlying the braid bars is locally exposed in cut banks related to the active channel (fig. 17.6; see also color plate). The post-1943 deposits mapped at this locality (shown in fig. 17.4) are 50 cm thick, although they are typically over 100 cm thick. This yields an average sedimentation rate of about 1.6 cm y^{-1} (100 cm per 60 years), although deposition was episodic, as discussed below. A cobble- to small boulder-size gravel is incised into the margin of the braid bar at this locality and at numerous other sites. As suggested by the aerial photography, this incision and deposition of gravel resulted from a flood or floods during the 1980s to early 1990s. The surface of the alluvium has a dark biological crust with rolling morphology (terminology of Belnap et al. 2001) that formed recently, certainly since 1943, if not since the flooding of the 1980s to early 1990s. The bars are classified as remnant braid bars, because they are not actively forming in the present discharge regimen.

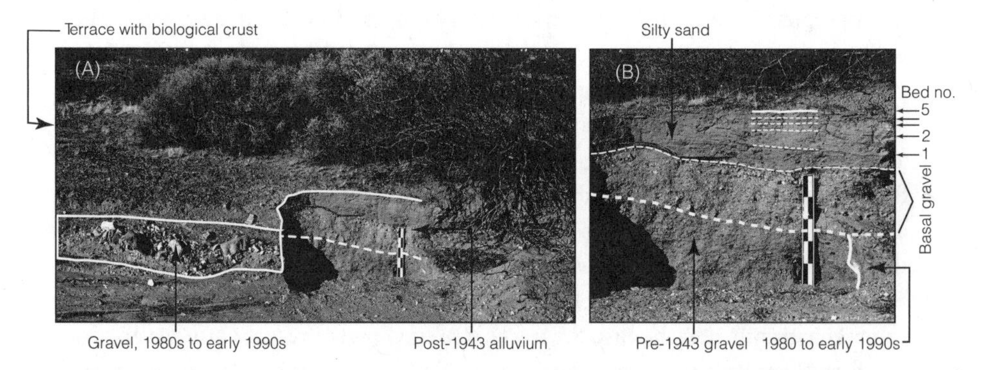

Fig. 17.6. Photographs showing a cutbank exposure of braid-bar alluvium. *A*, stratigraphic relations of post-1943 alluvium with older gravel (below dash line) and younger, incised gravel dating from the 1980s to early 1990s are visible in this photograph. *B*, shows the braid-bar alluvium consisting of basal gravel and silty sand overlying pre-1943 gravel. It is probable that five beds are present in the silty sand layer, representing five runoff events. The scale is 50 cm with 10-cm divisions. (See color plate following page 162.)

The alluvium of the remnant braid bars overlies sandy, pebble-size gravel along a gently sloping contact (dashed line in fig. 17.6a). Between the alluvium and the gravel lies a thin, resistant layer of silty clay. This resistant layer is weakly cemented, probably from repeated wetting and subaerial exposure. The post-1943 alluvium consists of basal gravel that is overlain gradationally by silty sand (fig. 17.6b). The silty sand has five beds separated by sharp, nonerosional contacts. Thus, including the basal gravel as part of the overlying silty sand, deposition of the post-1943 alluvium occurred during at least five floods from after 1943 until around 1979. These floods and associated alluviation resulted in visual removal of World War II–era military tracks. After 1979, the braid bars were evidently incised during one or more relatively large floods, which deposited gravel in higher parts of the channel adjacent to the braid bar, as well as on the braid bar surface.

Alluviation and Removal by Erosion in Tributary Washes

Discontinuous tracks and roads are also present outside the banks of Piute Wash. The alluvium covering the tracks occurs in tributary channels, on low-relief alluvial fans, and at the base of hillslopes adjacent to Piute Wash (fig. 17.2). The absence of tracks and the distinctive vegetation assemblage were used to map these post-1943 deposits (fig. 17.4). This alluvium is extensive, occupying 3.7 km^2 of the map area.

Brushy vegetation associated with the post-1943 sheetwash and tributary channel alluviums consists mainly of *Larrea* and *Acacia.* Older sheetwash and tributary channel surfaces lack *Acacia* and have widely scattered *Yucca schidigera* (Mojave yucca) and *Ferocactus cylindraceus* (barrel cactus). Where diagnostic vehicle tracks are absent, the presence of the latter two plants, along with an overall darker soil shade related to more developed biological soil crust, serves to distinguish the pre- and post-1943 surfaces.

The alluvium in tributary washes is sandy gravel with clasts ranging from pebble to cobble size. These deposits accumulated in channels incised 2–10 m into the piedmonts of Homer Mountain and the Dead Mountains (fig. 17.1). At the junction with Piute Wash, the deposits typically form low-relief alluvial fans that grade downslope into the main wash. Sheetwash alluvium is somewhat finer grained, consisting of sandy gravel with pebble to small cobble clasts. The sheetwash deposits occupy the area between tributary washes at the base of the 2–12 m high erosional (or perhaps fault) scarp that parallels both sides of Piute Wash in the study area north of latitude 34° 58' (fig. 17.4). It was not possible to map the sheetwash and tributary channel deposits separately at the scale of the map.

Where not recently active, the surface of the post-1943 sheetwash and tributary channel deposits has a biological crust resembling the crust on the deposits of Piute Wash (fig. 17.6a). The thickness of the channel and sheetwash sediment was estimated from the depth of burial of flood debris and World War II–era

Fig. 17.7. Photograph of alluvial deposits in a small pond on the upstream side of a camp road (location shown in fig. 17.4). The sedimentology indicates that runoff occurred twenty-two times between 1943 and 2001. The scale is 20 cm with 1 cm divisions. (See color plate following page 162.)

structures built for fortification and other purposes. Thickness varies widely from 5–50 cm.

Runoff Frequency of Tributary Streams

During construction of the road system at Camp Ibis, the cut-and-fill technique was used to cross the incised east-flowing washes. At these sites, the roadway dammed the channel, causing sediment to accumulate in small pond-like areas, as first noted by Prose and Metzger (1985). Sediment in one of the ponds (figs. 17.4 and 17.7; see also color plate 17.7) was excavated during the present study and analyzed in the field to determine the number of times sediment and water entered the pond.

The sediment in the ponds is thin, typically only 10–30 cm thick. The sedimentology consists of beds, or couplets, of material with particles that are coarser toward the bottom, and increasingly fine toward the top. Each couplet results from a ponding episode in which suspended sediment gradually settled through the water column at a velocity determined by grain size. These fining-upward couplets are bounded above and below by sharp contacts, which indicates a break in deposition and subsequent subaerial exposure (fig. 17.7). The transition between the relatively dark, coarse-grained basal unit (which is silty, very fine-grained sand) and the overlying, relatively light, fine-grained unit (which is silty clay) is gradational. Each couplet records a runoff-producing precipitation event; these events are thought to be the principal agents of landscape change and environmental recovery in the tributary washes at Camp Ibis. At the studied pond, 22 runoff events occurred from April 1943 to January 29, 2001, when the site was excavated. This gives an average runoff recurrence interval of 2.6 years, which is close to the average recurrence interval reported by Griffiths et al. (2006) for the Central, Northern, and Eastern Mojave Desert.

Precipitation History and Alluviation in Piute Valley

Precipitation in Piute Valley is recorded at only two weather stations: Needles, California, elevation 270 m, near the southern end of the valley, and Searchlight, Nevada, elevation 1038 m, near the northern end (fig. 17.1). Monthly precipitation totals were assembled from archival and commercially available databases (NOAA, Asheville, North Carolina, and EarthInfo, Inc.). The precipitation record begins in 1897 for Needles and 1914 for Searchlight. Although more weather stations would be useful, one may reasonably infer long-term precipitation patterns in the mapped area from the precipitation record of these stations.

Average annual precipitation at Needles and Searchlight is, correspondingly, 120 ± 14 mm and 196 ± 22 mm (± is the 95% confidence interval of the average). Precipitation is biseasonal and peaks in winter and summer. On a monthly basis, the correlation (i.e., Spearman correlation coefficient) between the two stations is $r = 0.72$. Assuming precipitation increases linearly with elevation, average annual precipitation in the Camp Ibis-Piute Wash area is 147 ± 17 mm.

The monthly precipitation time series of the two stations are shown in figure 17.8; the solid dots indicate precipitation (≥95th percentile, excluding those months with zero precipitation). These unusually wet periods probably had the potential to produce substantial runoff in Piute Wash and its tributaries in the Camp Ibis area. Although streamflow is not routinely measured in Piute Wash, a U.S. Geological Survey hydrographer (identified only by the initials "C.E.W.") stated in February 1940 that Piute Wash "carries about 10,000 sec. ft. [280 $m^3 s^{-1}$] when running full" where the wash crosses U.S. Highway 95 (fig. 17.4) (Nevada Historical Society, California Photographs, photograph number CA 576). The hydrographer probably estimated discharge from the high-water flood marks of September 1939, as discussed below.

The time between large precipitation events varies through the period of record. From roughly 1950 to 1975, the number of unusually wet months decreased, corresponding to an increase in the time between events (fig. 17.9). This episode of decreased precipitation in the mid-twentieth century, referred to as the mid-twentieth century drought, is typical of the Mojave Desert region (Hereford et al. 2004, 2006).

The relatively few large precipitation events from roughly 1950–1975 probably coincided with infrequent large floods in Piute Wash. This, in turn, would promote the spread of vegetation into the channel, development of braid bars, sediment accumulation during relatively rare flooding events, and burial of vehicle tracks. The condition of the channel in 1943 (figs. 17.3 and 17.5a) probably reflects the unusual rainfall of September 1939 (fig. 17.8), which likely produced a large flood or floods in Piute Wash with discharge as high as 280 $m^3 s^{-1}$. Flooding in the Mojave Desert at that time was widespread and severe and was caused by the passage of four tropical storms across the region, of which the storm of late September was the strongest (Smith 1986). Moreover, large rainfall events were relatively common before 1943, as compared with the 1950–1975 period. This quite likely increased the frequency of large floods, producing the wide, lightly vegetated channel present in 1943. The scoured condition of the braid channels in 1994 may reflect the numerous large rainfall events and resultant floods after around 1975.

Finally, runoff in the tributary channels of the study area occurred 22 times between 1943 and 2001, assuming the alluvial history of the studied pond site (fig. 17.7) is representative. The number of unusually wet months in this period is

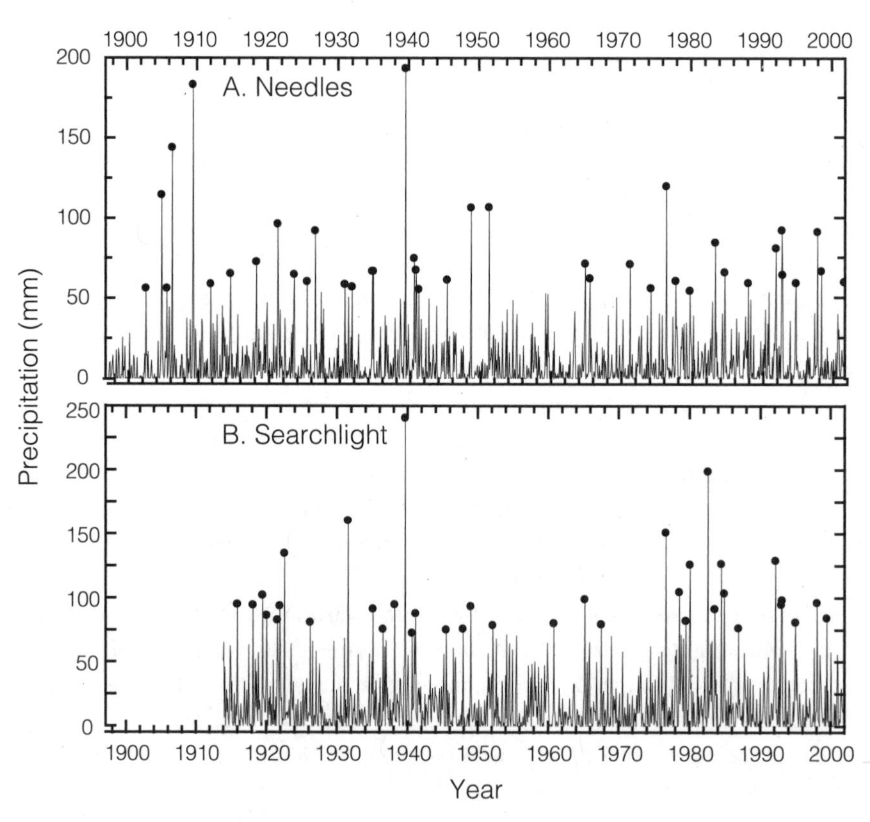

Fig. 17.8. Graphs of time series of monthly precipitation are presented for Needles, 1897–2002 (*A*), and Searchlight, 1914–2002 (*B*). Solid dots indicate months with precipitation ≥ 95th percentile of all months with precipitation.

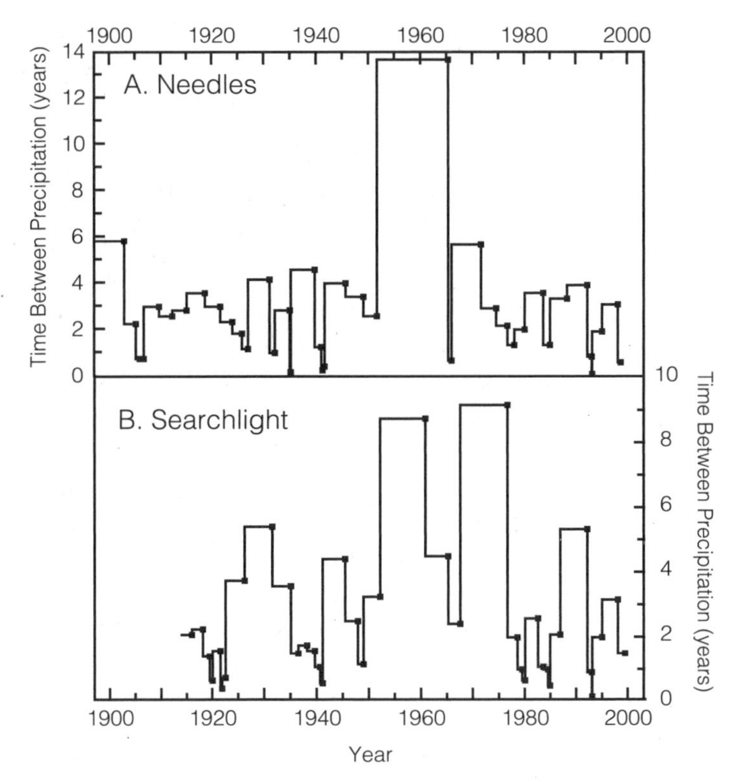

Fig. 17.9. Graphs of time series of years with precipitation ≥ 95th percentile of all months with measurable precipitation for Needles (*A*) and Searchlight (*B*). Note the increase in the time between events from roughly 1950–1975.

19 and 22 at Needles and Searchlight, respectively. This correspondence between wet conditions and recorded tributary runoff may not be entirely coincidental. If not coincidental, the rate of visual removal in the tributaries may have increased after around 1975, during the wet conditions of the 1980s and early 1990s.

CONCLUSIONS

Alluviation in Piute Wash and its tributaries after 1943 substantially altered the landscape of the study area, removing visible evidence of tank tracks and other World War II military activity in 6.2 km^2, which is about 30% of the area in and adjacent to the wash. This visual removal of anthropogenic disturbances restored the landscape on low-lying areas and in stream channels within 36–51 years. The removal of disturbances in Piute Wash probably resulted from a regional decrease in the frequency of runoff-producing precipitation, which enhanced establishment of vegetation in the channel and deposition of alluvium. Climate variation evidently controlled the rate of alluviation and visual removal.

Evidence supporting channel aggradation of Piute Wash is substantial. It is documented by the cross-cutting relation of numerous tank tracks with alluvium and by four sequential aerial photographic surveys of the area from 1943 to 1994 showing changes of channel planform and increased vegetation density (fig. 17.5). In the mapped area, sediment accumulated to an average depth of at least 1 m in 2.5 km^2 of channel during five floods with an average sedimentation rate of 1.6 cm y^{-1}. Alluviation, if not caused by the coincident spread of brushy vegetation into the channel, was at the very least enhanced by it. Surficial geologic mapping reveals that many large washes in the Central Mojave Desert have undergone a similar amount of alluviation from the 1950s to the late 1970s or early 1980s (R. Hereford *personal observation*).

Discontinuous tracks and roads show that aggradation also occurred outside of the main channel in tributary washes, on alluvial fans, and at the base of the hillslopes adjoining Piute Wash. These deposits vary widely in thickness from 5 to 50 cm, and they occupy 3.7 km^2 of the mapped area in Piute Valley. Visual removal by alluviation in the main channel and tributaries began between 1953 and 1979 throughout the study area, as shown in time-sequential aerial photography (fig. 17.5).

The road system remaining at Camp Ibis is 107 km in length, which is estimated to be 80% of its original length. Fluvial activity has fragmented the system, particularly at stream crossings. The roadbed at crossings is largely absent, having been eroded or covered by sediment at downstream localities. During construction of the camp, small washes crossing occupation sites were evidently filled in. Many of the washes have now reestablished channels across the cleared areas, which should eventually restore the original topography. These changes at Camp Ibis resulted from episodic runoff-producing precipitation, which affected the area 22 times since 1943, or about once every 2.6 years.

Finally, numerous large washes and valleys in the Mojave Desert are geomorphically and geologically similar to Piute Wash. A well-drained valley underlain by weakly consolidated sedimentary strata provides the potential source material for extensive visual removal by alluvial activity. Observations throughout the Central Mojave Desert show that fluvial erosion and alluviation on active geomorphic surfaces effectively removes evidence of disturbance.

ACKNOWLEDGMENTS

This work was supported by the Recoverability and Vulnerability of Desert Ecosystems Project of the U.S. Geological Survey. Robert H. Webb kindly supplied the archival aerial photography of Camp Ibis. Debra Block prepared the digital GIS database. Lee Amorosa and Kevin M. Schmidt reviewed the manuscript.

REFERENCES

Belnap, J., J. H. Kaltenecker, R. Rosentreter, J. Williams, S. Leonard, and D. Eldridge. 2001. Biological soil crusts: ecology and management. Technical Reference No.1730–2. U.S. Department of the Interior, Bureau of Land Management, Denver, Colorado.

Belnap, J., and S. D. Warren. 2002. Patton's tracks in the Mojave Desert, USA: an ecological legacy. *Arid Land Research and Management* **16**:245–258.

Bischoff, M. C. 2000. The desert training center/California-Arizona maneuver area, 1942–1944: historical and archeological contexts. Technical Series No. 75. Statistical Research, Tucson, Arizona.

Griffiths, P. G., R. Hereford, and R. H. Webb. 2006. Sediment yield and runoff frequency of small drainage basins in the Mojave Desert. *Geomorphology* **74**:232–244.

Hereford, R., R. H. Webb, and C. I. Longpré. 2004. Precipitation history of the Mojave Desert Region, 1893–2001. U.S. Geological Survey Fact Sheet No. 117–03. Reston, Virginia.

Hereford, R., R. H. Webb, and C. Longpré. 2006. Precipitation history and ecosystem response to multidecadal precipitation variability in the Mojave Desert and vicinity, 1893–2001. *Journal of Arid Environments* **67**:13–34.

Iwasiuk, R. J. 1979. Plant response parameters to General Patton's armored maneuvers in the eastern Mojave Desert of California. MS thesis. Loma Linda University, Loma Linda, California.

Lashlee, D., F. Briuer, W. Murphy, and E. V. McDonald. 2002. Geomorphic mapping enhances cultural resource management at the U.S. Army Yuma Proving Ground, USA. *Arid Land Research and Management* **16**:213–229.

Lathrop, E. W. 1983. Recovery of perennial vegetation in military maneuver areas. Pages 265–277 *in* R. H. Webb and H. G. Wilshire, editors. *Environmental effects of off-road vehicles: impacts and management in arid regions.* Springer-Verlag, New York, New York.

Lovich, J. E., and D. Bainbridge. 1999. Anthropogenic degradation of the southern California desert ecosystem and prospects for natural recovery and restoration. *Environmental Management* **24**:309–326.

Nichols, K. K., and P. R. Bierman. 2001. Forty-four years of ephemeral channel response to two years of intense World War II military activity, Camp Iron Mountain, Mojave Desert, California. *Reviews in Engineering Geology* **14**:123–136.

Prose, D. V. 1985. Persisting effects of armored military maneuvers on some soils of the Mojave Desert. *Environmental Geology Water Science* **7**:163–170.

Prose, D. V. 1986. Map showing areas of visible land disturbance caused by two military training operations in the Mojave Desert. U.S. Geological Survey Map No. MF-1855. Reston, Virginia.

Prose, D. V., and S. K. Metzger. 1985. *Recovery of soils and vegetation in World War II military base camps, Mojave Desert.* U.S. Geological Survey Open-File Report No. 85–234. Menlo Park, California.

Prose, D. V., S. K. Metzger, and H. G. Wilshire. 1987. Effects of substrate disturbance on secondary plant succession, Mojave Desert, California. *Journal of Applied Ecology* **24**:305–313.

Prose, D. V., and H. G. Wilshire. 2000. *The lasting effects of tank maneuvers on desert soils and intershrub flora.* U.S. Geological Survey Open File Report No. 00–512. Menlo Park, California.

Smith, W. 1986. The effects of eastern North Pacific tropical cyclones on the southwestern United States. Technical Memorandum No. NWS WR-197. National Oceanic and Atmospheric Administration, Salt Lake City, Utah.

Steiger, J. W., and R. H. Webb. 2000. *Recovery of perennial vegetation in military target sites in the eastern Mojave Desert, Arizona.* U.S. Geological Survey Open-File Report No. 00–355. Tucson, Arizona.

Webb, R. H. 2002. Recovery of severely compacted soils in the Mojave Desert, California, USA. *Arid Land Research and Management* **6**:292–305.

EIGHTEEN

A Review of Selected Long-Term Ecological Studies of the Mojave Desert Ecosystem

ROBERT H. WEBB, LESLEY A. DEFALCO, TODD C. ESQUE, AND PHILIP A. MEDICA

Understanding ecosystem processes in the Mojave Desert is a requirement of state and federal government agencies as well as a means for determining potential long-term changes induced by climatic fluctuations or land uses. Ecosystem function in the Mojave Desert is strongly influenced by both abiotic and biotic factors (Rundel and Gibson 1996), and an understanding of spatial variability induced by climatic variability and landform development is needed to determine where site-specific measurements should be made. Here, we examine the abundance of long-term studies among the three North American hot deserts in recently published literature. In addition, we explore what we have learned from the long-term studies in the Mojave Desert. Finally, we bring to light the types of long-term data that could help managers make better decisions, and we propose how the Nevada Test Site (NTS), with its historical data and ongoing research, serves as an example for understanding ecosystem quality and function in the Mojave Desert.

LONG-TERM STUDIES IN THE NORTH AMERICAN DESERTS

Long-term studies are essential for setting realistic management goals and developing appropriate plans for effectiveness monitoring. Information from long-term studies can also be used to determine the triggers for management action when resource conditions improve or decline (Elzinga et al. 1998) in an adaptive-management framework. Among their many responsibilities, managers are charged with maintaining populations of rare or sensitive plant and animal species, ensuring that human use of public lands does not degrade habitat for all species, and protecting the integrity of the resources that the land provides for human populations. Ecological phenomena occur over time spans and can have return intervals of several years (El Niño/Southern Oscillation, dramatic wildflower blooms, wildfires), decades (wet periods, drought periods), or centuries (vegetation succession, climate change). In the absence of a long-term temporal context, studies spanning only one to a few years may provide an incomplete picture of ecosystem dynamics in the absence of overt disturbance.

Here, we define one type of long-term study, which we will emphasize in this chapter, as one that repeatedly measures ecosystem attributes using the same techniques for extended time periods. This type is referred to as the iterative long-term study. An example of the quintessential long-term study is a storage rainfall gauge, where measurement of precipitation at a given point is made using the same technique for extended periods of time. The most common type of long-term study involving the terrestrial ecosystem of the Mojave Desert involves permanent plots, typically designed to repeatedly measure or census population changes in annual or perennial vegetation (e.g., Beatley 1980; Cody 2000; Webb et al. 2003). These plots may utilize minimal permanent fixtures, such as wooden stakes or rebar for permanent location, or they may involve considerable investment in infrastructure, such as at the Free-Air Carbon Dioxide Enrichment Facility or the Mojave Global Change Facility on NTS (Jordan et al. 1999; Smith et al. *this volume*).

Iterative studies provide the most useful data for developing and evaluating management objectives, but are the most uncommon. These studies document increasing, declining, or stable trends for resources in natural and disturbed areas. These studies generate data that can help managers determine the limits or thresholds for management goals because they (1) define the expected range of natural variability, and (2) describe the vulnerability and resiliency of the biota to climatic extremes. However, iterative studies require a commitment to maintaining the integrity of the sites or populations being studied, documenting and protecting the procedures after scientists retire, and preserving the data in reliable archival format. Examples of such studies include historical fluctuations in rainfall across the desert (Hereford et al. 2006); censuses of native and invasive annual plants (Beatley 1966, 1973; Hunter 1991); repeat photography used to estimate mortality, recruitment, and plant species compositional changes (Turner 1990; Bowers et al. 1995); measurement of perennial vegetation on permanent plots (Beatley 1980; Goldberg and Turner 1986; Cody 2000; Bowers 2005b, 2005c); long-term growth of tagged desert tortoises (Medica et al. 1975; Turner et al. 1987); density, body size and reproduction of lizards (Woodward 1994, 1995); and compositional changes in small mammal populations (Saethre 1995; Price et al. 2000).

In addition to iterative long-term studies, two other types of studies, reference and chronosequence, have been conducted in the Mojave Desert. Reference studies compare sites that were impacted at a known time in the past (typically by humans) with nearby control areas that had little or no impact. This type of study is commonly used to document the recovery of soils and vegetation following surface disturbances such as fire (Brooks and Matchett 2003), use of recreational or military vehicles (Webb 2002; Belnap and Warren 2002; Lei *this volume*), historical townsite occupation (Webb et al. 1988), and utility corridor construction (Webb et al., Natural Recovery, *this volume*). Reference studies, while widespread throughout the Mojave Desert due to their simplicity in design, do not alone provide sufficient information about recovery trajectories for managers because the range of natural variability cannot be estimated for a one-time comparison across the time scale measured. However, they are valuable for understanding the cumulative effects of conditions on plants and animals during the discrete time interval in which they are conducted.

Chronosequence studies use site-for-time comparisons as a proxy for examining trends (Webb et al. 1987, 1988; Hamerlynck et al. 2002; Bolling and Walker 2000, 2002). These studies provide a rapid appraisal of trends occurring in the past, and often can span a broader period of time with results obtained in a single field season. One excellent example of a chronosequence study is the analysis of packrat middens and other cave deposits for evaluating the changing distributions of plants and animals over the past 45,000 years (Wells and Jorgensen 1964; King 1976; Wells and Hunziker 1976; Mead 1981; Spaulding 1990; F. A. Smith et al. 2000). However, unless sites of the same age can be replicated, and additional factors (such as soil characteristics and local climate) are held constant across all sites, results of chronosequence studies may be confounded by site-specific characteristics. Also, this type of study generally does not document the trajectory of change, which may be important to understanding ecosystem processes (Webb et al., Natural Recovery, *this volume*).

Two other types of long-term studies contribute to resource management goals in the Mojave Desert. Evolutionary studies that examine the phylogenetic relationships for microbial, plant, and animal taxa ask how and why species have arrived or disappeared from the Mojave Desert over broad time scales and under fluctuating climates (Sanderson and Stutz 1994; Lee et al. 1996; Kelt 1999; Minckley et al. 2000; Riddle et al. 2000; Coleman et al. 2001; Hunter et al. 2001). Other evolutionary studies investigate how habitat fragmentation can influence population genetic structure of threatened or endangered plant and animal species (Lamb et al. 1989; Rainboth et al. 1989; Britten et al. 1997; Hickerson and Wolf 1998; Trepanier and Murphy 2001). In comparison, geologic studies also span time periods well beyond the careers of scientists and resource managers. Geologic studies have been useful for understanding how soil water dynamics and plant growth respond to soil horizon development (Smith et al. 1995; Hamerlynck et al. 2002) and how climate change has and will affect land forms such as sand dunes (Clarke and Rendell 1998) and the movement of vegetation communities over time (Spaulding 1990; Hockett 2000).

Length of Studies in the Mojave Desert

We queried the recent published literature (2002–2005) contained in Biological Abstracts and BIOSIS Reviews by searching for "desert" in the title and "North America" for geographic area, then further refined our search for iterative long-term studies of the Mojave Desert (1969–2005), as defined above and using "Mohave" and "Mojave" as subject terms. Our findings (fig. 18.1), which are preliminary and not exhaustive, illustrate the low proportion of long-term ecological studies in the hot deserts—Mojave, Sonoran, and Chihuahuan—of North America. Studies of longer duration, in particular, are fewer in the Mojave Desert compared with the Sonoran Desert.

Most iterative studies in the Mojave Desert were less than 5 years in duration (fig. 18.2), but a handful of studies were of longer duration. Most of the long-term studies are on what is now the Mojave National Preserve, Joshua Tree National Park, and the NTS (figs. 18.3 and 18.4). Collectively, these studies document long-term dynamics in vegetation (Cody 2000; Miriti et al. 2001; Webb et al. 2003), small mammals (Price et al. 2000), lizards (Medica and Turner 1976; Turner and Medica 1977; Turner et al. 1982), and desert tortoises (Turner et al. 1987; Longshore et al. 2003). This pattern has set a precedent for what is typically regarded as "long-term" for the Mojave Desert, even though studies spanning a single decade to multiple decades are common in the Chihuahuan (Geiger and McPherson 2005; Goheen et al. 2005; Valone and Kaspari 2005) and Sonoran (Goldberg and

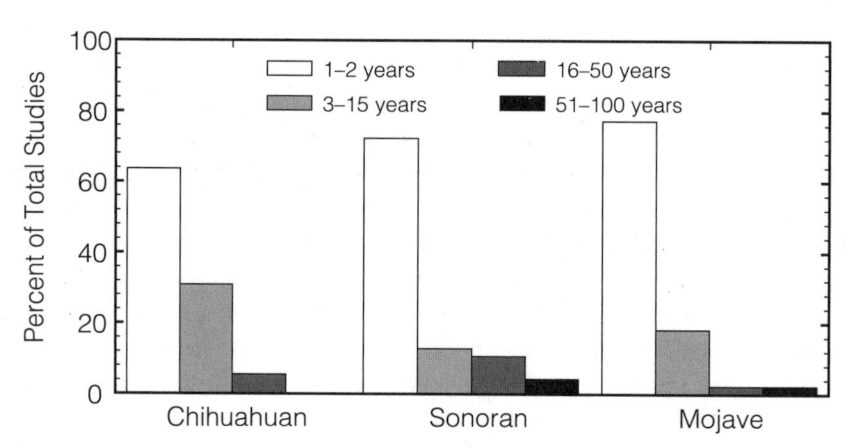

Fig. 18.1. Graph showing the percent of published studies versus length of the study period for North American warm deserts from 2002 through 2005. The graph clearly shows that there are few studies that extend beyond the past fifteen years.

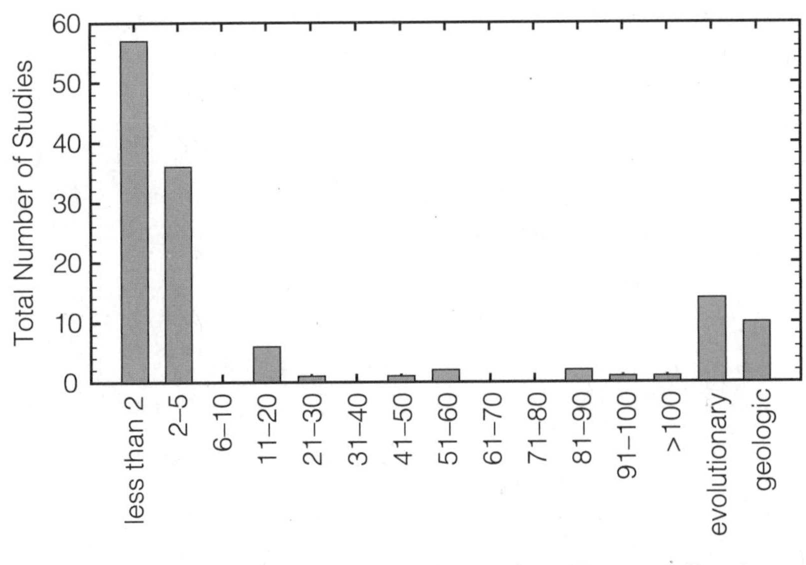

Fig. 18.2. Graph of a representative sample of 125 studies published from 1969 through 2005 for the Mojave Desert are plotted against the length of study period (years).

Turner 1986; Turner 1990; Bowers 2005a) deserts. The limited spatial distribution of long-term studies in the Mojave Desert suggests that one emphasis of future research might be the establishment of more long-term studies in areas representative of the other subregions of the Mojave Desert.

EXAMPLES OF LONG-TERM STUDIES IN THE MOJAVE DESERT

Variability in Regional Precipitation

Regional fluctuations in rainfall are a significant factor leading to large changes in desert ecosystems. Hereford et al. (2004, 2006) examined 52 long-term climate stations in the Mojave Desert and vicinity to assess long-term variability in annual, cold-season, and warm-season precipitation (also see Redmond *this volume*). Hereford et al. (2004) found that annual precipitation in the Mojave Desert region averages 137 mm y^{-1} and ranges from 34 to 310 mm y^{-1}. The driest year was 1953, and the wettest years were 1941 and 1983 (fig. 18.5). Most precipitation occurs in the cool season (October–April) and averages 95 mm y^{-1} with a range of 27–249 mm y^{-1} at the 52 long-term stations. West of longitude 117°W, approximately 82% of annual precipitation occurs from October through April. East of this longitude, 66% of annual precipitation occurs during those cool season months (Hereford et al. 2004).

A total of five climatic episodes are apparent in the time series of Figure 18.5:

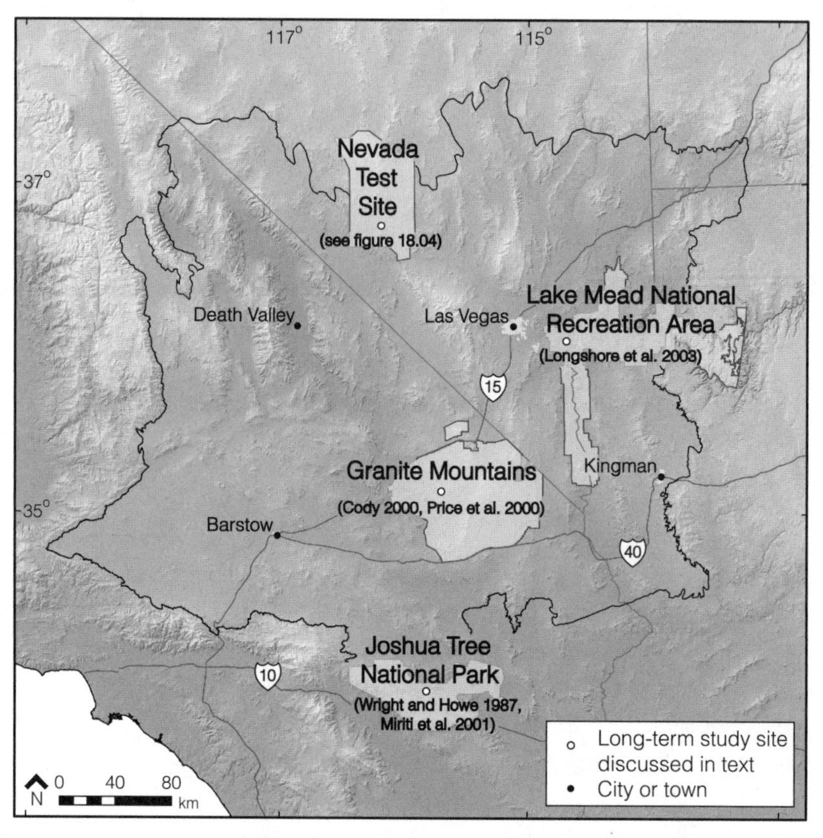

Fig. 18.3. Map of the greater Mojave Desert ecosystem showing the locations of the iterative study sites discussed in the text. A detailed map of the Nevada Test Site is provided in fig. 18.4.

(1) sustained drought occurred from 1893 through 1904; (2) above-average precipitation occurred from 1905 through 1941 with interruptions, especially in the 1930s; (3) the mid-century drought, a regional event, occurred from 1942 through 1975; (4) above-average precipitation, with some years at extremely high levels, occurred from 1976 through 1998 but was punctuated by the 1989 through 1991 drought; and (5) severe drought occurred from 1999 through 2004. The choice of bracketing dates for these episodes, particularly for the mid-century dry period, is somewhat subjective (Hereford et al. 2006), but the middle of the twentieth century was marked by drought and was bracketed by wetter episodes. In particular, the period from 1978 to 1998 was the wettest since the 1940s, if not the wettest in the entire the twentieth century (Hereford et al. 2006).

Wet and dry episodes in the Mojave Desert generally coincide with El Niño and La Niña conditions, respectively, which are part of the El Niño-Southern

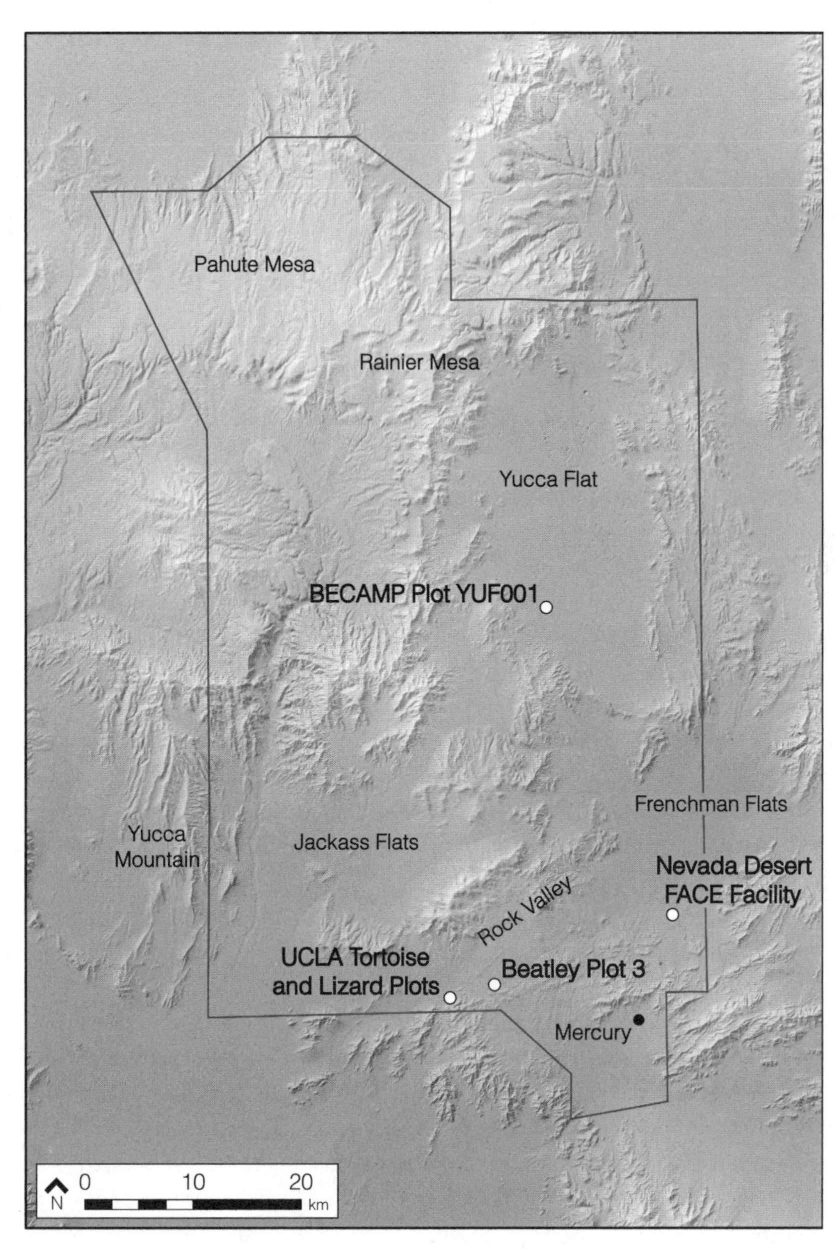

Fig. 18.4. Map of the Nevada Test Site in southern Nevada shows the locations of sites discussed in the text.

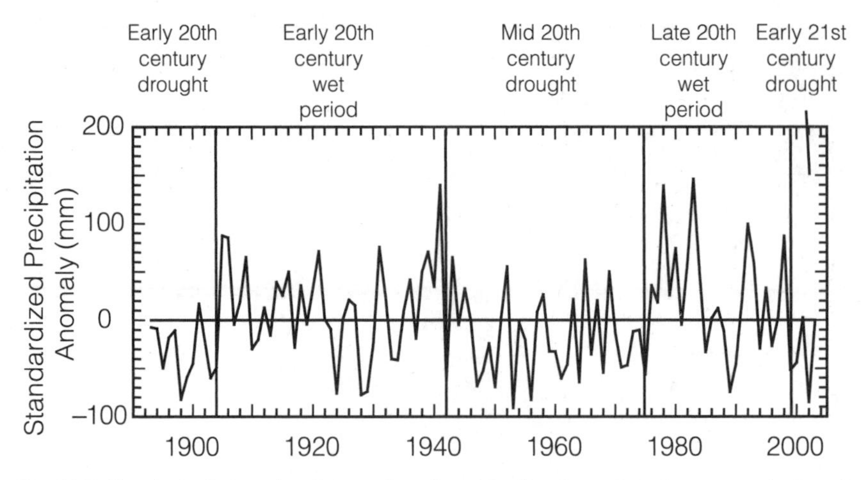

Fig. 18.5. Graph depicting the time series of standardized precipitation anomaly (mm), which represents the annual deviation from mean annual precipitation in the Mojave Desert, from 1893 through 2001 (Hereford et al. 2006). Mean precipitation = 137 mm.

Oscillation (ENSO) phenomenon of the Pacific Ocean (Rasmussen 1984). The most common measure of ENSO is the negative difference in sea-level air pressure between Tahiti and Darwin, Australia, which also is known as the Southern Oscillation Index (SOI). Annual precipitation in the Mojave Desert is correlated with the SOI, and above-normal annual precipitation occurred in 55% of years with El Niño conditions (Hereford et al. 2004). While this does not appear to be a strong relation, some of the highest rainfall years (e.g., 1978, 1983, 1993, 1998) had El Niño conditions.

The Pacific Decadal Oscillation (PDO) (Mantua and Hare 2002) is a climatic index related to sea-surface temperature and atmospheric pressure of the North Pacific Ocean and is broadly related to ENSO, affecting the spatial connection between ENSO and precipitation in the western United States (McCabe and Dettinger 1999). The PDO, which represents decadal-scale climatic variability, is thought to explain certain aspects of Mojave Desert climate because of its relation to twentieth century climate variability elsewhere in the western United States. Precipitation in the Mojave Desert region is significantly correlated with the PDO, which helps to explain the occurrence and timing of the five episodes of Mojave Desert precipitation and also establishes the global context of climate in this desert (Hereford et al. 2004, 2006).

Annual Plant Populations

Winter annual plants are one of the most dynamic components of the Mojave Desert as their abundance and diversity are intimately tied to winter rainfall

(Beatley 1967, 1969a; Bowers 1987). On a regional scale, the conventional wisdom is that heavy rainfall during ENSO years stimulates abundant winter annual plant production in the Mojave Desert. A recent study refines this relation by showing that the SOI is a better predictor of the abundance of winter annuals in the Mojave and Sonoran deserts when SOI is specifically analyzed on a monthly period of July through December (Bowers 2005b).

Plot 3, established on the NTS by Dr. Janice C. Beatley in 1963, provides the best example of long-term fluctuations in desert annuals (figs. 18.4 and 18.6). A long-term rainfall gauge provides precipitation data for Rock Valley, including shifting seasonal precipitation (fig. 18.6A). Winter precipitation, which averages 119.3 ± 78.3 mm at this station [coefficient of variation (CV) = 0.66], accounts for 74% of the annual precipitation from 1964 through 2005; summer rainfall has a

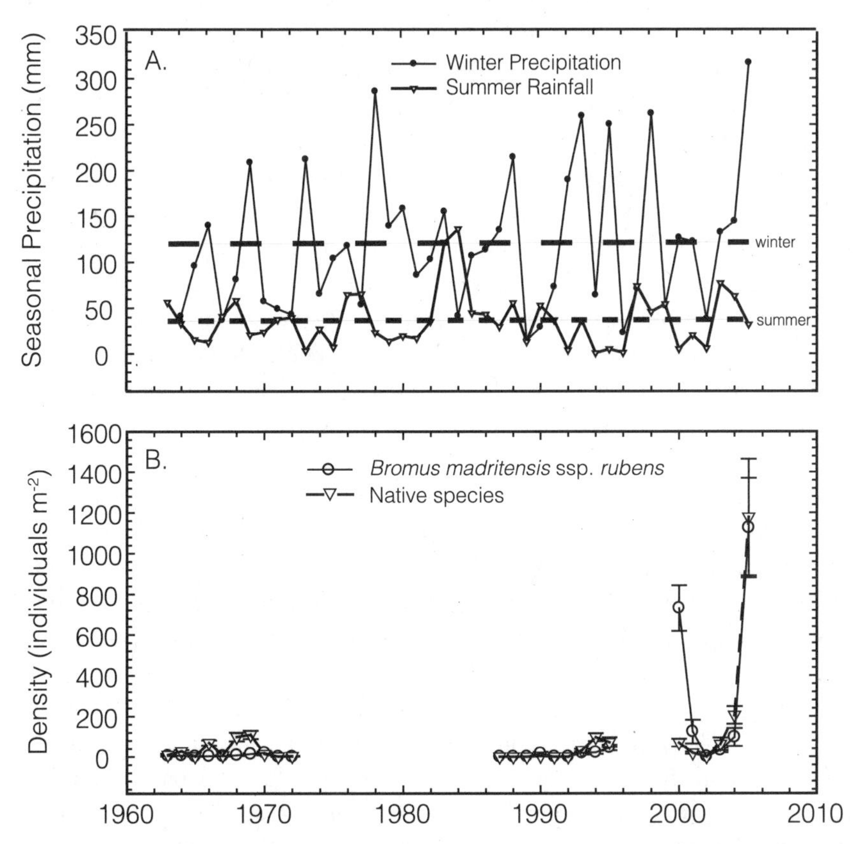

Fig. 18.6. Graphs showing seasonal precipitation (*A*) and annual plant densities (*B*) on Beatley Plot 3 at the Nevada Test Site from the 1960s to present (Hunter 1991, 1995a; L. A. DeFalco *unpublished data*). In *A*, the horizontal lines represent mean winter and summer precipitation, respectively.

less-variable average of 36.6 ± 29.6, albeit with a higher coefficient of variation (CV = 0.81). For this site, average winter precipitation during El Niño years (1966, 1969, 1973, 1978–1980, 1983, 1987, 1992–1993, 1995, 1998, and 2005) was 207 mm, or 74% above average for all years combined.

The abundance of native species from the early 1960s to the 1990s fluctuated with winter rainfall during the period of measurement, although censuses were not made in many of the El Niño years, such as 1998. Nonnative brome grasses, notably *Bromus madritensis* ssp. *rubens* (red brome), have become increasingly dominant in annual plant communities (Brown and Minnich 1986; Hunter 1991) coincident with the period of increased rainfall that predominated after the midcentury drought. Revisitation of an undisturbed plot (L. A. DeFalco et al. *unpublished data*), censused first by J. C. Beatley from 1963 through 1975, and then by R. B. Hunter from 1987 through 1995, suggests that although *B. madritensis* ssp. *rubens* was present from 1963 through the 1990s (fig. 18.7B), there is no indication that dominance by *B. madritensis* ssp. *rubens* in undisturbed areas is causing a decline in the density of all native species.

However, in 2000, *B. madritensis* ssp. *rubens* did occur on this plot at higher densities than native species. This likely resulted from the large seed bank established during the El Niño winter of 1997–1998, when *Bromus* densities were extremely high (S. D. Smith et al. 2000), and the fact that 1999 was a drought year with little germination of annual vegetation (fig. 18.6A). The 2000 high-density spring was followed by precipitous declines of all annuals during 2001 through 2003, and particularly during the extreme drought year of 2002 (fig. 18.6B).

The densities of both native and nonnative annual plants rebounded following the record wet winter of 2004 through 2005 (fig. 18.6). This demonstrates the complex interactions between climate, seed dormancy during drought, and seed dispersal. This recent increase in nonnative annual grasses such as *B. madritensis* ssp. *rubens* is highly relevant to land management because increased fine fuels increases fire frequency (Brooks et al. 2004). In addition, the relation among native and nonnative species is a topic of substantial interest to land managers (Brooks *this volume*) and can be better understood with long-term plot measurements.

Perennial Vegetation

Although perennial vegetation in the Mojave Desert was once considered to be static and unchanging, long-term studies demonstrate that variation in rainfall can have a dramatic impact on shrubland communities, notably during extended droughts (Reynolds et al. 1999; Lei 1999; Miriti et al. 2001; Breshears et al. 2005; Bowers 2005a). While the period of 1977 through 1984 was one of the wettest periods of the twentieth century, droughts occurred from 1989 through 1991 (Hunter 1994; Miriti et al. 2001), 1996, and 2000 through 2004. Many shrubs died during the 1989–1991 drought (Schultz and Ostler 1995a), suggesting that drought is a

B.

Fig. 18.7. Photographs showing Beatley Plot 3 at the Nevada Test Site on two different dates (Webb et al. 2003). *A,* on April 19, 1964, this view southeasterly across Plot 3 shows the typical mixed *Larrea tridentata* (creosote bush) community in that area and part of the Specter Mountain Range in the background (Janice Beatley Collection, 15-A). *B,* on May 9, 2000, the same view shows that biomass throughout the plot has increased considerably since the original photograph was taken. The lighter colored material in the foreground is bioturbation from digging by an animal of unknown species, possibly badger (R. H. Webb, Stake 4020B).

major mechanism for change in Mojave Desert ecosystems. With the resumption of rainfall, those plants that did not die during the severe drought added considerable biomass in the subsequent El Niño year of 1993 (Schultz and Ostler 1995b). In South Australia, Noble (1977) found that woody plant biomass was correlated with rainfall that occurred in the previous 12–42 months.

Reoccupation of plots first established on the NTS in 1963 demonstrates that, despite the droughts, the increase in the size and cover of plants between 1963 and 2000 is large for some species (Webb et al. 2003). Using Plot 3 as an example (fig. 18.7), the average height of *Larrea tridentata* (creosote bush) was 0.78 ± 0.22 m in 1975 and 1.06 ± 0.37 m in 2000, a statistically significant increase of 37%. The cover of *Larrea* and *Ephedra nevadensis* (Mormon tea) increased 93% and 145%, respectively. Few individuals of either species died in the 1989–1991 drought, but the cover of *Grayia spinosa* (spiny hopsage) decreased by 80% on this plot from 1989 through 2000, probably during the 1989–1991 drought. In general, plant associations on the NTS dominated by *Larrea* had large increases in the size of individual plants, as well as increases in the total number of plants (an indication of density) and total cover. Although biomass is not directly estimated, these phytoecological changes indicate that above-ground biomass has increased in *Larrea* assemblages on the NTS.

As indicated by the history of Plot 3, increases in *Larrea* and *Ambrosia dumosa* (white bursage), and decreases in *Grayia* and *Lycium andersonii* (wolfberry), indicate large changes in species composition as assessed by relative cover (fig. 18.8). Measurements made in 1989 indicate that the most significant changes occurred after this year, implicating the 1989–1991 drought (fig. 18.8). The drought year of 2002 caused additional changes to the perennial plant assemblage on Plot 3—total perennial cover declined, largely because of decreases in *Larrea* and *Ambrosia,* and *Lycium* and *Grayia,* which had not rebounded to their pre-1989 levels, remained at low levels.

Extreme cold plays a role in culling perennial plants as well. The freeze of December 24–25, 1990, was observed to damage *Menodora spinescens,* an endemic shrub of the Northern Mojave Desert (P. A. Medica *written communication*). The net effect was that the cover of *Menodora* on Beatley Plot 9 decreased from 10.0% in 1975 to 6.5% in 2000 (Webb et al. 2003). The same freeze caused damage to *Opuntia ramosissima* (pencil cholla) and *Opuntia basilaris* (beavertail cactus) in the Ivanpah Valley of eastern California. In 1937, an extreme freeze near St. George, Utah, killed above-ground branches of *Larrea* (Cottam 1937), but most individuals quickly resprouted from the root crowns (Fosberg 1938).

The amount of cover contributed by dead plants (either whole plants or frost- or drought-pruned parts) increased on the NTS in response to various extreme frosts, the 1989–1991 drought, and the 2002 drought year (Hereford et al. 2006). The climatically driven phenomena—both wet and dry periods—has

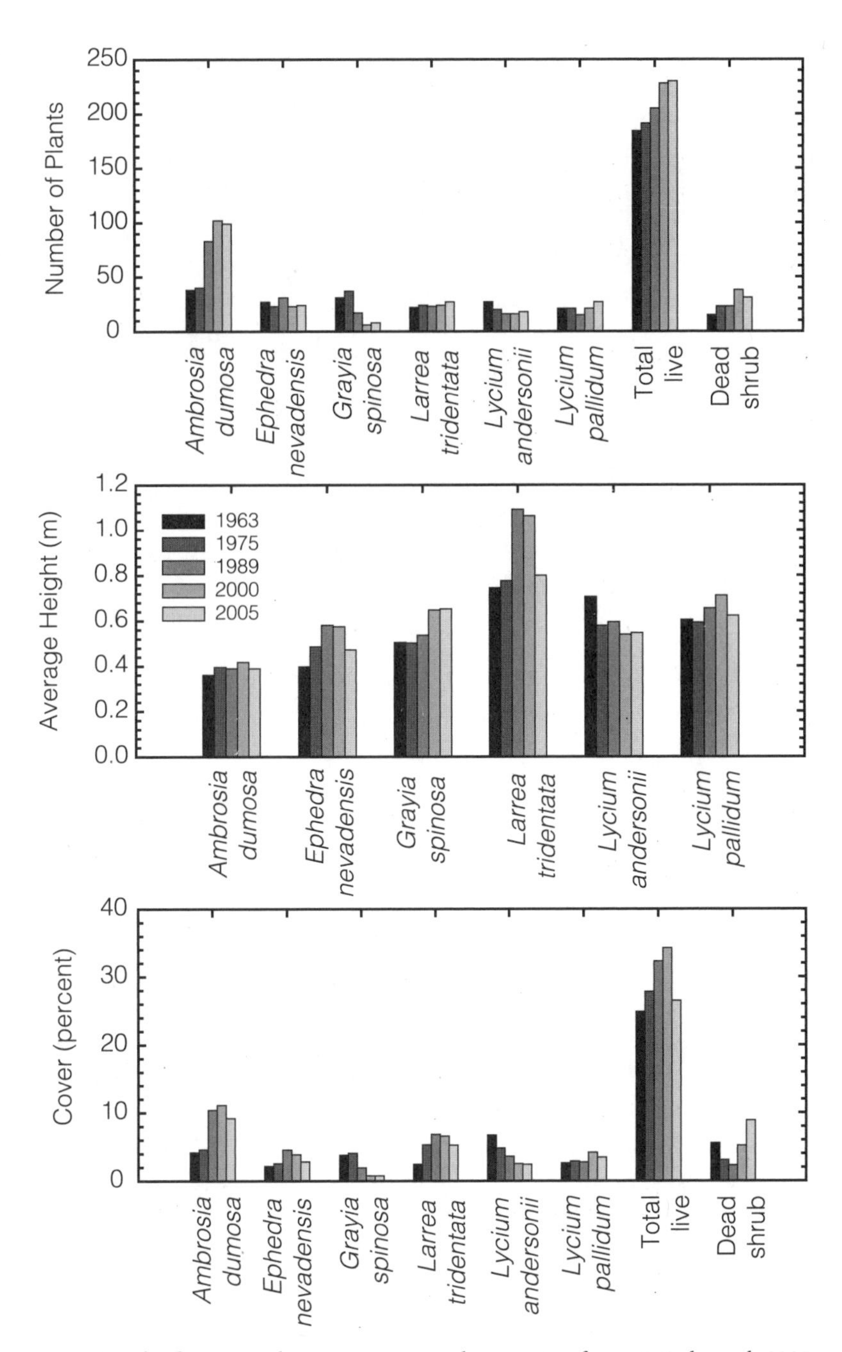

Fig. 18.8. Graphs depicting changes in perennial vegetation from 1963 through 2005 on Beatley Plot 3, located on the Nevada Test Site. Data from 1989, collected by R. B. Hunter, are not from the transect lines that Beatley established in 1963; the other data were collected using standardized methods (Webb et al. 2003).

Table 18.1 Change in density of selected reptile and rodent populations on the Nevada Test Site compared with winter (October–March) rainfall from 1987–1994 at Plot YUF001 (fig. 18.4)

Year	1987	1988	1989	1990	1991	1992	1993	1994
Rainfall (mm) Oct-Mar[a]	102.6	112.7	27.4	25.7	58.4	141.1	266.8	76.1
Spring *Uta* (no. ha^{-1})[b]	ND	9 1± 10	75 ± 10	48 ± 6	64 ± 6	114 ± 9	85 ± 11	118 ± 17
Number of *Uta* hatchlings[b]	113	68	10	38	102	188	137	19
Dipodomys merriami (no. ha^{-1})[c]	9.8 ± 0.3	5.2 ± 0	3.4 ± 0	5.0 ± 1.3	7.4 ± 0	15.1 ± 1.7	16.7 ± 1.4	14.4 ± 0.5
Dipodomys microps (no. ha^{-1})[c]	5.0 ± 0.7	5.2 ± 0.8	2.7 ± 0.7	2.3 ± 1.0	1.2 ± 0	5.4 ± 0.7	20.9 ± 1.2	34.8 ± 0.9
Perognathus longimembris (no. ha^{-1})[c]	27.8 ± 2.4	19.0 ± 1.8	9.0 ± 1.6	8.2 ± 4.7	13.2 ± 3.5	3.4 ± 1.9	3.4 ± 1.4	2.8 ± 0.2
Total rodent Density (no. ha^{-1})[c]	44.1	29.6	16.4	12.0	19.1	26.2	50.3	53.1

[a]Rainfall estimates are the average for the stations Buster Jangle and Yucca Lake, which are at elevations close to YUF001 and are less than 6.8 km away. The averages of Oct–Mar precipitation for these stations (1961–2003) are 100.6 and 108.8 mm, respectively.
[b]Density estimates and number of hatchlings are from Medica et al. (1994) and Woodward (1995).
[c]Density estimates from Saethre (1995).
ND = No data.
Density estimates are (numbers ha^{-1} ± 2 standard error) for animal populations.

changed the vegetation assemblage on Plot 3 from a *Lycium-Grayia* dominated assemblage in 1963 to a *Larrea-Ambrosia* assemblage in 2005. This underscores the potential for climatically driven processes to significantly alter the species composition of perennial vegetation in the Mojave Desert, thus it is important for management to take into account that the magnitude of the changes may rival changes induced by land-use practices such as livestock grazing, mining, and recreation.

Drought also affects other elements of perennial plant assemblages on the NTS. In the 1960s and 1970s, grasses [notably *Achnatherum speciosum* (desert needlegrass)] dominated the perennial vegetation in some parts of NTS, particularly Yucca Flat (Hunter 1995b). The 1989–1991 drought, one of the most severe in the history of southern Nevada, greatly reduced the density and cover of perennial grasses. These changes occurred without significant grazing pressure, which indicates that perennial plants are regulated by both anthropogenic and climatic processes.

Lizard Populations

Side-blotched lizard (*Uta stansburiana*) is ubiquitous in the Mojave Desert, and populations of this short-lived lizard respond rapidly to climatic fluctuations (Turner et al. 1970, 1982). At the NTS from 1987 through 1991, *Uta* populations on a baseline plot on Yucca Flat fluctuated tenfold (Woodward 1994, 1995). Initial densities of adult lizards in the spring of 1988 (91 ± 10 ha^{-1}) decreased to 75 ± 10 ha^{-1} by the spring of 1989, followed by a precipitous decline to 48 ± 6 ha^{-1} in the spring of 1990 and a rebound to 64 ± 6 ha^{-1} in 1991 (table 18.1). The initial decline in adult density was also matched by the lack of hatchlings during this drought, with 113 hatchlings enumerated in the summer of 1987, 10 in 1989, 38 in 1990, and 102 in 1991 (Medica et al. 1994). Subsequent *Uta* sampling from 1992 through 1994 documented an increase in density followed by another decline in 1993 and 1994 (Woodward, 1994, 1995).

Over a decade earlier, as part of the University of California at Los Angeles (UCLA) and the International Biome Program (IBP) studies at NTS, ten consecutive years (1964–1974) of lizard population studies in Rock Valley provided a wealth of information. Under favorable reproductive conditions, which coincide with high winter rainfall years, both clutch frequency and clutch size increased (Turner et al. 1974; Medica and Turner 1976; Turner et al. 1982), while under drought conditions *Uta* reproduction was greatly reduced or virtually nonexistent, as was observed in 1989 in Yucca Flat (table 18.1). These data show the collateral effect of drought (in this case, the 1989–1991 event) on reptile populations. On the other hand, these data are from a relatively short period (7–10 years) during which an extreme climatic event occurred. Longer-term data at the same location are needed to put these large population fluctuations into perspective.

Small Mammal Populations

Small mammal populations in the Mojave Desert are dependent upon winter precipitation and annual plant production (Beatley 1969b, 1976a, 1976b). In one 17-year study of kangaroo rats (genus *Dipodomys*) along a 615 m elevation gradient in the Granite Mountains, Price et al. (2000) studied interactions between two species, Merriam's kangaroo rat (*Dipodomys merriami*) and Panamint kangaroo rat (*Dipodomus panamintinus*) and the effect of climatic fluctuations on both small mammal species. Changes in abundance and species composition along their elevational gradient resulted from interannual variability in precipitation and subsequent changes in physical and biological conditions of their habitat. In particular, the larger *D. panamintinus* required a more dependable food supply inherent in higher elevation habitat, and *D. merriami* was restricted from moving to the higher elevations due to competition with *D. panamintinus* (Price et al. 2000).

On the NTS, small mammal populations fluctuated significantly from 1987 through 1991, which was mostly a period of drought. While some species of small mammals responded rapidly to climatic fluctuations (i.e., *D. merriami*), other longer-lived species did not respond as rapidly [i.e., Little pocket mouse (*Perognathus longimembris*) and Chisel-toothed kangaroo rat (*Dipodomys microps*)]. The population of *D. merriami* on Yucca Flat declined in 1989, increased slightly in 1990, and subsequently increased from 1991–1994 (Saethre 1995). Densities of *D. merriami* decreased from 9.8 ± 0.3 ha^{-1} in 1987 to a low of 3.4 ± 0 ha^{-1} in 1989 and then continued to increase to approximately 14–16 ha^{-1} in 1992–1994 (table 18.1). In 1987, *D. microps* was approximately half the density of *D. merriami.* Subsequently, *D. microps* decreased to 2.7 and 2.3 ha^{-1} in 1989 and 1990, respectively, to a low of 1.2 ± 0 ha^{-1} in 1991. In subsequent years, the density of *D. microps* surpassed that of *D. merriami* by a factor of two, reaching 34.8 ± 1 ha^{-1} by 1994 (table 18.1).

Another shift in the species composition of rodents at the NTS is illustrated by *P. longimembris.* This species, which was relatively numerous on Yucca Flat, declined precipitously from 27.8 ± 2.4 ha^{-1} in 1987 to 2.8 ± 0.2 ha^{-1} in 1994 (table 18.1), which was mostly a drought period. Other species of small rodents were present in most years but their numbers were generally low. A notable increase in deer mice (*Peromyscus maniculatus*) occurred in 1993, a relatively wet year, and the uncommon species Western harvest mouse (*Reithrodontomys megalotis*), which was only captured once in 1987–1992, increased in 1993 as well (Saethre 1995). In 2005, Yucca Flat was again sampled for small mammals, thus providing small-mammal population data for a period of 18 years. This long-term data set illustrates the complexities of the population dynamics of small mammals, which follow the variation in annual plant biomass induced by climatic fluctuations (Beatley 1969b).

Desert Tortoise Populations

Desert tortoise (*Gopherus agassizii*) is federally listed as a threatened species in the Mojave Desert, and recent declines in some populations have raised concerns about the long-term viability of this species. For its primary food supply, desert tortoise is dependent upon annual plant production, which strongly varies with the amount and timing of winter rainfall (Beatley 1974). Failure of annual plant production during prolonged droughts can lead to stress or mortality of these animals (Peterson 1996). Physiological health of *Gopherus* typically fluctuates with the seasonal and interannual changes in winter rainfall (Christopher et al. 1999). A six-year study that included the movements and home ranges of tortoises also demonstrated that *Gopherus* move less and remain close to shelters during years of low rainfall (Freilich et al. 2000).

The long-term study in Rock Valley (fig. 18.4) has provided the longest continuous growth data set from known-aged individuals of *Gopherus.* Between 1963 and 1965, researchers marked and measured 17 young (1- to 4-year old) tortoises within three plots. This cohort of tortoises has been repeatedly recaptured, measured, and weighed nearly every year since 1963 to document growth in relation to age. The growth of these tortoises in years of high winter rainfall (> 200 mm October–March) was rapid (12.3 ± 0.9 mm y^{-1}) in 1969, a wet El Niño year (fig. 18.6A), and slow in 1972 (1.8 ± 0.6 mm y^{-1}), a drought year (Medica et al. 1975). Subsequent studies of this cohort have continued to provide age-size relations over 24 years (Turner et al. 1987); physiological information pertaining to activity, drinking, water balance, and energy and salt budgets in contrasting wet and dry years (Nagy and Medica 1986); and details of long-term growth up through sexual maturity and greater than 45 years of age (P. A. Medica et al. *unpublished data*). The aggregate results show that *Gopherus* growth rates are understandably higher in wet years with higher annual plant production.

Gopherus populations not only exhibit dramatic temporal fluctuations associated with annual variability in climate, but are remarkably different across regions, especially in response to drought. The drought conditions that prevailed in the Mojave Desert (fig. 18.5) resulted in high mortality of *Gopherus* in Ivanpah Valley in the winter of 1981–1982 (Turner et al. 1984). The mortality of a cohort of radio-tagged tortoises (n = 69 in 1980, n = 76 in 1981) indicated that there was no difference between male and female survivorship. Mean mortality in 1980–1981 (4.4%), a relatively normal winter, and 1981–1982 (18.4%), a drought year, documented the influence of drought and the lack of ephemeral plant production. The combined dry weight of forbs and annual grasses was 0.07 g m^{-2} in the 1981 drought year, compared with 8.7 g m^{-2} in 1980 (Turner et al. 1984), and this apparently resulted in a four-fold increase in mortality.

Similarly, Longshore et al. (2003) presented data from two sites at Lake Mead National Recreation Area that showed large differences in mortality from 1994 through 2001. These two sites, Cottonwood Cove (290–360 m) and Grapevine Canyon (650–860 m), were 29 km apart, which should be well beyond the home ranges of these animals. Both sites have *Larrea-Ambrosia* assemblages, but annual productivity differences are unknown. Grapevine Canyon had an average of three times greater annual biomass than Cottonwood Cove during the measurement period (Longshore et al. 2003). *Gopherus* mortality was high at Cottonwood Cove and low at Grapevine Canyon; mortality at Cottonwood Cove was associated with low annual production during the 1996 drought and low precipitation in 1997 and 1999 (Longshore et al. 2003).

Localized summer rainstorms may influence the patchy distribution of desert tortoises. This was illustrated in the Eastern Mojave Desert in Ivanpah Valley when a localized mid-summer rainfall event of 48 mm occurred over an area of approximately 23 km^2 in August 1994. By mid-September, 95% of *Larrea* individuals produced new leaves and 55% produced flowers and fruit, and 88% of the *Ambrosia* and 94% of the *Lycium* shrubs responded by producing new leaves. The perennial grass *Pleuraphis rigida* (big galleta grass) greened up and produced new stems and seed. *Gopherus* within this greened-up area consumed the resprouted grass, as observed in fresh scats. In the nearby area (5 km northeast and ~ 70 m lower in elevation) that received only 0.8 mm of rainfall from the same storm, fresh tortoise scats contained cacti glochids (Medica et al. 1996).

The Turner et al. (1984) and Longshore et al. (2003) studies indicate that site-specific differences in annual productivity are important to *Gopherus* survival, even over short distances. This result underscores the need for population monitoring over extended periods at multiple sites. The spatial variability of winter precipitation may be an important consideration in understanding population dynamics of desert tortoise and other Mojave Desert species, particularly during dry years.

Nevada Test Site: A Model for Long-Term Studies in the Mojave Desert

As evident in the previous discussion, the Nevada Test Site (fig. 18.4), a facility now operated by the Department of Energy, has the largest number of long-term studies in the Mojave Desert and continues to provide valuable information for resource managers. Scientists from Brigham Young University and New Mexico Eastern University initiated ecological studies at the NTS in the late 1950s, focusing on vertebrate and invertebrate populations (Wills and Ostler 2001). Subsequent studies documented effects of above-ground nuclear detonation on vegetation (Shields and Wells 1962) and its subsequent recovery (Shields et al. 1963).

Beginning in the early 1960s, the Laboratory of Nuclear Medicine and Radiation Biology at UCLA initiated long-term ecological studies in Rock Valley (French 1964). This valley was not used for nuclear testing and was deemed a good location to conduct an ecological study of the effects of low-level, chronic exposure of gamma irradiation upon a natural community of plants and animals exposed to a ^{137}Cs source. In conjunction with this study, Dr. Beatley established the ecological study plots (Plots 2–5) adjacent to and within Rock Valley. The UCLA studies concluded with the International Biological Program (IBP) studies in the mid-1970s (Rundel and Gibson 1996), yet important literature about desert ecosystems continued to emerge from their work (Hunter 1991; Guo et al. 1998, 1999).

In the late 1960s, Rock Valley was selected as the Desert Biome, Mojave Desert site for the IBP. Baseline ecological data gathered by UCLA in Rock Valley was incorporated and expanded upon during the IBP studies. The study area was expanded by adding a 0.46 km^2 IBP validation site just north of the UCLA study plots (Turner 1973). The IBP studies were conducted between 1971 and 1976 to develop and validate an ecosystem model, as well as to provide basic data on ecological processes (Rundel and Gibson 1996).

If process information was lacking, then experimental manipulative research studies were developed to provide the necessary inputs for the modeling effort. These included October–November irrigation of acre plots in 1969 and 1971 to: (1) simulate winter rainfall, (2) induce the growth of ephemeral plants, and (3) document enhanced reproduction (clutch frequency) in *Uta stansburiana* (Turner et al. 1974). These data were subsequently incorporated into a lizard population model (Turner et al. 1982). Numerous ecological data were gathered and a series of reports and publications pertaining to both the validation and process studies were published by Utah State University. Summary and synthesized information pertaining to Rock Valley as well as to other desert validation sites in the Sonoran Desert (Silverbell), Chihuahuan Desert (Jornada), and Great Basin (Curlew Valley) were documented (Turner 1973; Goodall and Perry 1979, 1981; MacMahon and Wagner 1985).

In the mid-1980s, the Department of Energy had renewed interest in ecological studies and funded a monitoring program known as the Basic Environmental Compliance and Monitoring Program (BECAMP). This program was designed to track changes driven by climatic fluctuations as well as to document recovery from historical disturbances (Hunter and Medica 1992). Five baseline study areas (e.g., YUF001 on Yucca Flat) (fig. 18.4) that were approximately 500 m^2 in area with marked plots were established to monitor flora and fauna. Perennial vegetation was measured using belt transects of 2×25 m for evaluating the fate of individual plants. Lizard sampling plots (105 m^2), primarily designed to study *Uta* populations were established within the center of the 500 m^2. Additionally, five transects (500 m) traversed the entire baseline plot to monitor the presence of other lizard species, including whiptails [*Aspidoscelis* (=*Cnemidophorus*)], leopard lizard (*Gambelia*), horned lizard (*Phrynosoma*), and zebra-tailed lizard (*Callisaurus*). The lizard plot was surrounded by a 12 × 12 grid (144 m^2) used for small mammal census (Hunter and Medica 1989). Many plots were coincident with historical ecological study plots, and the most significant results involved the ecosystem response to the 1989–1991 drought (Hunter 1995a, 1995b).

The Legacy of Dr. Janice C. Beatley and the Nevada Test Site

The work of Dr. Janice C. Beatley, a UCLA researcher, well illustrates the legacy of long-term data on the NTS and within the Mojave Desert. In 1962, Beatley established her network of 68 permanent ecological plots ranging from 1000 to 2400 m in elevation (Beatley 1980; Webb et al. 2003). Although current study designs likely vary in detail, Beatley's plot layout provides a template for ongoing monitoring programs. Beatley constructed 100 ft × 100 ft (10,000 ft^2) permanent plots designed for monitoring perennial vegetation by line intercept. Outside this area, she measured winter annuals with 0.1 m^2 quadrat frames at 2 ft. intervals along a 100 ft. line during their reproductive peak in April–May from 1964 to 1972.

Associated with most of the plots, she established a rudimentary climate station to monitor precipitation, maximum and minimum temperatures, and soil moisture at three depths (monitored every two weeks) and she took large-format photographs of the plots in 1964 (usually four views). Within a grid encompassing the central plot, she measured rodent populations on a yearly basis. Her field research at the NTS ceased in 1975, and her numerous publications only documented and analyzed part of the data that was collected (Beatley 1969a, 1969b, 1979, 1980). Subsequent researchers (Hunter 1991, 1995b; L. A. DeFalco *unpublished data*) replicated the same techniques and equipment for consistency and fidelity of the long-term dataset.

Beatley's foresight has led to long-term sites of great historical value to the researchers that followed her. Measurements of perennial vegetation were made at several plots by BECAMP researchers in 1989 (Hunter 1995a, 1995b), and annuals adjacent to Plot 3 in Rock Valley were measured from 1987 through 1995 (Hunter 1991). Annual plant measurements have continued using Beatley's techniques from 2000 through 2005 on selected plots, including Plot 3. Perennial vegetation was remeasured on 66 surviving plots from 2000 through 2002 (Webb et al. 2003), with selected sites remeasured in 2003 and 2005 to document the impacts of drought (fig. 18.8).

The legacy of the NTS permanent plots has yielded some information that Beatley likely did not anticipate. Repeat photography (Webb et al. 2003) supports the results of plot monitoring and documents other important changes. For example, the increase in plant stature, density, and cover on plots, either documented or suggested by transect data, can clearly be seen. In some areas, notably Yucca Flat and Mid Valley, *Yucca brevifolia* (Joshua trees) have increased in height and density. The rate of change in Joshua trees is so high that identification of individuals between two views is difficult. Repeat photography on the NTS reinforces the concept that repeat photography supplements other types of long-term data and is an integral part of a long-term monitoring program.

BENEFITS OF LONG-TERM MONITORING

The extant long-term studies of the Mojave Desert ecosystem are relatively short and subsequently do not sufficiently capture many of the climatic fluctuations represented in the regional rainfall record. This is particularly important in analyzing the effects across broader climatic timescales, such as those generated by the Pacific Decadal Oscillation (Hereford et al. 2006), a commonly used index of climatic variation as it affects the western United States. Although the impact of drought has been of particular interest to researchers in other deserts (Swetnam and Betancourt 1998; Reynolds et al. 1999; Bowers 2005a; Breshears et al. 2005), the data we present in this paper indicate that wet periods, and particularly the extreme wet period of 1978 through 1998, are also of considerable importance.

Long-term data from permanent plots, such as those that Beatley established in 1963, represent monitoring information that is beneficial to researchers and land managers. They provide critical ecological information that can be collected in no other way. This chapter illustrates opportunities to generate and test hypotheses about environmental change far outside the normal time frame of 3- to 5-year funding cycles. From these plots, we are generating information about disturbance-recovery regimes, climate change, nonnative plant invasions, and plant/animal interactions through synthesis of data collected over a half century. Study plots have allowed the documentation of historic information used for other long-term studies at the NTS, such as understanding the impacts of elevated atmospheric carbon dioxide (e.g., the Nevada Desert FACE Facility: Jordan et al. 1999; Smith et al. *this volume*), nitrogen deposition, increasing rainfall, and disturbance of biological soil crusts (Mojave Global Change Facility) on the Mojave Desert ecosystem. What we learn from the Beatley plots and other long-term studies is stimulating discovery and insight that applies to a much larger geographic area and is of immediate use to managers, researchers, and the general public into the future.

ACKNOWLEDGMENTS

We wish to thank the U.S. Department of Energy, Nevada Operations Office, and Bechtel Nevada for their logistical and technical support for field work on the Nevada Test Site.

We especially thank Dr. Beatley for her foresight in setting up these ecological study plots. The authors thank James M. André and Stanley D. Smith for carefully reviewing the manuscript. This work was funded as part of the Recoverability and Vulnerability of Desert Ecosystems Project of the Priority Ecosystem Studies of the U.S. Geological Survey.

REFERENCES

Beatley, J. C. 1966. Ecological status of introduced brome grasses (*Bromus* ssp.) in desert vegetation of southern Nevada. *Ecology* **47**:548–554.

Beatley, J. C. 1967. Survival of winter annuals in the northern Mojave Desert. *Ecology* **48**:745–750.

Beatley, J. C. 1969a. Biomass of desert winter annual plant populations in southern Nevada. *Oikos* **20**:261–273.

Beatley, J. C. 1969b. Dependence of desert rodents on winter annuals and precipitation. *Ecology* **50**:721–724.

Beatley, J. C. 1973. Russian-thistle (*Salsola*) species in western United States. *Journal of Range Management* **26**:225–226.

Beatley, J. C. 1974. Phenological events and their environmental triggers in Mojave Desert ecosystems. *Ecology* **55**:856–863.

Beatley, J. C. 1976a. Environments of kangaroo rats (*Dipodomys*) and effects of environmental change on populations in southern Nevada. *Journal of Mammalogy* **57**:67–93.

Beatley, J. C. 1976b. Rainfall and fluctuating plant populations in relation to distributions and numbers of desert rodents in southern Nevada. *Oecologia* **24**:21–42.

Beatley, J. C. 1979. *Shrub and tree data for plant associations across the Mojave/Great Basin desert transition of the Nevada Test Site, 1963–1975.* Technical Report No. DOE/EV/2307–15 U-48. National Technical Information Service, U.S. Department of Energy, Springfield, Virginia.

Beatley, J. C. 1980. Fluctuations and stability in climax shrub and woodland vegetation of the Mojave, Great Basin and Transition Deserts of southern Nevada. *Israel Journal of Botany* **28**:149–168.

Belnap, J., and S. D. Warren. 2002. Patton's tracks in the Mojave Desert, USA: an ecological legacy. *Arid Land Research and Management* **16**:245–258.

Bolling, J. D., and L. R. Walker. 2000. Plant and soil recovery along a series of abandoned desert roads. *Journal of Arid Environments* **46**:1–24.

Bolling, J. D., and L. R. Walker. 2002. Fertile island development around perennial shrubs across a Mojave Desert chronosequence. *Western North American Naturalist* **62**:88–100.

Bowers, J. E. 2005a. Effects of drought on shrub survival and longevity in the northern Sonoran Desert. *Journal of the Torrey Botanical Society* **132**:421–431.

Bowers, J. E. 2005b. El Niño and displays of spring-flowering annuals in the Mojave and Sonoran deserts. *Journal of the Torrey Botanical Society* **132**:38–49.

Bowers, J. E. 2005c. Influence of climatic variability on local population dynamics of Sonoran Desert platyopuntia. *Journal of Arid Environments* **61**:193–210.

Bowers, J. E., R. H. Webb, and R. J. Rondeau. 1995. Longevity, recruitment, and mortality of desert plants in Grand Canyon, Arizona, USA. *Journal of Vegetation Science* **6**:551–564.

Bowers, M. A. 1987. Precipitation and the relative abundances of desert winter annuals: a

6-year study in the Northern Mohave Desert, Nevada, USA. *Journal of Arid Environments* **12**:141–150.

Breshears, D. D., N. S. Cobb, P. M. Rich, K. P. Price, C. D. Allen, R. G. Balice, W. H. Romme, J. H. Kastens, M. L. Floyd, J. Belnap, J. J. Anderson, O. B. Myers, and C. W. Meyer. 2005. Regional vegetation die-off in response to global-change-type drought. *Proceedings of the National Academy of Sciences* **102**:15144–15148.

Britten, H. B., B. R. Riddle, P. F. Brussard, R. Marlow, and T. E. Lee Jr. 1997. Genetic delineation of management units for the desert tortoise, *Gopherus agassizii,* in northeastern Mojave Desert. *Copeia* **1997**:523–530.

Brooks, M. L., C. M. D'Antonio, D. M. Richardson, J. B. Grace, J. E. Keeley, J. M. DiTomaso, R. J. Hobbs, M. Pellant, and D. Pyke. 2004. Effects of invasive alien plants on fire regimes. *BioScience* **54**:677–688.

Brooks, M. L., and J. R. Matchett. 2003. Plant community patterns in unburned and burned blackbrush (*Coleogyne ramosissima* Torr.) shrublands in the Mojave Desert. *Western North American Naturalist* **63**:283–298.

Brown, D. E., and R. A. Minnich. 1986. Fire and changes in creosote bush scrub of the western Sonoran Desert, California. *American Midland Naturalist* **116**:411–422.

Christopher, M. M., K. H. Berry, I. R. Wallis, K. A. Nagy, B. T. Henen, and C. C. Peterson. 1999. Reference intervals and physiologic alterations in hematologic and biochemical values of free-ranging desert tortoises in the Mojave Desert. *Journal of Wildlife Diseases* **35**:212–238.

Clarke, M., and H. M. Rendell. 1998. Climate change impacts on sand supply and the formation of desert sand dunes in the south-west USA. *Journal of Arid Environments* **39**:517–531.

Cody, M. L. 2000. Slow-motion population dynamics of Mojave Desert perennial plants. *Journal of Vegetation Science* **11**:351–358.

Coleman, M., D. G. Forbes, and R. J. Abbott. 2001. A new subspecies of *Senecio mohavensis* (Compositae) reveals old-new world species disjunction. *Edinburgh Journal of Botany* **58**:389–403.

Cottam, W. P. 1937. Has Utah lost claim to the lower Sonoran zone? *Science* **85**:563–564.

Elzinga, C. L., D. W. Salzer, and J. W. Willoughby. 1998. Measuring and monitoring plant populations. Technical Reference No. 1730–1. Bureau of Land Management, National Business Center, Denver, Colorado.

Fosberg, F. R. 1938. The lower Sonoran in Utah. *Science* **87**:39–40.

Freilich, J. E., K. P. Burnham, C. M. Collins, and C. A. Garry. 2000. Factors affecting population assessments of desert tortoises. *Conservation Biology* **14**:1479–1489.

French, N. R. 1964. *Description of a study of ecological effects on a desert area from chronic exposure to low level ionizing radiation.* Report No. UCLA 12–532. U.S. Atomic Energy Commission, Los Angeles, California.

Geiger, E. L., and G. R. McPherson. 2005. Response of semi-desert grasslands invaded by non-native grasses to altered disturbance regimes. *Journal of Biogeography* **32**:895–902.

Goheen, J. R., E. P. White, S. K. Morgan, and J. H. Brown. 2005. Intra-guild compensation regulates species richness in desert rodents. *Ecology* **86**:567–573.

Goldberg, D. E., and R. M. Turner. 1986. Vegetation change and plant demography in permanent plots in the Sonoran Desert. *Ecology* **67**:695–712.

Goodall, D. W., and R. A. Perry. 1979. *Arid-land ecosystems: structure, functioning and management.* Volume one. Cambridge University Press, Cambridge, England, UK.

Goodall, D. W., and R. A. Perry. 1981. *Arid-land ecosystems: structure, functioning and management.* Volume two. Cambridge University Press, Cambridge, England, UK.

Guo, Q., P. W. Rundel, and D. W. Goodall. 1998. Horizontal and vertical distribution of desert seed banks: patterns, causes, and implications. *Journal of Arid Environments* **38**:465–478.

Guo, Q., P. W. Rundel, and D. W. Goodall. 1999. Structure of desert seed banks: comparisons across four North American desert sites. *Journal of Arid Environments* **42**:1–14.

Hamerlynck, E. P., J. R. McAuliffe, E. V. McDonald, and S. D. Smith. 2002. Ecological responses of two Mojave Desert shrubs to soil horizon development and soil water dynamics. *Ecology* **83**:768–779.

Hereford, R., R. H. Webb, and C. I. Longpré. 2004. Precipitation history of the Mojave Desert region, 1893–2001. U.S. Geological Survey Fact Sheet No. 117–03. Reston, Virginia.

Hereford, R., R. H. Webb, and C. Longpré. 2006. Precipitation history and ecosystem response to multidecadal precipitation variability in the Mojave Desert and vicinity, 1893–2001. *Journal of Arid Environments* **67**:13–34.

Hickerson, L. L., and P. G. Wolf. 1998. Population genetic structure of *Arctomecon californica* Torrey and Fremont (Papaveraceae) in fragmented and unfragmented habitat. *Plant Species Biology* **13**:21–33.

Hockett, B. S. 2000. Paleobiogeographic changes at the Pleistocene-Holocene boundary near Pintwater Cave, southern Nevada. *Quaternary Research* **53**:263–269.

Hunter, K. L., J. L. Betancourt, B. R. Riddle, T. R. Van Devender, K. L. Cole, and G. W. Spaulding. 2001. Ploidy race distributions since the last glacial maximum in the North American desert shrub, *Larrea tridentata. Global Ecology and Biogeography* **10**:521–533.

Hunter, R. B. 1991. *Bromus* invasions on the Nevada Test Site: present status of *B. rubens* and *B. tectorum* with notes on their relationship to disturbance and altitude. *Great Basin Naturalist* **51**:176–182.

Hunter, R. B. 1994. Trends in perennial plant populations on the Nevada Test Site, 1989–1991. Pages 320–325 *in* R.B. Hunter, compiler. *Status of the flora and fauna on the Nevada Test Site, 1989–1991.* Technical Report No. DOE/NV/11432–57, Reynolds Electrical and Engineering Company, Las Vegas, Nevada.

Hunter, R. B. 1995a. Status of ephemeral plants on the Nevada Test Site, 1994. Pages 183–244 *in* R. B. Hunter, compiler. *Status of the flora and fauna on the Nevada Test Site, 1994.* Technical Report No. DOE/NV/11432–195. Reynolds Electrical and Engineering Company, Las Vegas, Nevada.

Hunter, R. B. 1995b. Status of perennial plants on the Nevada Test Site, 1994. Pages 245–348 *in* R. B. Hunter, compiler. *Status of the flora and fauna on the Nevada Test Site, 1994.* Technical Report No. DOE/NV/11432–195. Reynolds Electrical and Engineering Company, Las Vegas, Nevada.

Hunter, R. B., and P. A. Medica. 1989. *Status of the flora and fauna on the Nevada Test Site: results of continuing basic environmental research January through December 1987.* Technical Report No. DOE/NV/106030–2. Reynolds Electrical and Engineering Company, Las Vegas, Nevada.

Hunter, R. B., and P. A. Medica. 1992. Extent of land disturbance on the Nevada Test Site.

Pages 3–7 *in* R. B. Hunter, compiler. *Status of the flora and fauna on the Nevada Test Site, 1988*. Technical Report No. DOE/NV/106030–29. Reynolds Electrical and Engineering Company, Las Vegas, Nevada.

Jordan D. N., S. F. Zitzer, G. R. Hendrey, K. F. Lewin, J. Nagy, R. S. Nowak, S. D. Smith, J. S. Coleman, and J. R. Seemann. 1999. Biotic, abiotic and performance aspects of the Nevada Desert Free-Air CO_2 Enrichment Facility. *Global Change Biology* **5**:659–668.

Kelt, D. A. 1999. On the relative importance of history and ecology in structuring communities of desert small mammals. *Ecography* **22**:123–137.

King, T. J. Jr. 1976. Late Pleistocene-early Holocene history of coniferous woodlands in the Lucerne Valley region, Mojave Desert, California, USA. *Great Basin Naturalist* **36**: 227–238

Lamb, T., J. C. Avise, and J. W. Gibbons. 1989. Phylogeographic patterns in mitochondrial DNA of the desert tortoise *Xerobates agassizi* and evolutionary relationships among the North American gopher tortoises. *Evolution* **43**:76–87.

Lee, T. E. Jr., B. R. Riddle, and P. L. Lee. 1996. Speciation in the desert pocket mouse (*Chaetodipus penicillatus* Woodhouse). *Journal of Mammology* **77**:58–68.

Lei, S. A. 1999. Effects of severe drought on biodiversity and productivity in a creosote bush-blackbrush ecotone of southern Nevada. Pages 1–5 *in* E. D. McArthur, W. K. Ostler, and C. L. Wambolt, compilers. *Proceedings: Shrubland Ecotones, August 12–14, 1999, Ephraim, Utah*. Proceedings No. RMRS-P-00. US Department of Agriculture, Forest Service, Rocky Mountain Research Station, Ephraim, Utah.

Longshore, K. M., J. R. Jaeger, and J. M. Sappington. 2003. Desert tortoise (*Gopherus agassizii*) survival at two eastern Mojave Desert sites: death by short-term drought? *Journal of Herpetology* **37**:169–177.

MacMahon, J. A., and F. H. Wagner. 1985. The Mojave, Sonoran and Chihuahuan Deserts of North America. Pages 105–202 *in* M. Evenari, I. Noy-Meir, and D. W. Goodall, editors. *Hot deserts and arid shrublands. Ecosystems of the World.* Volume 12A. Elsevier Science Publications, Amsterdam, the Netherlands.

Mantua, N. J., and S. R. Hare. 2002. The Pacific Decadal Oscillation. *Journal of Oceanography* **58**:35–42.

McCabe, J. G., and M. D. Dettinger. 1999. Decadal variations in the strength of ENSO teleconnections with precipitation in the western United States. *International Journal of Climatology* **19**:1399–1410.

Mead, J. I. 1981. The last 30,000 years of faunal history within the Grand Canyon, Arizona. *Quaternary Research* **15**:311–326.

Medica, P. A., H. W. Avery, and J. E. Lovich. 1996. Localized mid-summer rainfall: its effects on perennial plant phenology, and importance to the desert tortoise (*Gopherus agassizii*). Page 50 *in* B. Bartholomew, editor. *Desert Tortoise Council, Proceedings of the Twenty-first Annual Symposium, March 29–31, 1996, Las Vegas, Nevada.* Desert Tortoise Council, Beaumont, California.

Medica, P. A., R. B. Bury, and F. B. Turner. 1975. Growth of the desert tortoise (*Gopherus agassizi*) in Nevada. *Copeia* **1975**:639–643.

Medica, P. A., M. B. Saethre, R. B. Hunter, and J. D. Drumm. 1994. Trends in reptile populations on the Nevada Test Site. Pages 1–49 *in* R. B. Hunter, compiler. *Status of the flora and fauna on the Nevada Test Site, 1989–91: results of continuing basic environmental moni-*

toring January through December 1989–1991. Technical Report No. DOE/NV/11432–57. Reynolds Electrical and Engineering Company, Las Vegas, Nevada.

Medica, P. A., and F. B. Turner. 1976. Reproduction by *Uta stansburiana* (Reptilia, Lacertilia, Iguanidae) in southern Nevada. *Journal of Herpetology* **10**:123–128.

Minckley, R. L., J. H. Cane, and L. Kervin. 2000. Origins and ecological consequences of pollen specialization among desert bees. *Proceedings of the Royal Society B: Biological Sciences* **267**:265–271.

Miriti, M. N., S. J. Wright, and H. F. Howe. 2001. The effects of neighbors on the demography of a dominant desert shrub (*Ambrosia dumosa*). *Ecological Monographs* **71**: 491–509.

Nagy, K. A., and P. A. Medica. 1986. Physiological ecology of desert tortoises in southern Nevada. *Herpetologica* **42**:73–92.

Noble, I. R. 1977. Long-term biomass dynamics in an arid chenopod shrub community at Koonamore, South Australia. *Australian Journal of Botany* **25**:639–653.

Peterson, C. C. 1996. Ecological energetics of the desert tortoise (*Gopherus agassizii*): effects of rainfall and drought. *Ecology* **77**:1831–1844.

Price, M. V., N. M. Waser, and S. A. McDonald. 2000. Elevational distributions of kangaroo rats (genus *Dipodomys*): long-term trends at a Mojave Desert site. *American Midland Naturalist* **144**:352–361.

Rainboth, W. J., D. G. Buth, and F. B. Turner. 1989. Allozyme variation in Mojave populations of the desert tortoise *Gopherus agassizi. Copeia* **1989**:115–123.

Rasmussen, E. M. 1984. El Niño, the ocean/atmosphere connection. *Oceanus* **27**:5–12.

Reynolds, J. F., R. A. Virginia, P. R. Kemp, A. G. de Soyza, and D. G. Tremmel. 1999. Impact of drought on desert shrubs: effects of seasonality and degree of resource island development. *Ecological Monographs* **69**:69–106.

Riddle, B. R., D. J. Hafner, and L. F. Alexander. 2000. Phylogeography and systematics of the *Peromyscus eremicus* species group and the historical biogeography of North American warm regional deserts. *Molecular Phylogenetics and Evolution* **17**:145–160.

Rundel, P. W., and A. C. Gibson. 1996. *Ecological communities and processes in a Mojave Desert ecosystem: Rock Valley, Nevada.* Cambridge University Press, New York, New York.

Saethre, M. B. 1995. Small mammal populations on the Nevada Test Site, 1994. Pages 75–148 *in* R. B. Hunter, compiler. *Status of the flora and fauna on the Nevada Test Site, 1994: results of continuing basic environmental monitoring January through December 1994.* Technical Report No. DOE/NV/11432–195. Reynolds Electrical and Engineering Company, Las Vegas, Nevada.

Sanderson, S. C., and H. C. Stutz. 1994. High chromosome numbers in Mojavian and Sonoran Desert *Atriplex canescens* (Chenopodiaceae). *American Journal of Botany* **81**:1045–1053.

Schultz, B. W., and W. K. Ostler. 1995a. Effects of prolonged drought on vegetation associations in the northern Mojave Desert. Pages 228–235 *in* B. A. Roundy, E. D. McArthur, E. Durant, J. S. Haley, and D. K. Mann, compilers. *Proceedings: Wildland Shrub and Arid Land Restoration Symposium, October 19–21, 1993, Las Vegas, Nevada.* General Technical Report No. INT-GTR-315. U.S. Department of Agriculture, Forest Service, Ogden, Utah.

Schultz, B. W., and W. K. Ostler. 1995b. Species and community response to above normal precipitation following prolonged drought at Yucca Mountain, Nevada. Pages 236–242 *in* B. A. Roundy, E. D. McArthur, E. Durant, J. S. Haley, D. K. Mann, compilers. *Proceedings: Wildland Shrub and Arid Land Restoration Symposium, October 19–21, 1993, Las Vegas, Nevada.* General Technical Report No. INT-GTR-315. U.S. Department of Agriculture, Forest Service, Ogden, Utah.

Shields, L. M., and P. V. Wells. 1962. Effects of nuclear testing on desert vegetation. *Science* **135**:38–40.

Shields, L. M., P. V. Wells, and W. H. Rickard. 1963. Vegetational recovery on atomic target areas in Nevada. *Ecology* **44**:697–705.

Smith, F. A., M. D. Matocq, K. E. Melandez, A. M. Ditto, and P. A. Kelly. 2000. How isolated are Pleistocene refugia? Results from a study on a relict woodrat population from the Mojave Desert. *Journal of Biogeography* **27**:483–500.

Smith, S. D., C. A. Herr, K. L. Leary, and J. M. Piorkowski. 1995. Soil-plant water relations in a Mojave Desert mixed shrub community: a comparison of three geomorphic surfaces. *Journal of Arid Environments* **29**:339–351.

Smith, S. D., T. E. Huxman, S. F. Zitzer, T. N. Charlet, D. C. Housman, J. S. Coleman, L. K. Fenstermaker, J. R. Seemann, and R. S. Nowak. 2000. Elevated CO_2 increases productivity and invasive species success in an arid ecosystem. *Nature* **408**:79–82.

Spaulding, W. G. 1990. Vegetational and climatic development of the Mojave Desert: the last glacial maximum to the present. Pages 166–199 *in* J. L. Betancourt, T. R. Van Devender, and P. S. Martin, editors. *Packrat middens.* University of Arizona Press, Tucson, Arizona.

Swetnam, T. W., and J. L. Betancourt. 1998. Mesoscale disturbance and ecological response to decadal climate variability in the American Southwest. *Journal of Climate* **11**:3128–3147.

Trepanier, T. L., and R. B. Murphy. 2001. The Coachella Valley fringe-toed lizard (*Uma inornata*): genetic diversity and phylogenetic relationships of an endangered species. *Molecular Phylogenetics and Evolution* **18**:327–334.

Turner, F. B., editor. 1973. Rock Valley validation site report. Research Memorandum No. RM 73-2. U.S. International Biological Program, Desert Biome, Ecology Center, Utah State University, Logan, Utah.

Turner, F. B., G. A. Hoddenbach, P. A. Medica, and J. R. Lannom. 1970. The demography of the lizard, *Uta stansburiana* Baird and Girard, in southern Nevada. *Journal of Animal Ecology* **39**:505–519.

Turner, F. B., and P.A. Medica. 1977. Sterility among female lizards *Uta stansburiana* exposed to continuous gamma irradiation. *Radiation Research* **70**:154–163.

Turner, F. B., P. A. Medica, and R. B. Bury. 1987. Age-size relationships of desert tortoises (*Gopherus agassizii*) in southern Nevada. *Copeia* **1987**:974–979.

Turner, F. B., P. A. Medica, K. W. Bridges, and R. I. Jennrich. 1982. A population model of the lizard *Uta stansburiana* in southern Nevada. *Ecological Monographs* **52**:243–259.

Turner, F. B., P. A. Medica, and C. L. Lyons. 1984. Reproduction and survival of the desert tortoise (*Scaptochelys agassizii*) in Ivanpah Valley, California. *Copeia* **1984**:811–820.

Turner, F. B., P. A. Medica, and D. D. Smith. 1974. Reproduction and survivorship of the lizard *Uta stansburiana,* and the effect of winter rainfall, density and predation on these processes. Pages 117–128 *in* Research Memorandum No. 74–26. U.S. International Biological Program, Desert Biome, Ecology Center, Utah State University, Logan, Utah.

Turner, R. M. 1990. Long-term vegetation change at a fully protected Sonoran Desert site. *Ecology* **71**:464–477.

Valone, T. J., and M. Kaspari. 2005. Interactions between granivorous and omnivorous ants in a desert grassland: results from a long-term experiment. *Ecological Entomology* **30**:116–121.

Webb, R. H. 2002. Recovery of severely compacted soils in the Mojave Desert, California, USA. *Arid Land Research and Management* **16**:291–305.

Webb, R. H., M. B. Murov, T. C. Esque, D. E. Boyer, L. A. DeFalco, D. F. Haines, D. Oldershaw, S. J. Scoles, K. A. Thomas, J. B. Blainey, and P. A. Medica. 2003. *Perennial vegetation data from permanent plots on the Nevada Test Site, Nye County, Nevada.* U.S. Geological Survey Open-File Report No. 03–336. Online. Tucson, Arizona. http://www.werc.usgs.gov/lasvegas/ofr-03-336.html.

Webb, R. H., J. W. Steiger, and E. B. Newman. 1988. The response of vegetation to disturbance in Death Valley National Monument, California. U.S. Geological Survey Bulletin No. 1793. U.S. Government Printing Office, Denver, Colorado.

Webb, R. H., J. W. Steiger, and R. M. Turner. 1987. Dynamics of Mojave Desert shrub vegetation in the Panamint Mountains. *Ecology* **50**:478–490.

Wells, P. V., and J. H. Hunziker. 1976. Origin of the creosote bush (*Larrea*) deserts of southwestern North America. *Annals of the Missouri Botanical Garden* **63**:843–861.

Wells, P. V., and C. D. Jorgensen. 1964. Pleistocene wood rat middens and climatic change in Mohave Desert—a record of juniper woodlands. *Science* **143**:1171–1174.

Wills, C. A., and W. K. Ostler. 2001. *Ecology of the Nevada Test Site: an annotated bibliography.* Technical Report No. DOE/NV/11718–594, Bechtel Nevada, Las Vegas, Nevada.

Woodward, B. 1994. Reptile populations on the Nevada Test Site in 1993. Pages 1–35 *in* R. B. Hunter, compiler. *Status of the flora and fauna on the Nevada Test Site, 1993: results of continuing basic environmental monitoring, January through December 1993.* Technical Report No. DOE/NV/11432–162. Reynolds Electrical and Engineering Company, Las Vegas, Nevada.

Woodward, B. 1995. Status of reptiles on the Nevada Test Site, 1994, and summary of work 1987–1994. Pages 1–73 *in* R. B. Hunter, compiler. *Status of the flora and fauna on the Nevada Test Site, 1994: results of continuing basic environmental monitoring, January through December 1994.* Technical Report No. DOE/NV/11432–195. Reynolds Electrical and Engineering Company, Las Vegas, Nevada.

NINETEEN

The Mojave Desert

Ecosystem Processes and Sustainability Revisited

ROBERT H. WEBB, LYNN F. FENSTERMAKER, JILL S. HEATON,
DEBRA L. HUGHSON, ERIC V. MCDONALD, AND DAVID M. MILLER

As discussed briefly in the preface, the Mojave Desert Science Symposium, held on November 16–19, 2004 (Desert Managers Group *undated*), was the third in a series of symposia designed to bring together scientists, managers, and the public to discuss common interests and concerns in the Mojave Desert. The third symposium provided a unique opportunity for these groups to discuss a wide variety of issues regarding the threats to and sustainability of the Mojave Desert. In this book, we have assembled several of the scientific contributions presented at the symposium that either provide new information or review previous investigations. The new information focuses on Mojave Desert resources, ecosystem processes, and sustainability. The reviews explore the significant scientific knowledge base that bears on management of this desert, which is unique for its climatic gradients and water resources. In this summary, we synthesize some of the overarching themes of the third symposium and the written contributions in this book.

HOW UNIQUE ARE THREATS TO THE MOJAVE DESERT?

One can hold the view that the threats to natural ecosystems are similar worldwide, and that only the specific details are unique in a region like the Mojave Desert. The amount of human use and population growth in this arid region are hardly unique, and, in fact, may be less significant, in terms of rates of change, than in other regions. In the Sahel region of northern Africa or the deserts of central Asia, for example, once-productive farm and range lands are being irretrievably converted, within the timeframe of human lifespans, into dunes and otherwise unproductive lands (Alsharhan et al. 2003), which is not happening on a large scale in most of the Mojave Desert. This distinction in no way diminishes the value of preserving the Mojave Desert and its resources, but it does provide some perspective on the nature of changes and our perception of those changes as "threats."

Desert landscapes change through time, as documented by historical ecology [e.g., analyses of pack rat middens, fire chronologies, aerial photographs (Swetnam et. al. 1999), and permanent plots (Webb et al., Long-Term Data, *this volume*)]. Climatic change and weather fluctuations guarantee that large-scale ecosystem changes will occur in the Mojave Desert, whether during major natural climatic shifts on the scale of thousands of years, decadal-scale droughts, or multiyear wet periods. One of the most critical questions raised among researchers looking at change in the Mojave Desert is whether the rate of inevitable change and its consequences to individual species is compounded by human effects. Climate is not constant (Redmond *this volume*), and some predictions, based on the concept of global change, hold that increases in temperature and interannual climate variability are to be expected (Smith et al. *this volume*). This expectation should influence, in a most fundamental way, how scientists and managers view this desert.

The effects of climate variability are compounded by human influences. One example is the occurrence of fire: wet years, which typically result from some incarnations of the global El Niño phenomenon in the Pacific Ocean (Redmond *this volume*), result in increased fuel on the landscape, typically in the form of annual plants. Humans have imported nonnative annuals to this desert, which tend to further increase the amount of fuel loading. Lightning and humans ignite wildfires with a frequency unknown in prehistoric times, but one that has increased in the last several decades (Brooks *this volume*). With global change, will the incidence of summer thunderstorms and lightning-caused wildfires increase in the future? If so, what are the eventual ecological and socioeconomic impacts?

The issue of unsustainable human populations, whose energy is provided by fossil fuels, water, and food products that all are imported from other regions (Hughson *this volume*), is not unique to the Mojave Desert—it is, in essence, a global issue with consequences at a regional scale and exacerbated by arid conditions. As Hughson (*this volume*) noted, a significant regional issue is the sustainability of water resources in the face of ever-expanding urban growth. While much of the water used in the Mojave Desert comes from outside of the ecosystem—northern California, northern Nevada, and the Colorado River—outside sources are used only after regional resources are either close to depletion or depleted. The reduced water resources will result in concomitant impacts on highly valued riparian areas (Lines 1999; Dudley *this volume*) and wildlife populations. At the heart of the issue of sustainability are questions that have not yet been addressed at a regional scale, including whether increasing human water use and sustainable wildlife populations are incompatible.

Another potential threat, which we are only beginning to understand, is the question of nutrient influxes into the ecosystem through atmospheric transport, both from the urban areas within the Mojave Desert and from those to the west and east. Deposition of nitrogen-bearing compounds, either through dry fallout or within precipitation, may alter the competitive interactions between native and nonnative annuals (Allen et al. *this volume*), again providing a positive-feedback mechanism that leads to more frequent fires at lower elevations in the desert. No single chapter in this book discusses fire as its major theme, yet many chapters deal with the large number of issues that contribute to this ongoing threat, which may be the most important single contributor to vegetation change in the Mojave Desert, particularly in areas that are otherwise protected from adverse land use impacts (e.g., Joshua Tree National Park, Mojave National Preserve).

The effects of roads on this ecosystem are both well understood and understudied. In their review of the ecological effects of roads, Brooks and Lair (*this volume*) discuss well-known impacts of roads, such as direct wildlife mortality, but they and Brooks (*this volume*) also emphasize less well-understood issues, such as the fact that roads can act as vectors for transport of nonnative vegetation. This question, at the forefront of research on invasive species, bears on a number of management-related issues, most notably control of the spread of *Brassica tournefortii* (Sahara mustard), a rapidly expanding nonnative with as-yet unknown ecosystem impacts. Road proliferation has been extreme in the Mojave National Preserve (Vogel and Hughson *this volume*), throughout the Mojave Desert, and particularly on the outskirts of urban areas. We are only now beginning to appreciate the additional threat of habitat fragmentation (Barrows and Allen *this volume*) induced by roads at the urban-wildland interface.

The proper management of roads and other disturbances is a major issue discussed in this book. Soil compaction, perhaps one of the best-understood effects of roads and other land use activities (Lei *this volume*), has been documented in the Mojave Desert, as well as in other ecosystems worldwide. Compaction lingers in the Mojave Desert for well over a century (Webb et al., Natural Recovery, *this volume*), making it a significant threat well past cessation of disturbances. While the negative effects of soil compaction on vegetation reestablishment can be reduced using an expensive mechanical ripping technique (Weigand and Rodgers *this volume*), we now understand all too well that such activities severely disrupt soil structure, which requires millennia to develop (Miller et al. *this volume;* Stevenson et al. *this volume*). What is unique about soil compaction in the Mojave Desert ecosystem is not how, why, or where it occurs, but how it can be controlled, the overall ecosystem effects, and the rate at which it responds to the abiotic and biotic processes that ameliorate its effects (Webb et al., Natural Recovery, *this volume*).

One question that looms large is whether soil compaction significantly influences the natural recovery process of annuals and perennials, and if so, to what degree. As previously discussed, increasing fire frequency in the desert involves the interplay of perennial species, nonnative annuals, climate change, and increased atmospheric nitrogen deposition. Nonnative species with dispersed root systems, such as the Mediterranean grasses now common in the Mojave Desert, may grow better than native species in slightly compacted soils (Adams et al. 1982). Therefore, even low levels of soil compaction could provide yet another positive feedback mechanism that promotes increased fire frequency. Because soil compaction may promote the productivity of nonnative plants, the subject is an important research question that illustrates the variety of threats faced by this desert.

The interrelations among natural processes, human impacts, and recovery rates raise scientific questions of great interest to land managers, for whom ecosystem restoration is a high priority. Given the expansive areas of disturbance in the Mojave Desert, and the large expense of active restoration (Weigand and Rodgers *this volume*), the viability of allowing natural recovery is gaining attention (Webb et al., Natural Recovery, *this volume*). Quaternary geology can be a useful tool in evaluating restoration potential and technique. From a purely visual perspective, which is often the primary consideration in restoration, young geomorphic surfaces with active fluvial or eolian processes may recover in less than a century (Hereford *this volume;* Webb et al., Natural Recovery, *this volume*), while disturbances on older surfaces may not recover in a timeframe acceptable to land managers.

Finally, recent research on recovery and restoration processes raises the potential for implementing both active and natural recovery in a single project (Weigand and Rodgers *this volume*). However, significant questions remain that, again, reflect the unique threats to, and remediation potential of, the Mojave Desert. For example, what are the goals of restoration? Most active restoration efforts aim to restore perennial vegetation cover, which may occur naturally in less than a century as described previously. Active restoration techniques can accelerate that recovery (Weigand and Rodgers *this volume*). Restoration of species composition is generally not an explicit goal of active restoration, nor is restoration of ecosystem function, nor the panoply of ecosystem resources, ranging from biological soil crusts and soil biota to wildlife populations. Current thinking seems to involve the cliché, "If you build it (perennial vegetation cover), they will come." Future research is needed to determine if this is, in fact, the case, or whether the real result of both active restoration and natural recovery in the Mojave Desert is the creation of transient new community types.

ECOSYSTEM PROCESSES: THE STATE OF THE SCIENCE

A major theme of this book is water, albeit not from the typical perspective of maintaining water resource sustainability for human use. Ecosystem processes in deserts are driven by water availability, in terms of both soil moisture and the interannual variability of seasonal precipitation. The Mojave Desert has long been classified as a high desert sustained primarily by winter precipitation, but we now know that the Mojave Desert contains an array of subregions with their own climatic and ecological characteristics (Redmond *this volume*). Once considered an ecosystem driven primarily by biological processes, where soil, in the conventional sense, was unimportant or did not exist, we now know that abiotic processes extending far beyond mere climatic considerations are perhaps of more importance to ecosystem stability in this desert. Many ecologists now believe that soil characteristics and hydrology may be the most important factors affecting how this ecosystem works spatially (Miller et al. *this volume;* Bedford et al. *this volume*). This change in thinking resulted from basic scientific research in the Mojave and other deserts, much of which was not driven by land-management questions, but instead, by simple curiosity.

In the years leading up to the landmark passage of the Federal Land Policy and Management Act of 1976, which largely focused on the California desert, the mantra "easily scarred and slowly healed" was bantered repeatedly to paint an image of a fragile Mojave Desert. Recent research has shown that these adjectives "easily" and "slowly" are relative and should not be universally applied to this desert, some aspects of which are more resilient than previously thought. The new, more resilient view of the Mojave Desert comes, in part, from the explicit recognition of the effects that geomorphic surfaces have on plant assemblages and ecosystem productivity (Miller et al. *this volume;* Bedford et al. *this volume;* Stevenson et al. *this volume*), as well as the fact that both natural climate variability (Redmond *this volume*) and anthropogenically driven changes in the atmosphere (Smith et al. *this volume*) have short- and long-term effects, which are compounded by land use practices (Webb et al., Natural Recovery, *this volume*). We should expect significant ecosystem changes to occur in the Mojave Desert and we should design those expectations into monitoring protocols.

Soil moisture appears to be another emerging key to ecosystem processes in the Mojave Desert. Older soils with vesicular horizons at the surface have greatly reduced infiltration rates compared with younger soils lacking such a horizon (Bedford et al. *this volume*). Subsurface soil development, notably calcic horizon formation, restricts root expansion (Shreve and Mallery 1933), thereby reducing the above-ground biomass of perennial species, notably *Larrea tridentata* (creosote bush), one of the most widespread species in this desert (Stevenson et al. *this volume*). That roots offer one of the most important explanations for ecosystem productivity (Schwinning and Hooten *this volume*) is hardly surprising, but the interrelation between below-ground root structure, soil moisture, and ecosystem processes is an emerging area of research that is discussed extensively in this book and is fertile ground for future research as well.

The value of previously established, long-term studies (Webb et al., Long-Term Data, *this volume*) and the need for future studies is another theme of this book. The longest-running ecosystem monitoring plots, on the Nevada Test Site, were established in an era when abiotic influences were underappreciated, and sampling designs mostly focused on obtaining vegetation types and did not account for the influence of geomorphic surfaces. As science and our understanding of the Mojave Desert advances, sampling designs and other methods need to be modified to address emerging questions.

The current emphasis on ecosystem monitoring among managers will likely yield a network of long-term study sites, but this network needs to be constructed with the explicit recognition that the Mojave Desert is a patchwork of abiotically controlled vegetation assemblages that may respond differently to the same stimuli. The amount of high-quality geospatial data depicting the patchwork nature of the Mojave is increasing (Miller et al. *this volume*), and all the information presented in this book indicates that ecosystem-monitoring programs of the future will be firmly rooted in such geospatial data. As discussed by Bedford et al. (*this volume*), geospatial modeling that includes both abiotic and biotic factors may provide answers to ecosystem productivity and stability at a broader scale than previously addressed using point and small-plot data.

For some, the question, "Don't we know enough?" is raised with respect to whether or not scientific research should continue. As several chapters in this volume imply, we may never know enough about the intricacies of the Mojave Desert. This book shows how basic science can guide management by offering new knowledge regarding ecosystem controls and ecosystem function on both larger and smaller scales, which may well have future bearing on the preservation of the uniqueness of this desert.

SUSTAINABILITY AND THE MOJAVE DESERT: EMERGING ISSUES FOR SCIENTISTS AND MANAGERS

Returning to questions raised by Wilkinson (*this volume*) and Hughson (*this volume*), the future of water is likely the most important question scientists and managers can address that bears directly on the sustainability of not only the ecosystem, but perhaps more importantly, the human presence in the Mojave Desert. As human water consumption increases, science undoubtedly will be used to address questions such as the effects of large-scale water importation and manipulation of the Mojave Desert's resources. These types of effects can already be seen around the margins of Lake Mead, where a national recreation area contains riparian vegetation that previously did not occur. Similarly, effluent discharge from waste-treatment plants creates artificial riparian areas such as the Las Vegas Wash while, conversely, ground-water extraction dries up springs, as has occurred in the Las Vegas Valley. Water and the urban-wildland interface are fast-emerging issues for both scientists and land managers, as exemplified by Las Vegas, as well as other fast-growing towns, such as Pahrump, Nevada, and Barstow, California.

The relation between science and management can be mutually beneficial. While basic research on ecosystem function is not often immediately useful to land managers, it should be recognized for its long-term value, and researchers should recognize its influence on both the evaluation of ecosystem disturbance and the design of inventory and monitoring programs. The advancement of scientific knowledge cannot occur in a vacuum; policy makers and managers play a crucial role in guiding scientists to evaluate threats to desert ecosystems, and in the management of those threats and their aftermath (e.g., mitigation of disturbances).

The Mojave Desert community must work collectively to seek ways to enhance the integration of science into management and policy, and to meet management needs through scientific inquiry. The comfort level of scientists for "engineering judgment," the euphemism commonly applied to decisions made with incomplete data, needs to be raised to facilitate communication with managers, who must use their judgment to deal with day-to-day decisions. Continuing dialogue and the ability to alter land-management plans as more data become available (i.e., adaptive management) is also essential. Equally important is the dissemination of scientific information in easily accessible formats. Scientists can help educate both the public and private sectors about how the Mojave Desert functions, the short- and long-term effects of land use practices, and the sustainability of its resources, which will help to promote the missions of scientists and managers alike.

ACKNOWLEDGMENTS

We thank Lesley DeFalco for information on the publication of the first Mojave Desert Science Symposium and her help with the third installment. Jeff Lovich provided the still-intact link to the web pages for the second Mojave Desert Science Symposium. The authors thank Jeffrey Lovich, Lesley DeFalco, and Jayne Belnap for critically reviewing this section.

REFERENCES

Adams, J. A., L. H. Stolzy, A. S. Endo, P. G. Rowlands, and H. B. Johnson. 1982. Desert soil compaction reduces annual plant cover. *California Agriculture* **36**:6–7.

Alsharhan, A. S., W. W. Wood, A. S. Goudie, A. Fowler, and E. M. Abdellatif. 2003. *Desertification in the third millennium.* A. A. Balkema Publishers, Lisse, the Netherlands.

Desert Managers Group. *Undated. Mojave Desert Science Symposium, November 16–18,*

2004, University of Redlands, California. Online. Desert Managers Group, Barstow, California.

Lines, G. C. 1999. *Health of native riparian vegetation and its relation to hydrologic conditions along the Mojave River, Southern California.* U.S. Geological Survey Water-Resources Investigations Report No. 99–4112. San Diego, California.

Shreve, R., and T. D. Mallery. 1933. The relation of caliche to desert plants. *Soil Science Society of America Journal* **35**:99–112.

Swetnam, T. W., C. D. Allen, and J. L. Betancourt. 1999. Applied historical ecology: using the past to manage for the future. *Ecological Applications* **9**:1189–1206.

CONTRIBUTORS

EDITORS

Dr. ROBERT H. WEBB has worked on long-term changes in natural ecosystems of the southwestern United States since 1976. He has degrees in engineering (B.S., University of Redlands, 1978), environmental earth sciences (M.S., Stanford University, 1980), and geosciences (Ph.D., University of Arizona, 1985). Since 1985, he has been a research hydrologist with the U.S. Geological Survey in Tucson, Arizona, and is an adjunct faculty member of the Departments of Geosciences and Hydrology and Water Resources at the University of Arizona. Webb has authored, coauthored, or edited 10 books, including *Environmental Effects of Off-Road Vehicles* (with Howard Wilshire); *Grand Canyon: A Century of Change; Floods, Droughts, and Changing Climates* (with Michael Collier); *The Changing Mile Revisited* (with Raymond Turner); *Cataract Canyon: A Human and Environmental History of the Rivers in Canyonlands* (with Jayne Belnap and John Weisheit); *The Ribbon of Green: Long-Term Change in Woody Riparian Vegetation in the Southwestern United States* (with Stanley Leake and Raymond Turner); and *Damming Grand Canyon: The 1923 U.S. Geological Survey Expedition on the Colorado River* (with Diane Boyer).

Dr. LYNN F. FENSTERMAKER has used remotely sensed data to map, monitor, and assess the effect of environmental stressors on vegetation at small and large scales since 1984. She has degrees in environmental resource management (B.S., The Pennsylvania State University, 1981), agronomy (M.S., The Pennsylvania State University, 1986) and biological sciences (Ph.D., University of Nevada, Las Vegas, 2003). Since 1990, she has been an ecological remote-sensing scientist (research associate and currently associate research professor) with the Desert Research Institute in Las Vegas, Nevada. She is also the director of two research efforts examining the effects of elevated CO_2 and other global change variables on the Mojave Desert ecosystem. Fenstermaker has edited two books: *Remote Sensing Classification Accuracy Assessment: A Compendium* and *Remote Sensing Applications for Acid Deposition.*

Dr. JILL S. HEATON has worked on spatial modeling of animal populations and their habitats, including threatened species, in the Mojave Desert since 1996. She has degrees in biology (B.S and M.S., University of North Texas, 1993 and 1996) and geography (Ph.D., Oregon State University, 2001). She has been an assistant professor in geography at the University of Nevada at Reno since 2004. Prior to that she held a three-year, post-doctoral research position at the University of Redlands in Redlands, California. Since completing her degree she has focused on issues surrounding the status of the threatened desert tortoise in the Mojave Desert, including population monitoring, habitat modeling, and decision support for implementation of management actions.

Dr. DEBRA L. HUGHSON worked in mining, oil and gas development, engineering, and nuclear-waste disposal in the western United States before joining the National Park Service in 2001 as the Science Advisor for Mojave National Preserve. She has earth science degrees from Arizona State University (B.S., geology, 1982), San Diego State University (M.S., hydrogeology, 1992), and New Mexico Institute of Mining and Technology (Ph.D., hydrology, 1997). She worked for several years on the Yucca Mountain Project at the Center for Nuclear Waste Regulatory Analyses in San Antonio, Texas, after completing a post-doctoral research appointment at the University of Arizona.

Dr. ERIC V. MCDONALD has conducted research on interactions among climate change, soil, geomorphic, and ecologic processes in arid and semiarid regions since 1983. Since 2000, he has focused work on environmental aspects of ecosystems in the Mojave Desert. He received degrees in geology (B.S., Humboldt State University, 1984), soil science (M.S., Washington State University, 1987), and earth and planetary sciences (Ph.D, University of New Mexico, 1994). McDonald has been a research professor with the Division of Earth and Ecosystem Sciences, Desert Research Institute (DRI), in Reno, Nevada, since 1998. He has also served as interim director for the Center for Arid Lands Environmental Management at DRI. He has edited a volume on desert environments entitled *Management and Mitigation of U.S. Department of Defense Lands Arid Lands* (with Steven Warren), which appeared as a special volume of *Arid Land Research and Management.*

Dr. DAVID M. MILLER has studied geology in all states of the intermountain west, but especially in the deserts of Utah and California. His studies span a wide range of geology, and include mapping and surface-process studies of the surficial geology, geomorphology, and ecology of the Mojave Desert region. He is currently engaged in studies of relations between desert ecology and soils, and active tectonics of the Mojave Desert. Miller has degrees in geology (B.S., Binghamton University, New York, 1973; Ph.D., University of California Los Angeles, 1978), and has been a research geologist for the U.S. Geological Survey in Menlo Park for 26 years. He has edited and contributed to several books on geology of the western United States, including *Tectonic and Stratigraphic Studies in the Eastern Great Basin* and *Jurassic Magmatism and Tectonics of the North American Cordillera.*

AUTHORS

EDITH ALLEN is a professor of plant ecology and cooperative extension specialist in the Department of Botany and Plant Sciences, University of California at Riverside. She received her Ph.D. in botany from the University of Wyoming in 1979.

MICHAEL ALLEN is director of the University of California at Riverside Center for Conservation Biology and is on the faculty in the Departments of Biology, Plant Pathology (where he is currently serving as department chair) and Environmental Sciences. He received his Ph.D. in botany from the University of Wyoming in 1980.

CAMERON BARROWS currently serves as the coordinator for the University of California at Riverside Center for Conservation Biology's Desert Studies Initiative in Palm Desert, California. He received his Ph.D. in environmental sciences-conservation biology at the University of California at Riverside in 2006.

DAVID BEDFORD is a geologist with the U.S. Geological Survey in Menlo Park, California. He has a B.S. in geology from Colorado State University (1996) and is currently pursuing a Ph.D. in geology with hydrology certificate at the University of Colorado.

JAYNE BELNAP has been a scientist with the Department of Interior since 1987 and currently is a research ecologist with the U.S. Geological Survey's Southwest Biological Science Center in Moab, Utah. She received her Ph.D. in botany and range science from Brigham Young University in 1991.

MATTHEW BROOKS is a research botanist for the U.S. Geological Survey's Western Ecological Research Center in Henderson, Nevada. His primary research focus is on the ecology and management of nonnative plants and fire in western North America. He received his Ph.D. in biology from the University of California at Riverside in 1998.

ANDRZEJ BYTNEROWICZ is an ecologist and senior scientist with the U.S. Department of Agriculture, Forest Service, Pacific Southwest Research Station in Riverside, California. He received a Ph.D. in natural sciences from Silesian University, Poland, in 1981.

THERESE CHARLET is currently the data manager of the Nevada Desert Research Center at the University of Nevada at Las Vegas. She has been involved in plant ecology research in the Mojave Desert for the past 10 years and received her M.S. in cellular and molecular biology from the University of Nevada, Reno in 1994.

LESLEY DEFALCO is a plant ecologist with the U.S. Geological Survey's Western Ecological Research Center in Henderson, Nevada. She received her Ph.D. in ecology, evolution and conservation biology from the University of Nevada, Reno in 2003.

TOM DUDLEY is currently a researcher with the Marine Science Institute of the University of California at Santa Barbara and adjunct professor in the Department of Natural Resource and Environmental Science at the University of Nevada, Reno. He received his Ph.D. in aquatic and population biology from the University of California at Santa Barbara in 1989.

TODD ESQUE is an ecologist with the U.S. Geological Survey's Western Ecological Research Center in Henderson, Nevada. He received his Ph.D. in ecology from the University of Nevada, Reno in 2003.

MARK FENN is a research plant pathologist with the U.S. Forest Service in Riverside, California. He received his Ph.D. in plant pathology from the University of California at Riverside in 1986.

RICHARD HEREFORD is a research geologist (emeritus) with the U.S. Geological Survey in Flagstaff, Arizona. His primary research interest is the effects of climate on landscape evolution, particularly alluvial processes. He received his M.S. degree in geology from Northern Arizona University in 1974.

MARK HOOTEN is a consulting ecologist and ecological risk assessor with the Neptune and Co., Inc. He has worked specifically on matters concerning Mojave Desert and Great Basin Desert ecology since 2000 and received his Ph.D. in biology from Montana State University at Bozeman in 1995.

BRIDGET LAIR was formerly an ecologist for the U.S. Geological Survey's Western Ecological Research Center, in Henderson, Nevada. She received M.S. degrees in geography and international affairs from Ohio University in 2002.

SIMON LEI is a biology and ecology instructor in the Department of Biology at Nevada State College in Henderson, Nevada. He received his Ph.D. in higher education teaching and leadership, with a minor in plant ecology from the University of Nevada, Las Vegas in 2003.

PHILIP MEDICA is a biologist/ecologist with the U.S. Geological Survey's Biological Resources Division in Henderson, Nevada. He has more than 40 years experience conducting research in the Mojave Desert, primarily involving desert reptiles and rodents. He has a B.S. in wildlife management (1964) and an M.S. in biology from New Mexico State University at Las Cruces (1966).

BETH NEWINGHAM is a plant community ecologist and is currently an assistant research faculty member at the University of Nevada, Las Vegas. She received her Ph.D. in organismal biology and ecology from the University of Montana in 2002.

GEOFF PHELPS is a geologist and geophysicist with the U.S. Geological Survey in Menlo Park, California. He has a B.S. in geology from the University of California at Berkeley (1990) and an M.S. in geographic information science from the University of London at Birkbeck (2003).

LEELA RAO is a Ph.D. candidate in the Department of Environmental Sciences at the University of California at Riverside. She has a B.A. in environmental studies from Scripps College (1997) and an M.E.M. (Master of Environmental Management) in resource ecology from Duke University (1999).

KELLY REDMOND serves as regional climatologist at the National Oceanic and Atmospheric Administration, Western Regional Climate Center, at the Desert Research Institute in Reno, Nevada. He received a Ph.D. (1982) and M.S. (1977) in meteorology from the University of Wisconsin and a B.S. (1974) in physics from MIT.

SARAH ROBINSON is an assistant professor of geospatial science with the Air Force Academy in Colorado Springs, Colorado. She received her Ph.D. in geology from Arizona State University in 2002.

JANE RODGERS is a vegetation ecologist with the National Park Service at Point Reyes National Seashore in California. She received her B.S. in forestry and natural resources management from the University of California at Berkeley in 1990.

SUSAN SCHWINNING is a plant ecologist at Texas State University in San Marcos, Texas. She received her Ph.D. in ecology and evolutionary biology from the University of Arizona in 1994.

KEVIN SCHMIDT is a geologist with the U.S. Geological Survey in Menlo Park, California. He received a Ph.D. in geology and geophysics from the University of Washington in 1999.

STANLEY SMITH is a physiological ecologist in the School of Life Science at the University of Nevada at Las Vegas. He received his Ph.D. in botany from the Arizona State University in 1981.

ROBERT STEERS is currently pursuing a Ph.D. in botany from the University of California at Riverside (expected in 2008). He received his B.S. in ecology and systematic biology from the California Polytechnic State University at San Luis Obispo in 2002.

BRYAN STEVENSON is currently a soil/ecosystem scientist with Landcare Research in New Zealand. He conducted post doctoral research at the Desert Research Institute, Nevada from 2003–2005 and received his Ph.D. in soil science from the Colorado State University in 1997.

KATHRYN THOMAS is an ecologist with the U.S. Geological Survey's Southwest Biological Science Center and is currently based in Tucson, Arizona. She received a Ph.D. in geography, with emphasis in vegetation and landscape studies, from the University of California at Santa Barbara in 1996.

JOHN VOGEL is a geographer with the U.S. Geological Survey's Southwest Geographic Science Team in Tucson, Arizona. He received a B.A. in geography, with an emphasis on cartographic techniques, from the University of California at Santa Barbara in 1978.

JAMES WEIGAND is with the Bureau of Land Management in Sacramento, California. He received his Ph.D. in forest ecology from Oregon State University in 1997.

CHARLES WILKINSON is Distinguished University Professor and the Moses Lasky Professor of Law at the University of Colorado. He graduated from Stanford Law School in 1966.

INDEX